Myb Transcription Factors: Their Role in Growth, Differentiation and Disease

PROTEINS AND CELL REGULATION

Volume 2

Series Editors: Professor Anne Ridley
Ludwig Institute for Cancer Research
and Department of Biochemistry and Molecular Biology
University College London
London
United Kingdom

Professor Jon Frampton
Professor of Stem Cell Biology
Institute for Biomedical Research,
Birmingham University Medical School,
Division of Immunity and Infection
Birmingham
United Kingdom

Aims and Scope

Our knowledge of the ways in which a cell communicates with its environment and how it responds to information received has reached a level of almost bewildering complexity. The large diagrams of cells to be found on the walls of many a biologist's office are usually adorned with parallel and interconnecting pathways linking the multitude of components and suggest a clear logic and understanding of the role played by each protein. Of course this two-dimensional, albeit often colourful representation takes no account of the three-dimensional structure of a cell, the nature of the external and internal milieu, the dynamics of changes in protein levels and interactions, or the variations between cells in different tissues.

Each book in this series, entitled "*Proteins and Cell Regulation*", will seek to explore specific protein families or categories of proteins from the viewpoint of the general and specific functions they provide and their involvement in the dynamic behaviour of a cell. Content will range from basic protein structure and function to consideration of cell type-specific features and the consequences of disease-associated changes and potential therapeutic intervention. So that the books represent the most up-to-date understanding, contributors will be prominent researchers in each particular area. Although aimed at graduate, postgraduate and principle investigators, the books will also be of use to science and medical undergraduates and to those wishing to understand basic cellular processes and develop novel therapeutic interventions for specific diseases.

Myb Transcription Factors: Their Role in Growth, Differentiation and Disease

Edited by

JON FRAMPTON
University of Birmingham, U.K.

KLUWER ACADEMIC PUBLISHERS
DORDRECHT / BOSTON / LONDON

A C.I.P. Catalogue record for this book is available from the Library of Congress.

ISBN 1-4020-2779-6 (HB)
ISBN 1-4020-2869-5 (e-book)

Published by Kluwer Academic Publishers,
P.O. Box 17, 3300 AA Dordrecht, The Netherlands.

Sold and distributed in North, Central and South America
by Kluwer Academic Publishers,
101 Philip Drive, Norwell, MA 02061, U.S.A.

In all other countries, sold and distributed
by Kluwer Academic Publishers,
P.O. Box 322, 3300 AH Dordrecht, The Netherlands.

Printed on acid-free paper

Printed in the Netherlands.

TABLE OF CONTENTS

Preface vii

List of contributors ix

Colour Section xiii

1 EVOLUTION OF MYB PROTEINS 1
Colin Davidson, Emily Ray and Joseph Lipsick

2 *DROSOPHILA* MYB 35
Lessons for the Understanding of Vertebrate Myb Proteins
Alisa L. Katzen

3 ESSENTIAL AND DIVERSE ROLES FOR C-MYB THROUGHOUT T CELL DEVELOPMENT 65
Kathleen Weston

4 POTENTIAL ROLES FOR C-MYB THOUGHOUT EARLY LYMPHOCYTE DEVELOPMENT 81
Timothy P. Bender

5 INVOLVEMENT OF C-MYB IN RED CELL AND MEGAKARYOCYTE DEVELOPMENT 107
Alexandros Vegiopoulos, Nikla R. Emambokus, Jon Frampton

6 C-MYB AS A KEY PLAYER IN THE CONTROL OF MYELOID CELL DIFFERENTIATION 133
Sandrine Sarrazin and Michael H. Sieweke

7 DOES C-MYB HAVE A ROLE IN HAEMOPOIETIC STEM CELLS AND MULTILINEAGE PROGENITORS? 145
Nikla R. Emambokus and Jon Frampton

8 A-MYB IN DEVELOPMENT AND CANCER 163
Ramana V. Tantravahi, Stacey J. Baker and E. Premkumar Reddy

9 B-MYB: A HIGHLY REGULATED MEMBER OF THE MYB TRANSCRIPTION FACTOR FAMILY 181
Roger J. Watson

10 REGULATION OF MAMMALIAN MYB GENE EXPRESSION 201
Fiona J. Tavner

11 THE C-MYB DNA BINDING DOMAIN 223
From Molecular Structure to Function
Kazuhiro Ogata, Tahir H. Tahirov and Shunsuke Ishii

12 MYB PARTNERSHIPS 239

Xianming Mo, Elisabeth Kowenz-Leutz and Achim Leutz

13 TARGET GENES OF V-MYB AND C-MYB 257

Karl-Heinz Klempnauer

14 THE MICROARRAY BIG BANG 271

Genome-Scale Identification of Myb Regulatd Genes

Scott A. Ness

15 THE V-MYB ONCOGENE 279

Two Models for Activation

Fan Liu and Scott A. Ness

16 C-MYB AND LEUKAEMOGENESIS 307

Juraj Bies and Linda Wolff

17 THE INVOLVEMENT OF MYB IN VASCULAR INJURY 331

Cathy M. Holt and Nadim Malik

18 REPRESSION OF MATRIX GENE EXPRESSION BY B-MYB 351

Claudia S. Hofmann and Gail E. Sonenshein

19 THE ROLE OF C-MYB IN GASTROINTESTINAL TRACT DEVELOPMENT AND CARCINOGENESIS 367

Robert G. Ramsay, Daniel Ciznadji and Gabriella Zupi

20 C-MYB AND CREB FUNCTION IN ADULT NEUROGENESIS 389

Theo Mantamadiotis, Sally Lightowler, Marijana Vanevski, Mark A. Rosenthal, Nikla R. Emambokus, Jon Frampton, Robert G. Ramsay

21 THE C-MYB GENE: A RATIONAL TARGET FOR TREATMENT OF HUMAN DISEASES 399

Susan E. Shetzline and Alan M. Gewirtz

Preface

MYB TRANSCRIPTION FACTORS

Jon Frampton
Institute for Biomedical Research, Birmingham University Medical School, Division of Immunity and Infection, Edgbaston, Birmingham B15 2TT, UK.

This volume brings together for the first time articles written by experts researching the c-Myb transcription factor and its related homologues B-Myb and A-Myb. The majority of chapters deal with vertebrate Myb proteins but discussion of related proteins from lower organisms is also included because of the light and understanding these shed on the structure and functioning of the proteins higher up the evolutionary ladder. The importance of Myb proteins is apparent from the wide range of cell types in which they are expressed and individual chapters describe their involvement in various haemopoietic cells and in the vasculature, gut and brain. The molecular mechanisms underlying transcriptional regulation by Myb proteins are explored in chapters dealing with the structure of these proteins, modulation of their activity through post-translational modifications or interaction with other proteins, and the identification of target genes. c-Myb has long been viewed as a potential therapeutic target in diseases involving cellular hyperproliferation and the exciting prospect that manipulation of its function could be used in the treatment of diseases as diverse as leukaemia and atherosclerosis is discussed. Researchers and students studying cellular regulation by transcription factors as well as clinical scientists in fields ranging from haematology and oncology to cardiology will benefit from this volume.

Contributors

Stacey J. Baker/ Fels Institute for Cancer Research and Molecular Biology/ Temple University School of Medicine/3307 N. Broad Street, Philadelphia, PA 19140, USA.

Timothy P. Bender /Department of Microbiology/ University of Virginia Health System/ Charlottesville, VA 22908-0734, USA/Phone +1 804 924 1246/Fax +1 804 982 1071/Email tpb3e@virginia.edu.

Juraj Bies/Laboratory of Molecular Virology/Cancer Research Institute, Slovak Academy of Sciences/833 91 Bratislava, Slovakia/Phone +421 7 59327 425/Fax +421 7 59327 250/Email Exonbies@savba.sk.

Daniel Ciznadji/Differentiation and Transcription Laboratory/Peter MacCallum Cancer Centre/St. Andrews Place, East Melbourne, 3002, Australia

Colin Davidson/Departments of Pathology and Genetics/Stanford University School of Medicine/Stanford, CA 94305-5324, USA.

Nikla R. Emambokus/Harvard Medical School/Childrens Hospital Boston NRB07.007B, 1 Blackfan Circle, Boston, MA 02115, USA/Phone +1 617 919 2049/Fax: +1 617 730 0222/E-mail emambokus@bloodgroup.tch.harvard.edu.

Jon Frampton/Institute for Biomedical Research/Birmingham University Medical School/Edgbaston, Birmingham, B15 2TT, UK/Phone +44 121 414 6812 Fax +44 121 415 8817 Email j.frampton@bham.ac.uk.

Alan M. Gewirtz/Departments of Internal Medicine, Pathology and Laboratory Medicine/University of Pennsylvania School of Medicine/421 Curie Boulevard, Philadelphia, PA 19104, USA/Phone +1 215 898 4499/Fax +1 215 573 2078/Email gewirtz@mail.med.upenn.edu.

Claudia S. Hofmann/Department of Biochemistry/Boston University School of Medicine/Boston Massachusetts 02118, USA.

Cathy M. Holt/University of Manchester/1.305 Stopford Building/Manchester, M13 9PT, UK/Phone +44 161 275 5671/Fax +44 161 275 5669/Email cathy.holt@man.ac.uk.

Shunsuke Ishii/Laboratory of Molecular Genetics/RIKEN Tsukuba Institite, 3-1-1 Koyadai, Tsukuba, Ibaraki 305-0074, Japan/Email sishii@rtc.riken.go.jp.

Alisa L. Katzen/Department of Biochemistry and Molecular Genetics/University of Illinois at Chicago/College of Medicine, Chicago, IL 60607-7170, USA/Phone +1 312 413 9215/Fax +1 312 413 0353/Email katzen@uic.edu.

Karl-Heinz Klempnauer/Institut für Biochemie/Universität Münster, Wilhelm Klemm Str. 2, D48149 Münster, Germany/Fax +49 251 8333206/Email klempna@uni-muenster.de.

Elisabeth Kowenz-Leutz/Max-Delbrück-Center for Molecular Medicine/Robert-Rössle-Str. 10, 13092 Berlin, Germany.

Achim Leutz/Max-Delbrück-Center for Molecular Medicine/Robert-Rössle-Str. 10, 13092 Berlin, Germany/Phone +49 30 9406 3735/Fax +49 30 9406 3298/Email aleutz@mdc-berlin.de.

Sally Lightowler/Differentiation and Transcription Laboratory/Peter MacCallum Cancer Centre/St. Andrews Place, East Melbourne, 3002, Australia.

Joseph S. Lipsick/Departments of Pathology and Genetics/Stanford University School of Medicine Stanford/CA 94305-5324, USA/Phone +1 650 723 1623/Fax +1 650 725 6902/Email lipsick@stanford.edu.

Fan Liu/Department of Molecular Genetics and Microbiology/915 Camino de Salud NE, MSC08 4660/University of New Mexico Health Sciences Center/Albuquerque, New Mexico 87131-0001, USA.

Nadim Malik/University of Manchester/1.305 Stopford Building Manchester, M13 9PT, UK.

Theo Mantamadiotis/Differentiation and Transcription Laboratory/Peter MacCallum Cancer Centre/St. Andrews Place, East Melbourne, 3002, Australia./Phone 03 9656 3715/Fax 03 9656 3738/Email t.mantamadiotis@pmci.unimelb.edu.au.

Xianming Mo/Max-Delbrück-Center for Molecular Medicine/Robert-Rössle-Str. 10, 13092 Berlin, Germany.

Scott A. Ness/Department of Molecular Genetics and Microbiology/915 Camino de Salud NE, MSC08 4660/University of New Mexico Health Sciences Center/Albuquerque, New Mexico 87131-0001, USA/Phone +1 505 272 9883/Fax +1 505 272 3463/Email ness@unm.edu.

Kazuhiro Ogata/Department of Biochemistry/Yokohama City University Graduate School of Medicine/3-9 Fukuura, Kanazawa-ku, Yokohama 236-0004, Japan/Phone +81 45 787 2589 2590/Fax +81 45 784 4530/Email: ogata@med.yokohama-cu.ac.jp.

Robert G. Ramsay/Differentiation and Transcription Laboratory/Peter MacCallum Cancer Centre/St. Andrews Place, East Melbourne, 3002, Australia/Phone 03 9656 1863/Fax03 9656 1411/Email r.ramsay@pmci.unimelb.edu.au.

Emily Ray/Departments of Pathology and Genetics/Stanford University School of Medicine Stanford/CA 94305-5324, USA.

E. Premkumar Reddy/Fels Institute for Cancer Research and Molecular Biology/Temple University School of Medicine/3307 N. Broad Street, Philadelphia, PA 19140, USA/Phone +1 215 707 4307/Fax +1 215 707 1454/Email reddy@unix.temple.edu.

Mark A. Rosenthal/Differentiation and Transcription Laboratory/Peter MacCallum Cancer Centre/St. Andrews Place, East Melbourne, 3002, Australia.

Sandrine Sarrazin/Centre d'Immunologie de Marseille Luminy/Campus de Luminy, Case 906, 13288 Marseille Cedex09, France.

Susan E. Shetzline/Departments of Internal Medicine/Pathology and Laboratory Medicine/University of Pennsylvania School of Medicine/421 Curie Blvd. BRB II/III Rm. 727, Philadelphia, PA 19104, USA/Phone: +1 215 898 5101/Fax +1 215 573 7049 Email shetzlin@mail.med.upenn.edu.

Michael Sieweke/Centre d'Immunologie de Marseille Luminy/Campus de Luminy, Case 906, 13288 Marseille Cedex09, France./Phone +33 4 91 26 94 38/Fax +33 4 91 26 94 30/Email Sieweke@ciml.univ-mrs.fr.

Gail E. Sonenshein/Department of Biochemistry/Boston University School of Medicine/Boston Massachusetts 02118, USA/Email gsonensh@bu.edu.

Tahir H. Tahirov/Department of Biochemistry/Yokohama City University Graduate School of Medicine/3-9 Fukuura, Kanazawa-ku, Yokohama 236-0004, Japan/Email tahir@med.yokohama-cu.ac.jp.

Ramana V. Tantravahi/Fels Institute for Cancer Research and Molecular Biology/Temple University School of Medicine/3307 N. Broad Street, Philadelphia, PA 19140, USA.

Fiona J. Tavner/Ludwig Institute for Cancer Research and Department of Virology/Faculty of Medicine/Imperial College London/Norfolk Place, London W2 1PG, UK/Phone +44 20 7503 7716/Fax +44 20 7724 8586/Email f.tavner@ic.ac.uk.

Marijana Vanevski/Differentiation and Transcription Laboratory/Peter MacCallum Cancer Centre/St. Andrews Place, East Melbourne, 3002, Australia.

Alexandros Vegiopoulos/Institute for Biomedical Research/Birmingham University Medical School/Edgbaston, Birmingham, B15 2TT, UK/Phone +44 121 414 6807/Fax +44 121 415 8817/Email vegiopoulos@web.de.

Roger J. Watson/Ludwig Institute for Cancer Research and Department of Virology/Faculty of Medicine/Imperial College London/Norfolk Place, London W2 1PG, UK/Phone +44 20 7563 7711/Fax +44 20 7724 8586/Email r.watson@ic.ac.uk.

Kathleen Weston/Cancer Research UK Centre for Cell and Molecular Biology Institute of Cancer Research/237 Fulham Road, London SW3 6JB, UK./Phone +44 20 7878 3846/Fax +44 20 7352 3299/Email kathy@icr.ac.uk.

Linda Wolff/Laboratory of Cellular Oncology/National Cancer Institute/NIH, Bethesda, MD 20892-4255, USA.

Gabriella Zupi/Experimental Chemotherapy Laboratory/Regina Elena Cancer Institute, Rome, Italy.

Color Version of Plates 1-21

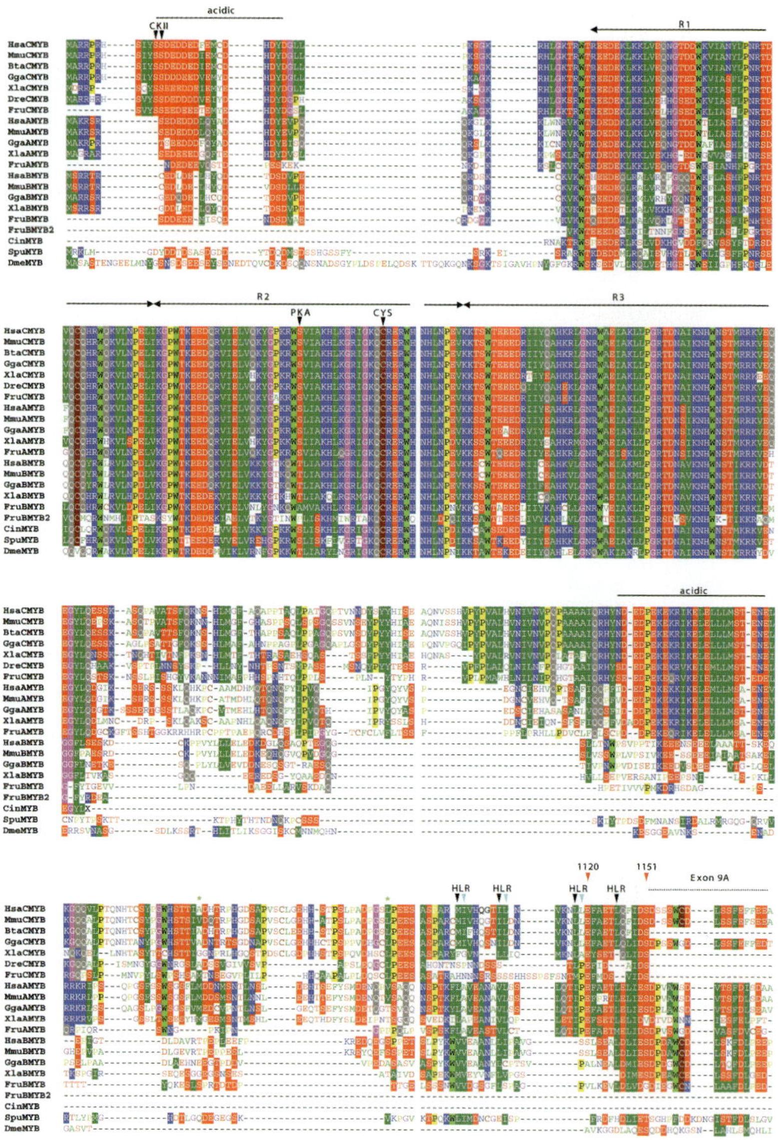

Figure 1

Amino acid alignment of invertebrate and vertebrate 3R Myb proteins. A multiple sequence alignment of complete and partial animal 3R Myb proteins was generated using ClustalW (Thompson et al., 1994) and edited with colour coding according to residue conservation using BioEdit (http://www.mbio.ncsu.edu/BioEdit/bioedit.html). The extreme amino terminal exonic regions were not determined for the Fugu and *Ciona* sequences. In addition, XX in the *Ciona* sequence indicates exonic regions that were indeterminable from the genomic sequence due to either poor conservation or genome assembly errors. The complete Fugu B-myb2 cDNA sequence could not be determined due to unfinished assembly of the genome sequence for this gene. The XX near the carboxyl terminus of *Xenopus* B-Myb indicates residues that were unaligned and not included to save space. Exon 9A encoded sequences were not available for bovine, *Xenopus*, Zebrafish and Fugu c-Myb sequences. CKII, Casein kinase II

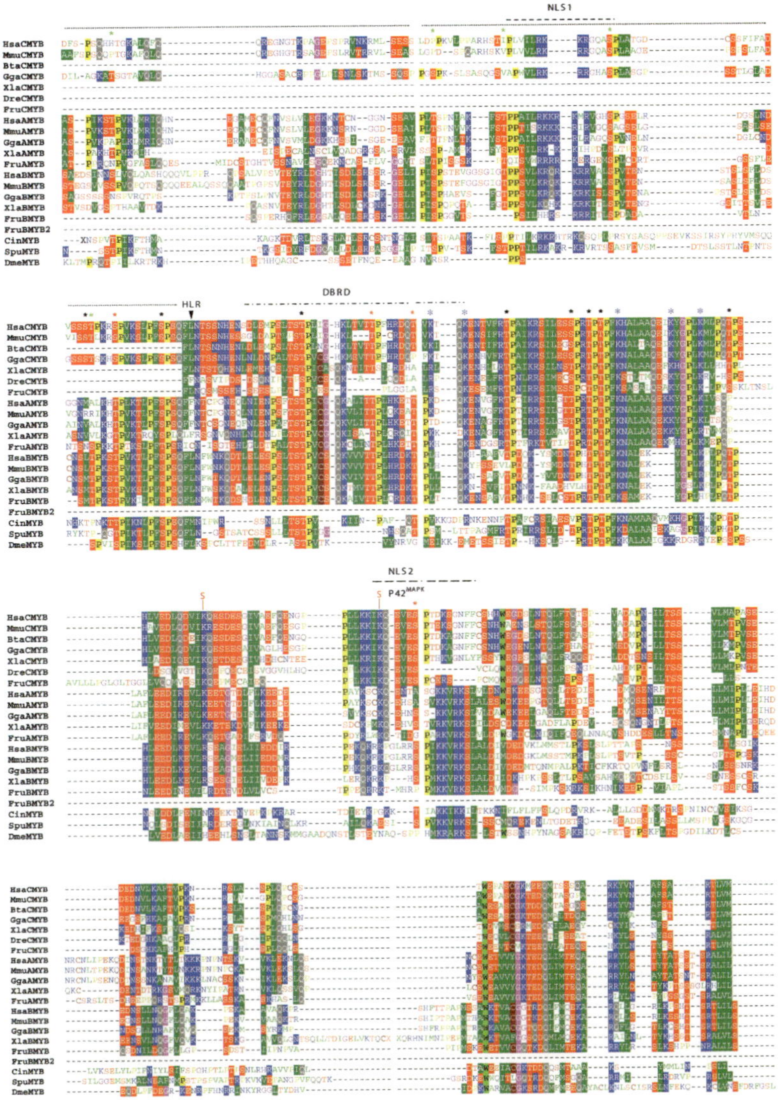

phosphorylation site; R1, R2 and R3 indicate the Myb repeats; PKA, protein kinase A phosphorylation site; CYS, redox modified cysteine residue; HLR, heptad leucine repeat, black and light blue arrowheads (the black arrowhead has been referred to as the "leucine zipper"); NLS1, nuclear localisation signal 1; 1120 and 1151, carboxy-terminal truncation mutants of AMV v-Myb; DBRD, DNA binding regulatory domain; NSL2, nuclear localisation signal 2; black asterisks, highly conserved putative phosphorylation sites for proline-directed kinases; putative cyclin A1 and A2/Cdk2 phosphorylation sites are indicated by the red asterisks and putative B-Myb phosphorylation sites are shown in green. Modified lysine residues are indicated by blue asterisks (acetylation) and **S** (sumoylation). Abbreviations: Hsa, *Homo sapiens* (Human); Mmu, *Mus musculus* (Mouse); Bta, *Bos taurus* (Cow); Gga, *Gallus gallus* (Chicken); Xla, *Xenopus laevis* (Xenopus); Dre, *Danio rerio* (Zebrafish); Fru, *Fugu rubripes* (Fugu); Cin; *Ciona intestinalis* (Ciona); Spu, *Stongylocentrotus purpuratus* (Sea Urchin); Dme, *Drosophila melanogaster* (Drosophila).

(see chapter 1, p.4 and 5)

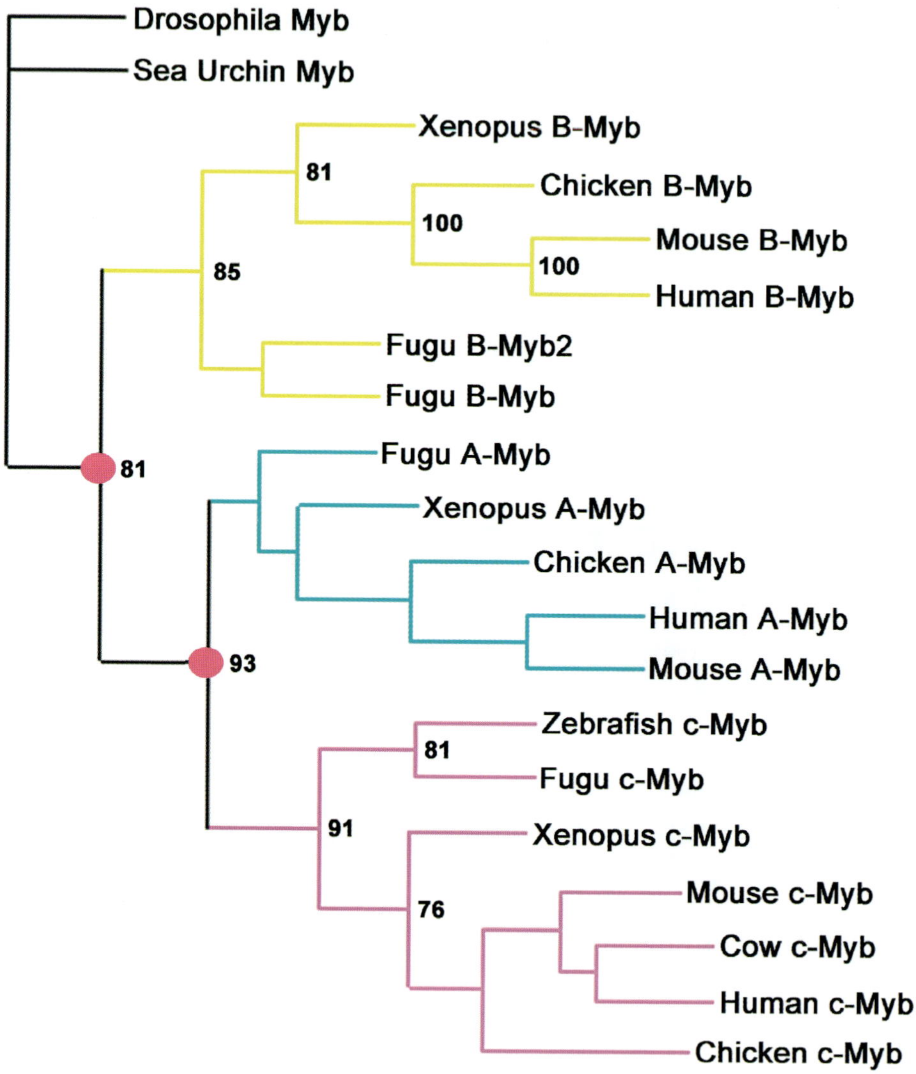

Figure 2

Consensus tree illustrating the phylogenetic relationship of animal 3R Myb proteins. The unrooted tree topology was estimated through neighbour joining using the distances calculated from an alignment of the R1, R2 and R3 domains using the Dayhoff-PAM substitution model in Phylip 3.5. The numbers at the nodes represent percent bootstrap support based on 1000 iterations. For purposes of clarity bootstrap values below 75% are not shown. Invertebrate 3R Myb sequences are shown in black, the B-Myb clade is shown in yellow, the A-Myb clade is shown in cyan and the c-Myb clade is shown in magenta. The pink circles indicate putative gene duplication events. *(see chapter 1, p.11)*

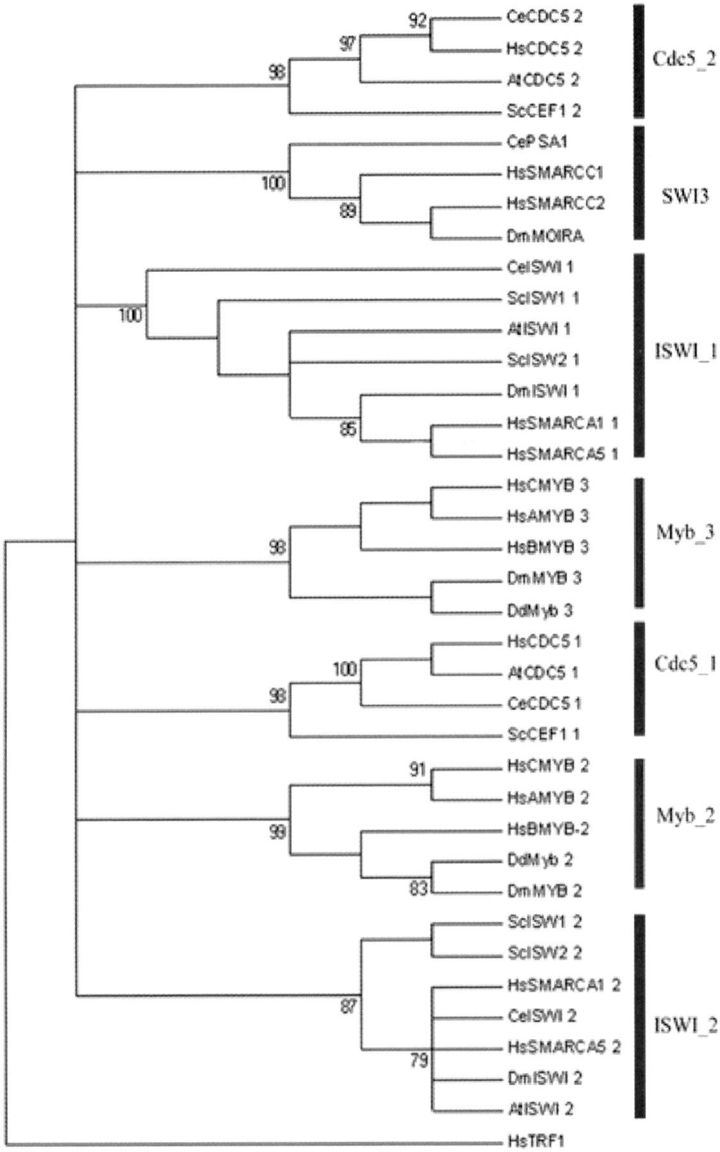

Figure 3

Phylogenetic analysis of individual Myb repeats. We performed unrooted minimum evolution analysis using a Poisson correction model with Molecular Evolutionary Genetics Analysis (MEGA) software version 2.1. This analysis generated a consensus tree from a Clustal X alignment containing Myb repeats from protein homologues of the Myb, CDC5, SWI3, and ISWI protein families. In the consensus tree shown, all Myb repeats within one of the seven groups can be inferred to share common ancestry based on bootstrap values greater than 75% (values lower than 75% are not shown). However, the order of branching which gives rise to the seven groups cannot be accurately inferred based on the alignment.

(see chapter 1, p.16)

```
                     ---helix1-----       -helix2-     --helix3---
cMyb_DNAB_R2                              WS            KQ    R  NH
cMyb_DNAB_R3    K S                       WA    K       N        NS
                    5    10   15          20   25        30   35   40
                    |    |    |           |    |         |    |    |
HsCMYB_1        KTRWTREEEDEKLKKLVEQ--5--WKVIANY-3--R-TDVQCQHRWQKVL
HsCMYB_2        KGPWTKEEDQRVIELVQK--5--WSVIAKH-3--R-IGKQCRERWHNHL
HsCMYB_3        KTSWTEEEDRIIYQAHKR--5--WAEIAKL-3--R-TDNAIKNHWNSTM
DmMYB_1         GKRWSKSEDVLLKQLVET--4--WEIIGPH-3--R-LEQQVQQRWAKVL
DmMYB_2         KGPWTRDEDDMVIKLVRN--5--WTLIARY-3--R-IGKQCRERWHNHL
DmMYB_3         KTAWTEKEDEIIYQAHLE--5--WAKIAKR-3--R-TDNAIKNHWNSTM
AtPCMyb1_1      KGQWTPEEDEVLCKAVER--5--WKKIAEC-3--R-TDVQCLHRWQKVL
AtPCMyb1_2      KGPWSKEEDNTIIDLVEK--5--WSTISQH-3--R-IGKQCRERWHNHL
AtPCMyb1_3      KNAWTQEEELTLIRAHQI--4--WAELMKF-3--R-SDNSIKNHWNSSV
AtMyb1_1        KGPWSKEEDDVLSELVKR--5--WSFIARS-3--R-SGKSCRLRWCNQL
AtMyb1_2        RNSFTEVEDQAIIAAHAI--4--WAVIAKL-3--R-TDNAIKNHWNSAL
CeCDC5_1        GGVWKNTEDEILKAAIMK--5--WSRIASL-2--RKSAKQCKARWFEWL
CeCDC5_2        KTEWSREEDEKLLHLAKL--4--WRTIAPI-2--R-TSAQCLERYEHLL
ScADA2          CPDWGADEELQLIKGAQT--5--WQDIADH-3--R-GKEEVKEHYLKYY
HsTADA2L        DPSWTAQEEMALLEAVMD--5--WQDVANQ-3--K-TKEECEKHYMKHF
ScRSC8          KKNWSDQEMLLLLEGIEM--4--WEKIADH-4--K-RVEDCIEKFLSLP
ScSWI3          DENWSKEDLQKLLKGIQE--4--WYKVAKN-3--K-SPEQCILRFLQLP
HsSMARCC1       GREWTEQETLLLLEALEM--4--WNKVSEH-3--R-TQDECILHFLRLP
HsSMARCC2       TREWTEQETLLLLEALEM--4--WNKVSEH-3--R-TQDECILHFLRLP
DmISWI_1        FTAWTKRDFNQFIKANEK--5--IDNIAKD-3--K-TPEEVIE-YNAVF
HsSMARCA1_1     FTNWTKRDFNQFIKANEK--5--IDNIARE-3--K-SPEEVME-YSAVF
HsSMARCA5_1     FTNWNKRDFNQFIKANEK--5--IENIARE-3--K-TPEEVIE-YSAVF
DmISWI_2        GKNYTEIEDRFLVCMLHK--8--YEELRAA-1--R-ASPQFRFDWFIKS
HsSMARCA1_2     GKNYTEEEDRFLICMLHK--8--YEELRQC-1--R-NAPQFRFDWFIKS
HsSMARCA5_2     GKNYTEEEDRFLICMLHK--8--YDELRQC-1--R-NSPQFRFDWFLKS
ScTFIIIB-B"     TDPWTVEEMIKFYKALSM--4--FNLISQL-3--R-SRKQVKAKFVNEE
HsNCOR1_1       MNVWTDHEKEIFKDKFIQ--4--FGLIASY-3--K-SVPDCVLYYYLTK
HsNCOR1_2       TSRWTEEEMEVAKKGLVE--4--WAAIAKM-3--K-SEAQCKNFYFNYK
ScTBF1          KRTWSKEEEEALVEGLKE--4--WSKILDL-13-R-TQVQLKDKARNWK
AtTRP1          RRPFSVAEVEALVQAVEK--5--WRDVKLC-7--R-TYVDLKDKWKTLV
HsTRF1          RQAWLWEEDKNLRSGVRK--5--WSKILLH-5--R-TSVMLKDRWRTMK
HsTRF2          KQKWTVEESEWVKAGVQK--5--WAAISKN-5--R-TAVMIKDRWRTMK
ScRAP1_1        KASFTDEEDEFILDVVRK--10-YDEISHY-3--H-TGNSIRHRFRVYL
ScRAP1_2        KRKFSADEDYTLAIAVKK--62-FKHFAEE-3--H-TENAWRDRFRKFL
HsRAP1          RIAFTDADDVALLTYKEN--12-WKAMEKS-4--H-SWQSLKDRYLKHL
HsSNAPC4_1      KQEWSREEEERLQAIAAA--5--WQKIAEE-4--R-SAFQCLQKFQQHN
HsSNAPC4_2      RKEWTEEEDRMLTQLVQE--8--YRRIVYY-3--R-DSMQLIYRWTKSL
HsSNAPC4_3      KGYWAPEEDAKLLQAVAK--4--WFKIREE-3--R-SDAQCRDRYLRRL
HsSNAPC4_4      KGRWNLKEEEQLIELIEK--4--WAKIASE-3--R-SGSQCLSKWKIMM
HsTTF1_1        KGRYSEGDTEKLKMYHSL--4--WKTIGEM-2--R-RSLSVALKFSQIS
HsTTF1_2        RGAWSKSETRKLIKAVEE--38-WVEVEAK-3--R-NWMQCKSKWTEIL
ScREB1_1        SKAFDKAEDAALERFINE--26-WNNIYKV-3--R-SSSSIYKHMRRKY
ScREB1_2        RGKWTAEEEQELAKLCAE--4--WAEIGKT-2--R-MPEDCRDRWRNYV
ScREB1_3        SNRWSVEEEELLKKVISD-148-WTIVSER-4--R-SRIQCRYKWNKLV
ScREB1_4        IAKIQTVKDDDMLWIFEK--13-WDELAAL-4--KLNGLELKLCYERMK
DvR2element     KARWSEEEALMMARKEVE--6--VRHLNKK-6--R-S-SDAISSYRKRG
Consensus       ...WT..E...L..........W..I.......R.........W
                ....S..................................Y/F
```

Figure 4

Amino acid alignments of Myb repeats identified high conservation of structural residues and an acidic patch in the first helix. We identified Myb repeats through analysis of the primary protein sequence and confirmed the boundaries of each domain using the Simple Modular Architechural Research Tool program (http://smart.embl-heidelberg.de/). We created multiple sequence alignments of Myb repeats using Clustal X and color coded the alignments based on conservation using the BioEdit program. Labelling of Myb repeats indicates genus and species with the first two letters, followed by the protein name. For proteins containing multiple Myb-motifs, repeats are numbered starting from the N-terminus and identified after an underscore. Residues known to bind DNA in c-Myb (labelled cMyb DNAB) are depicted in bold on the first two lines. Contributions from the second repeat are on the line labelled R2 and those from the third repeat are on the line labelled R3. A consensus of the most highly conserved residues is located on the last line emphasising the importance of the structural residues. Note the high conservation of acidic residues in the first helix compared to c-Myb DNA binding residues. Species and genus abbreviations: Hs, *Homo sapiens*, Dm, *Drosophila melanogaster*, Dv, *Drosophila varians*, Ce, *Caenorhabditis elegans*, At, *Arabidopsis thalia*, Sc, *Saccharomyces cerevisiae*. *(see chapter 1, p.21)*

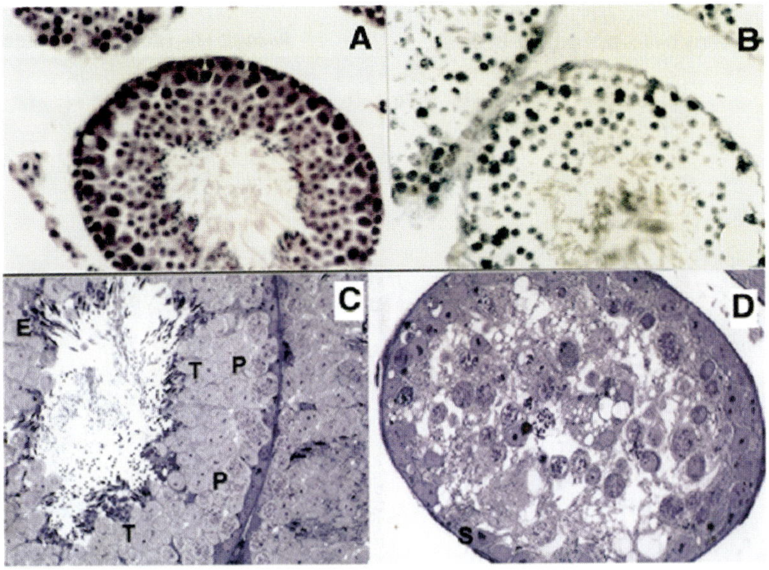

Figure 2

Expression pattern of A-*myb* and its role in the development of testis. (A) Haematoxylin and eosin-stained sections of adult mouse testis. (B) Adjacent section after *in situ* hybridisation to the anti-sense A-*myb* cRNA probe. (C and D) Sections of seminiferous tubules from either wild type or A-*myb*$^{-/-}$ mice. P, primary spermatocytes at pachytene; T, round spermatids.
(see chapter 8, p.167)

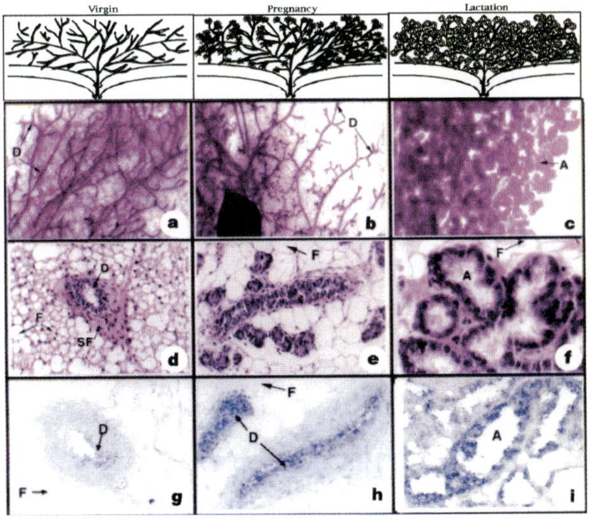

Figure 3

Expression of A-*myb* in mouse mammary tissue. The top panel shows a schematic representation of ductal branching in virgin, preganat and lactating mammary gland. (A-C) Whole mount preparations of mammary glands derived from a nulliparous, 10-day preganant and a lactating mouse two days after delivery. (D-F) Sections of the same tissues stained with haemotoxylin and eosin. (G-I) *In situ* hybridisation pattern of the breast sections with A-*myb* specific probe. A, alveoli; D, ductal epithelial cells; F, adipocytes; SF, fibroblasts.
(see chapter 8, p.171)

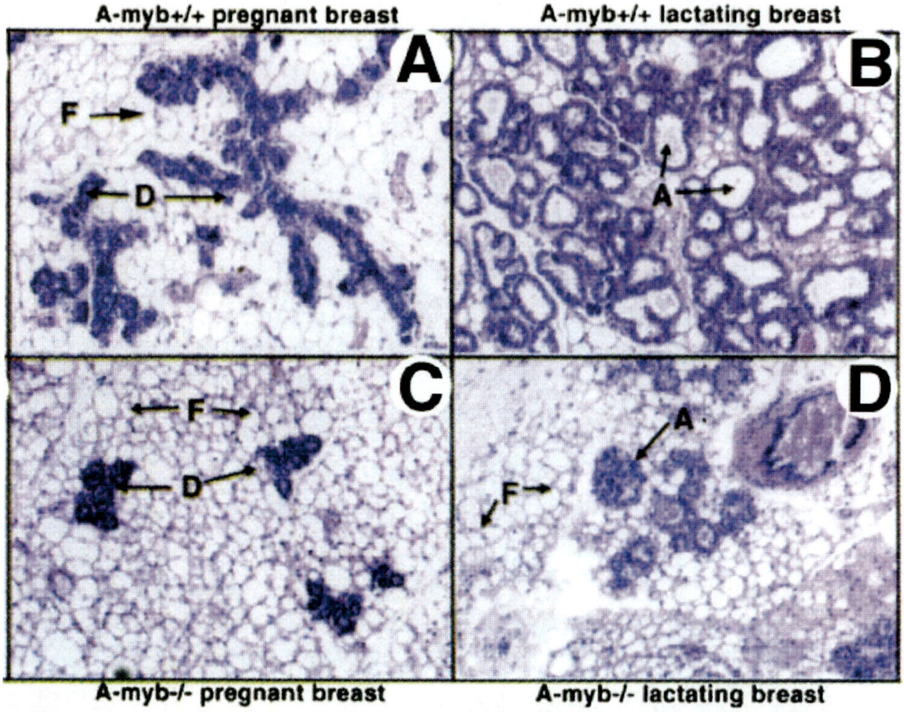

Figure 4

Defective breast development in mice lacking A-Myb. Mammary glands derived from 10 day pregnant and lactating (2 days after delivery of pups) wild-type and A-Myb-/- mice were used for histopathological analysis. Note the reduced proliferation of ductal cells and incompletely formed alveolar structure in A-Myb -/- mice, which leads to a failure of the A-Myb-/- mice to lactate. A, alveoli;D, ductal epithelial cells; F, adipocytes. *(see chapter 8, p.172)*

a **Functional maps of c-Myb and AMV v-Myb proteins**

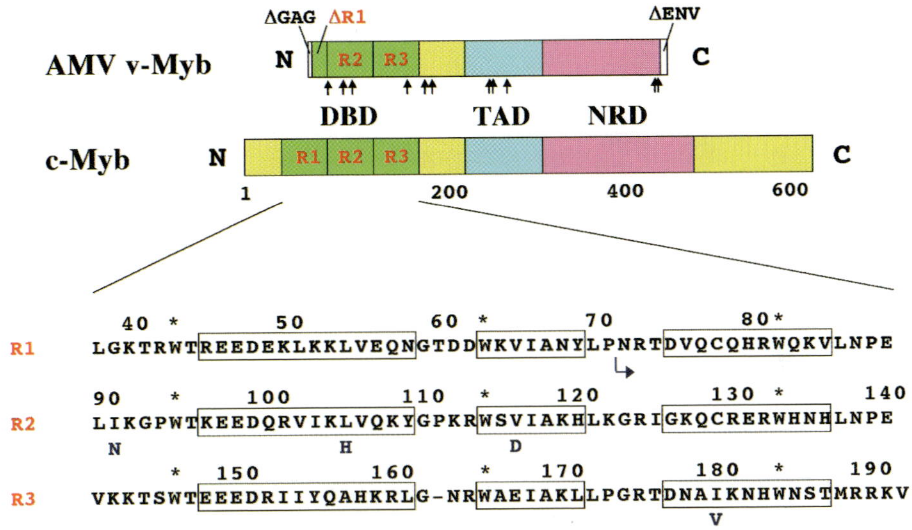

b **Consensus sequence for Myb protein**

	1	2	3	4	5	6	
5′-	T/C	A	A	C	G/T	G	-3′
3′-	A/G	T	T	G	C/A	C	-5′

Figure 1

Functional domain maps of c-Myb and AMV v-Myb (a) and consensus binding sequence for Myb proteins (b). (a) In c-Myb the amino acid sequence of the DBD is presented. Three helical regions of each repeat are boxed, and the periodically positioned tryptophans are marked with asterisks. The position of the N-terminal truncation and the four mutated residues in the DBD of AMV v-Myb are shown with an blue arrow and letters below the sequence, respectively. In AMV v-Myb, the truncated R1 and viral Gag and Env protein regions are shown as ΔR1, ΔGAG and ΔENV, respectively. The mutations are indicated by arrows. DBD; DNA-binding domain, TA; *trans*-activation domain, NRD; negative regulatory domain. *(see chapter 11, p.225)*

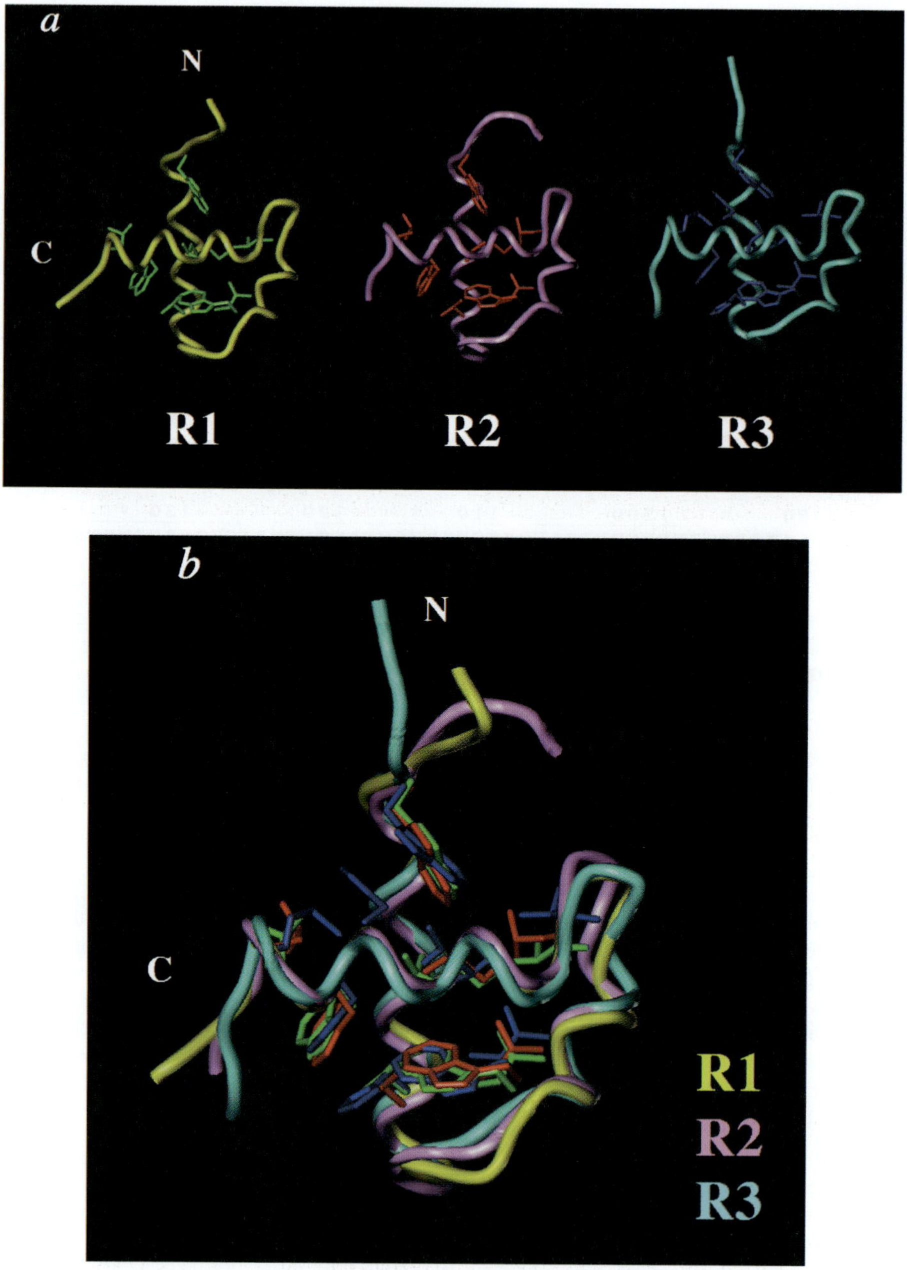

Figure 2

The NMR average structure of the c-Myb DBD consisting of the three subdomains, R1, R2 and R3 (a), and superimposition of them (b). The backbone of each subdomain is shown (R1 - yellow, R2 - magenta and R3 - cyan tubes) and residues in the hydrophobic core (R1 - green, R2 - red and R3 - blue). *(see chapter 11, p.226)*

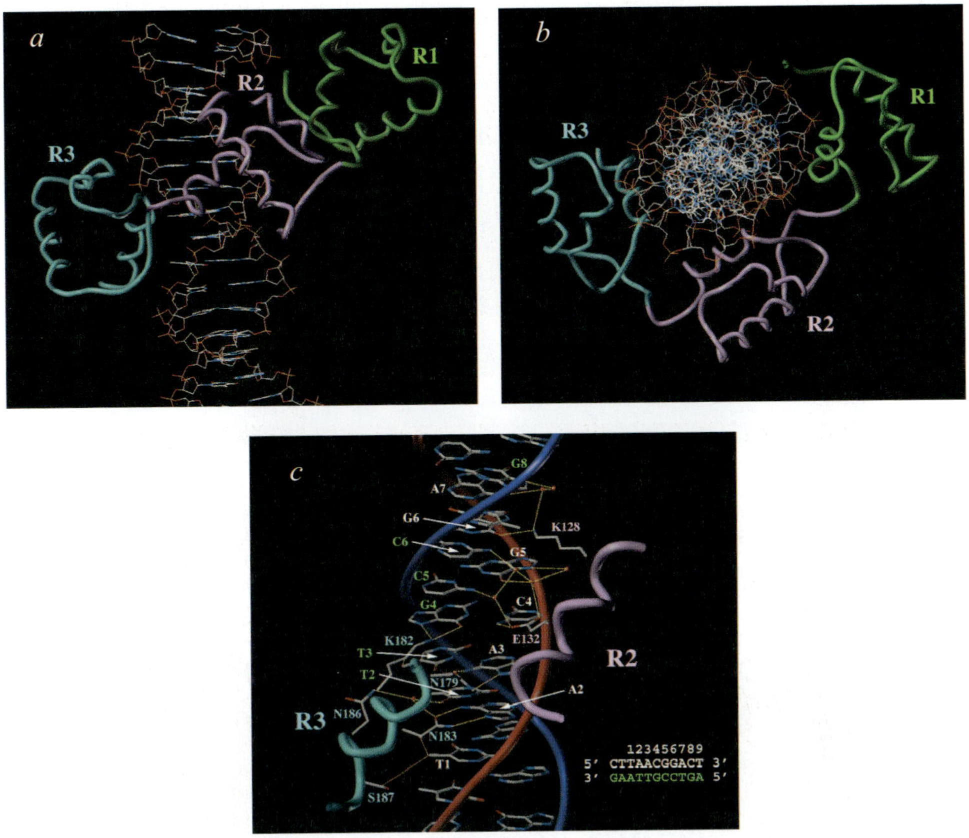

Figure 3

Side and top views of the crystal structure of the c-Myb DBD–DNA complex in the c-Myb–C/EBPβ–DNA ternary complex (a, b), and the specific interactions between c-Myb R2R3 and DNA (c). (a, b) For clarity, the C/EBPβ part has been omitted in these figures. In the c-Myb DBD, only the backbone structure is shown as a tube presentation coloured green, magenta and cyan for R1, R2 and R3, respectively. (c) In c-Myb, two recognition helices from R2 and R3 are shown as tubes coloured magenta and cyan, respectively. In the target DNA, the sugar-phosphate backbones are shown as red and blue tubes. The DNA bases and the side chains of c-Myb R2R3, which are involved in the specific protein–DNA interactions, are shown with capped stick presentations. The water molecules, which mediate the protein–DNA interactions, are shown as red spheres. In the specific protein–DNA interactions, hydrogen bonds and van der Waals interactions are indicated as yellow and orange dotted lines, respectively. The target DNA sequence is shown in the right-bottom corner. *(see chapter 11, p.229)*

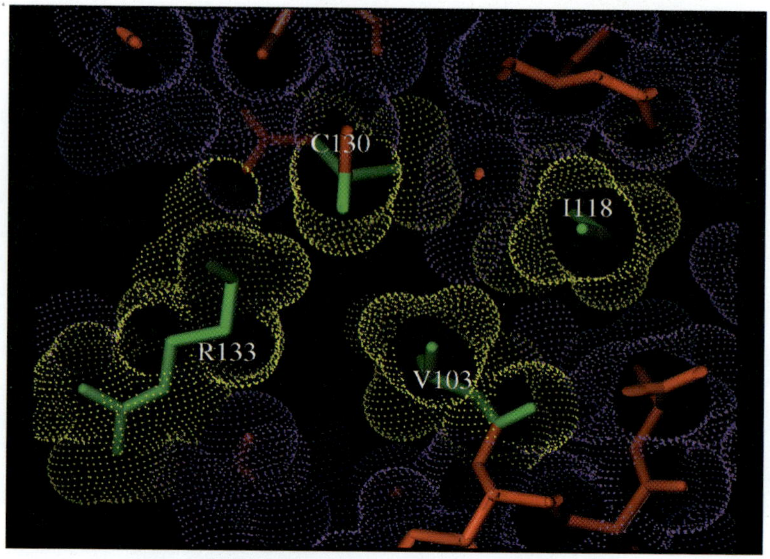

Figure 4

A cavity in the hydrophobic core of c-Myb R2. The side chains of some residues surrounding the cavity are shown as green capped sticks with labellings. The yellow dots represent the van der Waals surfaces of these residues. Other residues are shown as red capped sticks with purple dots of the van der Waals surfaces. *(see chapter 11, p.230)*

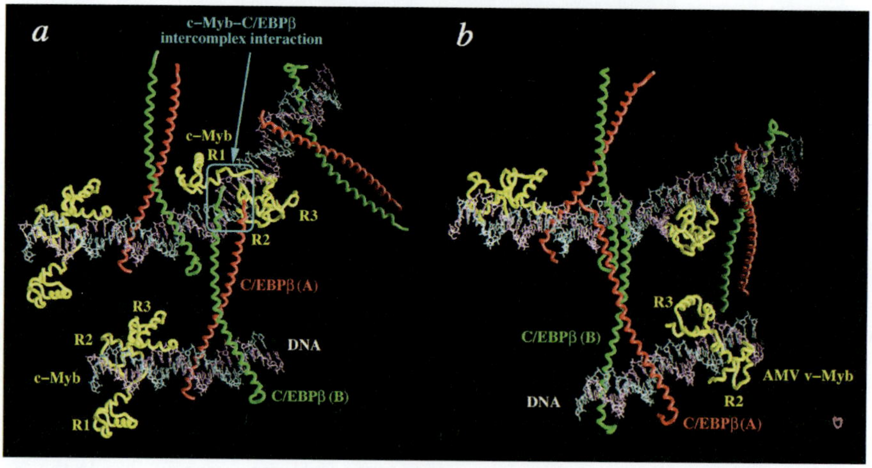

Figure 5

The structures of Myb–C/EBPβ–DNA complexes in crystals. (a) The crystal structure of the c-Myb–C/EBPβ–DNA complex. The backbone structures of c-Myb DBD and two peptide chains of the C/EBPβ homodimer (C/EBPβ(A) and C/EBPβ(B)) are shown as yellow, red and green tubes. The c-Myb–C/EBPβ intercomplex interaction is marked blue. (b) The crystal structure of the AMV v-Myb–C/EBPβ–DNA complex. The backbone structures of the AMV v-Myb DBD and two peptide chains of the C/EBPβ homodimer (C/EBPβ(A) and C/EBPβ(B)) are shown as yellow, red and green tubes. In this structure, intercomplex interaction is not observed. These figures were adopted from Ogata et al. (2003). *(see chapter 11, p.230)*

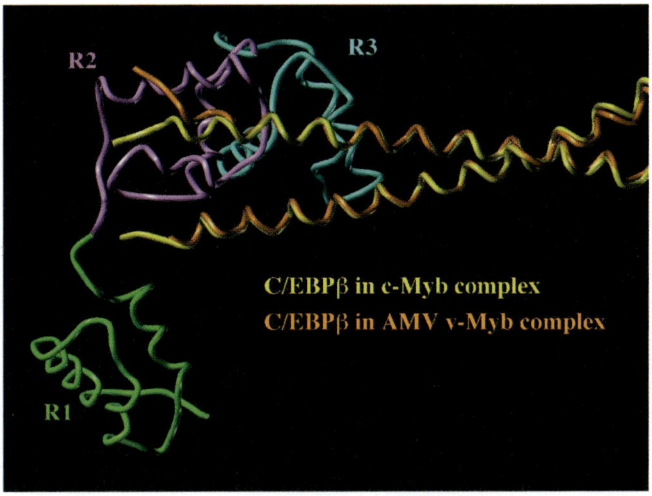

Figure 6

Superimposition of the C-terminal leucine-zipper parts of C/EBPβ in the crystal structures of the c-Myb–C/EBPβ–DNA and AMV v-Myb–C/EBPβ–DNA complexes. The backbones are shown as yellow and orange tubes, respectively. The backbone consisting of the R1, R2 and R3 subdomains of c-Myb DBD in the c-Myb–C/EBPβ–DNA complex is coloured green, magenta and cyan, respectively. The DNA part is excluded for clarity. One of the C-terminal positions of the leucine-zipper parts of C/EBPβ in the AMV v-Myb–C/EBPβ–DNA complex does not take a defined conformation and is not presented. This figure was adopted from Ogata et al. (2003). *(see chapter 11, p.233)*

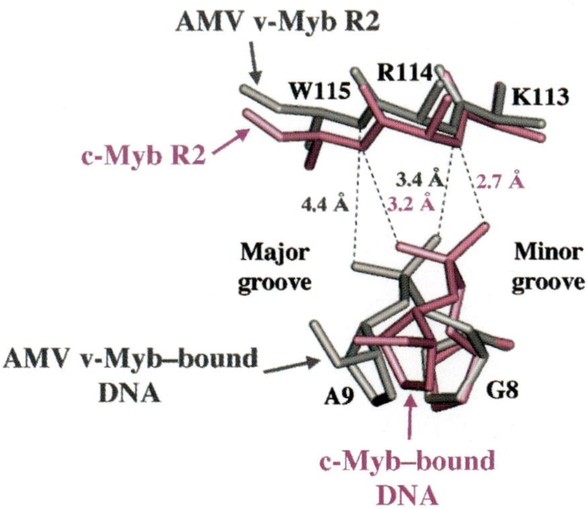

Figure 7

A close-up view of the hydrogen bonds between protein backbones and DNA phosphates in the R2 subdomains of the superimposed c-Myb–C/EBPβ–DNA and AMV v-Myb–C/EBPβ–DNA complex structures. c-Myb and AMV v-Myb are shown as magenta and silver capped sticks, respectively. The DNA base pair positions are labelled according to the numbering in Figure 3 (c). This figure was adopted from Tahirov et al. (2002).
(see chapter 11, p.233)

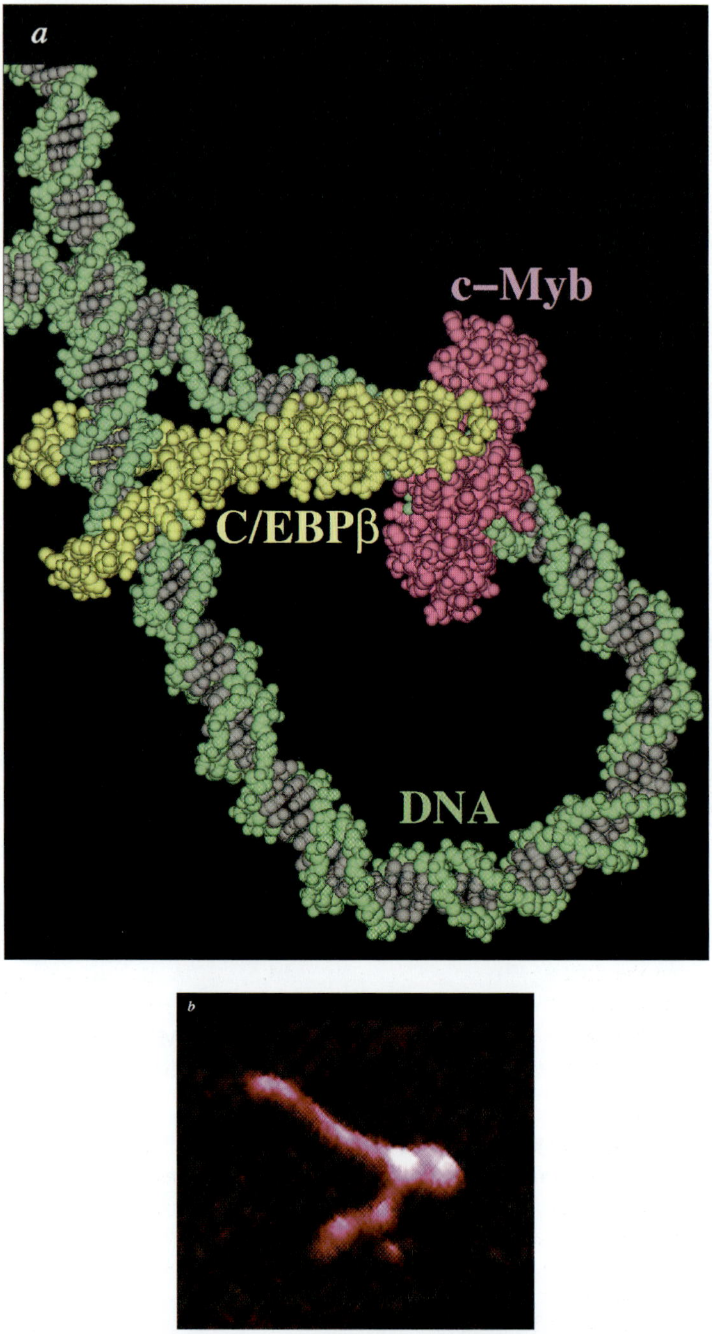

Figure 8

A modelled structure (a) and an AFM image (b) of the complex composed of c-Myb, C/EBPβ and the *mim-1* promoter DNA, showing DNA loop formation. These figures were adopted from Tahirov et al. (2002) (the issue cover) and Ogata et al. (2003). *(see chapter 11, p.234)*

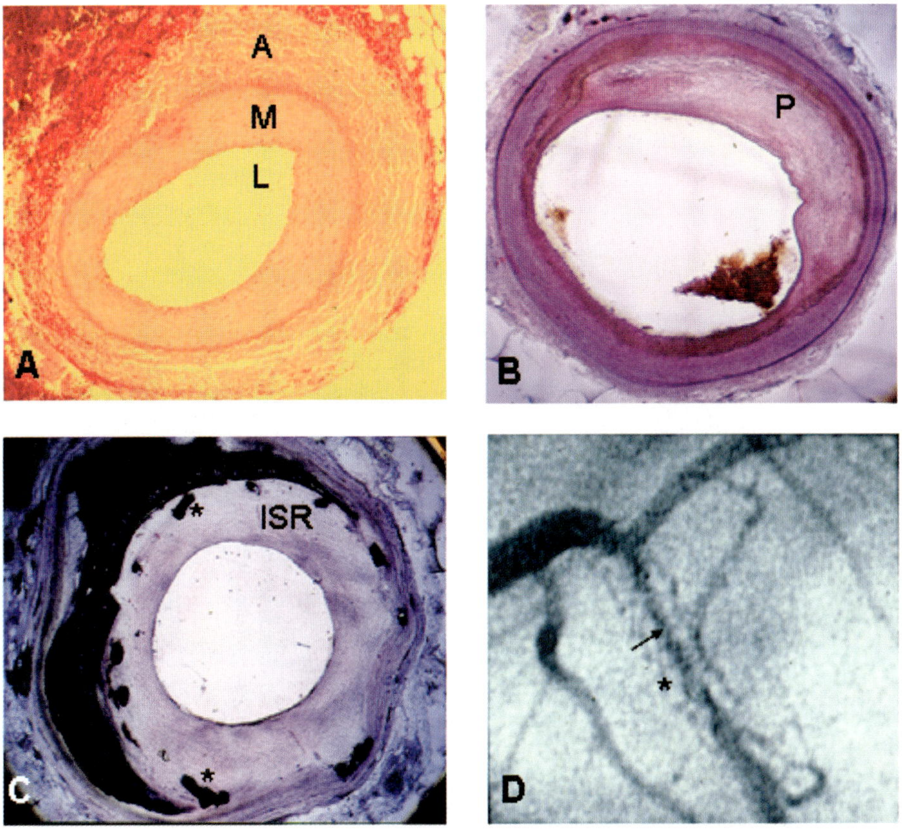

Figure 1

Histology of normal and diseased arteries. A. Transverse histological cross-section of normal artery showing the different layers of the vessel wall; lumen (L), media (M) and adventitia (A). In a normal artery, the intimal layer consists of a single layer of endothelial cells lining the vessel wall and is not visible in this photomicrograph. B. Transverse histological cross section of an artery showing features of atherosclerosis. The plaque (P) region of the vessel wall is contained within the thickened intima. Thinning of the medial layer is also present. C. Transverse histological cross section of an atherosclerotic artery that has been implanted with a stent. Asterisks represent stent struts and ISR represents in-stent restenosis encroaching on the vessel lumen. D. Coronary angiogram showing in-stent restenosis. For an angiogram image, the patient's artery is injected with a radio-opaque contrast media and visualised under X-ray. The vessel lumen appears in black. The arrow indicates restricted blood flow caused by narrowing of the blood vessel lumen. The asterisk represents a region of in-stent restenosis. The shaded appearance surrounding the vessel lumen represents the stent struts that are radio-opaque and indicates the original vessel lumen prior to onset of in-stent restenosis. *(see chapter 17, p.333)*

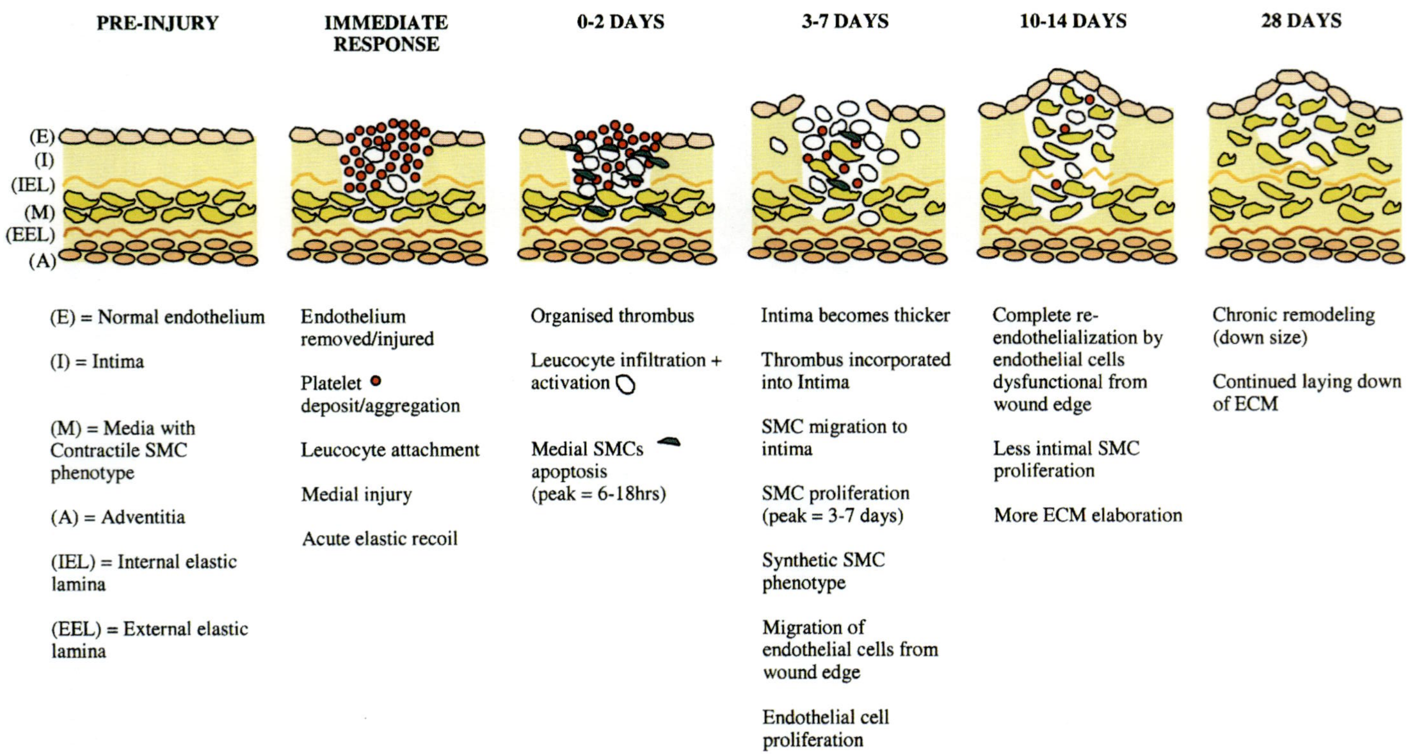

Figure 2

The possible mechanisms and time course of restenosis following percutaneous coronary intervention *(see chapter 17, p.336)*

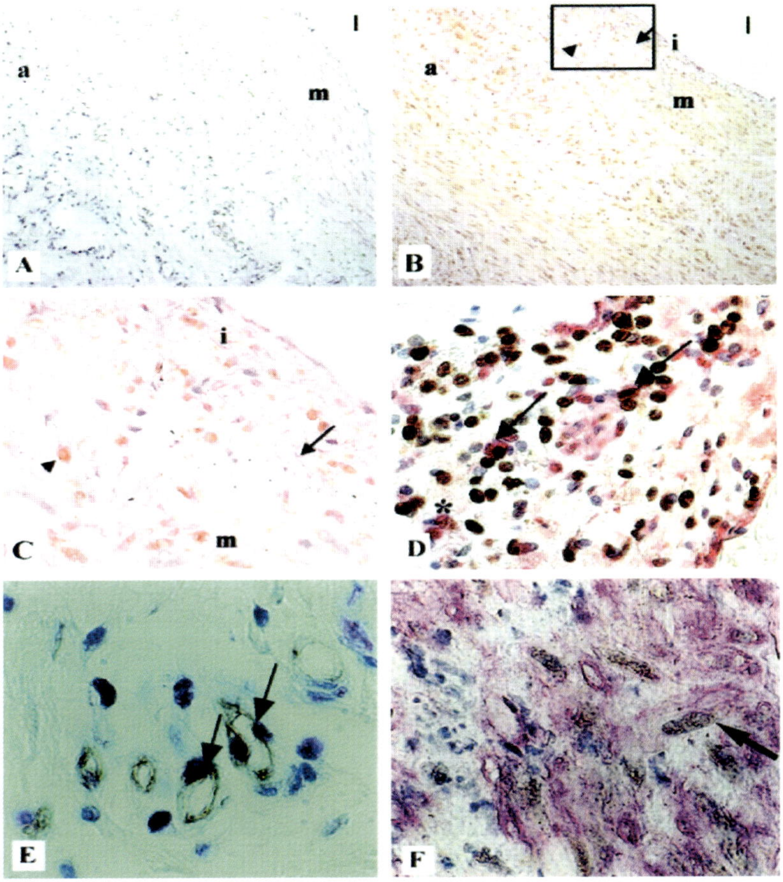

Figure 3

c-Myb expression and control in balloon injured pig coronary artery. A. Transverse histological section of a control pig coronary artery immunostained for c-Myb. Note the minimal positive staining. l indicates lumen; m, media; and a, adventitia (original magnification x20). B. Seven days after angioplasty. Numerous c-Myb-positive cells can be seen within the media (m, arrowhead) and are also present within the intima (i, brown). The arrow indicates internal elastic lamina (original magnification x20). C. High-power view of boxed area, shown in panel B, 7 days after angioplasty (original magnification x100). D. Six hours after angioplasty. A marked inflammatory infiltrate of CD68-positive cells (brown) with some positive for c-Myb staining is shown (red, arrow). Some c-Myb-positive cells are CD68 negative (*). The area of inflammation is localised to a region of trauma (original magnification x100). E. Three days after angioplasty showing adventitial microvessel stained positive for dolichos biflorus–lectin (brown). Some of the endothelial cells are c-Myb positive (red, arrows) (original magnification x100). (Taken from Lambert et al., 2001).
(see chapter 17, p.339)

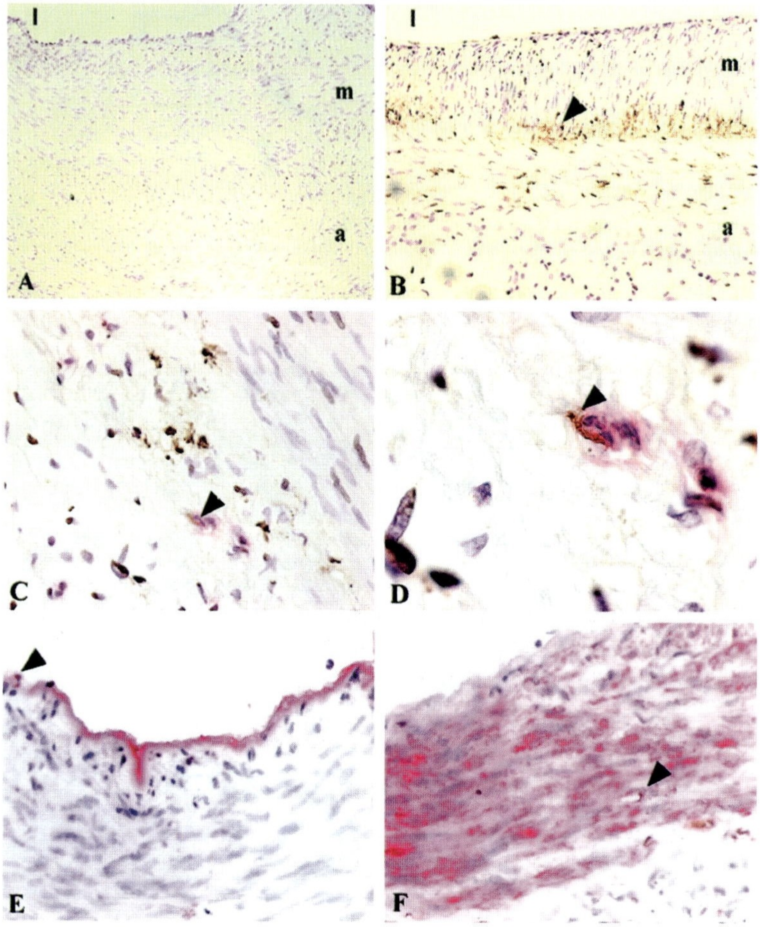

Figure 4

Pig coronary artery obtained 6 hours after balloon injury and delivery of c-myb antisense. A. Control artery that has undergone the TUNEL procedure. Note the lack of TUNEL-positive cells. l indicates lumen; m, media; and a, adventitia (original magnification x20). B. TUNEL-positive cells in the balloon-injured vessel showing brown staining and characteristic nuclear condensation. The majority of TUNEL-positive cells are located within the outer media (arrowhead) (original magnification x20). C. Macrophage stained with Mac387 (red, arrowhead) and TUNEL (brown) (original magnification x40). D. High-power view of area indicated by arrowhead in panel C (original magnification x100). E. von Willebrand factor antigen staining showing the vascular endothelial layer and a TUNEL-positive cell (arrowhead) (original magnification x40). F. α-actin-stained artery showing TUNEL-positive smooth muscle cell (arrowhead) (original magnification x40). (Taken from Lambert et al, 2001). *(see chapter 17, p.341)*

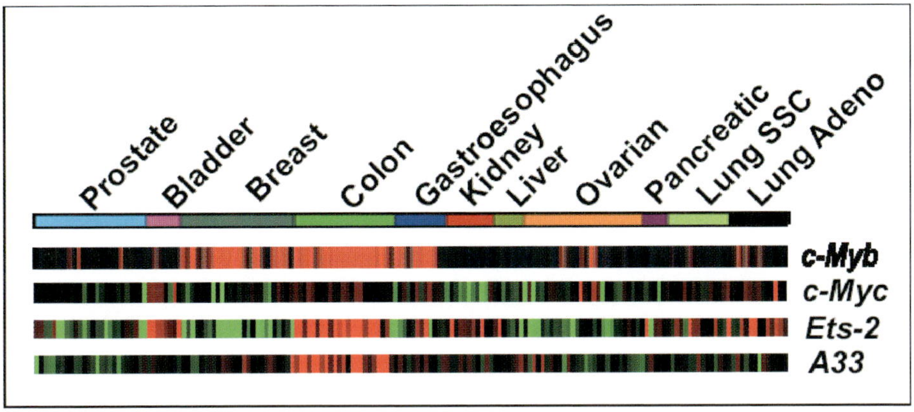

Figure 7

c-Myb over-expression is a feature of haemopoietic malignancies, however a recent survey of 174 epithelial tumours categorised by tissue of origin by Su et al (2001) demonstrated the relatively high expression (Red - high expression; Green - low expression) of *c-myb* in colon, gastroesophageal and breast cancers. For comparison two other transcription factors *c-myc* and *ets-2* and the intestine-specific gene *A33* are shown. *(see chapter 19, p.377)*

Chapter 1

EVOLUTION OF MYB PROTEINS

Colin Davidson*, Emily Ray* and Joseph Lipsick
Departments of Pathology and Genetics, Stanford University School of Medicine Stanford, CA 94305-5324, United States of America.
**These authors contributed equally.*

Abstract: The Myb domain is a highly conserved 50 amino acid motif found in all eukaryotes and often in tandem repeats within a single protein. DNA binding proteins with three such repeats (3R) are present in animals, plants, fungi and protists. Invertebrate animals have a single 3R *myb* gene, whereas all vertebrates studied thus far have at least three such genes. The 3R *myb* genes of vertebrates arose by two duplications from a B-*myb*/Dm-*myb*-like ancestral gene. The first duplication appears to have resulted in a neomorphic paralogue with a central transactivation domain. This new gene then duplicated again to give rise to A-*myb* and c-*myb*. An examination of a wide variety of more distantly related Myb domain-containing proteins suggests that the most highly conserved function of the Myb domain may be interaction with chromatin proteins rather than with DNA.

1. THREE REPEAT (3R) MYB PROTEINS

1.1 Introduction

Representative species from all kingdoms of eukaryotic life possess proteins that contain three repeated (3R) Myb domains (Table 1). In support of the ancient origin of 3R *myb* genes, a 3R *myb* gene has been identified from the draft genome sequence of the human intestinal parasite *Giardia lamblia* (Sun et al., 2002). *Giardia* are diplomonads, members of a deep branch of eukaryotes that lack mitochondria and peroxisomes and rely on a number of bacterial metabolic pathways (Hashimoto et al., 1995; Roger et al., 1998). Genes encoding 3R Myb proteins have been identified in other protists such as the cellular slime mold, *Dictyostelium discoideum*, and in all major plant lineages (Stober-Grasser et al., 1992; Braun et al., 1999; Kranz et al., 2000). In animals 3R Myb proteins have been identified in

J. Frampton (ed.), Myb Transcription Factors: Their Role in Growth, Differentiation and Disease, 1-33.

invertebrates (*Drosophila*), invertebrate chordates such as sea urchin (*Stongylocentrotus purpuratus*) and *Ciona intestinalis*, bony fish (Fugu and Zebrafish), amphibians (*Xenopus*), birds (chicken) and mammals (human, mouse and cow). Interestingly, not all animals possess 3R *myb* genes; analysis of the completely sequenced genome of the free-living nematode *C. elegans* indicates a 3R *myb* gene does not occur in this organism. Yet, the *C. elegans* genome does contain other Myb domain containing proteins such as CDC5. A 3R *myb* gene is not the only gene 'absent' from the genome of *C. elegans*; cancer genes such as *p53*, *neurofibromatosis type 1*, and the two genes implicated in tuberous sclerosis (*TSC1* and *TSC2*), are conserved between humans and *Drosophila* yet extraordinarily are not present in the *C. elegans* genome (Rubin et al., 2000). Whatever the function of Myb is in invertebrates, and by analogy B-Myb in vertebrates, it appears *C. elegans* has dispensed with the requirement for this protein.

It has been extensively reported that genome or large chromosomal region duplications may be responsible for the structure and evolution of vertebrate genomes (Abi-Rached et al., 2002; McLysaght et al., 2002). While controversial, it has been proposed that vertebrate genome evolution has occurred through two whole-genome duplication events that are thought to have taken place early in vertebrate evolution approximately 500 million years ago (Ohno, 1999). The high degree of sequence identity and the location of genes for human A-Myb (HSA 8q22), B-Myb (HSA 20q13.1) and c-Myb (HSA 6q22) on separate chromosomes suggest the vertebrate 3R Myb family benefited from large chromosomal, or possibly whole genome, duplication events. Other gene families that number three or greater in similar chromosomal regions as the 3R *myb* genes include the genes that encode the WNT1 inducible signaling pathway proteins (WISP1, 8q24.1-q24.3; WISP2, HSA20q12-g13.1; WISP3, 6q22-q23), D52 tumour-associated proteins (TPD52, HSA8q21; TPD52L2, HSA20q13.2-q13.3; TPD52L1, 6q22-q23), the pleiomorphic adenoma genes (PLAG1, 8q12; PLAGL2, HSA20q11.1; PLAGL1, HSA6q24-q25) and the eyes absent proteins (EYA1, HSA8q13.3; EYA2, HSA20q13.1; EYA4, HSA6q23; EYA3, HSA1q36). Interestingly, chromosomal regions 8q, 16q, 18q, 20q, 1p and 2p form a paralogon, or a series of paralogous regions that likely derived from a common ancestral region, in the human genome (http://195.220.67.166/paradb/) (Leveugle et al., 2003). In contrast, the genomes of invertebrates, including the urochrodate *Ciona intestinalis*, encode a single 3R *myb* gene that most closely resembles vertebrate B-Myb by sequence identity.

Outside of the repeated Myb domains, the plant *Giardia* and *Dictyostelium* 3R Myb proteins do not possess significant sequence identity to the 3R animal Myb proteins. For this reason, we have chosen to investigate the evolution of 3R animal proteins. In this section we review the domain structure conservation and evolution of animal 3R Mybs with special consideration given to newly isolated sequences from the draft genome sequence of *Fugu rubripes* (http://fugu.hgmp.mrc.ac.uk) and *Ciona intestinalis* (http://genome.jgi-psf.org/ciona4/ciona4.home.html). cDNA sequences from these newly sequenced organisms were predicted from the draft genome sequence by homology to previously isolated 3R *myb* sequences. In the case of the *Ciona myb* sequence and Fugu B-*myb*2 the complete cDNA sequence could not be determined due to incomplete assembly of the genome sequence for these genes. Based on phylogenetic analysis of vertebrate and invertebrate 3R *myb* sequences it is likely that the evolution of the vertebrate 3R *myb* gene family occurred via the proposed two rounds of vertebrate genome duplications thought to be responsible for the duplication and divergence of other vertebrate gene families. In addition, *Fugu* and possibly all bony fish have undergone further lineage-specific duplications generating additional B-*myb*-like genes.

1.2 Domain Structure of 3R Myb Proteins

Comparative sequence analysis, biochemical and molecular investigations of the v-Myb oncoprotein of avian myeloblastosis virus and its cellular counterpart c-Myb have identified important domains characteristic of 3R Myb proteins (Ganter et al., 1999).

1.2.1 Amino-terminal acidic domain

The amino-terminal region of 3R Myb proteins is highly acidic; this region is conserved across all animal 3R proteins investigated with the exception of the *Ciona* Myb sequence due to difficulty in unequivocally identifying this exonic region from the draft genome sequence (Figure 1, acidic). The function of this domain is unknown; however, its deletion by retroviral insertional mutagenesis in the murine and chicken c-*myb* genes results in oncogenic activation of c-Myb (Shen-Ong et al., 1984; Jiang et al., 1997). Immediately amino terminal to these acidic residues in c-Myb are casein kinase II (CKII) phosphorylation sites that appear to be conserved across all c-Myb sequences. There have been conflicting results reporting that phosphorylation of these sites can both reduce and increase the DNA-binding activity of c-Myb (Lüscher et al., 1990; Oelgeschlager et al., 1995; Ramsay et al., 1995). However, mutation of these sites is not sufficient for oncogenic activation (Dini et al., 1993).

1.2.2 Myb domain

The presence of three imperfect tandem Myb repeats (R1, R2 and R3) form the DNA binding domain of c-Myb and the other animal Myb proteins. Each repeat is approximately 50 amino acids in length and contains three conserved tryptophan resides that form a hydrophobic core important for the structural integrity of the DNA-binding domain (Anton and Frampton, 1988; Ogata et al., 1994). These tryptophan residues are conserved in all three Fugu orthologues and the novel partial Fugu B-Myb2 sequence (Figure 1, R1, R2 and R3). The second tryptophan residue in the R1 repeat of the

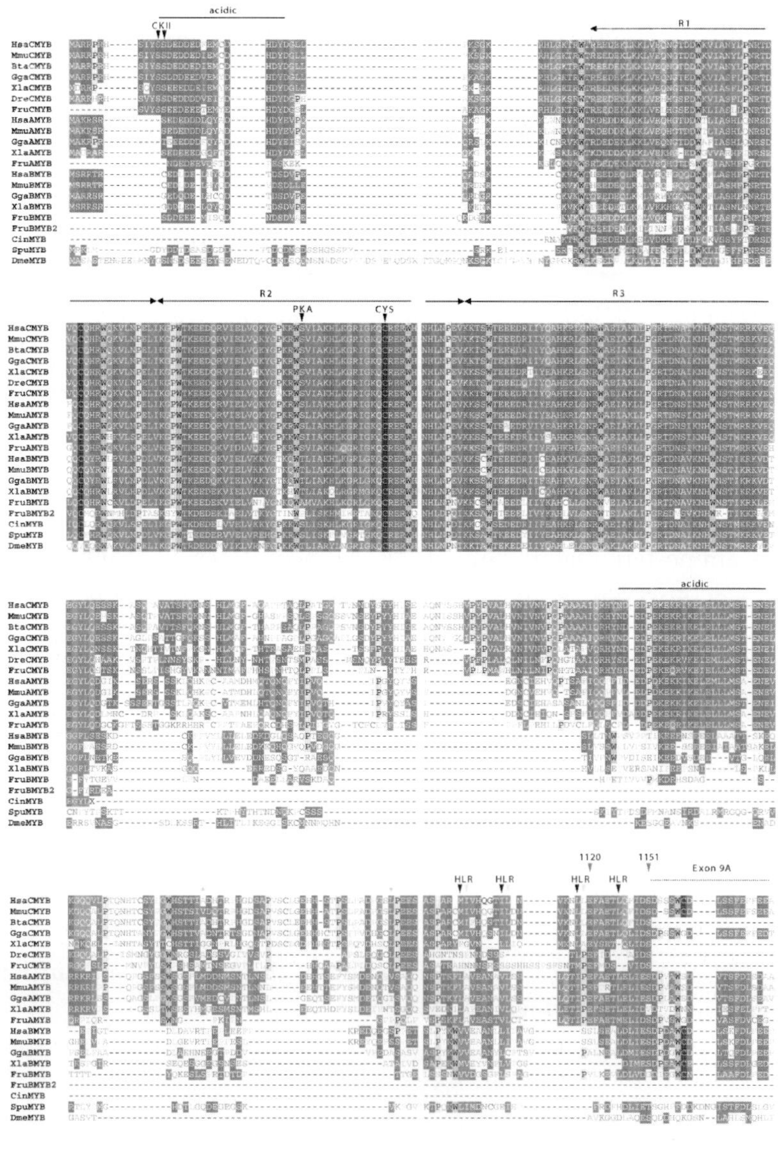

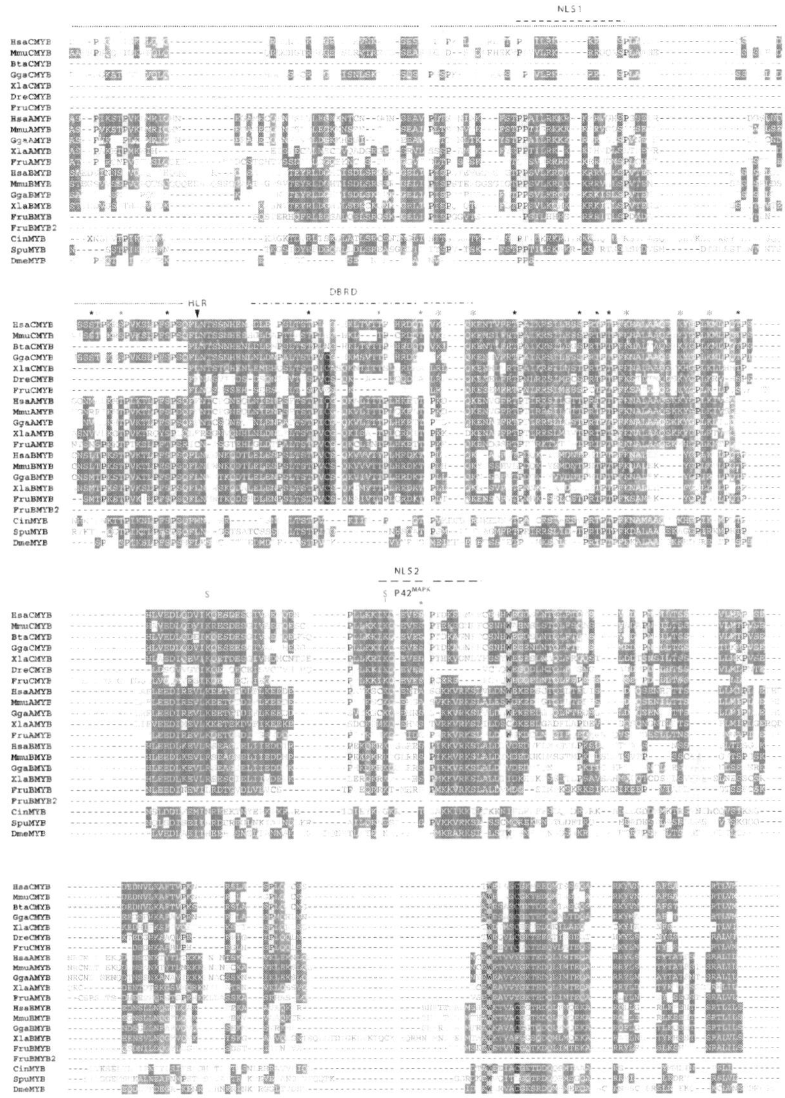

Figure 1

Amino acid alignment of invertebrate and vertebrate 3R Myb proteins. A multiple sequence alignment of complete and partial animal 3R Myb proteins was generated using ClustalW (Thompson et al., 1994) and edited with colour coding according to residue conservation using BioEdit (http://www.mbio.ncsu.edu/BioEdit/bioedit.html). The extreme amino terminal exonic regions were not determined for the Fugu and *Ciona* sequences. In addition, XX in the *Ciona* sequence indicates exonic regions that were indeterminable from the genomic sequence due to either poor conservation or genome assembly errors. The complete Fugu B-myb2 cDNA sequence could not be determined due to unfinished assembly of the genome sequence

for this gene. The XX near the carboxyl terminus of *Xenopus* B-Myb indicates residues that were unaligned and not included to save space. Exon 9A encoded sequences were not available for bovine, *Xenopus*, Zebrafish and Fugu c-Myb sequences. CKII, Casein kinase II phosphorylation site; R1, R2 and R3 indicate the Myb repeats; PKA, protein kinase A phosphorylation site; CYS, redox modified cysteine residue; HLR, heptad leucine repeat, black and light blue arrowheads (the black arrowhead has been referred to as the "leucine zipper"); NLS1, nuclear localisation signal 1; 1120 and 1151, carboxy-terminal truncation mutants of AMV v-Myb; DBRD, DNA binding regulatory domain; NSL2, nuclear localisation signal 2; black asterisks, highly conserved putative phosphorylation sites for proline-directed kinases; putative cyclin A1 and A2/Cdk2 phosphorylation sites are indicated by the red asterisks and putative B-Myb phosphorylation sites are shown in green. Modified lysine residues are indicated by blue asterisks (acetylation) and **S** (sumoylation). Abbreviations: Hsa, *Homo sapiens* (Human); Mmu, *Mus musculus* (Mouse); Bta, *Bos taurus* (Cow); Gga, *Gallus gallus* (Chicken); Xla, *Xenopus laevis* (Xenopus); Dre, *Danio rerio* (Zebrafish); Fru, *Fugu rubripes* (Fugu); Cin; *Ciona intestinalis* (Ciona); Spu, *Stongylocentrotus purpuratus* (Sea Urchin); Dme, *Drosophila melanogaster* (Drosophila).
(see colour section p. xiv-xv)

Ciona Myb sequence is conservatively substituted to a phenylalanine residue. The redox state of the cysteine residue at position 130 in R2 of c-Myb has been suggested to regulate DNA binding (Guehmann et al., 1992; Myrset et al., 1993). This cysteine residue is conserved in all the investigated sequences including the newly identified Fugu Myb sequences and the *Ciona* Myb sequence (Figure 1, CYS). Further, it has also been found that the serine residue at position 116 in R2 of c-Myb can be phosphorylated *in vitro* by cyclic AMP-dependent protein kinase A (PKA) (Ramsay et al., 1995). It is thought that phosphorylation of this site negatively influences the ability of c-Myb to bind DNA (Andersson et al., 2003). A serine residue is conserved at this position across all c-Myb, A-Myb, Sea Urchin Myb and *Ciona* Myb sequences (Figure 1, PKA). A conserved substitution of a phosphorylatable threonine residue occurs at this position in *Drosophila* Myb and all examined B-Myb sequences with exception of the two putative Fugu B-Myb sequences, where an alanine and proline occur at this position in Fugu B-Myb and B-Myb2, respectively.

1.2.3 Transcriptional activation domain, heptad leucine repeat, and exon 9A

The transcriptional activation (TA) domain of c-Myb has been mapped to the centre of the protein and is conserved within all c-Myb and A-Myb sequences identified. Both Fugu c-Myb and A-Myb display considerable conservation of the TA domain (Figure 1, acidic). However, this domain is poorly conserved and difficult to unambiguously align with the B-Myb sequences and the invertebrate Myb sequences. This is consistent with the failure to conclusively demonstrate transcriptional activation by B-Myb and the invertebrate Myb proteins except under special conditions or in particular cell lines (Mizuguchi et al., 1990; Foos et al., 1992; Watson et al., 1993; Tashiro et al., 1995; Hu et al., 1997; Lane et al., 1997; Ziebold et al., 1997). In addition, the central region of B-Myb is under far less evolutionary constraint than the central activation domains of A-Myb and c-Myb (Simon et al., 2002).

Further 3' of the central acidic domain of c-Myb is the heptad leucine repeat (HLR) region. Site-directed mutagenesis of specific leucine residues to proline activates c-Myb, possibly through inhibition of dimerisation, suggesting this region functions as a negative regulatory domain (Nomura et al., 1993). However, mutational analyses of v-Myb (mutant 1120 versus 1151) have demonstrated that this region, but not the leucine residues, is essential for both transcriptional activation and oncogenic transformation (Ibanez et al., 1988; Ibanez et al., 1990; Fu et al., 1996). Aliphatic amino acid residues can be aligned to important residues in the HLR in all c-Myb sequences with the exception of the *Xenopus*, Zebrafish and Fugu c-Myb sequences (Figure 1, HLR). There appears to be limited conservation of this region within A-Myb, B-Myb and invertebrate Myb sequences. However, the amino acids required for v-Myb transformation and transcriptional activation (EFAETLQLID) constitute a well-conserved portion of the HLR region across all animal Myb proteins (Figure 1, 1120-1151).

Immediately carboxy terminal to the HLR region is the alternatively spliced exon 9A of c-Myb. The larger protein, an approximately 120 amino acid increase, constitutes 20% or less of the total c-Myb protein but it has been reported that this alternative splicing event can increase the transcriptional activation of the c-Myb protein (Rosson et al., 1987; Shen-Ong et al., 1989; Woo et al., 1998). c-Myb sequences isolated from cow, *Xenopus*, zebrafish and Fugu lack this alternatively spliced exon, possibly due to incomplete sampling of the c-Myb transcripts from these species. However, examination of the Fugu c-*myb* gene indicates that exon 9A is very unlikely to occur as a splice variant in the Fugu c-Myb protein as a mere 483 base pairs separate the surrounding exons. The intronic and intergenic distances are greatly reduced in Fugu (Brenner et al., 1993) and it would appear that the compaction of the Fugu genome resulted in the loss of exon 9A from the Fugu c-*myb* gene. While alternatively spliced in c-Myb, the exon 9A sequence is conserved and rarely if ever spliced within A-Myb, B-Myb and invertebrate Myb sequences (Figure 1, exon 9A). Within this region a nuclear localisation signal has been mapped in *Xenopus* B-Myb (Figure 1, NLS1), mutation of lysine residues in this region abolished the nuclear localisation of *Xenopus* B-Myb (Humbert-Lan et al., 1999). With the exception of *Drosophila* Myb, this NLS domain, including the critical lysine residues, is conserved across all animal Myb sequences with identity to exon 9A (Figure 1, NLS1).

1.2.4 Negative regulatory region and post-translational modifications

The conserved carboxyl terminus of c-Myb is deleted in v-Myb and experimental evidence indicates this portion of the c-Myb protein acts to negatively regulate the transcriptional activation of c-Myb (Sakura et al., 1989; Hu et al., 1991; Dubendorff et al., 1992). The extreme carboxy terminus of all the animal 3R Myb proteins is well conserved (Figure 1). The last 88 residues of *Xenopus* B-Myb have been shown to negatively regulate the nuclear import of this protein presumably through intra- or intermolecular interaction with the putative nuclear localisation signals in the protein (Humbert-Lan et al., 1999). Previous experiments also indicate that the carboxyl terminus can inhibit or directly regulate DNA-binding of c-Myb (Ramsay et al., 1986; Tanaka et al., 1997).

In addition to the amino-terminal CKII and PKA phosphorylation sites, a number of other phosphorylation sites occur in c-Myb and v-Myb. Of the eight potential MAP kinase (proline-directed protein kinases) phosphorylation sites, four are conserved across all of the animal Myb proteins examined (Figure 1, black asterisks). Two-dimensional tryptic phosphopeptide mapping identified serine residue 533 of avian c-Myb as a phosphorylation target by $p42^{MAPK}$ kinase *in vitro* (Aziz et al., 1993). This serine residue is conserved in all identified c-Myb sequences (Figure 1, $p42^{MAPK}$). Conservation of phosphorylation sites across all animal 3R Myb proteins suggests a possible common regulation of 3R Myb proteins by MAP kinases through modulation of the conserved negative regulatory domain.

In *Xenopus* B-Myb the serine/threonine-rich region of proposed phosphorylation sites includes a second putative nuclear localisation signal shown to be required for the nuclear import and a region implicated in the negative regulation of DNA-binding of this protein (Humbert-Lan et al., 1999). The second nuclear localisation signal is well conserved in all A-Myb, B-Myb and invertebrate Myb sequences investigated with limited conservation in vertebrate c-Myb sequences (Figure 1, NLS2). The DNA-binding regulatory domain identified in *Xenopus* B-Myb spans a highly conserved phosphorylation motif consisting of three amino acids: theroine/serine, proline and aliphatic amino acids (valine, isoleucine or leucine) (Figure 1, DBRD). In addition, the B-Myb protein has been shown to be specifically phosphorylated during S-phase by cyclin A1 and A2/Cdk2 and cyclin E/Cdk2 at sites within the negative regulatory domain (Robinson et al., 1996; Sala et al., 1997; Saville et al., 1998; Bartsch et al., 1999; Johnson et al., 1999; Muller-Tidow et al., 2001). Threonines 447, 490 and 497 and serine 581 of murine B-Myb have been shown to be phosphorylated *in vitro* by cyclin A1 and A2/Cdk2 (Figure 1, red asterisks). A phosphorylatable residue is conserved at the position of threonine 447 across all Myb sequences for which sequence is available, while the remaining residues are highly conserved across the vertebrate Myb sequences. Further studies have identified additional putative B-Myb phosphorylatable residues with limited conservation across all investigated sequences (Figure 1, green asterisks) (Johnson et al., 1999; Johnson et al., 2002).

Recently, c-Myb has been shown to be acetylated within the negative regulatory domain by the CREB-binding protein (CBP) and p300 (Tomita et al., 2000; Sano et al., 2001). Both CBP and p300 possess histone acetyltransferase activity with CBP shown to acetylate five lysine residues in c-Myb (Figure 1, blue asterisks). Substitution of three lysine residues to arginine (K467R, K476R, and K481R) increased the transcriptional activity of c-Myb (Tomita et al., 2000), whereas substitution of all five residues by arginine (K438R, K441R, K467R, K476R, and K481R) decreased transcriptional activity (Sano et al., 2001). Only a single lysine is conserved at these positions across all identified 3R Myb sequences. However, the density of lysine residues in this region in all the investigated sequences suggest that modification of 3R Myb proteins by acetylation may be conserved across all 3R Myb paralogues and orthologues. Indeed, co-transfection experiments in a *Drosophila* cell line demonstrate an interaction between *Drosophila* CBP and Dm-Myb and that B-Myb is acetylated by p300 (Hou et al., 1997; Johnson et al., 2002).

Further lysine modifications occur to the c-Myb negative regulatory domain through SUMO-1 (small ubiquitin-related modifier) conjugation (Bies et al., 2002; Dahle et al., 2003). Two lysine residues have demonstrated SUMO-1 modification in c-Myb (Figure 1, indicated by a red "S") with substitution mutations of these residues to arginine (K503R and K527R) resulting in an enhancement of c-Myb transcriptional activity. Lysine residues at this position are conserved in all A- and c-Myb orthologues with the C-terminal lysine residue (K527) conserved in all vertebrate Myb sequences examined. However, the consensus sumoylation sequence (ΦKxE, Φ is hydrophobic) is not conserved in B-Myb and invertebrate 3R Myb sequences suggesting that sumoylation may be a mechanism of negatively regulating 3R Myb proteins with a transcriptional stimulatory activity.

1.3 Evolution of the 3R myb Gene Family

Despite the completion of a number of draft vertebrate genomes, the hypothesis that ancient genome duplications contributed to vertebrate genome evolution has yet to be rigorously tested (Skrabanek et al., 1998). The number and timing of genome duplications are still contentious issues, however, the most popular hypothesis supported by analysis of Hox gene clusters proposes two duplications: the first duplication occurred early in vertebrate evolution after the divergence of jawless vertebrates from non-vertebrate chordates such as amphioxus and a second duplication occurring subsequent to the divergence of jawless vertebrates from jawed vertebrates (Garcia-Fernandez et al., 1994; Holland et al., 1994). To assign identity to the newly isolated Fugu 3R *myb* sequences and determine the role of a gene duplication event in the generation of the 3R *myb* gene family we performed phylogenetic analysis on the 3R Myb domain from invertebrate and vertebrate Myb sequences (Figure 2).

The tree topology clearly supports two successive gene duplication events in the generation of the vertebrate 3R *myb* gene family (Figure 2, pink circles). The presence of A-, B- and c-*myb* genes in Fugu indicates that the gene duplication events that gave rise to the vertebrate 3R *myb* gene family occurred prior to the divergence of lobe- and ray-finned fish over 450 million years ago and lends support to the proposal that vertebrate genome evolution benefited from two rounds of whole-genome duplication. To more accurately date the timing of these gene duplications it will be of interest to determine whether jawless fish such as lamprey and hagfish possess all three 3R *myb* paralogues and if amphioxus, phylogenetically the closest non-vertebrate chordate, possesses an intermediate number of 3R *myb* genes or only a single gene. On the basis of the most common vertebrate genome duplication hypothesis one would predict that jawless vertebrates would possess two 3R *myb* genes and amphioxus would have a single 3R *myb* gene.

To account for the number of paralogous 3R *myb* genes found in jawed vertebrates following two rounds of gene or genome duplication some gene loss must have occurred. The second gene duplication event that gave rise to the A- and c-*myb* gene clades must have also generated a second B-*myb* paralogue. The loss of this second B-*myb* gene must have occurred early in vertebrate evolution as this gene is absent from all jawed vertebrates examined. Interestingly, the tree topology indicates that further Fugu, and possibly bony fish-specific, gene duplication events must have occurred more recently to account for the second Fugu B-*myb*-like gene (Figure 2). It will be of interest to determine the expression pattern and role of this additional B-*myb*-like gene in Fugu.

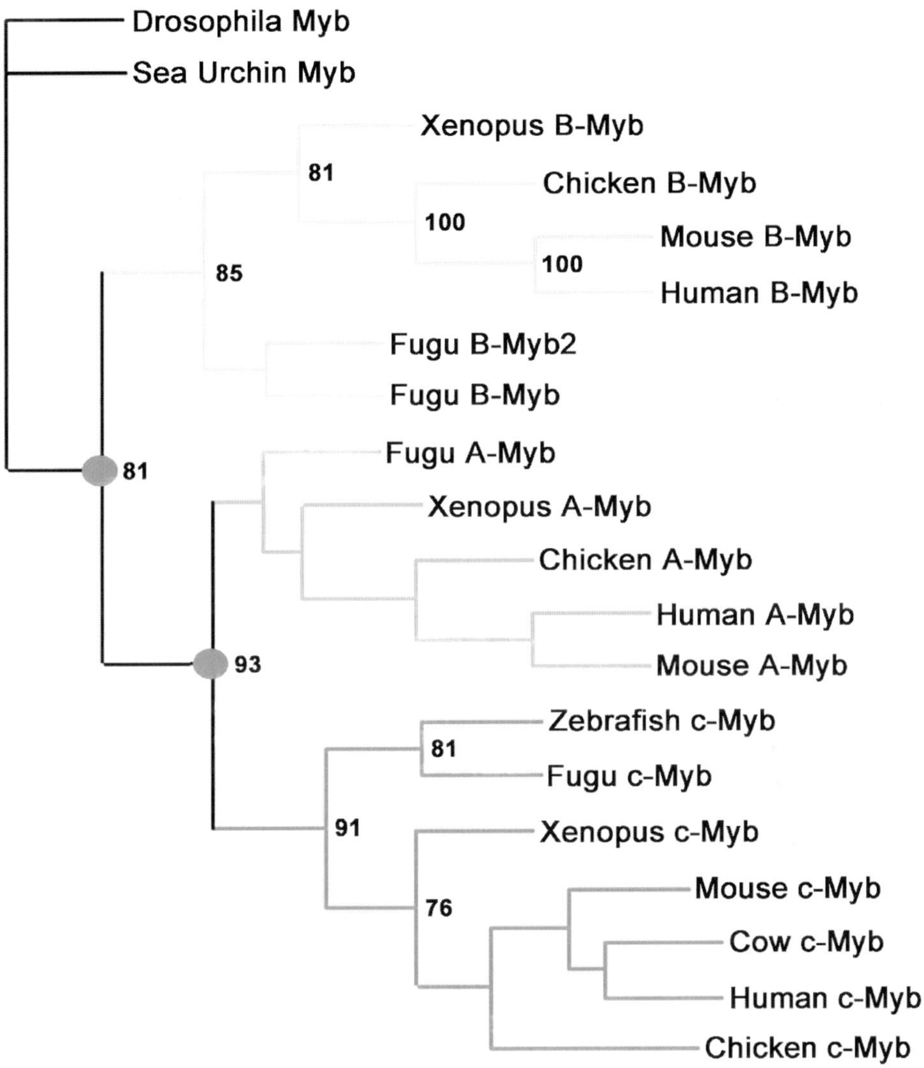

Figure 2

Consensus tree illustrating the phylogenetic relationship of animal 3R Myb proteins. The unrooted tree topology was estimated through neighbour joining using the distances calculated from an alignment of the R1, R2 and R3 domains using the Dayhoff-PAM substitution model in Phylip 3.5. The numbers at the nodes represent percent bootstrap support based on 1000 iterations. For purposes of clarity bootstrap values below 75% are not shown. Invertebrate 3R Myb sequences are shown in black, the B-Myb clade is shown in yellow, the A-Myb clade is shown in cyan and the c-Myb clade is shown in magenta. The pink circles indicate putative gene duplication events. (see colour section p. xvii)

The central acidic transactivation domain is the most striking difference between c-Myb and A-Myb versus B-Myb and the invertebrate Myb sequences (Figure 1, acidic). Analysis of vertebrate A-, B- and c-Myb sequences using a maximum likelihood based method for identifying evolutionarily constrained regions identified the transcriptional activation domain as a region under constraint in A- and c-Myb proteins but not in B-Myb (Simon et al., 2002). The evolutionary constraint on this region of A- and c-Myb proteins suggests that it is functionally important in these paralogues, whereas it is less important in B-Myb and the invertebrate Myb proteins. The accepted evolutionary relationship of *myb* genes predicts the origin of evolutionary constraint and the origin of the transcriptional activation function occurred in the ancestral A- and c-*myb* gene, prior to the second gene or genome duplication event, and subsequent to the separation from the B-*myb* lineage. It is also plausible that changes to the coding region of the ancestral A- and c-*myb* gene, such as the acquisition of the transactivation domain, also correlate with the changes that resulted in the tissue-specific expression of the A- and c-*myb* paralogues compared with the ubiquitous expression of B-*myb* and the invertebrate *myb* genes (Table 1).

The rapid progress in genome sequencing and analysis has revealed that eukaryotic genomes contain a surprisingly large number of duplicated genes (Prince et al., 2002). For example, at least 15% of the known human genes are recognisable as duplicates (Li et al., 2001). These duplicated genes can occur in tandem arrays (e.g. the *Hom/Hox* gene clusters), dispersed duplications (e.g. the three different immunoglobulin chain genes), or as whole genome duplications (e.g. the polyploid nature of maize). A comparison of genome sizes first led to the hypothesis that the prototypic vertebrate genome evolved by two rounds of genome-wide duplication (Ohno 1970). Consistent with this model, many vertebrate multi-gene families are represented by a single homologue in modern invertebrate species such as the sea urchin, fruit fly, and nematode (Holland et al., 1994). In some vertebrate species, including the salmonid fish, *Xenopus laevis* and the red viscacha rat, additional genome-wide duplications appear to have occurred much more recently (Bailey et al., 1978; Hughes et al., 1993; Gallardo et al., 1999). A major question in the field of gene and genome evolution concerns the mechanism by which duplicated genes are retained in the face of constant selective pressure. Current theories postulate three alternative fates for duplicated genes: (i) one copy is lost by genomic rearrangements or rendered non-functional by mutations (non-functionalisation); (ii) both copies are retained due to a rare mutation in one copy that creates a selective advantage (neo-functionalisation); (iii) both copies are retained due to complementary loss-of-function mutations (sub-functionalisation) (Prince et al., 2002). The sub-functionalisation model has recently gained in popularity, in large part due to comparative studies of the *Hox* genes of mice and zebrafish, the latter having undergone an additional genome-wide duplication followed by selective losses of duplicates. However, it remains unclear whether such studies of relatively recent duplications of an unusually large tandem array of duplicated genes (*Hox* clusters) can be generalised to explain the many gene duplications that occurred during vertebrate evolution from a common ancestor shared with modern invertebrates.

We propose that duplication of a B-*myb*-like ancestral gene was followed by the acquisition of the central activation domain in one gene copy, thereby imparting a neomorphic function to an A-*myb*/c-*myb*-like gene. The latter gene then duplicated resulting in the genesis of the closely related A-*myb* and c-*myb* genes that presumably were retained due to mutation and sub-functionalisation. While sub-functionalisation of the ancestral A- and c-*myb* genes likely contributed to their survival, it is tempting to speculate that some functions remain conserved between c-Myb and B-Myb (Campanero et al., 1999). In particular, the potential role of c-Myb in control of the cell cycle. Hence, both c-*myb* mRNA and protein expression begin in late G_1 and continue into S-phase when resting lymphocytes are stimulated to divide (Torelli et al., 1985; Lipsick et al., 1987). Similarly, B-*myb* mRNA is induced in late G_1 and early S-phase when quiescent cultured fibroblasts start cycling (Lam et al., 1992). Like other S-phase genes such as *cdc2*, *cyclin A*, *thymidylate synthetase*, *ribonucleotide reductase* and *E2F-1*, the B-*myb* promoter contains binding sites for E2F-1, a transcription factor responsible for negatively regulating gene expression in G_0 and early G_1 (Lam et al., 1993; Zwicker et al., 1996). Similarly, the c-*myb* promoter contains binding sites for E2F (Campanero et al., 1999). Although controversial, experiments with antisense oligonucleotides have suggested that c-*myb* expression is required for S-phase progression of haemopoietic cells (Gewirtz et al., 1989; Burgess et al., 1995). Potentially the cell cycle role is an ancestral function shared between c-Myb and B-Myb, with the acquisition of the transactivation domain leading to sub-functionalisation of the ancestral A- and c-Myb and a role as transcriptional activators. The B-Myb protein is the most closely related vertebrate Myb sequence to that of *Drosophila* Myb. Consistent with an ancestral cell-cycle role, *Drosophila* Myb (Dm-Myb) has been shown to localise to replicating DNA in mitotically dividing larval brain cells and endocycling larval fat body cells (Manak et al., 2002a). Loss of Dm-Myb function results in mitotic arrest and genomic instability (Fung et al., 2002; Manak et al., 2002a). Furthermore, Dm-Myb has been implicated in the unlicensed replication of the chorion loci in *Drosophila* ovarian follicle cells, with the protein shown to bind both *in vitro* and *in vivo* to site-specific DNA replication enhancer elements and to be required for amplification of the chorion gene loci (Beall et al., 2002).

In summary, comparative sequence analysis of animal 3R Myb proteins reveals the conservation of a number of important domains and motifs in the newly isolated sequences from Fugu and *Ciona*. Phylogenetic analysis suggests that the vertebrate 3R *myb* gene family benefited from two rounds of gene duplication prior to the divergence of a last common ancestor of the ray- and lobe-finned fish. In addition, further lineage-specific gene duplications have occurred in Fugu generating an additional B-*myb*-like sequence. The identification of 3R *myb* sequences from amphioxus and jawless vertebrates will further our understanding of the evolution of the 3R *myb* gene family and add to our knowledge of vertebrate genome evolution.

2. PLANT-SPECIFIC MYB PROTEINS

Myb repeat containing transcription factors are highly represented in plants, with more than two hundred 2R Myb protein genes in *Maize* (Rabinowicz et al., 1999; Dias et al., 2003) and over one hundred present in *Arabidopsis* (Riechmann et al., 2000; Stracke et al., 2001). 3R Myb-family transcription factors were recently identified in *Arabidopsis* (Braun et al., 1999). Detailed phylogenetic analysis of the large family of plant Myb proteins has allowed useful insight into the evolution of the plant genome (Dias et al., 2003). Interestingly, the 3R Myb proteins appear to be the oldest of the plant Myb family of transcription factors. The 2R Myb proteins resulted from loss of the first repeat which occurred before land plants and chlorophyte algae diverged. Duplication and divergence led to an atypical two repeat protein class that carries a mutation of the highly conserved tryptophan to a phenylalanine in the first helix of the second repeat (originating from the third repeat in 3R family members). The more typical plant 2R Myb family contains a leucine insertion in the first repeat (originating from the second repeat in the 3R Mybs) which appears to have occurred before mosses and angiosperms diverged. Many gene duplication events gave rise to a large family of typical 2R Myb family members. Further diversification of the C-terminal regions led to specific classes, such as the anthocyanin regulators and a class of 2R Mybs carrying a proline to alanine substitution between the two repeats. These large duplication events appear to be genome wide since the plant Myb genes are located throughout the genome as opposed to being in specific clusters (Rabinowicz et al., 1999).

3. OTHER MYB-RELATED PROTEINS

While this book focuses on the 3R Myb family of proteins, a thorough discussion of Myb evolution requires an examination of the origin and function of the extended family of Myb-related proteins. This section will discuss Myb-like proteins, defined as those proteins containing homology to at least one repeat from the Myb DNA-binding domain. An extensive list of Myb-related proteins includes factors involved in several biological processes all of which occur in the nucleus (Table 2) (Ganter et al., 1999). These include essential components of the mRNA splicing machinery (CDC5/CEF1), proteins involved in telomere regulation (TRF1, TRF2, Taz1, RAP1), factors involved chromatin remodelling and modifying complexes (ADA2, NCoR, SWI3, RSC8, ISWI), components of basal transcriptional initiation complexes (SNAPc4, TFIIIB-B"), and rDNA transcriptional initiation/termination factors (REB1, TTF1).

Traditionally, Myb-related repeats have been thought of as DNA-binding domains. However, Myb repeats occur in proteins that do not have specific DNA-binding properties, raising the possibility of an alternate, non-DNA-binding function for this domain. Therefore, the presence of a Myb-like motif in such a diverse group of proteins may provide clues to a function separate from DNA-binding yet to be elucidated.

3.1 The Relationship of Myb Repeats Among Different Protein Classes

Analysis of the evolution of the Myb repeat across such a diverse group of proteins is complex due to the small size of the conserved Myb repeat (~40 amino acids), thus, making it difficult to build robust phylogenetic trees. Nevertheless, phylogenetic analysis of highly conserved protein classes such as the Myb, CDC5/CEF1, SWI3 and ISWI families confidently places single repeats from these proteins into separate clades (Figure 3). Interestingly, in protein families with more than one repeat, when treated separately the single repeats clustered independently. For example, in the 3R Myb family the second repeat forms a separate clade from the third. The first repeat in the 3R Myb family proved to be too divergent for confident clustering based on bootstrap values. The lack of conservation suggests that the first repeat is not under the same selective constraint as the other two repeats that do bind DNA in a sequence-specific manner. Our phylogenetic analysis supports the idea that Myb repeats in the Myb, CDC5/CEF1, SWI3, and ISWI proteins each evolved separately from a common ancestral Myb repeat. However, the evolutionary relationships between Myb repeats from other protein families could not be inferred with strong statistical significance.

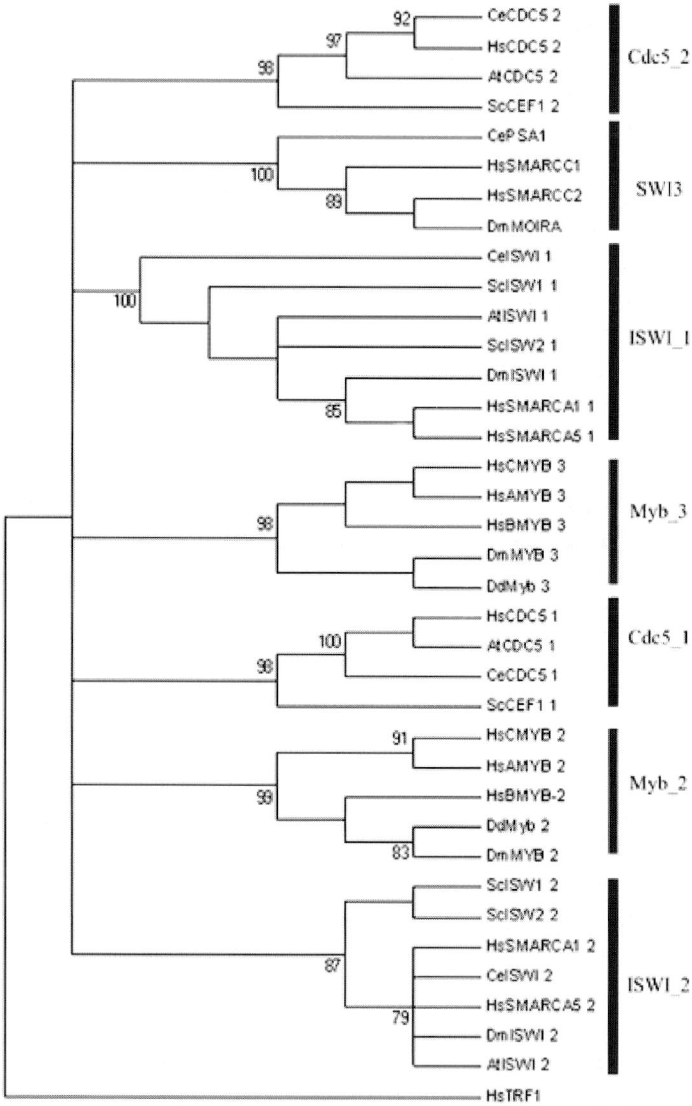

Figure 3

Phylogenetic analysis of individual Myb repeats. We performed unrooted minimum evolution analysis using a Poisson correction model with Molecular Evolutionary Genetics Analysis (MEGA) software version 2.1. This analysis generated a consensus tree from a Clustal X alignment containing Myb repeats from protein homologues of the Myb, CDC5, SWI3, and ISWI protein families. In the consensus tree shown, all Myb repeats within one of the seven groups can be inferred to share common ancestry based on bootstrap values greater than 75% (values lower than 75% are not shown). However, the order of branching which gives rise to the seven groups cannot be accurately inferred based on the alignment.
(see colour section p. xvii)

Previous work identified the telobox (Bilaud et al., 1996) and SANT domains (Aasland et al., 1996) as separate domains from the Myb repeat. In fact, the ISWI and SWI3 Myb repeats are both considered SANT domain proteins, but our analysis demonstrates that these repeats form independent clades, indicating that they should not be grouped together. In our analysis the telobox proteins did not cluster with the 3R family of Myb repeats and they could not be confidently placed in separate clade. Thus, our analysis demonstrates that phylogenetic support for some proposed subclasses of Myb repeats is weak.

The presence of Myb repeats in proteins with such diverse functions as well as the enormous expansion of the 2R Myb-family transcription factors in plants suggests an important role for Myb repeats in nuclear functions. A careful inspection of the overall alignment of Myb repeats should reveal residues important for maintaining the structure. A single Myb repeat consists of three helices maintained by a hydrophobic core (Ogata et al., 1994; Tahirov et al., 2002). Residues essential for maintaining the domain structure remain the most highly conserved (consensus in Figure 4). The greatest variation between Myb repeats from distant proteins occurs in the length of the helices and the turns between helices. The solved structure of yeast Rap1 bound to telomeric DNA constitutes a prime example. The yeast Rap1 DNA binding domain contains two distantly related Myb repeats. Analysis of the primary sequence does not readily predict the presence of Myb repeats given the existence of an insertion of 62 amino acids in the first helix of the second repeat. However, co-crystalisation with DNA demonstrated that this domain forms the classic three helix structure of a Myb repeat and binds DNA in a similar fashion (Konig et al., 1996) Also, secondary structure prediction and homology modelling of non-DNA-binding Myb repeats (ADA2, SWI3, NCoR1) strongly predict the presence of three helices in a similar orientation to those in solved Myb repeat structures (Aasland et al., 1996).

Examination of the Myb repeat sequences reveals that the DNA binding residues are not well conserved unless they are necessary for structural stability. For instance, the highly conserved arginine/lysine at position 26 (see numbering in Figure 4) binds DNA in c-Myb; surprisingly this residue is highly conserved in Myb repeats that do not bind DNA. NMR data on the second and third repeats of c-Myb demonstrates the involvement of this basic residue in the formation of a stabilising salt bridge with the highly conserved glutamic/aspartic acid residue at position 8 (Figure 4) in the first helix (Ogata et al., 1994). Thus, in addition to DNA binding, the conservation of residues at position 8 and 26 in the alignment in Figure 4 could be necessary for preservation of structural integrity.

Aside from residues essential for structural integrity, a stretch of acidic amino acids near the beginning of the first helix comprise the most conserved residues. Conservation begins with a polar threonine/serine at position 5 (Figure 4). While glutamic/aspartic acid residues are present between positions 6-11, the location of the highest concentration of acidic residues exists at positions 7, 8 and 9. The reason for such high conservation of the acidic character of the first helix remains unclear.

Conservation of the cysteine at position 32 in DNA-binding and non-DNA-binding Myb repeats may indicate a regulatory role for this residue. As previously mentioned, this cysteine residue has been proposed to mediate redox-regulated DNA-binding by c-Myb (Guehmann et al., 1992; Myrset et al., 1993). The role of this cysteine in other Myb repeats remains unclear, however its conservation suggests a possible role in functional regulation.

In some cases specific residues are conserved in a single protein class. For example, in RAP1 proteins a histidine is substituted for the highly conserved arginine/lysine at position 26. SWI3/RSC8 homologues exhibit conservation of a lysine at position 21 while ADA2 homologues show strong conservation of an acidic residue (glutamic or aspartic acid) at the same position. These residues are likely to have an important role in the function of the Myb repeat in these specific protein families, but do not have a more general structural role.

3.2 Myb Repeats as Protein-Protein Interaction Domains?

DNA-binding capacity of Myb-related proteins depends on the presence of two Myb repeats. Thus, even telomere binding proteins that contain a single repeat must homodimerise to achieve specific DNA-binding; (Bianchi et al., 1997; Spink et al., 2000). In c-Myb, recognition of DNA occurs through a basic face of the protein consisting of residues located primarily in the third helix, also known as the recognition helix (Lustig et al., 1995). The character of the third helix varies in many Myb-related proteins, being basic in many and acidic (ADA2, SWI3, RSC8) or hydrophobic (SNAPC4) in others, further suggesting that Myb repeats in these proteins are not utilised for specific DNA-binding.

Recent evidence has revealed a role for the second repeat of the c-Myb DNA-binding domain in protein-protein interactions. The co-crystal of c-Myb and C/EBPβ DNA-binding domains complexed with DNA provides evidence that physical interactions between cooperating transcription factors may play a role in their activity at promoters. Residues L106, Y110, R114, V117, K120, and H121 in the first turn and second helix of the second c-Myb repeat physically interact with DNA-bound C/EBPβ (Tahirov et al., 2002). The low conservation of these amino acids in the Myb repeats of other proteins is not surprising since one would expect residues involved in specific protein-protein interactions to be different depending on the two proteins involved. The high conservation of structural residues would be consistent with the Myb repeat acting as a structural scaffold for binding a diverse array of proteins.

A role for Myb repeats in protein-protein interactions may explain the observation that Myb-related proteins tend to function in complexes or must transiently interact with other proteins to function. For instance, some Myb-related proteins without specific DNA-binding capacity conserve the basic surface of the third helix. CDC5 functions in the splicesome, a complex of at least twenty-six proteins and the U2, U5, and U6 snRNAs (Ohi, 2002). CDC5 has two Myb repeats with a conserved basic face of helix 3. However, DNA-binding assays with CDC5 demonstrated a protein-DNA interaction only under low salt conditions, suggesting a lack of sequence specificity (Burns et al., 1999). Thus, Myb repeats in CDC5 are likely to serve a role separate from specific DNA recognition, perhaps mediating RNA or protein binding.

Another example of a Myb repeat containing protein that does not specifically bind DNA is SNAPc4. The SNAP complex (SNAPc) binds specifically to a proximal sequence element (PSE) in snRNA promoters to direct basal transcription through RNA polymerase II or III (Henry et al., 1995; Yoon et al., 1995). SNAPc4 (also known as SNAP190) contains four and a half tandemly arrayed Myb repeats (Rh, R1, R2, R3, R4). Interestingly, R3 and R4 are necessary and sufficient for sequence specific binding of the SNAPc to the PSE (Wong et al., 1998). However, the function of the other two and a half repeats remains unresolved. R1 and R2 have a more hydrophobic character, suggesting a role in protein-protein interactions. Whatever the function of R1 and R2 in SNAPc4, conservation of these repeats in fly, worm, and plant orthologues demonstrates that non-DNA-binding Myb repeats can have an important conserved function.

The transcriptional initiator/terminators TTF1 and REB1 bind to sequences that separate tandemly arrayed rDNA genes. TTF1 and REB1 terminate transcription of the upstream locus and activate transcription of the downstream rDNA promoter (Langst et al., 1997). Activation by TTF1 occurs by recruitment of the ISWI chromatin remodelling complex, NoRC (Strohner et al., 2001). The ability of TTF1 to recruit NoRC depends on protein elements upstream of the two conserved Myb repeats. TTF1-dependent promoter remodelling and subsequent activation require the Myb repeats (Langst et al., 1998). A necessity for the Myb repeats was attributed to their essential DNA-binding function, however contributions from protein-DNA and protein-protein interactions are not separable in these assays. Interestingly, insect R2 retrotransposons integrate at rDNA loci and contain a Myb-related repeat within their reverse transcriptase protein despite intense selection for minimisation of transposon size (Burke et al., 1999). It is therefore tempting to speculate that the Myb repeat functions to target the reverse transcriptase to rDNA repeats, perhaps by protein-protein interactions similar to those described for TTF1 and REB1.

3.3 Myb repeats as Chromatin-Interaction Domains?

While specific protein binding by Myb repeats may explain their presence in non-DNA binding proteins, a more general function may explain why this domain is present in so many nuclear factors. Most Myb-like proteins share the requirement to interact with chromatin in order to carry out their function. Myb domain proteins involved in chromatin remodelling complexes comprise the most obvious example. Mutational analysis of the single Myb repeat present in these proteins has yielded valuable insight into alternative functions for this domain.

ADA2 functions in yeast as an essential component of the SAGA and ADA histone acetylase complexes. Mutation of the conserved tryptophan and glutamic acid (residues 4 and 8 in Figure 4) and conserved aspartic acid and glutamic acids (residues 7-9 in Figure 4) to alanine leads to slow growth phenotypes similar to the *ada2* deletion mutant (Sterner et al., 2002). Although mutant protein assembled into SAGA complexes, the authors found a defect in histone acetyltransferase activity of mutant SAGA complexes. Mutation of other conserved residues to alanine did not result in slow growth or a defect in histone acetyltransferase activity. This suggests that these residues in the Myb repeat are essential for the acetyltransferase activity of the complex. Similarly, Boyer and colleagues found the Myb repeat of ADA2 to be necessary for transcription at the SAGA regulated HO endonuclease promoter (Boyer et al., 2002). The mutation of histidine 103 (residue 35 in Figure 4) to glutamic acid led to a 60% reduction in HO promoter activity, while deletion of residues 97-106 (position 29-43 in Figure 4) in helix 3 of the ADA2 Myb repeat led to complete loss of transcription from a HO reporter. Although the deletion mutant formed stable SAGA complexes, it demonstrated decreased binding *in vitro* to its histone H3 substrate. Further kinetic analysis indicated that mutations in the Myb repeat of ADA2 affect both histone substrate binding and catalysis by the SAGA complex. Thus, both the acidic residues in the first helix and amino acids in the third helix contribute to substrate recognition by ADA2.

```
                      ---helix1-----      -helix2-    --helix3---
cMyb_DNAB_R2                             WS              KQ    R  NH
cMyb_DNAB_R3     K S                     WA   K          N       NS
                     5    10   15        20   25         30   35   40
                     |    |    |         |    |          |    |    |
HsCMYB_1         KTRWTREEDEKLKKLVEQ--5--WKVIANY-3--R-TDVQCQHRWQKVL
HsCMYB_2         KGPWTKEEDQRVIELVQK--5--WSVIAKH-3--R-IGKQCRERWHNHL
HsCMYB_3         KTSWTEEEDRIIYQAHKR--5--WAEIAKL-3--R-TDNAIKNHWNSTM
DmMYB_1          GKRWSKSEDVLLKQLVET--4--WEIIGPH-3--R-LEQQVQQRWAKVL
DmMYB_2          KGPWTRDEDDMVIKLVRN--5--WTLIARY-3--R-IGKQCRERWHNHL
DmMYB_3          KTAWTEKEDEIIYQAHLE--5--WAKIAKR-3--R-TDNAIKNHWNSTM
AtPCMyb1_1       KGQWTPEEDEVLCKAVER--5--WKKIAEC-3--R-TDVQCLHRWQKVL
AtPCMyb1_2       KGPWSKEEDNTIIDLVEK--5--WSTISQH-3--R-IGKQCRERWHNHL
AtPCMyb1_3       KNAWTQEEELTLIRAHQI--4--WAELMKF-3--R-SDNSIKNHWNSSV
AtMyb1_1         KGPWSKEEDDVLSELVKR--5--WSFIARS-3--R-SGKSCRLRWCNQL
AtMyb1_2         RNSFTEVEDQAIIAAHAI--4--WAVIAKL-3--R-TDNAIKNHWNSAL
CeCDC5_1         GGVWKNTEDEILKAAIMK--5--WSRIASL-2--RKSAKQCKARWFEWL
CeCDC5_2         KTEWSREEDEKLLHLAKL--4--WRTIAPI-2--R-TSAQCLERYEHLL
ScADA2           CPDWGADEELQLIKGAQT--5--WQDIADH-3--R-GKEEVKEHYLKYY
HsTADA2L         DPSWTAQEEMALLEAVMD--5--WQDVANQ-3--K-TKEECEKHYMKHF
ScRSC8           KKNWSDQEMLLLLEGIEM--4--WEKIADH-4--K-RVEDCIEKFLSLP
ScSWI3           DENWSKEDLQKLLKGIQE--4--WYKVAKN-3--K-SPEQCILRFLQLP
HsSMARCC1        GREWTEQETLLLLEALEM--4--WNKVSEH-3--R-TQDECILHFLRLP
HsSMARCC2        TREWTEQETLLLLEALEM--4--WNKVSEH-3--R-TQDECILHFLRLP
DmISWI_1         FTAWTKRDFNQFIKANEK--5--IDNIAKD-3--K-TPEEVIE-YNAVF
HsSMARCA1_1      FTNWTKRDFNQFIKANEK--5--IDNIARE-3--K-SPEEVME-YSAVF
HsSMARCA5_1      FTNWNKRDFNQFIKANEK--5--IENIARE-3--K-TPEEVIE-YSAVF
DmISWI_2         GKNYTEIEDRFLVCMLHK--8--YEELRAA-1--R-ASPQFRFDWFIKS
HsSMARCA1_2      GKNYTEEEDRFLICMLHK--8--YEELRQC-1--R-NAPQFRFDWFIKS
HsSMARCA5_2      GKNYTEEEDRFLICMLHK--8--YDELRQC-1--R-NSPQFRFDWFLKS
ScTFIIIB-B"      TDPWTVEEMIKFYKALSM--4--FNLISQL-3--R-SRKQVKAKFVNEE
HsNCOR1_1        MNVWTDHEKEIFKDKFIQ--4--FGLIASY-3--K-SVPDCVLYYYLTK
HsNCOR1_2        TSRWTEEEMEVAKKGLVE--4--WAAIAKM-3--K-SEAQCKNFYFNYK
ScTBF1           KRTWSKEEEEALVEGLKE--4--WSKILDL-13-R-TQVQLKDKARNWK
AtTRP1           RRPFSVAEVEALVQAVEK--5--WRDVKLC-7--R-TYVDLKDKWKTLV
HsTRF1           RQAWLWEEDKNLRSGVRK--5--WSKILLH-5--R-TSVMLKDRWRTMK
HsTRF2           KQKWTVEESEWVKAGVQK--5--WAAISKN-5--R-TAVMIKDRWRTMK
ScRAP1_1         KASFTDEEDEFILDVVRK--10-YDEISHY-3--H-TGNSIRHRFRVYL
ScRAP1_2         KRKFSADEDYTLAIAVKK--62-FKHFAEE-3--H-TENAWRDRFRKFL
HsRAP1           RIAFTDADDVAILTYKEN--12-WKAMEKS-4--H-SWQSLKDRYLKHL
HsSNAPC4_1       KQEWSREEEERLQAIAAA--5--WQKIAEE-4--R-SAFQCLQKFQQHN
HsSNAPC4_2       RKEWTEEEDRMLTQLVQE--8--YRRIVYY-3--R-DSMQLIYRWTKSL
HsSNAPC4_3       KGYWAPEEDAKLLQAVAK--4--WFKIREE-3--R-SDAQCRDRYLRRL
HsSNAPC4_4       KGRWNLKEEEQLIELIEK--4--WAKIASE-3--R-SGSQCLSKWKIMM
HsTTF1_1         KGRYSEGDTEKLKMYHSL--4--WKTIGEM-2--R-RSLSVALKFSQIS
HsTTF1_2         RGAWSKSETRKLIKAVEE--38-WVEVEAK-3--R-NWMQCKSKWTEIL
ScREB1_1         SKAFDKAEDAALERFINE--26-WNNIYKV-3--R-SSSSIYKHMRRKY
ScREB1_2         RGKWTAEEEQELAKLCAE--4--WAEIGKT-2--R-MPEDCRDRWRNYV
ScREB1_3         SNRWSVEEEELLKKVISD-148-WTIVSER-4--R-SRIQCRYKWNKLV
ScREB1_4         IAKIQTVKDDDMLWIFEK--13-WDELAAL-4--KLNGLELKLCYERMK
DvR2element      KARWSEEEALMMARKEVE--6--VRHLNKK-6--R-S-SDAISSYRKRG
Consensus        ...WT..E...L...........W..I.......R.........W
                 ....S.......................................Y/F
```

Figure 4

Amino acid alignments of Myb repeats identified high conservation of structural residues and an acidic patch in the first helix. We identified Myb repeats through analysis of the primary protein sequence and confirmed the boundaries of each domain using the Simple Modular Architechural Research Tool program (http://smart.embl-heidelberg.de/). We created multiple sequence alignments of Myb repeats using Clustal X and color coded the alignments based on conservation using the BioEdit program. Labelling of Myb repeats indicates genus and species with the first two letters, followed by the protein name. For proteins containing multiple Myb-motifs, repeats are numbered starting from the N-terminus and identified after an underscore. Residues known to bind DNA in c-Myb (labelled cMyb DNAB) are depicted in bold on the first two lines. Contributions from the second repeat are on the line labelled R2 and those from the third repeat are on the line labelled R3. A consensus of the most highly conserved residues is located on the last line emphasising the importance of the structural residues. Note the high conservation of acidic residues in the first helix compared to c-Myb DNA binding residues. Species and genus abbreviations: Hs, *Homo sapiens*, Dm, *Drosophila melanogaster*, Dv, *Drosophila varians*, Ce, *Caenorhabditis elegans*, At, *Arabidopsis thalia*, Sc, *Saccharomyces cerevisiae*. (see colour section p. xviii)

Deletion of regions of the Myb repeat in SWI3 and RSC8 has also been informative. The SWI/SNF ATP-dependent chromatin remodelling complex requires the SWI3 subunit to activate transcription from inducible genes in yeast, including the HO endonuclease gene (Breeden et al., 1987). Deletion of a portion of helix 3, mutagenesis of at least two conserved structural residues to alanine, or substitution of arginine 564 (position 35 in Figure 4) for glutamic acid led to a loss of HO promoter activity (Boyer et al., 2002). However, mutants still assembled into SWI/SNF complexes and the deletions did not affect ATPase activity. Another ATP-dependent chromatin remodelling complex, RSC, is responsible for most chromatin restructuring in yeast and mutations in components of the complex are inviable. Like the SWI/SNF complex, deletion of residues 348-352 in helix 3 (position 29-33 in Figure 4) of the RSC8 Myb repeat led to loss of viability. The substrate for SWI/SNF and RSC complexes remains poorly defined but their activity depends on the presence of histone tails (Boyer et al., 2002).

Interestingly, *Drosophila* Myb and the p55 subunit of the chromatin assembly factor CAF1 copurify in a complex with three novel proteins that together associate with elements of the third chromosome chorion gene during amplification (Beall et al., 2002). The p55/RbAp48 histone chaperone also functions in nucleosome remodelling, histone deacetylation, and in methylated DNA binding complexes (reviewed in Ridgway et al., 2000). These data from *Drosophila* provide the first evidence that a 3R Myb family member functions in a chromatin modifying/assembly complex and may provide a functional link between Myb repeats in 3R Myb family proteins and Myb-related repeats in proteins involved in chromatin remodelling complexes.

More tantalizing evidence of a chromatin binding ability for Myb repeats comes from data on the TFIIIB complex. This complex contains three components: TBP, Brf1, and B" (also known as BDP1). Transcriptional initiation by RNA polymerase III requires the TFIIIB complex. Deletion of the single Myb repeat in B" leads to lethality in yeast (Ishiguro et al., 2002). Suppression of this mutant phenotype cannot be achieved by over-expression of other complex members, implying a function separate from complex assembly for the B" Myb repeat. Functional transcription can be attained *in vitro* on templates that are not assembled into chromatin by complexes containing B" Myb repeat mutant proteins, however, this domain is essential *in vivo* (Ishiguro et al., 2002). The requirement for an intact B" Myb repeat at promoters *in vivo* suggests an essential role for transcription of templates in the context of chromatin.

A general role for Myb repeats in chromatin interactions predicts a nucleosome binding capacity for Myb-related proteins that are not a part of chromatin modifying complexes. In fact, some Myb repeat containing proteins are known to have nucleosome binding capacity. It has been shown that chromatin reorganisation occurs in yeast upon yRap1p binding *in vivo* (Calvi et al., 1999; Yu et al., 2001) and that yRap1 has nucleosome binding capabilities *in vitro* (Rossetti et al., 2001). An interesting parallel exists in the mammalian TTF1 protein. TTF1 recruits the NoRC to remodel chromatin at rDNA promoters and has the ability to bind nucleosomes before remodelling occurs (Langst et al., 1998; Strohner et al., 2001). Nucleosome binding of both Rap1 and TTF1 depends on the presence of the DNA-binding Myb repeats.

Conservation of separate acidic and basic surfaces of the Myb repeat may facilitate interaction with chromatin, a structure comprised of basic histone tails and the acidic phosphate backbone of DNA. One possible role for conservation of a basic surface in non-DNA-binding Myb-related proteins could be a non-sequence specific interaction with the acidic phosphate backbone of DNA to stabilise the complex, while the highly conserved acidic residues (residues 6-11 in Figure 4) may function by interacting with the basic histone tails. In this regard, the basic face of the first c-Myb repeat non-specifically binds DNA in the cocrystal structure (Tahirov et al., 2002). Presumably, this interaction stabilises binding of the other two repeats. Interestingly, the first Myb repeat of ISWI proteins have a reversed arrangement, the first helix being relatively basic and the third helix being acidic, but the presence of an acidic and basic face is maintained. The SWI3, RSC8, and ADA2 homologues have conserved acidic residues in both the first (positions 7-9 in Figure 1) and third helix (positions 30, 31 in Figure 1) consistent with an ability to bind histone substrates. However, the scarcity of basic residues suggests that these Myb repeats may not interact directly with the DNA phosphate backbone.

3.4 Myb Repeats in General Regulatory Factors (GRFs)

Studies in yeast have identified a group of four proteins considered general regulatory factors (GRF) (Fourel et al., 2002). GRFs are abundant, essential proteins whose binding sites are frequent in the yeast genome (Lieb et al., 2001). Interestingly, these proteins generally collaborate with other factors and have a wide range of functions. For example, RAP1 acts as a transcriptional activator, facilitates DNA replication, functions in SIR-dependent silencing, regulates telomeres, and can establish boundaries between different chromatin states (reviewed in (Morse 2000)). REB1 assists RNA polI transcriptional activation and termination at rDNA loci as well as playing a role in silencing by acting as an insulator of chromatin boundaries (Ju et al., 1990; Packham et al., 1996; Reeder et al., 1999).

Although poorly characterised, TBF1 has an established role as an insulator of chromatin boundaries, binds to the telomeric repeat sequences recognised by TRF proteins, and can provide telomeric functions in budding yeast bearing these sequences at the termini of their chromosomes (Brigati et al., 1993; Fourel et al., 1999; Alexander et al., 2003). ABF1 functions in transcriptional regulation, silencing, DNA replication, and nucleotide excision repair (Diffley et al., 1989; Kang et al., 1995; Reid et al., 2000; Fourel et al., 2002). Interestingly, of the four GRFs known three (RAP1, REB1, and TBF1) contain Myb repeats. While the mechanism of GRF action remains unresolved it appears that these proteins work by a common means. In fact, the binding site of one GRF can substitute for another and swapping of protein domains leads to functional GRF chimaeric proteins. One common property of GRF proteins is their ability to act as protein insulators between different chromatin states. The hypothesis that GRFs function through local opening of chromatin structure could explain their multi-regulatory roles, however, this has yet to be tested conclusively.

While work on GRFs has been done in yeast, there are remarkable similarities to 3R Myb family proteins in higher eukaryotes. For example, both B-Myb and Dm-Myb are ubiquitously expressed, essential proteins. The consensus binding sites for these proteins are short and located throughout the genome. 3R Myb family proteins have been linked to diverse chromatin-mediated processes including DNA replication, transcriptional activation, and transcriptional repression. Interestingly, the loss of Dm-Myb results in disturbances of normal nuclear structure (Beall et al., 2002; Manak et al., 2002b). Thus, it is possible that the 3R Myb family constitutes an example of GRFs in higher eukaryotes.

ACKNOWLEDGEMENTS

EMR is supported by a USPHS Training Grant T32GM007365. Research in the Lipsick laboratory is supported by grants from the National Cancer Institute of the United States Public Health Service.

REFERENCES

Aasland, R., Stewart, A.F. and Gibson, T. (1996) The SANT domain: a putative DNA-binding domain in the SWI-SNF and ADA complexes, the transcriptional co-repressor N-COR and TFIIIb. Trends Bio. Sci. *21*, 87-88.

Abi-Rached, L., Gilles, A., Shiina, T., Pontarotti, P. and Inoko, H. (2002) Evidence of en bloc duplication in vertebrate genomes. Nat. Genet. *31*, 100-105.

Alexander, M.K. and Zakian, V.A. (2003) Rap1p telomere association is not required for mitotic stability of a C(3)TA(2) telomere in yeast. EMBO J. *22*, 1688-1696.

Andersson, K.B., Kowenz-Leutz, E., Brendeford, E.M., Tygsett, A.H., Leutz, A. and Gabrielsen, O.S. (2003) Phosphorylation-dependent down-regulation of c-Myb DNA binding is abrogated by a point mutation in the v-myb oncogene. J. Biol. Chem. *278*, 3816-3824.

Anton, I.A. and Frampton, J. (1988) Tryptophans in myb proteins. Nature *336*, 719.

Aziz, N., Wu, J., Dubendorff, J.W., Lipsick, J.S., Sturgill, T.W. and Bender, T.P. (1993) c-Myb and v-Myb are differentially phosphorylated by p42mapk in vitro. Oncogene *8*, 2259-2265.

Bailey, G.S., Poulter, R.T. and Stockwell, P.A. (1978) Gene duplication in tetraploid fish: model for gene silencing at unlinked duplicated loci. Proc. Natl. Acad. Sci. USA *75*, 5575-5579.

Bartsch, O., Horstmann, S., Toprak, K., Klempnauer, K.H. and Ferrari, S. (1999) Identification of cyclin A/Cdk2 phosphorylation sites in B-Myb. Eur. J. Biochem. *260*, 384-391.

Beall, E.L., Manak, J.R., Zhou, S., Bell, M., Lipsick, J.S. and Botchan, M.R. (2002) Role for a Drosophila Myb-containing protein complex in site-specific DNA replication. Nature *420*, 833-837.

Bianchi, A., Smith, S., Chong, L., Elias, P. and de Lange, T. (1997) TRF1 is a dimer and bends telomeric DNA. EMBO J. *16*, 1785-1794.

Bies, J., Markus, J. and Wolff, L. (2002) Covalent attachment of the SUMO-1 protein to the negative regulatory domain of the c-Myb transcription factor modifies its stability and transactivation capacity. J. Biol. Chem. *277*, 8999-9009.

Bilaud, T., Koering, C.E., Binet-Brasselet, E., Ancelin, K., Pollice, A., Gasser, S.M. and Gilson, E. (1996) The telobox, a Myb-related telomeric DNA binding motif found in proteins from yeast, plants and human. Nucl. Acids Res. *24*, 1294-1303.

Boyer, L.A., Langer, M.R., Crowley, K.A., Tan, S., Denu, J.M. and Peterson, C.L. (2002) Essential role for the SANT domain in the functioning of multiple chromatin remodeling enzymes. Mol. Cell *10*, 935-942.

Braun, E.L. and Grotewold, E. (1999) Newly discovered plant c-myb-like genes rewrite the evolution of the plant myb gene family. Plant Physiol. *121*, 21-24.

Breeden, L. and Nasmyth, K. (1987) Cell cycle control of the yeast HO gene: cis- and trans-acting regulators. Cell *48*, 389-97.

Brenner, S., Elgar, G., Sandford, R., Macrae, A., Venkatesh, B. and Aparicio, S. (1993) Characterization of the pufferfish (fugu) genome as a compact model vertebrate genome. Nature *366*, 265-268.

Brigati, C., Kurtz, S., Balderes, D., Vidali, G. and Shore, D. (1993) An essential yeast gene encoding a TTAGGG repeat-binding protein. Mol. Cell. Biol. *13*, 1306-1314.

Burgess, T.L., Fisher, E.F., Ross, S.L., Bready, J.V., Qian, Y.X., Bayewitch, L.A., Cohen, A.M., Herrera, C.J., Hu, S.S., Kramer, T.B., et al. (1995) The antiproliferative activity of c-myb and c-myc antisense oligonucleotides in smooth muscle cells is caused by a nonantisense mechanism. Proc. Natl. Acad. Sci. USA *92*, 4051-4055.

Burke, W.D., Malik, H.S., Jones, J.P. and Eickbush, T.H. (1999) The domain structure and retrotransposition mechanism of R2 elements are conserved throughout arthropods. Mol. Biol. Evol.*16*, 502-511.

Burns, C.G., Ohi, R., Krainer, A.R. and Gould, K.L. (1999) Evidence that Myb-related CDC5 proteins are required for pre-mRNA splicing. Proc. Natl. Acad. Sci. USA *96*, 13789-13794.

Calvi, B.R. and Spradling, A.C. (1999) Chorion gene amplification in Drosophila: A model for metazoan origins of DNA replication and S-phase control. Methods *18*, 407-417.

Campanero, M.R., Armstrong, M. and Flemington, E. (1999) Distinct cellular factors regulate the c-myb promoter through its E2F element. Mol. Cell. Biol. *19*, 8442-8450.

Dahle, O., Andersen, T.O., Nordgard, O., Matre, V., Del Sal, G. and Gabrielsen, O.S. (2003) Transactivation properties of c-Myb are critically dependent on two SUMO-1 acceptor sites that are conjugated in a PIASy enhanced manner. Eur. J. Biochem. *270*, 1338-1348.

Dias, A.P., Braun, E.L., McMullen, M.D. and Grotewold, E. (2003) Recently Duplicated Maize R2R3 Myb Genes Provide Evidence for Distinct Mechanisms of Evolutionary Divergence after Duplication. Plant Physiol. *131*, 610-620.

Diffley, J.F. and Stillman, B. (1989) Similarity between the transcriptional silencer binding proteins ABF1 and RAP1. Science *246*, 1034-1038.

Dini, P.W. and Lipsick, J.S. (1993) Oncogenic truncation of the first repeat of c-Myb decreases DNA binding in vitro and in vivo. Mol. Cell. Biol. *13*, 7334-7348.

Dubendorff, J.W., Whittaker, L.J., Eltman, J.T. and Lipsick, J.S. (1992) Carboxy-terminal elements of c-Myb negatively regulate transcriptional activation in cis and in trans. Genes Dev. *6*, 2524-2535.

Foos, G., Grimm, S. and Klempnauer, K.H. (1992) Functional antagonism between members of the myb family: B-myb inhibits v-myb-induced gene activation. EMBO J. *11*, 4619-4629.

Fourel, G., Miyake, T., Defossez, P.A., Li, R. and Gilson, E. (2002) General regulatory factors (GRFs) as genome partitioners. J. Biol. Chem. *277*, 41736-41743.

Fourel, G., Revardel, E., Koering, C.E. and Gilson, E. (1999) Cohabitation of insulators and silencing elements in yeast subtelomeric regions. EMBO J. *18*, 2522-2537.

Fu, S.L. and Lipsick, J.S. (1996) FAETL motif required for leukemic transformation by v-Myb. J. Virol. *70*, 5600-5610.

Fung, S.M., Ramsay, G. and Katzen, A.L. (2002) Mutations in Drosophila myb lead to centrosome amplification and genomic instability. Development *129*, 347-359.

Gallardo, M.H., Bickham, J.W., Honeycutt, R.L., Ojeda, R.A. and Kohler, N. (1999) Discovery of tetraploidy in a mammal. Nature *401*, 341.

Ganter, B. and Lipsick, J.S. (1999) Myb and oncogenesis. Adv Cancer Res *76*, 21-60.

Garcia-Fernandez, J. and Holland, P.W. (1994) Archetypal organization of the amphioxus Hox gene cluster. Nature *370*, 563-566.

Gewirtz, A.M., Anfossi, G., Venturelli, D., Valpreda, S., Sims, R. and Calabretta, B. (1989) G1/S transition in normal human T-lymphocytes requires the nuclear protein encoded by c-myb. Science *245*, 180-183.

Guehmann, S., Vorbrueggen, G., Kalkbrenner, F. and Moelling, K. (1992) Reduction of a conserved Cys is essential for Myb DNA-binding. Nucl. Acids Res. *20*, 2279-2286.

Hashimoto, T., Nakamura, Y., Kamaishi, T., Nakamura, F., Adachi, J., Okamoto, K. and Hasegawa, M. (1995) Phylogenetic place of mitochondrion-lacking protozoan, Giardia lamblia, inferred from amino acid sequences of elongation factor 2. Mol. Biol. Evol. *12*, 782-793.

Henry, R.W., Sadowski, C. L., Kobayashi, R. and N. Hernandez. (1995) A TBP-TAF complex required for transcription of human snRNA genes by RNA polymerase II and III. Nature *374*, 653-657.

Holland, P.W., Garcia-Fernandez, J., Williams, N.A. and Sidow, A. (1994) Gene duplications and the origins of vertebrate development. Development *Supplement*, 125-133.

Hou, D.X., Akimaru, H. and Ishii, S. (1997) Trans-activation by the Drosophila myb gene product requires a Drosophila homologue of CBP. FEBS Lett, *413*, 60-64.

Hu, Y.H., Cheng, L., Hochleitner, B.W., and Xu, Q.B. (1997) Activation of mitogen-activated protein kinases (ERK/JNK) and AP- 1 transcription factor in rat carotid arteries after balloon injury. Arterioscl. Thromb. Vasc. Biol.*17*, 2808-2816.

Hu, Y.L., Ramsay, R.G., Kanei-Ishii, C., Ishii, S. and Gonda, T.J. (1991) Transformation by carboxyl-deleted Myb reflects increased transactivating capacity and disruption of a negative regulatory domain. Oncogene *6*, 1549-1553.

Hughes, M.K. and Hughes, A.L. (1993) Evolution of duplicate genes in a tetraploid animal, Xenopus laevis. Mol. Biol. Evol. *10*, 1360-1369.

Humbert-Lan, G. and Pieler, T. (1999) Regulation of DNA binding activity and nuclear transport of B-Myb in Xenopus oocytes. J. Biol. Chem. *274*, 10293-10300.

Ibanez, C.E. and Lipsick, J.S. (1988) Structural and functional domains of the myb oncogene: requirements for nuclear transport, myeloid transformation, and colony formation. J. Virol. *62*, 1981-1988.

Ibanez, C.E. and Lipsick, J.S. (1990) trans activation of gene expression by v-myb. Mol. Cell. Biol. *10*, 2285-2293.

Ishiguro, A., Kassavetis, G.A. and Geiduschek, E.P. (2002) Essential roles of Bdp1, a subunit of RNA polymerase III initiation factor TFIIIB, in transcription and tRNA processing. Mol. Cell. Biol. *22*, 3264-3275.

Jiang, W., Kanter, M.R., Dunkel, I., Ramsay, R.G., Beemon, K.L. and Hayward, W.S. (1997) Minimal truncation of the c-myb gene product in rapid-onset B-cell lymphoma. J. Virol. *71*, 6526-6533.

Johnson, L.R., Johnson, T.K., Desler, M., Luster, T.A., Nowling, T., Lewis, R.E. and Rizzino, A. (2002) Effects of B-Myb on gene transcription: phosphorylation-dependent activity ans acetylation by p300. J. Biol. Chem. *277*, 4088-4097.

Johnson, T.K., Schweppe, R.E., Septer, J. and Lewis, R.E. (1999) Phosphorylation of B-Myb regulates its transactivation potential and DNA binding. J. Biol. Chem. *274*, 36741-36749.

Ju, Q.D., Morrow, B.E. and Warner, J.R. (1990) REB1, a yeast DNA-binding protein with many targets, is essential for growth and bears some resemblance to the oncogene myb. Mol. Cell. Biol. *10*, 5226-5234.

Kang, J.J., Yokoi, T.J. and Holland, M.J. (1995) Binding sites for abundant nuclear factors modulate RNA polymerase I-dependent enhancer function in Saccharomyces cerevisiae. J. Biol. Chem. *270*, 28723-28732.

Konig, P., Giraldo, R., Chapman, L. and Rhodes, D. (1996) The crystal structure of the DNA-binding domain of yeast RAP1 in complex with telomeric DNA. Cell *85*, 125-136.

Kranz, H., Scholz, K. and Weisshaar, B. (2000) c-MYB oncogene-like genes encoding three MYB repeats occur in all major plant lineages. Plant J. *21*, 231-235.

Lam, E.W., Robinson, C. and Watson, R.J. (1992) Characterization and cell cycle-regulated expression of mouse B-myb. Oncogene *7*, 1885-1890.

Lam, E.W. and Watson, R.J. (1993) An E2F-binding site mediates cell-cycle regulated repression of mouse B-myb transcription. EMBO J. *12*, 2705-2713.

Lane, S., Farlie, P. and Watson, R. (1997) B-Myb function can be markedly enhanced by cyclin A-dependent kinase and protein truncation. Oncogene *14*, 2445-2453.

Langst, G., Becker, P.B. and Grummt, I. (1998) TTF-I determines the chromatin architecture of the active rDNA promoter. EMBO J. *17*, 3135-3145.

Langst, G., Blank, T.A., Becker, P.B. and Grummt, I. (1997) RNA polymerase I transcription on nucleosomal templates: the transcription termination factor TTF-I induces chromatin remodeling and relieves transcriptional repression. EMBO J. *16*, 760-768.

Leveugle, M., Prat, K., Perrier, N., Birnbaum, D. and Coulier, F. (2003) ParaDB: a tool for paralogy mapping in vertebrate genomes. Nucl. Acids Res. *31*, 63-67.

Li, W.H., Gu, Z., Wang, H. and Nekrutenko, A. (2001) Evolutionary analyses of the human genome. Nature *409*, 847-849.

Lieb, J.D., Liu, X., Botstein, D. and Brown, P.O. (2001) Promoter-specific binding of Rap1 revealed by genome-wide maps of protein-DNA association. Nat. Genet. *28*, 327-334.

Lipsick, J.S. and Boyle, W.J. (1987) c-myb protein expression is a late event during T-lymphocyte activation. Mol. Cell. Biol. *7*, 3358-3360.

Luscher, B., Christenson, E., Litchfield, D.W., Krebs, E.G. and Eisenman, R.N. (1990) Myb DNA binding inhibited by phosphorylation at a site deleted during oncogenic activation. Nature *344*, 517-522.

Lustig, B. and Jernigan, R.L. (1995) Consistencies of individual DNA base-amino acid interactions in structures and sequences. Nucl. Acids Res. *23*, 4707-4711.

Manak, J.R., Mitiku, N. and Lipsick, J.S. (2002a) Mutation of the Drosophila homologue of the Myb protooncogene causes genomic instability. *99*, 7438-7443.

Manak, J.R., Mitiku, N., Lipsick, J.S. (2002b) Mutation of the Drosophila homologue of the Myb protooncogene causes genomic instability. Proc. Natl. Acad. Sci. USA *99*, 7438-7443.

McLysaght, A., Hokamp, K. and Wolfe, K.H. (2002) Extensive genomic duplication during early chordate evolution. Nat. Genet. *31*, 200-204.

Mizuguchi, G., Nakagoshi, H., Nagase, T., Nomura, N., Date, T., Ueno, Y. and Ishii, S. (1990) DNA binding activity and transcriptional activator function of the human B-myb protein compared with c-MYB. J. Biol. Chem. *265*, 9280-9284.

Morse, R.H. (2000) RAP, RAP, open up! New wrinkles for RAP1 in yeast. Trends Genet. *16*, 51-53.

Muller-Tidow, C., Wang, W., Idos, G.E., Diederichs, S., Yang, R., Readhead, C., Berdel, W.E., Serve, H., Saville, M., Watson, R. and Koeffler, H.P. (2001) Cyclin A1 directly interacts with B-myb and cyclin A1/cdk2 phosphorylate B-myb at functionally important serine and threonine residues: tissue-specific regulation of B-myb function. Blood *97*, 2091-2097.

Myrset, A.H., Bostad, A., Jamin, N., Lirsac, P.N., Toma, F. and Gabrielsen, O.S. (1993) DNA and redox state induced conformational changes in the DNA-binding domain of the Myb oncoprotein. EMBO J. *12*, 4625-4633.

Nomura, T., Sakai, N., Sarai, A., Sudo, T., Kanei-Ishii, C., Ramsay, R.G., Favier, D., Gonda, T.J. and Ishii, S. (1993) Negative autoregulation of c-Myb activity by homodimer formation through the leucine zipper. J. Biol. Chem. *268*, 21914-21923.

Oelgeschlager, M., Krieg, J., Luscher-Firzlaff, J.M. and Luscher, B. (1995) Casein kinase II phosphorylation site mutations in c-Myb affect DNA binding and transcriptional cooperativity with NF-M. Mol. Cell. Biol. *15*, 5966-5974.

Ogata, K., Morikawa, S., Nakamura, H., Sekikawa, A., Inoue, T., Kanai, H., Sarai, A., Ishii, S. and Nishimura, Y. (1994) Solution structure of a specific DNA complex of the Myb DNA-binding domain with cooperative recognition helices. Cell *79*, 639-648.

Ohi, M.D., Link, A.J., Ren, L., Jennings, J.L., McDonald, W.H., and Gould, K.L. (2002) Proteomics analysis reveals stable multiprotein complexes in both fission and budding yeasts containing Myb-related Cdc5p/Cef1p, novel pre-mRNA splicing factors, and snRNAs. Mol Cell Biol *22*, 2011-24.

Ohno, S. 1970. *Evolution by Gene Duplication*. Springer-Verlag, Berlin and New York.

Ohno, S. (1999) Gene duplication and the uniqueness of vertebrate genomes circa 1970-1999. Semin. Cell Dev. Biol. *10*, 517-522.

Packham, E.A., Graham, I.R. and Chambers, A. (1996) The multifunctional transcription factors Abf1p, Rap1p and Reb1p are required for full transcriptional activation of the chromosomal PGK gene in Saccharomyces cerevisiae. Mol. Gen. Genet. *250*, 348-356.

Prince, V.E. and Pickett, F.B. (2002) Splitting pairs: the diverging fates of duplicated genes. Nat. Rev. Genet. *3*, 827-837.

Rabinowicz, P.D., Braun, E.L., Wolfe, A.D., Bowen, B. and Grotewold, E. (1999) Maize R2R3 myb genes. Sequence analysis reveals amplification in the higher plants. Genetics *153*, 427-444.

Ramsay, R.G. (1995) DNA-binding studies using in vitro synthesized Myb proteins. Meth. Mol. Biol. *37*, 369-377.

Ramsay, R.G., Ikeda, K., Rifkind, R.A. and Marks, P.A. (1986) Changes in gene expression associated with induced differentiation of erythroleukemia: protooncogenes, globin genes, and cell division. Proc. Natl. Acad. Sci. USA *83*, 6849-6853.

Ramsay, R.G., Morrice, N., Van Eeden, P., Kanagasundaram, V., Nomura, T., De Blaquiere, J., Ishii, S. and Wettenhall, R. (1995) Regulation of c-Myb through protein phosphorylation and leucine zipper interactions. Oncogene *11*, 2113-2120.

Reeder, R.H., Guevara, P. and Roan, J.G. (1999) Saccharomyces cerevisiae RNA polymerase I terminates transcription at the Reb1 terminator in vivo. Mol. Cell. Biol. *19*, 7369-7376.

Reid, J.L., Iyer, V.R., Brown, P.O. and Struhl, K. (2000) Coordinate regulation of yeast ribosomal protein genes is associated with targeted recruitment of Esa1 histone acetylase. Mol. Cell. *6*, 1297-1307.

Ridgway, P. and Almouzni, G. (2000) CAF-1 and the inheritance of chromatin states: at the crossroads of DNA replication and repair. J. Cell Sci. *113*, 2647-2658.

Riechmann, J.L., Heard, J., Martin, G., Reuber, L., Jiang, C., Keddie, J., Adam, L., Pineda, O., Ratcliffe, O.J., Samaha, R.R., Creelman, R., Pilgrim, M., Broun, P., Zhang, J.Z., Ghandehari, D., Sherman, B.K. and Yu, G. (2000) Arabidopsis transcription factors: genome-wide comparative analysis among eukaryotes. Science *290*, 2105-2110.

Robinson, C., Light, Y., Groves, R., Mann, D., Marias, R. and Watson, R. (1996) Cell-cycle regulation of B-Myb protein expression: specific phosphorylation during the S phase of the cell cycle. Oncogene *12*, 1855-1864.

Roger, A.J., Svard, S.G., Tovar, J., Clark, C.G., Smith, M.W., Gillin, F.D. and Sogin, M.L. (1998) A mitochondrial-like chaperonin 60 gene in Giardia lamblia: evidence that diplomonads once harbored an endosymbiont related to the progenitor of mitochondria. Proc. Natl. Acad. Sci. USA *95*, 229-234.

Rossetti, L., Cacchione, S., De Menna, A., Chapman, L., Rhodes, D. and Savino, M. (2001) Specific interactions of the telomeric protein Rap1p with nucleosomal binding sites. J. Mol. Biol. *306*, 903-913.

Rosson, D., Dugan, D. and Reddy, E.P. (1987) Aberrant splicing events that are induced by proviral integration: implications for myb oncogene activation. Proc. Natl. Acad. Sci. USA *84*, 3171-3175.

Rubin, G.M., Yandell, M.D., Wortman, J.R., Gabor Miklos, G.L., Nelson, C.R., Hariharan, I.K., Fortini, M.E., Li, P.W., Apweiler, R., Fleischmann, W., Cherry, J.M., Henikoff, S., Skupski, M.P., Misra, S., Ashburner, M., Birney, E., Boguski, M.S., Brody, T., Brokstein, P., Celniker, S.E., Chervitz, S.A., Coates, D., Cravchik, A., Gabrielian, A., Galle, R.F., Gelbart, W.M., George, R.A., Goldstein, L.S., Gong, F., Guan, P., Harris, N.L., Hay, B.A., Hoskins, R.A., Li, J., Li, Z., Hynes, R.O., Jones, S.J., Kuehl, P.M., Lemaitre, B., Littleton, J.T., Morrison, D.K., Mungall, C., O'Farrell, P.H., Pickeral, O.K., Shue, C., Vosshall, L.B., Zhang, J., Zhao, Q., Zheng, X.H. and Lewis, S. (2000) Comparative genomics of the eukaryotes. Science *287*, 2204-2215.

Sakura, H., Kanei-Ishii, C., Nagase, T., Nakagoshi, H., Gonda, T.J. and Ishii, S. (1989) Delineation of three functional domains of the transcriptional activator encoded by the c-myb protooncogene. Proc. Natl. Acad. Sci. USA *86*, 5758-5762.

Sala, A., Kundu, M., Casella, I., Engelhard, A., Calabretta, B., Grasso, L., Paggi, M.G., Giordano, A., Watson, R.J., Khalili, K. and Peschle, C. (1997) Activation of human B-MYB by cyclins. Proc. Natl. Acad. Sci. USA *94*, 532-536.

Sano, Y. and Ishii, S. (2001) Increased affinity of c-Myb for CREB-binding protein (CBP) after CBP- induced acetylation. J. Biol. Chem. *276*, 3674-3682.

Saville, M.K. and Watson, R.J. (1998) The cell-cycle regulated transcription factor B-Myb is phosphorylated by cyclin A/Cdk2 at sites that enhance its transactivation properties. Oncogene *17*, 2679-2689.

Shen-Ong, G.L., Lüscher, B. and Eisenman, R.N. (1989) A second c-myb protein is translated from an alternatively spliced mRNA expressed from normal and 5'-disrupted myb loci. Mol. Cell. Biol. *9*, 5456-63.

Shen-Ong, G.L., Potter, M., Mushinski, J.F., Lavu, S. and Reddy, E.P. (1984) Activation of the c-myb locus by viral insertional mutagenesis in plasmacytoid lymphosarcomas. Science *226*, 1077-1080.

Simon, A.L., Stone, E.A. and Sidow, A. (2002) Inference of functional regions in proteins by quantification of evolutionary constraints. Proc. Natl. Acad. Sci. USA *99*, 2912-2917.

Skrabanek, L. and Wolfe, K.H. (1998) Eukaryote genome duplication - where's the evidence? *8*, 694-700.

Spink, K.G., Evans, R.J. and Chambers, A. (2000) Sequence-specific binding of Taz1p dimers to fission yeast telomeric DNA. Nucl. Acids Res. *28*, 527-533.

Sterner, D.E., Wang, X., Bloom, M.H., Simon, G.M. and Berger, S.L. (2002) The SANT domain of Ada2 is required for normal acetylation of histones by the yeast SAGA complex. J. Biol. Chem. *277*, 8178-8186.

Stober-Grasser, U., Brydolf, B., Bin, X., Grasser, F., Firtel, R.A. and Lipsick, J.S. (1992) The Myb DNA-binding domain is highly conserved in Dictyostelium discoideum. Oncogene *7*, 589-596.

Stracke, R., Werber, M. and Weisshaar, B. (2001) The R2R3-MYB gene family in Arabidopsis thaliana. Curr. Opin. Plant Biol. *4*, 447-456.

Strohner, R., Nemeth, A., Jansa, P., Hofmann-Rohrer, U., Santoro, R., Langst, G. and Grummt, I. (2001) NoRC - a novel member of mammalian ISWI-containing chromatin remodeling machines. EMBO J. *20*, 4892-4900.

Sun, C.H., Palm, D., McArthur, A.G., Svèard, S.G. and Gillin, F.D. (2002) A novel Myb-related protein involved in transcriptional activation of encystation genes in Giardia lamblia. Mol. Microbiol. *46*, 971-984.

Tahirov, T.H., Sato, K., Ichikawa-Iwata, E., Sasaki, M., Inoue-Bungo, T., Shiina, M., Kimura, K., Takata, S., Fujikawa, A., Morii, H., Kumasaka, T., Yamamoto, M., Ishii, S. and Ogata, K. (2002) Mechanism of c-Myb-C/EBP beta cooperation from separated sites on a promoter. Cell *108*, 57-70.

Tanaka, Y., Nomura, T. and Ishii, S. (1997) Two regions in c-myb proto-oncogene product negatively regulating its DNA-binding activity. FEBS Lett. *413*, 162-168.

Tashiro, S., Takemoto, Y., Handa, H. and Ishii, S. (1995) Cell type-specific trans-activation by the B-myb gene product: requirement of the putative cofactor binding to the C-terminal conserved domain. Oncogene *10*, 1699-1707.

Thompson, J.D., Higgins, D.G. and Gibson, T.J. (1994) CLUSTAL W: improving the sensitivity of progressive multiple sequence alignment through sequence weighting, postion-specific gap penalties and weight matrix choice. Nucl. Acids Res. *22*, 4673-4680.

Tomita, A., Towatari, M., Tsuzuki, S., Hayakawa, F., Kosugi, H., Tamai, K., Miyazaki, T., Kinoshita, T. and Saito, H. (2000) c-Myb acetylation at the carboxyl-terminal conserved domain by transcriptional co-activator p300. Oncogene *19*, 444-451.

Torelli, G., Selleri, L., Donelli, A., Ferrari, S., Emilia, G., Venturelli, D., Moretti, L. and Torelli, U. (1985) Activation of c-myb expression by phytohemagglutinin stimulation in normal human T lymphocytes. Mol. Cell. Biol. *5*, 2874-2877.

Watson, R.J., Robinson, C. and Lam, E.W. (1993) Transcription regulation by murine B-myb is distinct from that by c-myb. Nucl. Acids Res. *21*, 267-272.

Wong, M.W., Henry, R.W., Ma, B., Kobayashi, R., Klages, N., Matthias, P., Strubin, M. and Hernandez, N. (1998) The large subunit of basal transcription factor SNAPc is a Myb domain protein that interacts with Oct-1. Mol. Cell. Biol. *18*, 368-377.

Woo, C.H., Sopchak, L. and Lipsick, J.S. (1998) Overexpression of an alternatively spliced form of c-Myb results in increases in transactivation and transforms avian myelomonoblasts. J. Virol. *72*, 6813-6821.

Yoon, J.-B., Murphy, S., Bai, L., Wang, Z. and R. G. Roeder. (1995) Proximal sequence element-binding transcription factor (PTF) is a multisubunit complex required for transcription of both RNA polymerase II- and RNA polymerase III-dependent small nuclear RNA genes. Mol. Cell. Biol. *15*, 2019-2027.

Yu, L., Sabet, N., Chambers, A. and Morse, R.H. (2001) The N-terminal and C-terminal domains of RAP1 are dispensable for chromatin opening and GCN4-mediated HIS4 activation in budding yeast. J. Biol. Chem. *276*, 33257-33264.

Ziebold, U., Bartsch, O., Marais, R., Ferrari, S. and Klempnauer, K.-H. (1997) Phosphorylation and activation of B-Myb by cyclin A-Cdk2. Curr. Biol. *7*, 253-260.

Zwicker, J., Liu, N., Engeland, K., Lucibello, F.C. and Muller, R. (1996) Cell cycle regulation of E2F site occupation in vivo. Science *271*, 1595-1597.

Table 1. Eukaryotic 3R Myb Proteins

3R Myb	**Organism**	**Expression Pattern**	**Transcriptional Activation**	**Function**
c-Myb	Human, Mouse, Chicken, *Xenopus*, *Fugu*	Tissue specific - Human and mouse: immature haemopoietic cells of all lineages (Chen, 1980; Westin et al., 1982; Gonda and Metcalf, 1984; Duprey and Boettiger, 1985; Kirsch et al., 1986; Ramsay et al., 1986), immature epithelial cells and other tissues such as colon, respiratory tract, skin and retina (91, 113); *Xenopus*: throughout development, highest levels in the intestine, heart, liver, lung and ovary (Amaravadi and King, 1994).	Yes	Mouse: definitive haemopoiesis (Mucenski et al., 1991).
B-Myb	Human, Mouse, Chicken, *Xenopus*, *Fugu*	Ubiquitous - Mouse and man: expressed through out mouse development (Sitzmann et al., 1996); *Xenopus*: specific to the developing central nervous system (Humbert-Lan and Pieler, 1999).	Conditional*	Mouse: early embryo E4.5 - 6.5 (Tanaka et al., 1999)
A-Myb	Human, Mouse, Chicken, *Xenopus*, *Fugu*	Tissue-specific - Human and mouse: central nervous system, germinal centre B lymphocytes, mammary gland epithelium and testes (Mettus et al., 1994; Trauth et al., 1994; Golay et al., 1998); *Xenopus*: mitotic spermatogonial cells (Sleeman, 1993)	Yes	Mouse: spermatogenesis and mammary gland proliferation (Toscani et al., 1997)
Invertebrates	*Ciona intestinalis*, Sea Urchin, *Drosophila*, *Anopheles gambiae*	Ubiquitous - *Drosophila*: expression through out embryonic development and in both larval mitotic and endocycling cells (Katzen et al., 1985; Manak et al., 2002)	Conditional*	*Drosophila*: cell cycle regulation; *S. purpuratus* :transcriptional repression (Coffman et al., 1997)
Plants	*Arabidopsis*, Rice (*Oryza sativa*), Moss (*Physcomitrella patens*), Delta Maidenhair Fern (*Adiantum raddianum*),	ND	ND	ND
Fungi	*Neurospora crassa*	ND	ND	ND
Early Eukaryotes (Protists)	*Giardia lamblia*, *Dictyostelium discoideum*	ND	Yes	*G. lamblia* - regulation of encystation genes (Sun et al., 2002)

* B-Myb and Dm-Myb transcriptional activation has been shown predominantly in exceptional conditions in which genes are over-expressed, particularly in specific cancer cell lines. These proteins do not score as transcriptional activators in budding yeast (unpublished data).

Table 2: Myb-related proteins

Myb related protein	Function	Complexes	Bind DNA	Loss of Function
A- and c-Myb 2R Plant Mybs	Transcriptional Regulation	?	Yes	c-Myb-lethal, defect in blood cell development A-Myb-defect in mammary gland development and spermatogenesis 2R Plant Mybs- variety of developmental and signalling defects
B-Myb and Dm-Myb	Cell Cycle Regulation	Dm-Myb/p55 CAF1 complex	Yes	Nonviable-proliferation defect in mice; cell cycle defects in flies
Telobox (Taz1, TBF1, TRFs)	Telomere binding and regulation	Telomere binding protein complex	Yes	TBF1-lethal TRF1-increased length of telomeres TRF2-end to end fusion of chromosomes Taz1-defects in meiosis and recombination
Rap1	Telomere regulation, transcription, silencing	Yeast-SIR; Human-TRFs	Yes	yRAP1-lethal
ADA2	Histone acetylation	SAGA, ADA	No	yADA2-viable, slow growth, transcriptional defects
N-CoR and SMRT	Histone Deacetylation	HDAC complexes	No	Yeast (SNT1)-viable, meiotic specific repression defect
SWI3 / RSC8	ATP-dependent chromatin remodeling	SWI/SNF (SWI3) RSC (RSC8)	No	ySWI3-viable, slow growth, transcriptional defects yRSC8-lethal
ISWI	ATP-dependent chromatin remodeling	CHRAC, NURF, ACF, NoRC	No	yISW1-altered chromatin structure at specific promoters, enhanced haploid invasive growth yISW2-defects in spindle formation, transcription, and altered chromatin patterns
CDC5/CEF1	mRNA splicing	Splicesome	?	Lethal due to G2 arrest, depletion leads to accumulation of unspliced mRNAs
SNAPC4	Transcription of snRNA by pol III	SNAP complex	Yes	-
TFIIIB-B"	RNA polymerase III transcription	TFIIIB component of RNA pol III holoenzyme	Yes	Lethal in yeast
TTF1 / REB1	Transcriptional termination	?	Yes	REB1-lethal

Chapter 2

DROSOPHILA MYB

Lessons for the Understanding of Vertebrate Myb Proteins

Alisa L. Katzen
Department of Biochemistry and Molecular Genetics, University of Illinois at Chicago, College of Medicine, Chicago, IL 60607-7170, United States of America.

Abstract: The fruit fly, *Drosophila melanogaster*, provides a powerful genetic and developmental system in which to dissect cellular and biochemical processes, making it an attractive model system for investigating the function of evolutionarily conserved genes. The DMyb protein encoded by *Dm myb*, the single *myb* gene in *Drosophila*, shares several biochemical properties with the vertebrate Myb proteins. Genetic studies have demonstrated the physiological relevance of previously identified biochemical interactions with CBP and Cyclin A. The consequences of altering DMyb activity within the developing animal demonstrate that it plays multiple roles in the cell cycle; promoting both S-phase and M-phase in diploid cells, acting to preserve diploidy by suppressing endoreduplication and, within at least one developmental setting, participating directly in the initiation of DNA replication. Recent findings suggest that DMyb may also participate in the regulation of some developmental patterning or differentiation decisions.

1. INTRODUCTION

The fruit fly, *Drosophila melanogaster* provides a powerful genetic system in which to dissect developmental processes and elucidate biochemical pathways. During the past fifteen years, it has become increasingly evident that the majority of biochemical pathways that regulate important developmental decisions have been highly conserved during evolution. Several of these pathways that were refractory to analysis in vertebrate systems, have been studied in *Drosophila*. For example, genetic screens for secondary mutations that enhance or suppress phenotypes caused by mutations in receptor tyrosine kinase genes, played a major role in elucidating the pathways by which this class of proteins send information from the cell surface to the nucleus (reviewed in Blenis, 1993; Perrimon,

J. Frampton (ed.), Myb Transcription Factors: Their Role in Growth, Differentiation and Disease, 35-64.

1994). Perhaps the most surprising examples, however, have been the discoveries that even though vertebrate and insect hearts and eyes look and function quite differently from one another, the "master control genes" that specify these organs have been conserved (Bodmer and Venkatesh, 1998; Gehring and Ikeo, 1999; Baker, 2001). Sequencing of the *Drosophila* genome has confirmed the extent of conservation with mammalian genes as well as previous findings that the number of members representing each gene family tends to be smaller in *Drosophila* than in mammals (Adams et al., 2000; Rubin et al., 2000). A systematic search for more than 900 known "human disease genes" revealed that greater than 75% of them were represented in the *Drosophila* genome (Reiter et al., 2001).

Unlike the three members of the *myb* gene family in vertebrates, *Drosophila* contains a single gene, *Dm myb*, encoding the DMyb protein (Katzen et al., 1985; Peters et al., 1987; Adams et al., 2000). Since publication of the first results from an analysis of loss-of-function mutants (Katzen and Bishop, 1996), additional mutations in *Dm myb* have been isolated and this approach to studying Myb protein function has gained momentum. Here, the published literature will be reviewed and potential implications for mammalian gene function will be discussed.

2. COMPARISON OF THE DROSOPHILA AND VERTEBRATE MYB PROTEINS

As shown in Figure 1, the DNA-binding domain (DBD) is highly conserved between DMyb and the vertebrate Myb proteins. Moreover, unlike more distantly Myb-related proteins, DMyb shares three additional regions of conservation with the vertebrate family of Myb proteins (Bishop et al., 1991). Within the four conserved domains, the overall identity is greater between DMyb and c-Myb than between DMyb and A-Myb or B-Myb (Table 1). This also holds true individually for regions I, II and III. In contrast, DMyb region IV is considerably more homologous to B-Myb than to c-Myb or A-Myb. This analysis therefore suggests that the single *Dm myb* gene is the ancestral gene, prior to the duplications that produced the three vertebrate *myb* genes. However, it does not determine whether the function of DMyb corresponds to one of the vertebrate Myb proteins or encompasses their combined functions.

Aside from the four conserved domains described above, the c-Myb and A-Myb proteins contain a conserved "acidic domain" that is required for transcriptional activation, which is not found in B-Myb or DMyb (see Figure 1). In addition, while this region of c-Myb and A-Myb has diverged slowly,

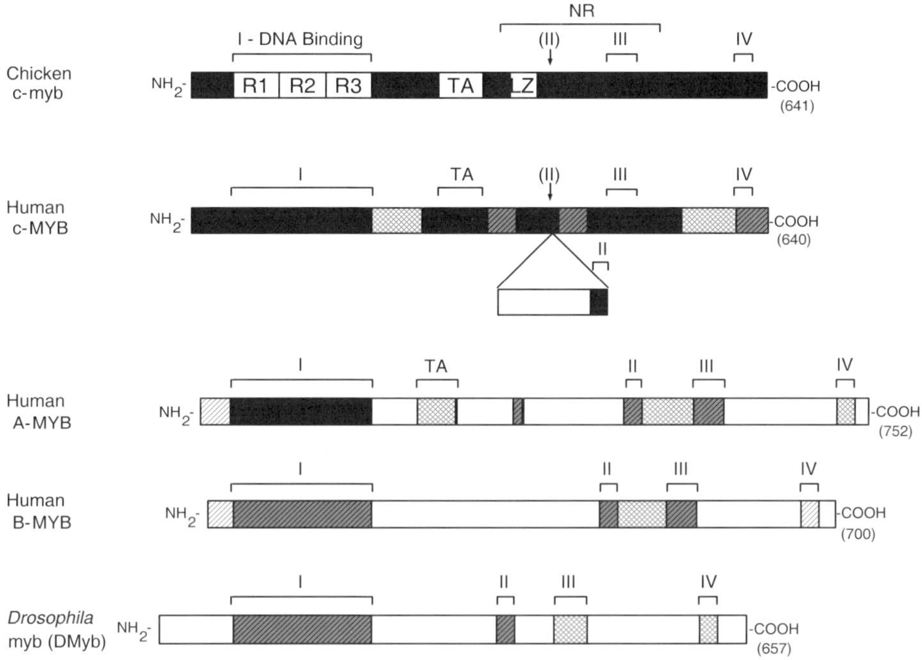

Figure 1

Topographies of vertebrate and *Drosophila* Myb proteins. Shown at top is a schematic representation of the prototypic chicken c-Myb protein. The four regions of conservation shared between vertebrate and *Drosophila* Myb proteins are indicated by Roman numerals. Also indicated are the three imperfect tandem repeats (R1, R2, and R3) that comprise the DNA-binding domain (region I), the transcriptional activator domain (TA), the leucine zipper region (LZ), and the negative regulatory domain (NR). Below the chicken c-Myb protein are schematics for the human c-Myb, A-Myb and B-Myb proteins and the *Drosophila* myb (DMyb) protein. For the human c-Myb protein, an additional region encoded by an alternatively spliced exon (exon 9A) that contains the majority of conserved region II is also depicted. The levels of amino acid conservation (percent identity) relative to the chicken c-Myb protein are indicated as follows: ■ = 81-100%; ▨ = 61-80%; ▩ = 41-60%; ▨ = 26-40%; □ = 0-25%.

the equivalent sequence in B-Myb has changed more significantly during evolution, indicating that it may not serve as an important functional domain. This phylogenetic analysis suggests that B-*myb* is the ancestral gene from which A-*myb* and c-*myb* arose by successive gene duplications, the first of which was accompanied by the acquisition of a central transcriptional activation domain (Simon et al., 2002). Therefore, DMyb would be more closely related to B-Myb than to A-Myb or c-Myb. Some support for this hypothesis was provided by early reports of difficulties in demonstrating the transcriptional activation abilities of B-Myb and DMyb (Foos et al., 1992; Watson et al., 1993; Hou et al., 1997). Subsequent studies have demonstrated that both B-Myb and DMyb are able to activate transcription, although compared to A-Myb and c-Myb, their activities appear to be more sensitive to regulation by phosphorylation or coactivators such as CBP/p300 (Ansieau et al., 1997; Lane et al., 1997; Sala et al., 1997; Ziebold et al., 1997; Saville and Watson, 1998; Johnson et al., 1999; Bessa et al., 2001; Jackson et al., 2001; Johnson et al., 2002; Li and McDonnell, 2002; Fung et al., 2003; S-M. Fung and A.L.K., manuscript in preparation).

In addition to sequence analysis, there are other aspects of the *Drosophila* and vertebrate *myb* genes that bear consideration when trying to determine the vertebrate orthologue of *Dm myb*. *In situ* analysis revealed that *Dm myb* RNA is expressed in all proliferating cells throughout development (Katzen and Bishop 1996). This is similar to B-*myb*, which is expressed in cycling cells from all lineages that have been examined, but different from both c-*myb* and A-*myb* that are expressed in a restricted set of cell types (Lipsick and Wang, 1999; Oh and Reddy, 1999). Consistent with these RNA expression patterns, genetic ablation of mouse B-*myb* results in very early embryonic lethality whereas A-*myb* or c-*myb* knockout mice exhibit defects in proliferation or development of specific cell types (Tanaka et al., 1999; Mucenski et al., 1991; Toscani et al., 1997). Further evidence that B-Myb function may be required in all dividing cells comes from studies showing that *B-myb* antisense oligonucleotides can inhibit proliferation of myeloid, lymphoid, glioblastoma, fibroblast and neuroblastoma cell lines (Arsura et al., 1992; Sala and Calabretta, 1992; Raschella et al., 1995). Similarly, a requirement for DMyb has been demonstrated in a variety of cell types (see below and Katzen et al., 1998; Fung et al., 2002; Manak et al., 2002; Okada et al., 2002).

Table 1. Homology between conserved domains of *Drosophila* Myb and each of the human Myb proteins.

	Conserved Region [No. of amino acids]				
Human protein	I [158]	II [19]	III [35]	IV [19]	I-IV [231]
c-Myb	66% (88%)[a]	74% (89%)	60% (71%)	42% (68%)	64% (84%)
A-Myb	65% (87%)	47% (79%)	49% (63%)	52% (74%)	60% (81%)
B-Myb	60% (84%)	58% (84%)	46% (57%)	63% (89%)	58% (80%)

[a] Shown is the percentage of amino acid identity (similarity) of each conserved region in the human proteins in comparison to the DMyb sequence.

Finally, recent studies in our laboratory have revealed genetic interactions between mutations in *Dm myb* and mutations in *cyclin A* (see section 4.4.2*)*. We have demonstrated that co-expression of Cyclin A with DMyb in tissue culture cells enhances the ability of DMyb to transactivate expression of a reporter construct (Fung et al., 2003; S-M. Fung and A.L.K., manuscript in preparation). On the other hand, while B-Myb and A-Myb have been shown to be activated by Cyclin A/Cyclin dependent kinase (Cdk) phosphorylation, c-Myb does not appear to be regulated in this manner (Lane et al., 1997; Sala et al., 1997; Ziebold et al., 1997; Ziebold and Klempnauer, 1997). Therefore, it seems likely that B-Myb is the functional counterpart of DMyb, although the evidence does not exclude the possibility that DMyb function may also encompass the roles of A-Myb and c-Myb.

3. THE DMYB PROTEIN: EXPRESSION AND BIOCHEMICAL CHARACTERISTICS

3.1 Expression and Intracellular Localisation

Northern and *in situ* analysis revealed that maternally expressed *Dm myb* mRNA is present at high levels in early embryos, that *Dm myb* transcripts are abundant in virtually all proliferating cells throughout development, and that the levels do not appear to vary appreciably during the cell cycle, at least in embryos (Katzen, 1990; Katzen and Bishop, 1996). This last finding was somewhat surprising since expression of each of the vertebrate *myb* genes has been shown to be tightly regulated in a cell cycle dependent or tissue-specific manner (reviewed in Lipsick and Wang, 1999; Oh and Reddy, 1999). Recently, we identified a transcription factor involved in regulating *Dm myb* expression. Although E2F has been implicated in the regulation of B-*myb* and c-*myb* expression (Sala et al., 1994; Oh and Reddy, 1999; Humbert et al., 2000), no E2F binding sites are located upstream of the *Dm myb* gene. However, two perfect binding sites for the DNA replication-related element binding factor (DREF) are located within 150 base pairs upstream of the start of the *Dm myb* transcript. DREF binds to these sites and they have been shown to be required for efficient activity of the *Dm myb* promoter in cutured cells (Sharkov et al., 2002). DREF regulates a number of *Drosophila* genes involved in DNA replication or cell cycle progression, including the *dE2F* gene and genes known to be regulated by E2F in mammals (Hirose et al., 1996; Ohno et al., 1996; Takahashi et al., 1996; Yamaguchi et al., 1996; Sawado et al., 1998; Lefai et al., 2000; Ruiz De Mena et al., 2000). Since no homologue of DREF had been identified in vertebrates, it was thought to perform an "E2F-like" role in *Drosophila*. Recently, however, a putative human DREF-like protein has been identified and, intriguingly, one of it candidate target genes is c-*myb* (Ohshima et al., 2003).

Preliminary *in situ* analysis of the DMyb protein in embryos and developing adult tissues (imaginal discs) indicates that its level is not directly correlated with that of *Dm myb* mRNA (G. Ramsay and A.L.K., unpublished observations). Whether or not regulation occurs at the level of DMyb protein stability has not yet been investigated, but there is some evidence supporting regulation at the level of translational efficiency (Sharkov et al., 2002). The *Dm myb* transcription unit has an unusually long 5'-UTR with the AUG initiation codon at position +605 of the spliced transcript. Surprisingly, the AUG for the DMyb protein is the eighth AUG in the transcript, a situation that is likely to lead to poor efficiency of translation since initiation sites in eukaryotic mRNAs are usually reached via a scanning mechanism that begins at the 5' end (Kozak, 1999). None of the eight AUGs is in an optimal context for the initiation of translation, although the *Dm myb* AUG does display appropriate bases at the two positions that have been shown to be most critical for efficient initiation (Kozak, 1999). Supporting the possibility that this upstream region is important in the regulation of DMyb protein levels, the efficiency of *in vitro* translation or expression from a construct in cultured eukaryotic cells were both dramatically increased when it was removed (Sharkov et al., 2002). It is possible that the *Dm myb* AUG is reached by a combination of leaky scanning and translational reinitiation, or that it may depend upon the presence of an Internal Ribosome Entry Site (IRES), an RNA element that directs internal initiation (Kozak, 1999).

Immunocytochemical analysis of the DMyb protein in cultured cells showed that it is localised in the nucleus as has been reported for all of the vertebrate Myb proteins (Jackson et al., 2001). However, preliminary studies in embryos and developing adult tissues suggest that the intracellular localisation of DMyb may be more complex (A.L.K. and colleagues, unpublished observations). Most, if not all of the DMyb protein is observed in the nucleus in pre-cellularised *Drosophila* embryos and in tissues where DMyb is ectopically expressed in transgenic animals. On the other hand, in later embryos and developing adult tissues, the DMyb protein appears to be at higher levels in the cytoplasm than the nucleus in some cells. At present, it is not clear what mechanisms might be regulating the intracellular distribution of DMyb or whether the changes in localisation correspond to various stages of the cell division cycle. However, it is of interest that there are also reports of the v-Myb, c-Myb and B-Myb proteins being observed outside the nucleus in certain circumstances (Klempnauer et al., 1984; Bading et al., 1988; Bouwmeester et al., 1994).

3.2 DNA Binding Activity

The DBDs of all three vertebrate Myb proteins bind to the consensus Myb binding site (MBS), Py**AAC**(G/T)**G**, the bold residues interacting specifically with the protein (Biedenkapp et al. 1988; Ness et al. 1989; Mizuguchi et al. 1990; Howe and Watson 1991; Golay et al. 1994). Individual Myb proteins show some specific preferences for nucleotides flanking the core binding site (Mizuguchi et al. 1990; Howe and Watson 1991). The DBD of DMyb had been shown to bind a double-stranded oligonucleotide containing a consensus MBS, but not to an oligonucleotide containing a mutated motif (Oehler et al., 1990; Madan et al., 1995). Studies performed in our laboratory using CASTing (cyclic amplification of selected targets (Pollock and Treisman, 1990) and electrophoretic mobility shift analyses (EMSA), indicated that the best consensus sequence for *in vitro* DMyb binding is AACGGPyPyG/T (Jackson et al., 2001).

3.3 Transcriptional Activation

An initial study reported that DMyb was only capable of activating transcription from a reporter construct in cultured insect (Schneider) cells when it was co-expressed with dCBP, the *Drosophila* homologue of the co-activator CBP (Hou et al., 1997). However, we observed a ten-fold activation of a reporter in the same cells (Jackson et al., 2001). One possibility is that the cell lines were subtley different, and indeed, we found that after many months in continuous culture, Schneider cells appeared to lose the ability to respond to DMyb (J. Jackson and A.L.K., unpublished observations). On the other hand, a notable difference between the experiments was the use of reporter constructs which differ at the positions 4 and 6 of the MBS: 5'-TAAC**G**G**T**TT-3' in pT81luc-3xA used by Jackson et al. (2001), and 5'-TAAC**T**G**A**CA-3' in pADHCAT6MBS-1 used by Hou et al. (1997). Interestingly, substitution of the G at position 4 with a T results in a dramatic decrease in binding affinity of DMyb while a pyrimidine appears to be favoured at position 6 (Jackson et al., 2001). So it seems that the more limited ability of DMyb to transactivate expression in the experiments of Hou et al. (1997) may be due to the use of a lower affinity binding site for DMyb.

DMyb is more efficient than c-Myb at activating transcription in Schneider cells but is a less effective transactivator than c-Myb in several mammalian cell lines (Jackson et al., 2001). This difference may be explained by the presence of one or more specific co-factors in each cell line that interact with a poorly conserved region of the Myb protein. This putative factor does not appear to be dCBP since co-expression of dCBP with DMyb in mammalian cells produced only a mild increase in transcriptional activation by DMyb (J. Jackson and A.L.K., unpublished observations). A more mundane possibility, and the one we favour at present, is that each Myb protein functions most efficiently at the normal body temperature for its organism of origin, that is 25°C in *Drosophila* compared to 37°C in mammals.

4. WHAT HAS BEEN LEARNED ABOUT THE PHYSIOLOGICAL ROLES OF DMYB?

4.1 DMyb Provides an Essential Function During Drosophila Development

Based on its evolutionary conservation and the presence of a single *myb* gene in *Drosophila*, it seemed likely that DMyb would provide an essential function to the organism. Therefore, a classical genetic screen was carried out to isolate recessive lethal mutations in the vicinity of the *Dm myb* gene located on the X-chromosome at position 13F14 (Katzen and Bishop, 1996). Confirming that DMyb does provide an essential function, two temperature-sensitive recessive lethal alleles, myb^1 and myb^2, were isolated. Additional recessive lethal alleles (including amorphs or nulls) have been reported subsequently (Manak et al., 2002; Okada et al., 2002). Furthermore, temperature shift studies with the temperature-sensitive alleles and examination of mutant phenotypes revealed that DMyb is important for both embryonic and imaginal development and that it serves a role in oogenesis and the development of many tissues (Katzen and Bishop, 1996; Manak et al., 2002). In addition, when the temperature sensitive mutants were raised at temperatures that were permissive for viability, phenotypic defects were observed in wings, abdominal cuticle, and flight ability (Katzen and Bishop, 1996).

4.2 DMyb Promotes Mitosis and Suppresses Endoreduplication

The next question to address was the cellular basis of the mutant *myb* phenotypes. This issue was first pursued for one of the "viable phenotypes". Of these, the wing phenotype was the most consistent, and was also of particular interest because the phenotype itself already suggested possibilities for the underlying cellular basis of the resulting cuticular defect. Mutant wings were approximately the same size as wild type, but had about half the number of hairs that were considerably larger than normal (Katzen et al., 1998). In wild type wings, each cell that is not specialised for another purpose is represented by a single hair (Postlethwait, 1978). Therefore, the reduced density of hairs on mutant wings suggested two possibilities; either these wings had fewer cells, each of which was larger, or they had the same number of cells as wild type, but only some of the cells produced hairs.

As is true for the rest of the thoracic and head epidermal tissue, adult wings are formed from imaginal discs, which are specified during embryogenesis. Imaginal disc cells are diploid and proliferate throughout larval development, completing only their final one or two cell divisions during early pupation (Postlethwait, 1978; Cohen, 1993). In the case of the wing, it has been shown that by shortly after puparium formation (APF) the majority of cells become arrested in G2 and remain so until 12 hours APF. The cells then divide and progress through their final cell cycle before becoming postmitotic at 24 hours APF (Schubiger and Palka, 1987). When developing wings from wild type and mutant *myb*^{1} and *myb*^{2} animals were examined, no differences were apparent through early pupation when the cells are arrested in G2. In contrast, in postmitotic pupal wings, the density of nuclei was approximately half of that in wild type wings and there was a one-to-one correspondence between the number of nuclei and the number of developing hairs in both mutant and wild type wings. No apoptotic nuclei were observed in any of the pupal samples. These findings demonstrate that the mutant wings have fewer, larger cells which each produce a hair (Katzen et al., 1998).

The finding raised the question of why the mutant wings are essentially normal in size even though they are composed of approximately half the number of cells as wild type wings? Injection of 5-bromo-2-deoxyuridine (BrdU) into developing pupae revealed that mutant *myb*^{1} wing cells enter into their final S-phase, but apparently cannot progress through their final division. Additional support for this conclusion was provided by the finding that ectopic expression of either of two regulators of the G_2/M transition, Cyclin dependent kinase 1 (Cdk1) or String (*Drosophila* homologue of Cdc25, the protein tyrosine/threonine phosphatase that regulates Cdk1 activity), was able to partially suppress the mutant *myb* phenotype in adult wings. Together, these findings indicated that the mutant wings were of normal size because the wing cells were arrested in G2, and therefore had DNA contents of 4C instead of 2C, which then led to the enlargement of the nuclei, cells and hairs (Katzen et al., 1998).

A final experimental approach used to confirm that the mutant cells were arrested in G2 of their final cell cycle, revealed additional complexity. The relative DNA contents of wild type and mutant nuclei were compared using high resolution, three-dimensional wide-field fluorescence microscopy after staining with the DNA-binding dye, DAPI (4'6-Diamidine-2-phenylindole) (Katzen et al., 1998). Unexpectedly, the results revealed heterogeneity within the population of mutant cells. Although most of the mutant cells had nuclear DNA contents that were consistent with their being arrested in G2, there was also a population of cells with the same DNA content as wild type cells. This suggested that some of the cells, perhaps those that divide earliest, retain enough DMyb activity to complete the final division. More surprisingly, some of the mutant cells (the majority when the mutants were raised at temperatures that are non-permissive for adult viability) had DNA contents in excess of 4C, indicating that instead of remaining in G2, a fraction of the wing cells that were unable to divide entered into endoreduplication. Taken together these results led us to conclude that in proliferating cells, DMyb functions to promote mitosis while simultaneously acting to suppress endoreduplication.

The finding that diploid cells with reduced levels of DMyb activity enter into endoreduplication cycles correlates well with the earlier observation that DMyb is not expressed in larval tissues that undergo endoreduplication (Katzen and Bishop, 1996). Ectopic expression of full length and C-terminally truncated (ΔDMyb) versions of DMyb in developing salivary glands suppressed endoreduplication, although the ΔDMyb protein was considerably more effective (Fitzpatrick et al., 2002). As predicted from the mitotic block in the mutant *myb* wing cells, ectopically expressed DMyb induced higher levels of mitosis in imaginal disc cells as visualised by immunostaining with an antibody (PH3) for a mitotic-specific phospho-epitope on histone H3. No significant difference was observed between the effects of the full-length and C-terminally truncated proteins, suggesting that the negative regulatory effects meditated by the C-terminus may be at least partially "relieved" in these cells (Fitzpatrick et al., 2002). Further evidence that DMyb may play an important role in regulating the G2/M transition was provided by the finding that it is a transcriptional regulator of the mitotic cyclin, *cyclin B*. Ectopic expression of a C-terminally truncated DMyb protein in eye imaginal discs induced *cyclin B* expression, and in clones of cells that were homozygous for loss-of-function mutations in *Dm myb*, *cyclin B* expression was lost (Okada et al., 2002). Seven potential DMyb-binding sites were identified in a 733 bp fragment encompassing the *cyclin B* transcription start site, which was able to direct high level expression of a luciferase reporter gene in Schneider cells. This expression was greatly decreased when either endogenous DMyb levels were reduced by RNA interference or when all of the putative DMyb-binding sites were mutated (Okada et al., 2002).

4.3 Additional Roles for DMyb in the Cell Cycle

The abdominal phenotype in temperature-sensitive *myb* mutants that are raised at temperatures permissive for viability includes missing bristles and patches of undifferentiated and unpigmented cuticle (Katzen and Bishop, 1996). This phenotype is characteristic of situations in which there are not enough adult epidermal cells to replace all larval cells (Poodry, 1975), and is therefore suggestive of a defect in cell proliferation. Although this phenotype superficially appears to represent another example of the cell division defect analysed in the wing, there are several differences between wing and abdominal phenotypes at the cuticular level (Fung et al., 2002). In addition, the developmental programs of these two epidermal tissues are distinct. For example, unlike the epidermis of the head and thorax, the abdominal epidermis is formed from small nests of imaginal cells called abdominal histoblasts, which do not divide during larval development but undergo rapid proliferation after puparium formation (Madhavan and Madhavan, 1980). These differences suggested that an analysis of the cellular basis of the abdominal phenotype might reveal more about DMyb function (Fung et al., 2002). As was the case for the wing imaginal discs, no apparent defects were observed in the larval abdominal histoblast nests of the *myb* mutants. However, when the histoblasts started to proliferate shortly after puparium formation, it became evident that the mutant *myb* cells proliferated more slowly than wild type controls, with the differential in proliferation rates increasing when the animals were incubated at higher, non-permissive temperatures. Although, the mutant abdominal epidermal cells continued to divide slowly as pupal development proceeded, they were able to partially compensate for this by continuing to proliferate well beyond the timepoint when wild type cells have become postmitotic (circa 41 hours APF at 25°C). Despite the reduced rate of proliferation, the mitotic index of the mutant *myb* abdominal epidermal cells during pupation was higher than wild type controls, indicating that mutant cells were progressing especially slowly through mitosis.

Analysis of the distribution of cells among the various stages of mitosis using the PH3 antibody revealed that in *myb* mutants a higher percentage were in the early stages of mitosis, prior to metaphase, compared to wild type controls. Most notably, there was a dramatic increase in the percentage of cells with a distinctive DAPI/PH3-staining pattern that is characteristic of cells in "pre-prophase", a stage corresponding to the initiation of histone H3 phosphorylation at the G2/M transition (Hendzel et al., 1997). This staining pattern is rarely observed in wild type cells (1% or less of PH3 staining cells), indicating that the cells progress through this stage rapidly. In contrast, in mutant *myb* samples the frequency of pre-prophase cells was consistently higher and increased as pupation progressed, reaching levels of more than one third of all PH3-staining cells. An increase in the percentage of pre-prophase cells was not observed in abdominal epidermal cells in the presence of *Minute* mutations, which are a class of mutations that slow cell proliferation throughout the animal. These results indicate that in *myb* mutants the abdominal epidermal cells progress through the early stages of prophase at an abnormally slow rate.

The finding that mutant *myb* abdominal histoblasts progress slowly through mitosis prompted a closer examination of the mitotic cells to determine whether cytological defects could be detected. Although histoblast proliferation was slower throughout pupal development, mitotic defects were not evident in early divisions but, after first being detected at about 30 hours APF (at 25ºC), became increasingly commonplace later. The majority of mitotic abnormalities were associated with aberrant numbers of centrosomes, ranging from one to eight (instead of the normal two), with the most common numbers being three or four. Cells with single centrosomes organising monopolar spindles were occasionally seen, but cells with extra centrosomes were much more common. Most, if not all, of the supernumerary centrosomes nucleated mitotic spindles with variable consequences for the dividing cells. When the additional centrosomes were located at, or near each pole, a bipolar spindle was still formed allowing for chromosome separation. More commonly, extra centrosomes formed multipolar spindles, pulling chromosomes in multiple directions and occasionally organizing more than a single metaphase plate. When a large number of centrosomes were distributed throughout the cell, a proper spindle apparatus was not formed. Chromosomal abnormalities, such as lagging chromosomes during metaphase and anaphase, were common in mitotic cells with aberrant centrosome numbers, but were also observed in mitotic cells containing two centrosomes. These mitotic defects appear to be specific to mutations in *Dm myb* since abnormalities in centrosome numbers were not observed when cell proliferation was slowed down via either *Minute* mutations or ectopic expression of the *Drososphila* retinoblastoma family protein (RBF; Du and Dyson, 1999).

How are the abnormal mitoses in *myb* mutants that have extra centrosomes forming multipolar spindles resolved? Do they apoptose, get trapped in metaphase indefinitely, complete division and form multiple daughter cells with unbalanced chromosome segregation, or return to an interphase state? A definitive answer is difficult since mitoses in abdominal histoblasts cannot be readily monitored *in vivo*. However, the first two possibilities appear to be rare, since neither elevated levels of apoptosis nor ever increasing numbers of metaphase chromosomes were observed. The lack of increased apoptosis was surprising given the severity of mitotic defects and evidence that vertebrate Myb can prevent apoptosis (Frampton et al., 1996; Taylor et al., 1996), thereby raising the possibility that apoptosis may be suppressed in abdominal epidermal cells.

Most mutant *myb* abdominal epidermal cells were larger than wild type, and these larger cells generally contained either larger nuclei, which were often abnormally shaped or multilobed, or two fused or separated nuclei. There was also a minor population of mutant *myb* cells that were smaller than wild type and contained correspondingly undersized nuclei. The variation in nuclear size and morphology was confirmed by quantitative microscopic analysis, showing that DNA content ranged from subdiploid levels to as high as 7-fold normal diploid levels. The mitotic defects observed in the mutant *myb* cells imply that even some of the cells within the normal range of DNA content may be aneuploid rather than diploid. This latter possibility was supported by fluorescent *in situ* hybridisation (FISH) analysis, which indicated that the number of hybridisation signals in mutant cells was variable.

What is the origin of such an array of mitotic defects? One possibility is that the primary defect causes some cells to fail to complete mitosis or cytokinesis. The resulting cells would inevitably contain extra centrosomes, which could then actively contribute to additional mitotic defects in subsequent divisions. This scenario is unlikely since there was no evidence of polyploidy or abnormalities in nuclear morphology (binucleate or multilobed) prior to the first centrosomal defects. However, a clue may be provided by a mild centrosome abnormality in which two centrosomes are present at or near each mitotic spindle pole. This defect was observed much more frequently in cells that were in late anaphase or telophase than in cells that were in metaphase or early anaphase, indicating it may arise after metaphase. This defect, which is never seen in wild type samples, appears to reflect a precocious separation of the centriole pairs associated with each centrosome, and is likely to be an early sign of a breakdown in the coordination of nuclear and centrosome cell cycles. This initial breakdown in coordination could then be compounded in subsequent divisions, a possibility that fits well with the findings that the percentage of mutant *myb* histoblasts with visible mitotic abnormalities increases dramatically as pupal development proceeds. Furthermore, the spectrum of mitotic defects, which include aberrant numbers of centrosomes, grossly abnormal DNA morphology, aneuploidy, and polyploidy, are characteristic of situations in which the coordination of centrosome and nuclear cycles has been disturbed (Sluder and Hinchcliffe, 1999).

Why do the cellular defects in *myb* mutants indicate that DMyb is required for both promotion of the G2/M transition and suppression of endoreduplication in wings (Katzen et al., 1998), whereas it is required for appropriate progression through mitosis in the abdomen (Fung et al., 2002)? One possibility is that new aspects of DMyb function may be revealed in abdominal histoblasts because the demands of their developmental program make them more sensitive to reductions in the protein's activity. Histoblasts proliferate three to four times more rapidly than wing cells (Madhavan and Madhavan, 1980), and the levels of *Dm myb* mRNA are lower in histoblasts than in wing discs (Katzen and Bishop, 1996). During pupation, abdominal development is significantly more delayed than wing development in *myb* mutants (10-12 hours versus 1.5 hour - Katzen et al., 1998; Fung et al., 2002), indicating that abdominal histoblasts are indeed more sensitive to reductions in DMyb function. It is also possible that additional mitotic functions for DMyb are revealed in abdominal cells because regulation of the G2/M transition is not as restrictive as it is in wing cells. However, in subsequent studies with null alleles of *Dm myb,* similar defects to those observed in abdominal cells with the temperature sensitive alleles of *myb* (slow proliferation, abnormal centrosome numbers and mitotic spindle sturctures, aneuploidy and polyploidy), have been observed in imaginal disc and neural cells (Manak et al., 2002). Furthermore, similar mitotic defects have recently been observed in early myb^1 embryos when they were collected from homozygous myb^1 females that had been incubated at non-permissive temperatures (G. Scaria and A.L.K., unpublished observations). Taken together, these results demonstrate that the requirement for DMyb function in order to progress through as well as enter mitosis, is not limited to abdominal epidermal cells.

Manak and colleagues also reported that some cells stained simultaneously for BrdU incorporation and with the PH3 antibody, suggesting that they had entered into mitosis while they were still in the process of replicating their chromosomes (Manak et al., 2002). The authors proposed that DMyb functions in DNA replication rather than in regulating mitosis, and that the mitotic defects observed in *myb* mutants are an indirect consequence of the inability of the mutant cells to completely replicate their chromosomes, followed by an aberrant S to M-phase transition. Data from two subsequent papers supported the hypothesis that DMyb plays a role in S-phase. Firstly, ectopic expression of DMyb in imaginal discs induced an increase in the overall percentage of S-phase cells and pushed cells into S-phase even when developmental signals normally dictate cell cycle arrest (Fitzpatrick et al., 2002). Secondly, the DMyb protein was shown to have a direct, non-transcriptional role in the amplification of the chorion loci in follicle cells (Beall et al., 2002 - see Section 4.4.3). Nevertheless, the findings that ectopic DMyb drives cells into M-phase as well as S-phase (Fitzpatrick et al., 2002), and that *cyclin B* is a target of transcriptional regulation by DMyb (Okada et al., 2002), support the earlier evidence that

DMyb has a *bonafide* role in regulating entry into, and progression through, mitosis (Katzen et al., 1998; Fung et al., 2002).

Since ectopic DMyb activity was found to induce increased levels of both S- and M-phase in diploid cells, it should lead to massive overgrowth of imaginal disc tissue. Surprisingly, although wing discs were malformed and appeared to be somewhat overgrown (or bulging) in the areas where DMyb was ectopically expressed, massive overgrowth was not observed (Fitzpatrick et al., 2002). In cases where the animals survived to adulthood, the wings were not obviously enlarged (C.A. Fitzpatrick and A.L.K., unpublished results). Further analysis revealed that the increases in cellular proliferation induced by ectopic DMyb were accompanied by increases in apoptosis (Fitzpatrick et al., 2002). It is likely that the increased apoptosis is an indirect consequence of the ectopic DMyb activity, since expression of other cell cycle regulators produced similar results (C.A. Fitzpatrick and A.L.K., unpublished results), and previous studies have demonstrated that cell death is often induced when the cell cycle is deregulated in imaginal discs (Asano et al., 1996; Du et al., 1996; Milan et al., 1997; Neufeld et al., 1998). However, there are also indications that ectopic DMyb may have a more direct influence on developmental signalling pathways, which could also contribute to the increased levels of apoptosis (see Section 4.5). Recently, we have found that the wing discs are greatly enlarged when cell death is inhibited by coexpression of the baculovirus P35 caspase inhibitor protein and a DMyb transgene (C. A. Fitzpatrick and A.L.K., unpublished results). This confirms our conclusions from earlier studies that DMyb activity promotes cell proliferation and growth in imaginal tissues. This is notably different from the situation when E2F, which promotes cell proliferation but not growth in the presence of P35, is over-expressed in that more cells are produced, but they are smaller and the overall size of the tissue is essentially unaffected (Neufeld et al. 1998).

4.4 Genetic and Biochemical Interactions with Other Genes and Their Products

4.4.1 dCBP

As discussed in Section 3.3, the *Drosophila* CBP protein, dCBP, enhances the ability of DMyb to activate transcription in cultured cells (Hou et al., 1997; Fung et al., 2002). DMyb and dCBP were shown to physically interact *in vitro* using a bacterially expressed GST-dCBP fusion protein to "pull-down" full-length *in vitro* translated DMyb protein (Hou et al., 1997). The dCBP-binding domain of DMyb was mapped to amino acids 230-406, immediately C-terminal of the DMyb DNA-binding domain. When this region was *in vitro* translated, it interacted as strongly with GST-dCBP as did the full-length DMyb protein. The dCBP-binding domain also appears to encompass the transcriptional activating domain, since in combination with the DMyb DNA-binding domain, it is capable of activating transcription from a reporter construct in transient transfection assays (Hou et al., 1997). Therefore, the authors conclude that dCBP binds to the transactivating domain of DMyb, which is similar to what has been demonstrated for the mammalian A-Myb and c-Myb proteins (Dai et al., 1996; Facchinetti et al., 1997).

In *Drosophila*, studies of chromosomal deletions have shown that few loci have observable haploid phenotypes, and that for most loci, the level of expression is proportional to gene copy number (Muller et al., 1931; Lindsley et al., 1972). Therefore, for most genes, a mutation in a single copy (with the other copy being wild type) will not result in a phenotypic defect unless the biochemical pathway in which it participates has already been "sensitised" by a mutation in another gene that functions in the same pathway. The hypomorphic alleles of *Dm myb* discussed above, myb^1 and myb^2, provide appropriately sensitised backgrounds in which to test the phenotypic consequences of reducing the levels of another gene product. Reducing the levels of dCBP (also known as *nejire*) by introduction of one copy of an amorphic allele dramatically reduced the viability of myb^1 and myb^2 homozygotes at temperatures that are normally permissive for viability, but had no significant effects on the viability of control animals that were not mutant for *Dm myb* (Fung et al., 2003). In addition, phenotypic defects observed in *myb* mutants (especially myb^1) were enhanced by reduced levels of dCBP. Hence, in wings, the density of hairs was further reduced and the presence of multiple hairs protruding from a single position was greatly increased. In abdomens, the presence of undifferentiated cuticle was much more pronounced in certain areas, especially on the dorsal surface where white undifferentiated cuticle was present between most segments and along the dorsal midline (Fung et al., 2003).

The cellular basis of the enhanced phenotypes in wings and abdomens was also investigated (Fung et al., 2003). In wings, the variability in nuclear and cellular sizes and shapes became more pronounced when dCBP levels were reduced within the context of a myb^1 mutant, with enlarged, misshapen or multi-lobed nuclei or multiple separated nuclei within a single cell being commonly observed (note that in this experiment, a few cells with bi-lobed nuclei were also observed in myb^1 wing cells with wild type levels of dCBP, an abnormality not detected in previous experiments, presumably because of its rarity). The enhanced cellular defects associated with decreased dCBP levels generated more extreme variability in the number, size and orientation of prehairs. Most notably, single cells producing two or more prehairs were common, which correlated with the adult phenotype. The presence of multi-lobed or multiple separated nuclei in *myb*^{1} cells with decreased dCBP levels indicates that the cells entered, but did not complete, mitosis or cytokinesis, a phenotype that is qualitatively different from most of the *myb*^{1} mutant wing cells with wild type levels of dCBP. Initially, this result appears to be counter-intuitive because it suggests that the *myb*^{1} mutant cells with reduced dCBP levels are progressing further into the cell cycle than the *myb*^{1} cells with normal levels of dCBP. However, the presence of fewer, larger wing cells when dCBP levels are reduced indicates that at least a portion of these cells are failing to complete cell division in the previous (second to last) cell cycle, thereby accounting for the enhanced phenotype. This finding is also consistent with the earlier incidence of defects observed in imaginal disc cells in animals that are homozygous for amorphic alleles of *Dm myb* (Manak et al., 2002).

As described above (see Section 4.3), abdominal histoblasts that are mutant for *Dm myb* proliferate more slowly than wild type cells, which leads to a delay in the replacement of the polyploid larval cells by adult epidermal cells. When dCBP levels are reduced in *myb*^{1} mutants, proliferation and replacement are severely retarded and it appears that significant regions of larval cells are never replaced, which is likely to account for the undifferentiated cuticle observed between segments and along the dorsal midline in adults. The mitotic index, which is already abnormally high in mutant *myb* cells, is nearly doubled when dCBP levels are reduced. When considered in combination with the reduced rate in proliferation, this finding demonstrates that these cells are severely delayed in their progression through mitosis. In the later cell cycles of abdominal epidermal cells that are mutant for *Dm myb*, abnormal mitoses associated with multiple functional centrosomes, unequal chromosome segregation, formation of micronuclei, and/or failure to complete cell division are common (see Section 4.3 and Fung et al., 2002). It seemed likely that the mitotic abnormalities and slowed rates of cellular proliferation in *myb* mutants were directly related to each other. However, this supposition is contradicted by data from the analysis of mutant *myb* cells with reduced levels of dCBP. Although cell proliferation was dramatically slower in these cells, no obvious effects on the size or morphology of individual cells and nuclei were observed, and no changes in the timing or rate of mitotic defects were detected. These findings are consistent with observations in adults that in regions where differentiated cuticle has formed, the phenotype is not appreciably different between *myb* mutants with normal and reduced dCBP levels, suggesting that the centrosomal and chromosomal abnormalities may be at least partially independent of the reduced rate of proliferation.

The documentation of a genetic interaction between *Dm myb* and *dCBP* provides direct evidence that the biochemical interaction between CBP and Myb proteins is physiologically relevant within the context of a developing animal, and identified DMyb as a one of the transcription factors that is sensitive to decreases in dCBP levels. In addition, although other studies had implicated dCBP in several signal transduction pathways that regulate developmental patterning (Akimaru et al., 1997; Waltzer and Bienz, 1998; Waltzer and Bienz, 1999; Takaesu et al., 2002), the finding that reduced levels of dCBP enhance the cell division defects of mutations in *Dm myb*, provides the first explicit evidence that dCBP is directly involved in regulating cell proliferation (Fung et al., 2003).

4.4.2 Cyclin A

Based on the experimental evidence that *Dm myb* participates in several aspects of cell cycle regulation, mutations in several other genes known to regulate the cell cycle were tested for genetic interactions with the myb^1 and myb^2 mutants. Of the genes tested, decreased levels of the mitotic cyclin, Cyclin A, produced the strongest effects, significantly reducing the viability of the *myb* mutants and enhancing the wing and abdominal phenotypes. These phenotypic modifications appear to be similar to those associated with decreased levels of dCBP within a mutant *myb* background, and indeed, the cellular defects and cuticular phenotypes in the wing are quite comparable. In the abdomen, however, there are distinct differences between the effects of reduced Cyclin A and dCBP levels. For example, cuticular defects in adult abdomens are much more severe when Cyclin A levels are decreased in that all bristles on the ventral surface and most bristles on the dorsal surface are missing and segments are often malformed or fused. During pupal development, abdominal cells that are mutant for *myb* proliferate more slowly and exhibit a higher mitotic index when Cyclin A levels are lowered. This is similar to the consequences of reducing dCBP, except that both effects are milder with reductions in Cyclin A (S-M. Fung and A.L.K., manuscript in preparation). The main difference seems to be that while the timing and rate of mitotic defects is not affected by decreased dCBP levels, this is not the case with Cyclin A. In comparison to mutant abdominal histoblasts with normal levels of Cyclin A, mitotic defects (predominantly abnormal centrosome numbers) arise earlier and occur at considerably higher rates when Cyclin A levels are reduced. In addition, among the population of cells in mitosis, the percentage of cells in anaphase and early telophase is increased (with a reciprocal decrease in the percentage of cells in metaphase), indicating that mutant *myb* cells with lowered Cyclin A levels have difficulty exiting from mitosis. The resulting cells are greatly enlarged and frequently contain either multilobed or multiple nuclei, which is indicative of a failure to complete mitosis or cytokinesis. Although these abnormal "adult cells" eventually replace most of the larval cells in the abdominal epidermis, the severe cuticular defects displayed in adult abdomens indicate that they are unable to differentiate appropriately.

The physiological relevance of an interaction between DMyb and Cyclin A is further supported by another genetic approach. In agreement with previously published results (Okada et al., 2002), we found that ectopic expression of the C-terminally truncated DMyb protein in the eye imaginal disc causes the adult eye to be roughened and decreased in size (due to an apoptotic response to the disruption of cell cycle regulation). This phenotype was almost completely suppressed when Cyclin A levels were reduced by the introduction of one mutant gene (C.A. Fitzpatrick and A.L.K., unpublished observations). The ability of decreased levels of Cyclin A to suppress the effects of ectopic DMyb activity in the eye are similar to the previously published effects of reducing Cyclin B in this context (Okada et al., 2002). However, unlike Cyclin A, reducing the levels of Cyclin B in *myb*^{1} or *myb*^{2} mutants did not affect any aspect of the mutant phenotypes (S-M. Fung and A.L.K., unpublished observations).

This finding of a physiologically relevant interaction between DMyb and Cyclin A is of particular interest in light of published reports from several laboratories that show that the vertebrate B-Myb protein can be phosphorylated by a Cyclin A/Cdk complex, leading to stimulation of its transactivation capacity (Lane et al., 1997; Sala et al., 1997; Ziebold et al., 1997). The transactivation potential of the A-Myb protein has also been reported to be activated by Cyclin A/Cdk mediated phosphorylation (Ziebold and Klempnauer, 1997). Hence, the underlying biochemical basis of the genetic interaction between mutant *Dm myb* and *cyclin A* may be that a smaller proportion of the DMyb protein is phosphorylated by Cyclin A/Cdk when the levels of Cyclin A are decreased, reducing the already compromised activities of the mutant DMyb proteins. The possibility that the *cyclin A* gene could be a target of DMyb and that the genetic interaction reflects a reduction in *cyclin A* RNA expression was excluded by showing that *cyclin A* transcript levels were the same in cells that were either mutant for or ectopically expressing *Dm myb*.

To begin to address the possibility that DMyb activity may be regulated by Cyclin A/Cdk phosphorylation, the effects of co-expressing Cyclin A with DMyb in transient transfection assays were examined. As with the B-Myb and A-Myb proteins, co-expression of Cyclin A enhanced the transcriptional activation ability of DMyb. Co-expression of Cyclin A and Cdk1 with DMyb in larval salivary glands, which do not normally express these proteins, resulted in a decrease in the mobility of a portion of the

DMyb protein in Western blots. Interestingly, of the four consensus sites for Cyclin A/Cdk phosphorylation in DMyb, two (S381 and T447) are located within evolutionarily conserved domains and correspond to sites (T447 and T524) that have been shown to be phosphorylated by Cyclin A/Cdk in B-Myb (Bartsch et al., 1999). Therefore, it seems most likely that the basis of the genetic interaction between *Dm myb* and *cyclin A* is that the DMyb protein is a target for phosphorylation by a Cyclin A/Cdk complex, and that the phosphorylated DMyb protein is a more potent transcriptional activator.

4.4.3 Novel interactions

Recently, Beall and colleagues isolated and identified five proteins that bound to *ACE3* or *Ori-β* sequences, two *cis*-regulatory elements required for site-specific gene amplification of the chorion loci in follicle cells (Beall et al., 2002). These five proteins which include DMyb, Caf1 (Chromatin assembly factor 1 subunit), and three other proteins about which little is known (p40, p120 and p130, encoded by computed genes CG15119, CG6061 and CG3480 or *twilight*, respectively), form a complex which is not dependent on the presence of DNA. Direct interaction between the DMyb-containing complex and the Orc proteins was demonstrated by co-immunoprecipitated of the Orc complex proteins by anti-DMyb antibodies and of the DMyb-containing complex by anti-Orc2 antibodies.

Of the five proteins in the DMyb-containing complex, only DMyb and p120 were shown to interact directly with the *ACE3* DNA fragment. The 40 bp region protected by DMyb in DNase I protection assays contains two Myb-binding site consensus sequences (5'-AACGG and 5'-ACCTG) in opposite orientations. When either the DMyb or one of the p120-protected regions were deleted, amplification of the transgenic reporter sequences was dramatically reduced, indicating the functional importance of these sequences. To test for the functional requirement of the DMyb protein at the endogenous amplification loci, mitotic clones of follicle cells were generated that were homozygous for one of the amorphic alleles of *Dm myb* (MH107 or MH30 (Manak et al., 2002). Unlike the surrounding cells, no subnuclear foci of BrdU incorporation at the chorion gene amplification sites were observed in cells that were mutant for *Dm myb*, indicating that DMyb is indeed required for amplification. However, DMyb is apparently not required for localisation of the Orc complex, since Orc2 continued to be localised to the subnuclear foci in the mutant *myb* cells (Beall et al., 2002).

If the DMyb protein is not recruiting the Orc complex to the chorion gene amplification origin, what is its function? Some insights can be gleaned by considering what is known about the other members of the complex. The best characterised of these is Caf1, which along with its vertebrate homologue, hCAF-1 p48 (RbAp48), is a member of an evolutionarily conserved subfamily of WD-repeat proteins (Taunton et al., 1996; Tyler et al., 1996). Caf1 had already been shown to be a member of the chromatin assembly factor 1 (CAF1) and nucleosome remodelling factor (NURF) complexes (Krude, 1995; Roth and Allis, 1996; Martinez-Balbas et al., 1998). As a member of CAF1, Caf1 acts as a chaperone for Histones H3 and H4, accompanying them from the cytoplasm (where they are also associated with histone acetyltransferase) to the nucleus (where they are also associated with histone deacetylase), and is subsequently involved in assembling these histones into nucleosomes at the DNA replication fork. The nucleosome remodelling factor complex participates in transcription factor-mediated chromatin remodelling and contains three other subunits including ISWI, a protein related to the *Drosophila* Brahma and yeast SWI2/SNF2 transcriptional regulators. These findings indicate that Caf1 may function as a common platform for the assembly of protein complexes involved in chromatin construction or remodelling, which would suggest a similar role for the newly identified DMyb-containing Caf1 complex.

The presence of the p130 *twilight* (*twit*) gene product in this complex is also notable. Sequence analysis indicates that *twit* is one of two *Drosophila* genes that encode homologues of the *Caenorhabditis elegans lin-9* gene (White-Cooper et al., 2000). The protein product of the *always early (aly)* gene, the other *lin-9* homologue in *Drosophila*, is associated with chromatin and is required for the maintenance of normal chromatin structure in primary spermatocytes (White-Cooper et al., 2000). The authors propose that *aly* regulates transcription of cell cycle and terminal differentiation genes during spermatogenesis through a chromatin remodelling complex. Therefore, the presence of both Caf1 and p130 is a strong indication that this DMyb-containing complex may play an essential role in remodelling the chromatin at the origin of replication for chorion gene amplification.

Can this role for DMyb at the origin of replication for chorion gene amplification be generalised to less specialised forms of DNA replication? The same DMyb-containing complex of five proteins was found in nuclear extracts prepared from ovaries, embryos, and two tissue culture cell lines, indicating that it may have a ubiquitous function at replication origins. On the other hand, *Dm myb* mutant follicle cells appear to progress normally through the endoreplication cycles that precede chorion gene amplification (Beall et al., 2002). Furthermore, nuclei that normally become polyploid during larval development, such as salivary gland nuclei, successfully undergo endoreplication in animals that are homozygous for either hypomorphic or null alleles of *Dm myb*, and neither *Dm myb* mRNA nor protein are detectable in larval salivary glands (Katzen and Bishop, 1996; Katzen et al., 1998; Fung et al., 2002; S-M. Fung and A.L.K., unpublished observations). Together, these findings indicate that DMyb is not required for the initiation of DNA replication at origins during endoreduplication.

What about the situation in proliferating cells? In animals that are homozygous for null mutations in *Dm myb*, DNA replication and cell proliferation do occur, albeit slowly, and many mitotic defects are observed (see below), indicating that the DMyb protein could be functioning to "enable" DNA replication origins. Embryos containing temperature-sensitive alleles of *Dm myb* display mitotic defects when their mothers have been exposed to the restrictive temperature for at least 24 hours. In contrast, when mutant embryos are collected at the permissive temperature and are then shifted to the restrictive temperature no defect is observed, indicating that the critical period for these defects occurs during oogenesis (G. Scaria and A.L.K., unpublished observations). This finding is consistent with DMyb functioning as a transcriptional regulator during oogenesis rather than being required for initiation of replication at origins during early embryogenesis. However, this data does not rule out the possibility that DMyb is involved in initiating replication at a subset of origins in the nuclei of proliferating cells.

4.5 Does DMyb have a Role Outside the Cell Cycle?

Recent studies in our laboratory have revealed that changes in the levels of DMyb activity can disturb several developmental processes, including wing vein formation, wing margin specification, appendage determination, and dorsal closure of the thorax and abdomen (C.A. Fitzpatrick, G. Ramsay and A.L.K., unpublished results). These findings suggest that DMyb participates in some differentiation or developmental patterning decisions. Furthermore, the phenotypic defects indicate that in some cases its "patterning functions" may be completely independent of its role in the cell cycle, whereas in others DMyb may function in the cross-talk that occurs between the regulation of developmental patterning and cell proliferation. At a superficial level, it seems surprising that changes in the levels of DMyb can elicit such pleiotropic effects. However, developmental patterning repeatedly uses a small set of essential signalling pathways (Wingless or Wnt; Hedgehog, Dpp or TGF-β/BMP; receptor tyrosine kinases, represented in these processes by the EGF receptor; Jun N-terminal kinase and Notch - see Gerhart, 1999 and Curtiss et al., 2002) and it is possible that all of the observed phenotypic defects could reflect the participation of DMyb in a subset of these pathways.

5. PERSPECTIVES AND FUTURE DIRECTIONS

Figure 2 represents a summary of the processes in which DMyb has been implicated.

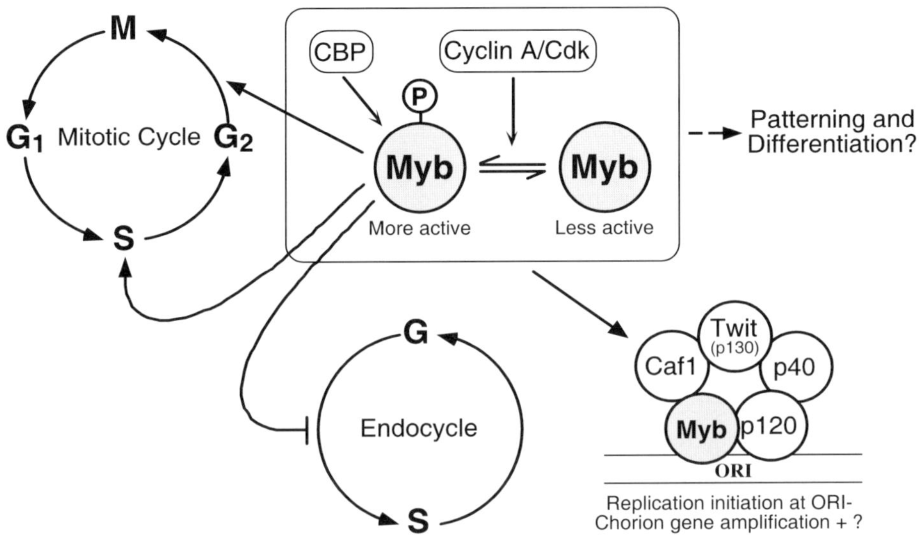

Figure 2

Schematic showing the various functions and biochemical interactions ascribed to the DMyb protein. Refer to the text for descriptions and references. Note that for the DMyb-containing complex that binds to the *cis*-regulatory elements (ORI) required for site-specific gene amplification of the chorion loci in follicle cells, the specific protein-protein contacts between the five proteins have not been defined. However, DMyb and p120 have been shown to make direct contacts with the DNA.

The abundance of activities in which DMyb has been implicated prompts the question of whether it is really functioning independently in all of these different processes or whether some of the apparent functions are actually indirect consequences of other primary activities? A definitive answer to this question is not possible at present, but since the primary biochemical activities of DMyb are to bind DNA and regulate transcription, it is conceivable that DMyb could be binding directly to some origins of replication while simultaneously regulating the expression of a variety of genes that are required for multiple aspects of the cell cycle. It is also not clear whether all of the functions in which DMyb has been implicated represent the function of a single vertebrate Myb protein (with the likely candidate being B-Myb), the combined functions of all three vertebrate Myb proteins, or something in between.

The availability of genetic tools together with improvements in technology should make it possible to greatly expand our understanding of DMyb during the next several years. For example, it is now possible to manipulate DMyb activity levels (up and down) in temporal or tissue specific patterns. The affected cells can be "marked" with green fluorescent protein and isolated as living cells using FACS. Combined with microarray analysis such manipulations should enable identification of the genes that are transcriptionally regulated by DMyb. It should also become apparent whether DMyb can act as both a repressor and an activator of transcription (transgenes encoding the DMyb DNA-binding domain fused to either the VP16 activating domain or the Engrailed repressor domain should be useful for this). Given the situation with vertebrate Mybs, it will be interesting to determine whether the target genes differ depending on developmental stage or tissue type. The small size and somewhat degenerate nature of the MBS consensus sequence makes bioinformatics of little use in identifying candidate target genes, however, the identification of a series of *bonafide* DMyb target genes combined with the availability of the complete sequence for the *Drosophila* genome will facilitate such analyses. In this way it should become apparent whether a target gene requires multiple MBS, whether there are distance constraints on the position of the MBS relative to the transcription start site, and whether there is a requirement for neighboring binding sites for other transcription factor(s).

In addition to the identification of target genes, a greater understanding of DMyb function will involve further investigation of its role in cell cycle regulation, developmental patterning and differentiation processes. This will include determination of the signalling pathways with which DMyb interacts and how these can influence its activity. DMyb could be regulated by modification of its ability to bind DNA, to regulate transcription, by altering its subcellular localisation, or by affecting its stability. For example, as described above, there is evidence that the cell cycle is regulating the activity of DMyb via Cyclin A/Cdk phosphorylation. Each site within the DMyb protein that is subject to phosphorylation by Cyclin A/Cdk could be assessed for its *in vivo* relevance by testing whether transgenic lines in which sites has been individually mutated to prevent phosphorylation, can still rescue mutant alleles of *Dm myb*.

Although, it is unlikely that every target gene, biochemical interaction, and physiological role of DMyb is performed by one or more of the vertebrate Myb proteins, the demonstrated similarities make it likely that many of the specific activities will also be conserved. Therefore, knowledge gained from studies of the *Drosophila myb* gene during the next few years can be expected to continue to shed light on vertebrate Myb protein function and suggest tangible lines of investigation.

ACKNOWLEDGEMENTS

I thank Gary Ramsay for critical reading of this manuscript and other members of my laboratory, including Siau-Min Fung, Carrie Fitzpatrick and George Scaria for sharing their ideas and unpublished data. Research conducted in our laboratory is supported by a National Institutes of Health grant to A.L.K. (GM68961).

REFERENCES

Adams, M.D., Celniker, S.E., et al. (2000) The genome sequence of Drosophila melanogaster. Science *287*, 2185-2195.

Akimaru, H., Chen, Y., Dai, P., Hou, D.X., Nonaka, M., Smolik, S.M., Armstrong, S., Goodman R.H. and Ishii, S. (1997) Drosophila CBP is a co-activator of cubitus interruptus in hedgehog signalling. Nature *386*, 735-738.

Ansieau, S., Kowenz-Leutz, E., Dechend, R. and Leutz, A. (1997) B-Myb, a repressed trans-activating protein. J Mol Med *75*, 815-819.

Arsura, M., Introna, M., Passerini, F., Mantovani, A. and Golay, J. (1992) B-myb antisense oligonucleotides inhibit proliferation of human hematopoietic cell lines. Blood *79*, 2708-2716.

Asano, M., Nevins, J.R. and Wharton, R.P. (1996) Ectopic E2F expression induces S phase and apoptosis in Drosophila imaginal discs. Genes Dev *10*, 1422-1432.

Bading, H., Gerdes, J., Schwarting, R., Stein, H. and Moelling, K. (1988) Nuclear and cytoplasmic distribution of cellular myb protein in human haematopoietic cells evidenced by monoclonal antibody. Oncogene *3*, 257-265.

Baker, N.E. (2001) Master regulatory genes; telling them what to do. Bioessays *23*, 763-766.

Bartsch, O., Horstmann, S., Toprak, K., Klempnauer, K-H. and Ferrari, S. (1999) Identification of cyclin A/Cdk2 phosphorylation sites in B-Myb. Eur J Biochem *260*, 384-391.

Beall, E.L., Manak, J.R., Zhou, S., Bell, M., Lipsick, J.S. and Botchan, M.R. (2002) Role for a Drosophila Myb-containing protein complex in site-specific DNA replication. Nature *420,* 833-837.

Bessa, M., Saville, M.K. and Watson, R.J. (2001) Inhibition of cyclin A/Cdk2 phosphorylation impairs B-Myb transactivation function without affecting interactions with DNA or the CBP coactivator. Oncogene *20*, 3376-3386.

Biedenkapp, H., Borgmeyer, U., Sippel, A.E. and Klempnauer, K-H. (1988) Viral myb oncogene encodes a sequence-specific DNA-binding activity. Nature *335*, 835-837.

Bishop, J.M., Eilers, M., Katzen, A.L., Kornberg, T., Ramsay, G. and Schirm, S. (1991) MYB and MYC in the cell cycle. Cold Spring Harb Symp Quant Biol *56*, 99-107.

Blenis, J. (1993) Signal transduction via the MAP kinases: proceed at your own RSK. Proc Natl Acad Sci U S A *90*, 5889-5892.

Bodmer, R. and Venkatesh, T.V. (1998) Heart development in Drosophila and vertebrates: conservation of molecular mechanisms. Dev Genet *22*, 181-186.

Bouwmeester, T., van Wijk, I., Wedlich, D. and Pieler, T. (1994) Functional aspects of B-Myb in early Xenopus development. Oncogene *9*, 1029-1038.

Cohen, S.M. (1993) Imaginal disc development. The Development of Drosophila melanogaster. ed. 747-841. Cold Spring Harbor Laboratory Press, Cold Spring Harbor.

Curtiss, J., Halder, G. and Mlodzik, M. (2002) Selector and signalling molecules cooperate in organ patterning. Nat Cell Biol *4*, E48-51.

Dai, P., Akimaru, H., Tanaka, Y., Hou, D.X., Yasukawa, T., Kanei-Ishii, C., Takahashi, T. and Ishii, S. (1996) CBP as a transcriptional coactivator of c-Myb. Genes Dev *10*, 528-540.

Du, W. and Dyson, N. (1999) The role of RBF in the introduction of G1 regulation during Drosophila embryogenesis. EMBO J *18*, 916-925.

Du, W., Xie, J.E. and Dyson, N. (1996) Ectopic expression of dE2F and dDP induces cell proliferation and death in the Drosophila eye. EMBO J *15*, 3684-3692.

Facchinetti, V., Loffarelli, L., Schreek, S., Oelgeschlager, M., Luscher, B., Introna, M. and Golay, J. (1997) Regulatory domains of the A-Myb transcription factor and its interaction with the CBP/p300 adaptor molecules. Biochem J *324*, 729-736.

Fitzpatrick, C.A., Sharkov, N.V., Ramsay, G. and Katzen, A.L. (2002) Drosophila myb exerts opposing effects on S phase, promoting proliferation and suppressing endoreduplication. Development *129*, 4497-4507.

Foos, G., Grimm, S. and Klempnauer, K-H. (1992) Functional antagonism between members of the myb family: B-myb inhibits v-myb-induced gene activation. EMBO J *11*, 4619-4629.

Frampton, J., Ramqvist, T. and Graf, T. (1996) v-Myb of E26 leukemia virus up-regulates bcl-2 and suppresses apoptosis in myeloid cells. Genes Dev *10*, 2720-2731.

Fung, S.-M., Ramsay, G. and Katzen, A.L. (2002) Mutations in Drosophila myb lead to centrosome amplification and genomic instability. Development *129*, 347-359.

Fung, S.-M., Ramsay, G and Katzen, A.L. (2003) Myb and CBP: Physiological relevance of a biochemical interaction. Mech Dev *120*, 711-720.

Gehring, W.J. and Ikeo, K. (1999) Pax 6: mastering eye morphogenesis and eye evolution. Trends Genet *15*, 371-377.

Gerhart, J. (1999) 1998 Warkany lecture: signaling pathways in development. Teratology *60*, 226-239.

Golay, J., Loffarelli, L., Luppi, M., Castellano, M. and Introna, M. (1994) The human A-myb protein is a strong activator of transcription. Oncogene *9*, 2469-2479.

Hendzel, M.J., Wei, Y., Mancini, M.A., Van Hooser, A., Ranalli, T., Brinkley, B.R., Bazett-Jones, D.P. and Allis, C.D. (1997) Mitosis-specific phosphorylation of histone H3 initiates primarily within pericentromeric heterochromatin during G2 and spreads in an ordered fashion coincident with mitotic chromosome condensation. Chromosoma *106*, 348-360.

Hirose, F., Yamaguchi, M., Kuroda, K., Omori, A., Hachiya, T., Ikeda, M., Nishimoto, Y. and Matsukage, A. (1996) Isolation and characterization of cDNA for DREF, a promoter-activating factor for Drosophila DNA replication-related genes. J Biol Chem *271*, 3930-3937.

Hou, D.X., Akimaru, H. and Ishii, S. (1997) Trans-activation by the Drosophila myb gene product requires a Drosophila homologue of CBP. Febs Lett *413*, 60-64.

Howe, K.M. and Watson, R.J. (1991) Nucleotide preferences in sequence-specific recognition of DNA by c-myb protein. Nucleic Acids Res *19*, 3913-3919.

Humbert, P.O., Verona, R., Trimarchi, J.M., Rogers, C., Dandapani, S. and Lees, J.A. (2000) E2f3 is critical for normal cellular proliferation. Genes Dev *14*, 690-703.

Jackson, J., Ramsay, G., Sharkov, N.V., Lium, E. and Katzen, A.L. (2001) The role of transcriptional activation in the function of the Drosophila myb gene. Blood Cells Mol Dis *27*, 446-455.

Johnson, L.R., Johnson, T.K., Desler, M., Luster, T.A., Nowling, T., Lewis, R.E. and Rizzino, A. (2002) Effects of B-Myb on gene transcription: phosphorylation-dependent activity ans acetylation by p300. *J Biol Chem 277*, 4088-4097.

Johnson, T.K., Schweppe, R.E., Septer, J. and Lewis, R.E. (1999) Phosphorylation of B-Myb regulates its transactivation potential and DNA binding. J Biol Chem *274*, 36741-36749.

Katzen, A.L. (1990) Proto-oncogenes in Drosophila: molecular and genetic analysis. University of California at San Francisco.

Katzen, A.L. and Bishop, J.M. (1996) myb provides an essential function during Drosophila development. Proc Natl Acad Sci U S A *93*, 13955-13960.

Katzen, A.L., Jackson, J., Harmon, B.P., Fung, S-M., Ramsay, G. and Bishop, J.M. (1998) Drosophila myb is required for the G2/M transition and maintenance of diploidy. Genes Dev *12*, 831-843.

Katzen, A.L., Kornberg, T.B. and Bishop, J.M. (1985) Isolation of the proto-oncogene c-myb from D. melanogaster. Cell *41*, 449-456.

Klempnauer, K.H., Symonds, G., Evan, G.I. and Bishop, J.M. (1984) Subcellular localization of proteins encoded by oncogenes of avian myeloblastosis virus and avian leukemia virus E26 and by chicken c-myb gene. Cell *37*, 537-547.

Kozak, M. (1999) Initiation of translation in prokaryotes and eukaryotes. Gene *234*, 187-208.

Krude, T. (1995) Chromatin assembly factor 1 (CAF-1) colocalizes with replication foci in HeLa cell nuclei. Exp Cell Res *220*, 304-311.

Lane, S., Farlie, P. and Watson, R. (1997) B-Myb function can be markedly enhanced by cyclin A-dependent kinase and protein truncation. Oncogene *14*, 2445-2453.

Lefai, E., Fernandez-Moreno, M.A., Alahari, A., Kaguni, L.S. and Garesse, R. (2000) Differential regulation of the catalytic and accessory subunit genes of Drosophila mitochondrial DNA polymerase. J Biol Chem *275*, 33123-33133.

Li, X. and McDonnell, D.P. (2002) The transcription factor B-Myb is maintained in an inhibited state in target cells through its interaction with the nuclear corepressors N-CoR and SMRT. Mol Cell Biol *22*, 3663-3673.

Lindsley, D.L., Sandler, L., Baker, B.S., Carpenter, A.T., Denell, R.E., Hall, J.C., Jacobs, P.A., Miklos, G.L., Davis, B.K., Gethmann, R.C., Hardy, R.W., Steven, A.H., Miller, M., Nozawa, H., Parry, D.M., Gould-Somero, M. and Gould-Somero, M. (1972) Segmental aneuploidy and the genetic gross structure of the Drosophila genome. Genetics *71*, 157-184.

Lipsick, J.S. and Wang, D.M. (1999) Transformation by v-Myb. Oncogene *18*, 3047-3055.

Madan, A., Radha, P.K., Hosur, R.V. and Padhy, L.C. (1995) Bacterial expression, characterization and DNA binding studies on Drosophila melanogaster c-Myb DNA-binding protein. Eur J Biochem *232*, 150-158.

Madhavan, M.M. and Madhavan, K. (1980) Morphogenesis of the epidermis of adult abdomen of Drosophila. J Embryol Exp Morphol *60*, 1-31.

Manak, J.R., Mitiku, N. and Lipsick, J.S. (2002) Mutation of the Drosophila homologue of the Myb protooncogene causes genomic instability. Proc Natl Acad Sci U S A *99*, 7438-7443.

Martinez-Balbas, M.A., Tsukiyama, T., Gdula, D. and Wu, C. (1998) Drosophila NURF-55, a WD repeat protein involved in histone metabolism. Proc Natl Acad Sci U S A *95*, 132-137.

Milan, M., Campuzano, S. and Garcia-Bellido, A. (1997) Developmental parameters of cell death in the wing disc of Drosophila. Proc Natl Acad Sci U S A *94*, 5691-5696.

Mucenski, M.L., McLain, K., Kier, A.B., Swerdlow, S.H., Schreiner, C.M., Miller, T.A., Pietryga, D.W., Scott Jr., W. and Potter, S.S. (1991) A functional c-myb gene is required for normal murine fetal hepatic hematopoiesis. Cell *65*, 677-689.

Muller, H.J., League, B.B. and Offerman, C.A. (1931) Effects of dosage changes of sex-linked genes, and the compensatory effects of other gene differences between male and female. Anat Rec *51 (Suppl.),* 110.

Ness, S.A., Marknell, A. and Graf, T (1989) The v-myb oncogene product binds to and activates the promyelocyte- specific mim-1 gene. Cell *59*, 1115-1125.

Neufeld, T.P., de la Cruz, A.F., Johnston, L.A. and Edgar, B.A. (1998) Coordination of growth and cell division in the Drosophila wing. Cell *93*, 1183-1193.

Oehler, T., Arnold, H., Biedenkapp, H. and Klempnauer, K-H. (1990) Characterization of the v-myb DNA binding domain. Nucleic Acids Res *18*, 1703-1710.

Oh, I.H. and Reddy, E.P. (1999) The myb gene family in cell growth, differentiation and apoptosis. Oncogene *18*, 3017-3033.

Ohno, K., Hirose, F., Sakaguchi, K., Nishida, Y. and Matsukage, A. (1996) Transcriptional regulation of the Drosophila CycA gene by the DNA replication-related element (DRE) and DRE binding factor (DREF). Nucleic Acids Res *24*, 3942-3946.

Ohshima, N., Takahashi, M. and Hirose, F. (2003) Identification of a human homologue of the DREF transcription factor with a potential role in regulation of the histone H1 gene. J Biol Chem *278*, 22928-22938.

Okada, M., Akimaru, H., Hou, D.X., Takahashi, T. and Ishii, S. (2002) Myb controls G(2)/M progression by inducing cyclin B expression in the Drosophila eye imaginal disc. EMBO J *21*, 675-684.

Perrimon, N. (1994) Signalling pathways initiated by receptor protein tyrosine kinases in Drosophila. Curr Opin Cell Biol *6*, 260-266.

Peters, C.W., Sippel, A.E., Vingron, M. and Klempnauer, K-H. (1987) Drosophila and vertebrate myb proteins share two conserved regions, one of which functions as a DNA-binding domain. EMBO J *6*, 3085-3090.

Pollock, R. and Treisman, R. (1990) A sensitive method for the determination of protein-DNA binding specificities. Nucleic Acids Res *18*, 6197-6204.

Poodry, C.A. (1975) Autonomous and non-autonomous cell death in the metamorphosis of the epidermis of *Drosophila*. Wilhelm Roux Archives *178*, 333-336.

Postlethwait, J.H. (1978) Clonal analysis of *Drosophila* cuticular patterns. The Genetics and Biology of *Drosophila*. ed. 359-441. Academic Press, London.

Raschella, G., Negroni, A., Sala, A., Pucci, S., Romeo, A. and Calabretta, B. (1995) Requirement of b-myb function for survival and differentiative potential of human neuroblastoma cells. J Biol Chem *270*, 8540-8545.

Reiter, L.T., Potocki, L., Chien, S., Gribskov, M. and Bier, E (2001) A systematic analysis of human disease-associated gene sequences in Drosophila melanogaster. Genome Res *11*, 1114-1125.

Roth, S.Y. and Allis, C.D. (1996) Histone acetylation and chromatin assembly: a single escort, multiple dances? Cell *87*, 5-8.

Rubin, G.M., Yandell, M.D., et al. (2000) Comparative genomics of the eukaryotes. Science *287*, 2204-2215.

Ruiz De Mena, I., Lefai, E., Garesse, R. and Kaguni, L.S. (2000) Regulation of mitochondrial single-stranded DNA-binding protein gene expression links nuclear and mitochondrial DNA replication in drosophila. J Biol Chem *275*, 13628-13636.

Sala, A. and Calabretta, B. (1992) Regulation of BALB/c 3T3 fibroblast proliferation by B-myb is accompanied by selective activation of cdc2 and cyclin D1 expression. Proc Natl Acad Sci U S A *89*, 10415-10419.

Sala, A., Kundu, M., Casella, I., Engelhard, A., Calabretta, B., Grasso, L., Paggi, M.G., Giordano, A., Watson, R.J., Khalili, K. and Peschle, C. (1997) Activation of human B-MYB by cyclins. Proc Natl Acad Sci U S A *94*, 532-536.

Sala, A., Nicolaides, N.C., Engelhard, A., Bellon, T., Lawe, T.C., Arnold, A., Grana, X., Giordano, A. and Calabretta, B. (1994) Correlation between E2F-1 requirement in the S phase and E2F-1 transactivation of cell cycle-related genes in human cells. Cancer Res *54*, 1402-1406.

Saville, M.K. and Watson, R.J. (1998) The cell-cycle regulated transcription factor B-Myb is phosphorylated by cyclin A/Cdk2 at sites that enhance its transactivation properties. Oncogene *17*, 2679-2689.

Sawado, T., Hirose, F., Takahashi, Y., Sasaki, T., Shinomiya, T., Sakaguchi, K., Matsukage, A. and Yamaguchi, M. (1998) The DNA replication-related element (DRE)/DRE-binding factor system is a transcriptional regulator of the Drosophila E2F gene. J Biol Chem *273*, 26042-26051.

Schubiger, M. and Palka, J. (1987) Changing spatial patterns of DNA replication in the developing wing of Drosophila. Dev Biol *123*, 145-153.

Sharkov, N.V., Ramsay, G. and Katzen, A.L. (2002) The DNA replication-related element-binding factor (DREF) is a transcriptional regulator for the Drosophila myb gene. Gene *297*, 209-219.

Sluder, G. and Hinchcliffe, E.H. (1999) Control of centrosome reproduction: the right number at the right time. Biol Cell *91*, 413-427.

Simon, A.L., Stone, E.A. and Sidow, A. (2002) Inference of functional regions in proteins by quantification of evolutionary constraints. Proc. Natl. Acad. Sci. USA *99*, 2912-2917.

Takaesu, N.T., Johnson, A.N., Sultani, O.H. and Newfeld, S.J. (2002) Combinatorial signaling by an unconventional Wg pathway and the Dpp pathway requires Nejire (CBP/p300) to regulate dpp expression in posterior tracheal branches. Dev Biol *247*, 225-236.

Takahashi, Y., Yamaguchi, M., Hirose, F., Cotterill, S., Kobayashi, J., Miyajima, S. and Matsukage, A. (1996) DNA replication-related elements cooperate to enhance promoter activity of the drosophila DNA polymerase alpha 73-kDa subunit gene. J Biol Chem *271*, 14541-14547.

Tanaka, Y., Patestos, N.P., Maekawa, T. and Ishii, S (1999) B-myb is required for inner cell mass formation at an early stage of development. J Biol Chem *274*, 28067-28070.

Taunton, J., Hassig, C.A. and Schreiber, S.L. (1996) A mammalian histone deacetylase related to the yeast transcriptional regulator Rpd3p. Science *272*, 408-411.

Taylor, D., Badiani, P. and Weston, K. (1996) A dominant interfering Myb mutant causes apoptosis in T cells. Genes Dev *10*, 2732-2744.

Toscani, A., Mettus, R.V., Coupland, R., Simpkins, H., Litvin, J., Orth, J., Hatton, K.S. and Reddy, E.P. (1997) Arrest of spermatogenesis and defective breast development in mice lacking A-myb. Nature *386*, 713-717.

Tyler, J.K., Bulger, M., Kamakaka, R.T., Kobayashi, R. and Kadonaga, J.T. (1996) The p55 subunit of Drosophila chromatin assembly factor 1 is homologous to a histone deacetylase-associated protein. Mol Cell Biol *16*, 6149-6159.

Waltzer, L. and Bienz, M. (1998) Drosophila CBP represses the transcription factor TCF to antagonize Wingless signalling. Nature *395*, 521-525.

Waltzer, L. and Bienz, M. (1999) A function of CBP as a transcriptional co-activator during Dpp signalling. EMBO J *18*, 1630-1641.

Watson, R.J., Robinson, C. and Lam, E.W. (1993) Transcription regulation by murine B-myb is distinct from that by c-myb. Nucleic Acids Res **21**, 267-272.

White-Cooper, H., Leroy, D., MacQueen, A. and Fuller, M.T. (2000) Transcription of meiotic cell cycle and terminal differentiation genes depends on a conserved chromatin associated protein, whose nuclear localisation is regulated. Development *127*, 5463-5473.

Yamaguchi, M., Hirose, F. and Matsukage, A. (1996) Roles of multiple promoter elements of the proliferating cell nuclear antigen gene during Drosophila development. Genes Cells *1*, 47-58.

Ziebold, U., Bartsch, O. Marais, R., Ferrari, S. and Klempnauer, K-H. (1997) Phosphorylation and activation of B-Myb by cyclin A-Cdk2. Curr Biol *7*, 253-260.

Ziebold, U. and Klempnauer, K-H. (1997) Linking Myb to the cell cycle: cyclin-dependent phosphorylation and regulation of A-Myb activity. Oncogene *15*, 1011-1019.

Chapter 3

ESSENTIAL AND DIVERSE ROLES FOR C-MYB THROUGHOUT T CELL DEVELOPMENT

Kathleen Weston
Cancer Research UK Centre for Cell and Molecular Biology Institute of Cancer Research, 237 Fulham Road, London SW3 6JB, United Kingdom.

Abstract: The complex process of T cell development provides an ideal model system in which to probe how the c-Myb transcription factor functions to regulate the many cellular events in which it has been implicated as a vital player. Thymopoiesis can be investigated both in and out of the body, and mature, peripheral T cells can be easily manipulated in tissue culture, allowing detailed biochemical investigation of T cell activation in response to antigenic stimulation. Through the use of mouse model systems, c-Myb activity has been shown to be required at many points in the life of a T cell; depending on the developmental stage, cell death, the cell cycle, and cell fate determination can all be affected by perturbation of c-Myb function. This review will discuss the key regulatory events during T cell ontogeny which involve c-Myb, and will then attempt to provide a molecular explanation for how c-Myb might be working.

1. T CELL DEVELOPMENT AND THE EXPRESSION OF C-MYB

In contrast to other haemopoietic lineages, T cells require the unique cellular environment provided by the thymus in which to mature. A schematic of thymopoiesis is shown in Figure 1. Early T cell lineage progenitors entering the thymus from the bone marrow are double negative for the later T cell markers CD4 and CD8 ($CD4^-CD8^-$; DN). Their progress towards maturity can be tracked through the DN compartment by their expression of CD44 and CD25 in the sequence: $CD44^+CD25^-$ (DN1); $CD44^+CD25^+$ (DN2); $CD44^-CD25^+$ (DN3) and $CD44^-CD25^-$ (DN4). DN1 cells are oligopotent, generating not only classical T cells but also B cells,

J. Frampton (ed.), Myb Transcription Factors: Their Role in Growth, Differentiation and Disease, 65-80.

natural killer (NK) cells and thymic dendritic (DC) cells, although they cannot make other haemopoietic lineages. By the DN2 stage, the capacity to produce B and NK cells is lost. DN3 cells have lost DC potential, and it is in this subset that the T cell receptor (TCR) β, γ and δ chains are actively rearranged, leading to a final divergence into either the αβ lineage or the minority γδ T cell lineage (reviewed in Di Santo et al., 2000).

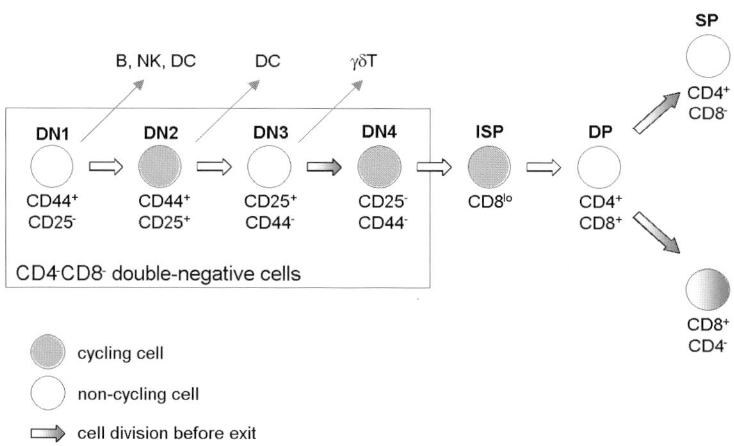

Figure 1. T cell development

Following the DN4 stage, cells expressing a functional TCRβ chain become transiently single positive for CD8 (the Immature Single Positive (ISP) subset), and then progress to the $CD4^+CD8^+$ double positive (DP) stage, where rearrangement of the TCRα chain takes place. Many T cells are lost at this point, either deleted by negative selection if their TCR shows too high an affinity for self-MHC, or killed by their failure to produce a functional rearrangement. Surviving positively selected cells down-regulate either the CD4 or the CD8 antigen, become single positive (SP), and are exported to the periphery. Upon antigen activation, they become functionally mature $CD4^+CD8^-$ helper cells or $CD4^-CD8^+$ cytotoxic cells (reviewed in Janeway et al., 2001)

During foetal development, c-*myb* is expressed in the earliest detectable day 12 thymic rudiment. By day 18, as the thymus matures, c-*myb* is maintained at high levels in the thymic cortex, where the more immature thymocytes reside, but is completely absent from the medulla, the site of more mature cells; this pattern is carried through into adulthood (Ess et al., 1999). Analysis of c-*myb* gene expression in sorted thymocytes confirms the *in situ* pattern. DN thymocytes express high levels of c-*myb* mRNA (Pearson and Weston, 2000), as do DP cells prior to positive selection. No or very low c-*myb* expression is observed in SP thymocytes or $CD69^{hi}$ (i.e. post-selection) DP cells (D. Maurice, J. Hooper, unpublished). However, when resting T cells are stimulated to divide in response to antigen, they re-express c-*myb* at high levels in an interleukin-2 (IL-2)-dependent fashion (Stern and Smith, 1986). In sorted thymocytes and T cells from adult mice, B-*myb* (MYBL2) is detectable in a similar pattern to c-*myb*, and A-*myb* (MYBL1) is absent from all but resting peripheral T cells (Golay et al., 1991; J. Hooper, R. Pearson, unpublished).

2. DISRUPTION OF C-MYB EXPRESSION IN T CELLS

An assessment of the effects on T cell development of losing c-Myb expression was initially hampered by the embryonic lethality of the homozygous c-*myb* null (*c-myb*$^{-/-}$) phenotype (Mucenski et al., 1991). A number of approaches have now been used to circumvent this problem. Using chimaeric *c-myb*$^{-/-}$/*Rag1*$^{-/-}$ chimaeric mice, c-Myb was shown to be essential for the earliest stage of T cell development, as no *c-myb*$^{-/-}$ cells were able to progress past the DN1 oligopotent precursor stage (Allen et al., 1999). To examine what happens during T cell development post-DN1, our laboratory made an active repressor specific for Myb target sites on DNA. This dominant negative protein, termed MENT, was made by fusing the c-Myb DNA binding domain to the repressor domain of the *Drosophila* Engrailed protein. MENT is able to switch off Myb target genes at protein levels equivalent to physiological levels of c-Myb, and it behaves similarly to a grossly over-expressed simple competitive inhibitor comprising the c-Myb DNA binding domain alone. When targetted specifically to the T cell lineage in transgenic mice, the MENT protein caused multiple defects in T cell development, with the defects becoming more severe with increasing transgene expression level. The progress of double negative cells from the DN3 to the DN4 compartment was inhibited, and there was also a build up of $CD8^+$ ISPs. DP cells were present at lower levels, as were SP cells. There was also a marked negative skew in the ratio of $CD4^+$SP to $CD8^+$ SP cells, in both the thymus and the spleen, caused by a greater loss of $CD4^+$SP than $CD8^+$ SP cells. In addition to these defects during T cell ontogeny, the ability of mature T cells to activate *in vitro* in response to anti-CD3 stimulation was also compromised (Badiani et al., 1994).

More recently, in the process of making a conditional knockout *c-myb* mouse, a hypomorphic allele of *c-myb*, termed *c-myb*^{loxP}, has been developed. Homozygous *c-myb*$^{loxP/loxP}$ mice are viable, and preliminary analysis of their T cells shows that they have similar albeit more severe defects to those described using the MENT mouse model. There is a pronounced block to DN development beyond the DN3 stage, and those few cells able to become SP are skewed with respect to the CD4:CD8 ratio, changing from a wild type figure of 3:1 to 1.4:1 (Emambokus et al., 2003). The c-Myb-specific DN3 block is also observed in a second conditional knockout, in which c-*myb* is specifically deleted in DN2 cells and beyond by the use of an lck-Cre transgenic mouse strain crossed to *c-myb*$^{loxP/loxP}$ animals (T. Bender, personal communication).

To complement these loss-of-function experiments, our laboratory has also made transgenic mice that over-express oncogenic v-Myb in a T-lineage-specific fashion. These mice have enlarged thymuses that fail to involute with age, mainly caused by the persistence of $CD4^+$ SP cells, which eventually results in slow onset $CD4^+$ SP tumours. Somewhat strangely, activation assays on peripheral T cells from these animals prior to their developing lymphomas show that, as for MENT animals, activation in response to anti-CD3 is inhibited (Badiani et al., 1996; A. Lauder and K.W, unpublished).

How might c-Myb be causing these diverse phenotypes? Although no data is yet available on the DN1 block, we have studied the blocks seen in MENT mice. As seems to be standard for c-Myb, what cellular process is being affected most is context-dependent. The DN3 block appears to be mostly caused by inhibition of the cell cycle (Pearson and Weston, 2000), whilst the loss of DP cells and failure of peripheral T cells to proliferate is due to enhanced apoptosis, with little change in the cell cycle (Lauder et al., 2001; Taylor et al., 1996). The observed apoptosis can only be partially rescued by the c-Myb target gene *bcl-2*, implying that c-Myb is also regulating Bcl-2 independent death pathways (Lauder et al., 2001). Regarding the CD4:CD8 SP ratio skew, it appears that whilst MENT biases against $CD4^+$ SP cells, forced expression of v-Myb on a CD8-selecting transgenic background is able to turn cells towards the $CD4^+$ fate, implying some effect on selection (R. Pearson, unpublished).

Our current knowledge of c-Myb-regulated genes is insufficient to explain these observations. Although *bcl-2*, a well-characterised target gene for c-Myb (Frampton et al., 1996; Taylor et al., 1996), is able to partially rescue the survival defect seen in DP cells and in resting T cells, it is unlikely that this is a normal event, as there is no coincidence of expression of c-Myb and Bcl-2 at these times in wild-type animals. Therefore, it is still necessary to develop a picture of the relevant c-Myb target genes, and to establish their biological relevance by fitting c-Myb itself into the network of regulatory pathways specifying T cell development. The remainder of this review is devoted to discussion and speculation about these two crucial issues.

3. EARLY THYMOPOIESIS AND THE REQUIREMENT FOR C-MYB

3.1 DN1 Cells

The reason for the inability of DN1 cells to mature if they lack c-Myb (Allen et al., 1999) has not been investigated, but it seems highly probable that aberrant regulation of the likely c-Myb target gene c-*kit* may play some part. The c-*kit* promoter contains Myb consensus binding sites, and is responsive to c-Myb in reporter assays (Ratajczak et al., 1998). In experimental systems where c-Myb is inactivated, a reduction in expression levels of c-*kit* mRNA has been reported (Hogg et al., 1997; White and Weston, 2000), although c-*kit* is still expressed in c-*myb*$^{-/-}$ cells during early haemopoiesis (Clarke et al., 2000; Sumner et al., 2000). Loss of c-Kit, the receptor for stem cell factor (SCF), together with defective interleukin-7 (IL-7) signalling, results in a complete block to thymopoiesis beyond the DN1 stage (Di Santo et al., 1999). The phenotype is less severe in c-*kit* mutant mice when the IL-7 signalling pathway is intact; although thymopoiesis occurs relatively normally in young animals, the adult thymus is almost completely depleted of all post DN1 thymocyte subsets (Waskow et al., 2002). Therefore, although clearly not a complete explanation for the c-*myb*$^{-/-}$ phenotype, down-regulation of c-*kit* expression may well play a part.

A second potential player in the DN1 defect seen in c-*myb*$^{-/-}$ animals is α4-integrin (ITGA4). This gene is co-regulated by c-Myb and c-Ets-1, whose ability to activate the α4-integrin promoter is inhibited by the zinc finger-homeodomain repressor ZEB (Postigo et al., 1997). In MENT mice, some down-regulation of α4-integrin is observed in DN3 cells (A. Castellanos, unpublished), and so it is possible that loss of c-Myb activity in earlier thymic precursors also results a deficit in α4-integrin expression. An anti-α4-integrin antibody has been shown to inhibit adhesion of haemopoietic precursor cells isolated from foetal liver to foetal thymic lobes (Kawakami et al., 1999), and knockout of the α4-integrin gene results in thymic atrophy shortly after birth (Arroyo et al., 2000), so population and maintenance of the thymus may be fundamentally flawed in the absence of c-Myb.

3.2 DN3 Cells and β Selection

The DN3 stage of thymocyte differentiation is an important control point. Productive TCRβ rearrangement and generation of a functional preTCR, comprising the TCRβ chain complexed to the invariant pTα chain, is known as β-selection. Triggering of the signalling pathways lying downstream of the preTCR complex is essential for DN3 cells to blast and undergo a massive proliferative burst as they mature through to the DP stage (see Figure 1). As described above, DN3 thymocytes from MENT mice are unable to cycle properly, resulting in far fewer cells being generated during the proliferative burst, and hence far fewer DP thymocytes (Pearson and Weston, 2000). Recent data have indicated that there may also be a partial defect in preTCR signalling in MENT mice at the level of complex assembly. Although TCRβ rearrangements do occur (Pearson and Weston, 2000), both *Rag2*, one of the genes required for TCRβ rearrangement , and the *pTα* gene itself, may both be c-Myb target genes (Reizis and Leder, 2001; Wang et al., 2000), and hence their expression may be attenuated in the MENT mice.

What signalling pathways might c-Myb lie on during β-selection? Loss of both IL-7R signalling and preTCR signalling has no effect on expression of c-*myb* (Pearson and Weston, 2000), suggesting either that there is no extracellular regulation, or that another pathway is required. It now seems likely that the former explanation is true. Recently, the c-*myb* gene has been described as a downstream target of the WNT signalling pathway (van de Wetering et al., 2002). In thymocytes, WNT signals are transduced via the TCF1 transcription factor, and $Tcf1^{-/-}$ mice display a very similar phenotype to MENT animals, being defective in cell cycling during the DN3 and ISP stages of thymopoiesis (Schilham et al., 1998; Verbeek et al., 1995). This intriguing link is currently under investigation.

A second potential link to extracellular signalling comes via the serine/threonine kinase Pim1, which has been shown to potentiate the transcription activation function of c-Myb (Leverson et al., 1998). Pim1 is commonly induced in response to many mitogens and cytokines (reviewed in Domen et al., 1993) and has been proposed to be an effector of the IL-7 signalling pathway during DN development (Jacobs et al., 1999). Pim1 appears to play an important role in proliferative signalling during β-selection (Schmidt et al., 1998). When over-expressed, it causes an increase in the number of cycling cells, and it can also rescue the DN3 block in $Rag^{-/-}$

thymocytes, allowing development of DP cells (Jacobs et al., 1999; Leduc et al., 2000). c-Myb is required for both of these functions of Pim1, as they are blocked in the presence of an MENT transgene (Pearson and Weston, 2000). Interestingly, Pim1 is up-regulated at late times in DN development (Schmidt et al., 1998), perhaps acting to fine-tune MYB activity post β-selection.

The crucial downstream targets of c-Myb during β-selection still remain to be elucidated. However, one intriguing candidate is the c-*myc* oncogene, described some years ago as a c-Myb regulated gene (Cogswell et al., 1993; Nakagoshi et al., 1992; Zobel et al., 1991). Loss of c-*myc* during DN development leads to a DN defect (Douglas et al., 2001) bearing some similarities to that seen in MENT mice. Given the importance of c-Myc for stimulation of the cell cycle in multiple tissues, regulation of c-*myc* by c-Myb in DN development may be of great importance.

4. C-MYB IN MATURE T CELLS

4.1 c-Myb Lies on the IL-2 Signalling Pathway to Cell Survival

Triggering of the TCR on resting T cells in response to antigen leads to assembly of the high-affinity IL-2 receptor (IL-2R) and synthesis of IL-2, and subsequent progression of cells from G1 into S phase of the cell cycle (reviewed in Janeway et al., 2001). Studies of the molecules involved in signalling from the IL-2R have shown that activation of phosphoinositide 3-kinase (PI3K) and its downstream effector protein kinase B (PKB), appear to be most important for the survival functions mediated by IL-2 (reviewed in Datta et al., 1999). PI3K, via PKB, has been shown to provide a survival signal for resting and activated T cells (Jones et al., 2002; Jones et al., 2000). c-Myb is necessary for this survival pathway, as expression of the MENT transgene is able to block the enhanced survival during T cell activation seen in the presence of an activated PKB transgene (Lauder et al., 2001). Moreover, c-Myb can be transcriptionally up-regulated by signalling from the IL-2R (Rohwer et al., 1996; Stern and Smith, 1986), and this occurs via the PI3K pathway. The specific PI3-K inhibitor Ly294002 or a dominant negative PI3K molecule is able to block *Myb* mRNA induction, in contrast to the MEK inhibitor PD98059, which has no effect. Protein kinase B (PKB) appears to be the principal transducer of the PI3K signal, as a dominant negative PKB molecule is able to completely inhibit PI3K-mediated activation of the c-*myb* promoter. The c-*myb* promoter contains a region of 119bp (-304/-185) which is conserved between humans, mice and chickens

(Urbanek et al., 1988), and is therefore likely to contain regulatory sequences of importance. There are two elements within this region, an NF-κB site (-266/-256), and an E2F site (-278/-271), which are required both for promoter activity and transduction of the PI3K/PKB stimulus (Campanero et al., 1999; Lauder et al., 2001; Muller et al., 2001; Sala et al., 1994).

Although during peripheral T cell activation, c-Myb can be switched on by PI3K, this is not the case earlier on in thymopoiesis. Interleukin-7 (IL-7) is a major cytokine required for DN development, and it too signals a survival function via PI3K (Pallard et al., 1999). DN thymocytes from IL-$7R^{-/-}$ mice still express ample c-*myb* mRNA, and similarly, c-*myb* mRNA levels are not affected by treatment of wild-type DN3 thymocytes with the PI3K inhibitor LY294002 (Pearson and Weston, 2000). Therefore, in one context (activated T cells), c-Myb is required for protection from apoptosis and is transcriptionally upregulated by a known anti-apoptotic PI3K signal, and in another (DN3 cells), it seems to have no effect on apoptosis and is not PI3K responsive.

4.2 c-Myb and FAS

The proliferation of T cells in response to antigen is curtailed by a process known as activation-induced cell death (AICD) (for review, see Lenardo et al., 1999). Ligand-induced activation of death receptors, primarily FAS, leads to the recruitment of adapter molecules, which in turn recruit procaspases which are rapidly cleaved and activated, triggering effector caspases which kill the host cell. The whole process of receptor-mediated AICD is independent of nascent protein synthesis (Itoh et al., 1991), and cannot be prevented by expression of Bcl-2 (Strasser et al., 1995). MENT T cells appear to be more susceptible to AICD (Lauder et al., 2001), leading to speculation that c-Myb may be able to negatively regulate the FAS pathway. A previous connection between c-Myb and FAS has been postulated, as T cells from *lpr/lpr* mice, which carry an inactivating mutant of FAS, contain extremely high levels of c-*myb* mRNA (Mountz and Steinberg, 1989; Mountz et al., 1984; Yokota et al., 1987). Therefore, there is a potential feedback mechanism between c-Myb and the FAS pathway, in which lack of a FAS signal promotes c-*myb* expression, and c-Myb suppresses FAS-mediated apoptosis. Recently, we have been able to show that the MENT transgene is dominant over the *lpr/lpr* phenotype; MENT:*lpr/lpr* mice have reduced numbers of T cells and there is no excessive lymphoproliferation (D.Rate, unpublished). We are currently investigating whether c-Myb is able to regulate a specific inhibitor of FAS killing, or acts at a level in the apoptotic pathway below the point at which Bcl-2-dependent and independent death signals converge.

4.3 A Role for c-MYB Beyond Activation?

Given the well-documented effects of c-Myb on differentiation in myeloid and erythroid cells (reviewed in Ness, 1996), it seems rather likely that it may also affect the maturation process by which activated $CD4^+$ and $CD8^+$ cells respectively differentiate into effector helper (Th) or killer (Tc) cells. Currently, no data addressing this issue have been published. One topic which might merit further investigation for some involvement of c-Myb is the differentiation of $CD4^+$ Th precursor cells into either Type 1 Th (Th1) or Type 2 Th (Th2) cells. The T cell specific GATA factor GATA3 is necessary and sufficient to dictate Th2 cell fate, and the transcription factor c-Maf is essential for transcription of the interleukin 4 (IL-4) gene (Kim et al., 1999), which encodes the hallmark cytokine of Th2 cells (reviewed in Ho and Glimcher, 2002). Intriguingly, both GATA3 and c-Maf may be potential interacting partners with c-Myb. C-Maf and c-Myb have been reported to interact with each other during myeloid differentiation, forming inhibitory complexes on the CD13 promoter (Hegde et al., 1998), and another GATA family member, GATA1, competes with c-Myb for interaction with the co-activator CBP/p300, causing mutual inhibition of their transactivation activity (Takahashi et al., 2000). This raises the highly speculative but interesting possibility that during helper cell differentiation, in addition to its normal role as a transcription activator, c-Myb might be able to act as a context-dependent repressor of gene activation and hence influence cell fate decision-making.

5. OTHER PROVOCATIVE C-MYB T CELL TARGETS

5.1 CD4

As discussed above, cooperation between c-Myb and other transcription factors is a common feature of many c-Myb-regulated promoters. The repressor HES1 is able to bind c-Myb and create a complex on the *CD4* silencer leading to down-regulation of *CD4* gene transcription (Allen et al., 2001). However, c-Myb has also been reported by the same laboratory to up-regulate *CD4* transcription (Siu et al., 1992), and which of these mechanisms predominate at any time is currently unclear. Data from c-*myb* mutant mice are equivocal. The transgenic MENT and v-Myb models from our laboratory show an effect consistent with suppression of CD4 SP development when c-Myb activity is curtailed, and an increase in CD4 SP cells when c-Myb is overactive (see above), but when DP and CD4 SP cells are analysed, levels of CD4 expression appear normal (K.W., unpublished). This is also true of splenocytes that carry the c-*myb*$^{loxP/loxP}$ hypomorphic allele generated by Jon Frampton's laboratory (J. Frampton, personal communication). Regulation of CD4 expression is, however, a tempting explanation for the effects of c-Myb on positive selection described above.

5.2 TCR δ and γ

The enhancers of both the TCR δ and γ chain loci are positively regulated by c-Myb in concert with a member of the Runx family of transcription factors (Hernandez-Munain and Krangel, 1994; Hernandez-Munain and Krangel, 1995; Hernandez-Munain et al., 1996; Hsiang et al., 1995; Redondo et al., 1995). *In vivo* footprinting of the TCR δ enhancer shows that the Runx protein plays primarily a structural role, inducing a conformational change in the enhanceosome and thereby increasing c-Myb binding; in contrast, c-Myb binding has no apparent reciprocal effect on Runx binding, but is required for transcriptional activation (Hernandez-Munain and Krangel, 2002). Runx proteins can act as context-dependent activators or repressors of transcription (reviewed in Lund and van Lohuizen, 2002), and all three family members are important for thymopoiesis. Runx1 (also known as PEBP2αB/AML1) and Runx3 (PEBP2αC) function to repress *CD4* expression via the *CD4* silencer at two separate times during thymopoiesis, and Runx3 is also required for the specification of functional T cells (Taniuchi et al., 2002). Although Runx2 (PEBP2αA/CBFA1) has no obvious T cell phenotype when deleted (Taniuchi et al., 2002), over-expression results in an expanded DN4 and CD8 ISP population with reduced capacity for proliferation (Vaillant et al., 2002). As these results are reminiscent of those found with MENT mice, it is worth investigating whether c-Myb is acting solely to switch on genes in the DN subset; perhaps expression of MENT is simply exacerbating a repressor effect of c-Myb, which can be mimicked by over-expression of a potential partner in repression.

5.3 Adenosine Deaminase

Regulation of the adenosine deaminase (*Ada*) gene by c-Myb has potentially important implications for T cell development. The *Ada* gene has a thymocyte-specific locus control region (LCR), within which is a Myb binding site absolutely required for function of the LCR in transgenic mice (Ess et al., 1995). The key role of ADA in humans in the development and function of the immune system is demonstrated by the impaired lymphoid development and severe combined immunodeficiency syndrome (SCID) associated with a congenital defect in the enzyme (reviewed in Hershfield and Mitchell, 1995). A mouse model for examining ADA deficiency in lymphoid tissues has been developed, and once again, the T cell defects observed in these animals are similar but not identical to those seen in mice deficient in c-Myb activity, as *Ada*$^{-/-}$ mice have hypocellular thymuses and spleens, a block in DN to DP differentiation, and significantly lower numbers of DP and SP thymocytes. An increase in levels of the ADA substrates adenosine and 2'-deoxyadenosine is observed, and it is these metabolites which are thought to mediate the effects of ADA loss on thymopoiesis (Blackburn et al., 1998). Accumulation leads to apoptosis: either FAS-mediated, by up-regulation of AdoHcy hydrolase, which inhibits transmethylation reactions (Ratter et al., 1996); or by affecting DNA strand break repair, via disruption of deoxynucleotide pools, or by accumulation of dATP which triggers activation of caspase 3 (Liu et al., 1996). *Ada*$^{-/-}$ thymopoiesis can be rescued by introduction of Bcl-2, the anti-apoptotic APAF1 protein, or caspase inhibitors, further demonstrating that the defect at this stage is primarily apoptotic (Thompson et al., 2000). In mature *Ada*$^{-/-}$ T cells, proliferation in response to antigen is inhibited, as ADA deficiency reduces tyrosine phosphorylation of TCR-associated signaling molecules and blocks TCR-triggered calcium increases. Interestingly, there is no effect on apoptosis (Apasov et al., 2001).

6. FUTURE PROSPECTS

Despite the effort expended thus far, we are still a long way from having an informative picture of how c-Myb works during T cell development. Fortunately, a far greater battery of techniques now exists with which to approach the problem. T cell targets can be identified using microarray analysis, and their relevance tested in foetal thymic organ culture, or using the recently described *in vitro* thymocyte differentiation protocol (Schmitt and Zuniga-Pflucker, 2002). Mice with conditional deletions of the c-*myb* and B-*myb* genes will allow issues of specificity to be addressed, and also will be invaluable in formulating genetic approaches to delineating signalling pathways dependent on c-Myb activity. Within the next few years, it seems likely that many important issues will be resolved.

REFERENCES

Allen, R.D., 3rd, Bender, T.P. and Siu, G. (1999) c-Myb is essential for early T cell development. Genes Dev *13*, 1073-1078.

Allen, R.D., 3rd, Kim, H.K., Sarafova, S.D. and Siu, G. (2001) Negative regulation of CD4 gene expression by a HES-1-c-Myb complex. Mol Cell Biol *21*, 3071-3082.

Apasov, S.G., Blackburn, M.R., Kellems, R.E., Smith, P.T. and Sitkovsky, M.V. (2001) Adenosine deaminase deficiency increases thymic apoptosis and causes defective T cell receptor signaling. J Clin Invest *108*, 131-141.

Arroyo, A.G., Taverna, D., Whittaker, C.A., Strauch, U.G., Bader, B.L., Rayburn, H., Crowley, D., Parker, C.M. and Hynes, R.O. (2000) In vivo roles of integrins during leukocyte development and traffic: insights from the analysis of mice chimeric for alpha 5, alpha v, and alpha 4 integrins. J Immunol *165*, 4667-4675.

Badiani, P., Corbella, P., Kioussis, D., Marvel, J. and Weston, K. (1994) Dominant interfering alleles define a role for c-Myb in T cell development. Genes Dev *8*, 770-782.

Badiani, P., Kioussis, D., Swirsky, D., Lampert, I. and Weston, K. (1996) T-cell lymphomas in v-Myb transgenic mice. Oncogene *13*, 2205-2212.

Blackburn, M.R., Datta, S.K. and Kellems, R.E. (1998) Adenosine deaminase-deficient mice generated using a two-stage genetic engineering strategy exhibit a combined immunodeficiency. J Biol Chem *273*, 5093-5100.

Campanero, M.R., Armstrong, M. and Flemington, E. (1999) Distinct cellular factors regulate the c-myb promoter through its E2F element. Mol Cell Biol *19*, 8442-8450.

Clarke, D., Vegiopoulos, A., Crawford, A., Mucenski, M., Bonifer, C. and Frampton, J. (2000) In vitro differentiation of c-myb(-/-) ES cells reveals that the colony forming capacity of unilineage macrophage precursors and myeloid progenitor commitment are c-Myb independent. Oncogene *19*, 3343-3351.

Cogswell, J.P., Cogswell, P.C., Kuehl, W.M., Cuddihy, A.M., Bender, T.P., Engelke, U., Marcu, K.B. and Ting, J.P. (1993) Mechanism of c-myc regulation by c-Myb in different cell lineages. Mol Cell Biol *13*, 2858-2869.

Datta, S.R., Brunet, A. and Greenberg, M.E. (1999) Cellular survival: a play in three Akts. Genes Dev *13*, 2905-2927.

Di Santo, J.P., Aifantis, I., Rosmaraki, E., Garcia, C., Feinberg, J., Fehling, H.J., Fischer, A., von Boehmer, H. and Rocha, B. (1999) The common cytokine receptor gamma chain and the pre-T cell receptor provide independent but critically overlapping signals in early alpha/beta T cell development. J Exp Med *189*, 563-574.

Di Santo, J.P., Radtke, F. and Rodewald, H.R. (2000) To be or not to be a pro-T? Curr Opin Immunol *12*, 159-165.

Domen, J., van der Lugt, N.M., Laird, P.W., Saris, C.J. and Berns, A. (1993) Analysis of Pim-1 function in mutant mice. Leukemia *7 Suppl 2*, S108-112.

Douglas, N.C., Jacobs, H., Bothwell, A.L. and Hayday, A.C. (2001) Defining the specific physiological requirements for c-Myc in T cell development. Nat Immunol *2*, 307-315.

Emambokus, N.R., Vegiopoulos, A., Harman, B., Jenkinson, E.J., Anderson, G. and Frampton, J. (2003) Progression through key stages of the hematopoietic hierarchy is dependent on distinct threshold levels of c-Myb. EMBO J *22*, 4478-4488.

Ess, K. C., Whitaker, T.L., Cost, G.J., Witte, D.P., Hutton, J.J. and Aronow, B.J. (1995) A central role for a single c-Myb binding site in a thymic locus control region. Mol Cell Biol *15*, 5707-5715.

Ess, K.C., Witte, D.P., Bascomb, C.P. and Aronow, B.J. (1999) Diverse developing mouse lineages exhibit high-level c-Myb expression in immature cells and loss of expression upon differentiation. Oncogene *18*, 1103-1111.

Frampton, J., Ramqvist, T. and Graf, T. (1996) v-Myb of E26 leukemia virus up-regulates bcl-2 and suppresses apoptosis in myeloid cells. Genes Dev *10*, 2720-2731.

Golay, J., Capucci, A., Arsura, M., Castellano, M., Rizzo, V. and Introna, M. (1991) Expression of c-myb and B-myb, but not A-myb, correlates with proliferation in human hematopoietic cells. Blood *77*, 149-158.

Hegde, S.P., Kumar, A., Kurschner, C. and Shapiro, L.H. (1998) c-Maf interacts with c-Myb to regulate transcription of an early myeloid gene during differentiation. Mol Cell Biol *18*, 2729-2737.

Hernandez Munain, C. and Krangel, M.S. (1994) Regulation of the T-cell receptor delta enhancer by functional cooperation between c-Myb and core-binding factors. Mol Cell Biol *14*, 473-483.

Hernandez-Munain, C. and Krangel,M. (1995) C-Myb and core binding factor PEPB2 display functional synergy but bind independently to adjacent sites in the T-cell receptor delta-enhancer. Molecular and Cellular Biology *15*, 3090-3099.

Hernandez-Munain, C. and Krangel, M.S. (2002) Distinct roles for c-Myb and core binding factor/polyoma enhancer-binding protein 2 in the assembly and function of a multiprotein complex on the TCR delta enhancer in vivo. J Immunol *169*, 4362-4369.

Hernandez-Munain, C. Lauzurica, P. and Krangel, M.S. (1996) Regulation of T cell receptor delta gene rearrangement by c-Myb. J Exp Med *183*, 289-293.

Hershfield, M.S. and Mitchell, B.S. (1995) Immunodeficiency diseases caused by adenosine deaminase deficiency and purine nucleoside phosphorylase deficiency. In The Metabolic and Molecular Basis of Inherited Disease, C.R. Scriver, A.L. Beaudet, W.S. Sly and D. Valle, eds. (New York, McGraw-Hill, Inc.), pp. 1725-1768.

Ho, I.C. and Glimcher, L.H. (2002) Transcription: tantalizing times for T cells. Cell *109 Suppl*, S109-120.

Hogg, A., Schirm, S., Nakagoshi, H., Bartley, P., Ishii, S., Bishop, J.M. and Gonda, T.J. (1997) Inactivation of a c-Myb/estrogen receptor fusion protein in transformed primary cells leads to granulocyte/macrophage differentiation and down regulation of c-*kit* but not c-*myc* or *cdc2*. Oncogene *15*, 2885-2898.

Hsiang, H.H., Goldman, J.P. and Raulet, D.H. (1995) The role of c-Myb or a related factor in regulating the T cell receptor gamma gene enhancer. J Immunol *154*, 5195-5204.

Itoh, N., Yonehara, S., Ishii, A., Yonehara, M., Mizushima, S., Sameshima, M., Hase, A., Seto, Y. and Nagata, S. (1991) The polypeptide encoded by the cDNA for human cell surface antigen Fas can mediate apoptosis. Cell *66*, 233-243.

Jacobs, H., Krimpenfort, P., Haks, M., Allen, J., Blom, B., Demolliere, C., Kruisbeek, A., Spits, H. and Berns, A. (1999) PIM1 reconstitutes thymus cellularity in interleukin 7- and common gamma chain-mutant mice and permits thymocyte maturation in Rag- but not CD3gamma-deficient mice. J Exp Med *190*, 1059-1068.

Janeway, C.A., Travers, P., Walport, M. and Shlomchik, M.J. (2001) Immunobiology: the immune system in health and disease - 5th ed. (New York, Garland Publishing).

Jones, R.G., Elford, A.R., Parsons, M.J., Wu, L., Krawczyk, C.M., Yeh, W.C., Hakem, R., Rottapel, R., Woodgett, J.R. and Ohashi, P.S. (2002) CD28-dependent activation of protein kinase B/Akt blocks Fas-mediated apoptosis by preventing death-inducing signaling complex assembly. J Exp Med *196*, 335-348.

Jones, R.G., Parsons, M., Bonnard, M., Chan, V.S., Yeh, W.C., Woodgett, J.R. and Ohashi, P.S. (2000) Protein kinase B regulates T lymphocyte survival, nuclear factor kappaB activation, and Bcl-X(L) levels in vivo. J Exp Med *191*, 1721-1734.

Kawakami, N., Nishizawa, F., Sakane, N., Iwao, M., Tsujikawa, K., Ikawa, M., Okabe, M. and Yamamoto, H. (1999) Roles of integrins and CD44 on the adhesion and migration of fetal liver cells to the fetal thymus. J Immunol *163*, 3211-3216.

Kim, J.I., Ho, I.C., Grusby, M.J. and Glimcher, L.H. (1999) The transcription factor c-Maf controls the production of interleukin-4 but not other Th2 cytokines. Immunity *10*, 745-751.

Lauder, A., Castellanos, A. and Weston, K. (2001) c-Myb transcription is activated by protein kinase B (PKB) following interleukin 2 stimulation of Tcells and is required for PKB-mediated protection from apoptosis. Mol Cell Biol *21*, 5797-5805.

Leduc, I., Karsunky, H., Mathieu, N., Schmidt, T., Verthuy, C., Ferrier, P. and Moroy, T. (2000) The Pim-1 kinase stimulates maturation of TCRbeta-deficient T cell progenitors: implications for the mechanism of Pim-1 action. Int Immunol *12*, 1389-1396.

Lenardo, M., Chan, K.M., Hornung, F., McFarland, H., Siegel, R., Wang, J. and Zheng, L. (1999) Mature T lymphocyte apoptosis--immune regulation in a dynamic and unpredictable antigenic environment. Annu Rev Immunol *17*, 221-253.

Leverson, J.D., Koskinen, P.J., Orrico, F.C., Rainio, E.M., Jalkanen, K.J., Dash, A.B., Eisenman, R.N. and Ness, S.A. (1998) Pim-1 kinase and p100 cooperate to enhance c-Myb activity. Mol Cell *2*, 417-425.

Liu, X., Kim, C.N., Yang, J., Jemmerson, R. and Wang, X. (1996) Induction of apoptotic program in cell-free extracts: requirement for dATP and cytochrome c. Cell *86*, 147-157.

Lund, A.H. and van Lohuizen, M. (2002) RUNX: a trilogy of cancer genes. Cancer Cell *1*, 213-215.

Mountz, J.D. and Steinberg, A.D. (1989) Studies of c-myb gene regulation in MRL-lpr/lpr mice. Identification of a 5' c-myb nuclear protein binding site and high levels of binding factors in nuclear extracts of lpr/lpr lymph node cells. J Immunol *142*, 328-335.

Mountz, J.D., Steinberg, A.D., Klinman, D.M., Smith, H.R. and Mushinski, J.F. (1984) Autoimmunity and increased c-myb transcription. Science *226*, 1087-1089.

Mucenski, M.L., McLain, K., Kier, A.B., Swerdlow, S.H., Schreiner, C.M., Miller, T.A., Pietryga, D.W., Scott, W.J., Jr. and Potter, S.S. (1991) A functional c-*myb* gene is required for normal murine fetal hepatic hematopoiesis. Cell *65*, 677-689.

Muller, H., Bracken, A.P., Vernell, R., Moroni, M.C., Christians, F., Grassilli, E., Prosperini, E., Vigo, E., Oliner, J.D. and Helin, K. (2001) E2Fs regulate the expression of genes involved in differentiation, development, proliferation, and apoptosis. Genes Dev *15*, 267-285.

Nakagoshi, H., Kanei-Ishii, C., Sawazaki, T., Mizuguchi, G. and Ishii, S. (1992) Transcriptional activation of the c-myc gene by the c-myb and B-myb gene products. Oncogene *7*, 1233-1240.

Ness, S.A. (1996) The Myb oncoprotein: regulating a regulator. Biochim Biophys Acta *1288*, F123-139.

Pallard, C., Stegmann, A.P., van Kleffens, T., Smart, F., Venkitaraman, A. and Spits, H. (1999) Distinct roles of the phosphatidylinositol 3-kinase and STAT5 pathways in IL-7-mediated development of human thymocyte precursors. Immunity *10*, 525-535.

Pearson, R. and Weston, K. (2000) c-Myb regulates the proliferation of immature thymocytes following beta-selection. EMBO J *19*, 6112-6120.

Postigo, A.A., Sheppard, A.M., Mucenski, M.L. and Dean, D.C. (1997) c-Myb and Ets proteins synergize to overcome transcriptional repression by ZEB. EMBO J *16*, 3924-3934.

Ratajczak, M.Z., Perrotti, D., Melotti, P., Powzaniuk, M., Calabretta, B., Onodera, K., Kregenow, D.A., Machalinski, B. and Gewirtz, A. M. (1998) Myb and ets proteins are candidate regulators of c-kit expression in human hematopoietic cells. Blood *91*, 1934-1946.

Ratter, F., Germer, M., Fischbach, T., Schulze-Osthoff, K., Peter, M.E., Droge, W., Krammer, P.H. and Lehmann, V. (1996) S-adenosylhomocysteine as a physiological modulator of Apo-1-mediated apoptosis. Int Immunol *8*, 1139-1147.

Redondo, J.M., Hernandez-Munain, C. and Krangel, M.S. (1995) Transcriptional regulation of the human T cell receptor delta gene. Immunobiology *193*, 288-292.

Reizis, B. and Leder, P. (2001) The upstream enhancer is necessary and sufficient for the expression of the pre-T cell receptor alpha gene in immature T lymphocytes. J Exp Med *194*, 979-990.

Rohwer, F., Todd, S. and McGuire, K.L. (1996) The effect of IL-2 treatment on transcriptional attenuation in proto- oncogenes pim-1 and c-myb in human thymic blast cells. J Immunol *157*, 643-649.

Sala, A., Nicolaides, N.C., Engelhard, A., Bellon, T., Lawe, D.C., Arnold, A., Grana, X., Giordano, A. and Calabretta, B. (1994) Correlation between E2F-1 requirement in the S phase and E2F-1 transactivation of cell cycle-related genes in human cells. Cancer Res *54*, 1402-1406.

Schilham, M.W., Wilson, A., Moerer, P., Benaissa-Trouw, B. J., Cumano, A. and Clevers, H. C. (1998) Critical involvement of Tcf-1 in expansion of thymocytes. J Immunol *161*, 3984-3991.

Schmidt, T., Karsunky, H., Rodel, B., Zevnik, B., Elsasser, H.P. and Moroy, T. (1998) Evidence implicating Gfi-1 and Pim-1 in pre-T-cell differentiation steps associated with beta-selection. EMBO J *17*, 5349-5359.

Schmitt, T.M. and Zuniga-Pflucker, J.C. (2002) Induction of T cell development from hematopoietic progenitor cells by delta-like-1 in vitro. Immunity *17*, 749-756.

Siu, G., Wurster, A.L., Lipsick, J.S. and Hedrick, S.M. (1992) Expression of the CD4 gene requires a Myb transcription factor. Mol Cell Biol *12*, 1592-1604.

Stern, J.B. and Smith, K.A. (1986) Interleukin-2 induction of T-cell G1 progression and c-myb expression. Science *233*, 203-206.

Strasser, A., Harris, A.W., Huang, D.C.S., Krammer, P.H. and Cory, S. (1995) Bcl-2 and Fas/APO-1 regulate distinct pathways to lymphocyte apoptosis. EMBO J *14*, 6136-6147.

Sumner, R., Crawford, A., Mucenski, M. and Frampton, J. (2000) Initiation of adult myelopoiesis can occur in the absence of c-Myb whereas subsequent development is strictly dependent on the transcription factor. Oncogene *19*, 3335-3342.

Takahashi, T., Suwabe, N., Dai, P., Yamamoto, M., Ishii, S. and Nakano, T. (2000) Inhibitory interaction of c-Myb and GATA-1 via transcriptional co- activator CBP. Oncogene *19*, 134-140.

Taniuchi, I., Osato, M., Egawa, T., Sunshine, M.J., Bae, S.C., Komori, T., Ito, Y. and Littman, D.R. (2002) Differential Requirements for Runx Proteins in CD4 Repression and Epigenetic Silencing during T Lymphocyte Development. Cell *111*, 621-633.

Taylor, D., Badiani, P. and Weston, K. (1996) A dominant interfering Myb mutant causes apoptosis in T cells. Genes Dev *10*, 2732-2744.

Thompson, L.F., Van de Wiele, C.J., Laurent, A.B., Hooker, S.W., Vaughn, J.G., Jiang, H., Khare, K., Kellems, R.E., Blackburn, M.R., Hershfield, M.S. and Resta, R. (2000) Metabolites from apoptotic thymocytes inhibit thymopoiesis in adenosine deaminase-deficient fetal thymic organ cultures. J Clin Invest *106*, 1149-1157.

Urbanek, P., Dvorak, M., Bartunek, P., Pecenka, V., Paces, V. and Travnicek, M. (1988) Nucleotide sequence of chicken *myb* proto-oncogene promoter region: detection of an evolutionarily conserved element. Nucl Acids Res *16*, 11521-11531.

Vaillant, F., Blyth, K., Andrew, L., Neil, J.C. and Cameron, E.R. (2002) Enforced expression of Runx2 perturbs T cell development at a stage coincident with beta-selection. J Immunol *169*, 2866-2874.

van de Wetering, M., Sancho, E., Verweij, C., de Lau, W., Oving, I., Hurlstone, A., van der Horn, K., Batlle, E., Coudreuse, D., Haramis, A.P., Tjon-Pon-Fong, M., Moerer, P., van den Born, M., Soete, G., Pals, S., Eilers, M., Medema, R. and Clevers, H. (2002) The beta-catenin/TCF-4 complex imposes a crypt progenitor phenotype on colorectal cancer cells. Cell *111*, 241-250.

Verbeek, S., Izon, D., Hofhuis, F., Robanus-Maandag, E., te Riele, H., van de Wetering, M., Oosterwegel, M., Wilson, A., MacDonald, H.R. and Clevers, H. (1995) An HMG-box-containing T-cell factor required for thymocyte differentiation. Nature *374*, 70-74.

Wang, Q.F., Lauring, J. and Schlissel, M.S. (2000) c-Myb binds to a sequence in the proximal region of the RAG-2 promoter and is essential for promoter activity in T-lineage cells. Mol Cell Biol *20*, 9203-9211.

Waskow, C., Paul, S., Haller, C., Gassmann, M. and Rodewald, H. (2002) Viable c-Kit(W/W) mutants reveal pivotal role for c-kit in the maintenance of lymphopoiesis. Immunity *17*, 277-288.

White, J.R. and Weston, K. (2000) Myb is required for self-renewal in a model system of early hematopoiesis. Oncogene *19*, 1196-1205.

Yokota, S., Yuan, D., Katagiri, T., Eisenberg, R., Cohen, P.L. and Ting, J.P.-Y. (1987) The expression and regulation of c-*myb* transcription in B6/*lpr* Lyt-2^- , L3T4^- T lymphocytes. J Immunol *139*, 2810-2817.

Zobel, A., Kalkbrenner, F., Guehmann, S., Nawrath, M., Vorbrueggen, G. and Moelling, K. (1991) Interaction of the v-and c-Myb proteins with regulatory sequences of the human c-myc gene. Oncogene *6*, 1397-1407.

Chapter 4

POTENTIAL ROLES FOR C-MYB THROUGHOUT EARLY LYMPHOCYTE DEVELOPMENT

Timothy P. Bender
Department of Microbiology, University of Virginia Health System, Charlottesville, VA 22908-0734, United States of America.

Abstract: Expression of the c-Myb transcription factor is primarily associated with the immature stages of haemopoietic differentiation and expression is down regulated to low or undetectable levels as haemopoietic maturation progresses. During lymphocyte development, c-Myb is abundantly expressed in early precursors and down regulation of c-Myb expression appears to occur near or during the time of repertoire selection. This pattern of expression has long suggested a significant role for c-Myb during lymphocyte development, which has been provocatively reinforced by reports of putative c-Myb target genes that are crucial for lymphocyte development. However, gaining insight has been greatly impeded by the embryonic lethality of traditional null c-*myb* mutations and the lack of a tractable genetic model to study c-Myb function. This chapter will discuss the relationship between c-Myb and lymphocyte development and discuss the prospect of gaining insight into c-Myb activity using conditional approaches to gene targetting.

1. INTRODUCTION

The *c-myb* protooncogene was first identified as the cellular homologue of the transforming element identified in two replication-defective acute transforming avian retroviruses, AMV and E26, that cause myeloblastic and erythroblastic leukaemias (Oh et al., 1999). Expression of *c-myb* has historically been associated with immature haemopoietic cells. In each haemopoietic lineage, *c-myb* mRNA and protein expression is greatest during the immature stages of differentiation and down regulation of expression is associated with cellular maturation (Graf, 1992). The pattern of *c-myb* expression suggested that it would play an important role during

J. Frampton (ed.), Myb Transcription Factors: Their Role in Growth, Differentiation and Disease, 81-105.

haemopoietic maturation, and this has been supported in several model systems. Most significantly, mice that lack a functional *c-myb* gene die by day 15 of embryonic development due to a defect in definitive haemopoiesis (Mucenski et al., 1991). More recent experiments using tissue-specific expression of a dominant interfering Myb allele (MEnT) or c-*myb*$^{-/-}$/Rag1$^{-/-}$ chimaeric mice have suggested a role for c-Myb during the early stages of T cell development (Allen et al., 1999; Badiani et al., 1994). While these experiments demonstrate the importance of c-Myb to normal haemopoiesis, relatively little is understood about the physiologic role(s) played by c-Myb or the downstream effectors of c-Myb activity. In addition, it is now apparent that c-Myb expression is more widely distributed than originally appreciated in both embryonic and adult tissues including gut epithelium, hair follicles, breast duct epithelium and vascular smooth muscle tissue (Brown et al., 1992; Ess et al., 1999).

Vertebrates possess two additional genes that are closely related to c-Myb, A-Myb and B-Myb (Nomura et al., 1988). While all three mammalian Myb proteins can bind the consensus c-Myb DNA-binding sequence, they have distinct tissue and cell type patterns of expression. A-*myb* is highly expressed in male germ cells, female breast duct epithelium and subsets of B lymphocytes (Foos et al., 1994; Mettus et al., 1994; Toscani et al., 1997; Trauth et al., 1994) while B-*myb* expression appears to be nearly ubiquitous (Golay et al., 1991; Nomura et al., 1988; Reiss et al., 1991). Phenotypes of the individual *myb* family member null mutations generally reflect the pattern of A-, B- and c-*myb* expression. c-*myb* null mice die at d15 due to a severe anaemia although other organs appeared normal (Mucenski et al., 1991) and the B-*myb* null mutation is lethal at the pre-implantation stage (Tanaka et al., 1999). In contrast, A-*myb* null mice are viable, though A-*myb* null males are infertile due to a block during spermatogenesis (Toscani et al., 1997). A-*myb* null females are fertile but do not nurse as proliferation of breast tissue does not occur during pregnancy. The phenotypes of individual *myb* family null mutations in mice suggest that these proteins do not simply mediate redundant functions. For example, the ubiquitously expressed B-*myb* does not substitute for c-*myb* in c-*myb* null mice. However, the embryonic lethality of c- and B-*myb* null mutations and the relatively complex phenotype of the A-*myb* null mutation have proved to be strong impediments to understanding the physiologic roles played by the Myb proteins in vertebrates. This chapter will discuss a conditional gene targetting approach to make tissue specific c-*myb* mutations in mice and the relationship between c-*myb* expression and the early stages of lymphocyte development as well as potential roles for c-Myb in each lymphoid lineage.

2. CONDITIONAL MUTAGENESIS TO STUDY C-MYB FUNCTION

Our recent experiments using the RAG blastocyst complementation assay clearly demonstrated that c-Myb is required for production of mature B and T cells (Allen et al., 1999). Close inspection of bone marrow from c-*myb*$^{-/-}$/*Rag1*$^{-/-}$ chimaeric mice did not identify B220^{+} or CD19^{+} B cell precursors. The thymi from c-*myb*$^{-/-}$/*Rag1*$^{-/-}$ chimaeric mice contained a population of CD44^{lo} CD25^{-} thymocyte precursors, suggesting that c-Myb is not required for haemopoietic stem cell (HSC) function *per se* but likely does play a significant role at the early precursor stage during both B and T cell development. However, these results also preclude gaining insight into c-Myb function during the later stages of B and T cell development. To circumvent the embryonic lethality of traditional c-*myb* null alleles we have used the Cre/loxP approach to create conditional c-*myb* deficient mice (T.P. Bender and K. Rajewsky, unpublished data). A targetting vector was built based on a 7.6 kbp EcoRI genomic DNA fragment encoding the 5' untranslated region, exon I, intron I, exon II and a portion of intron II from the mouse c-*myb* gene (Bender et al., 1987; Toth et al., 1995). A single *lox*P site was inserted approximately 1.9 kbp upstream of exon II and a *lox*P flanked neomycin cassette was inserted approximately 0.6 kbp downstream of exon II (Figure 1). This strategy was used because the exonI/exon II splice is in the first reading frame while the remaining downstream splices take place in reading frames two and three. Thus, removal of exon II results in out-of-frame downstream splicing events. In addition, the region of the c-*myb* locus that is targetted by our construct contains the cryptic exon II promoter that has been reported to be active in a number of leukaemic cell lines (Jacobs et al., 1994). The linearised targetting construct was introduced into IB10 mouse embryonic stem cells using standard procedures (Torres and Kuhn, 1997). To produce a minimally perturbed targetted c-*myb* locus the neomycin cassette was removed by transiently transfecting a Cre producing plasmid into the targetted ES cell lines and identifying subclones that had deleted the neomycin cassette but retained exon II. The c-*myb* locus containing exon II flanked by *lox*P sites is referred to as the "floxed" or c-*myb*^{f} allele. Further deletion of the *lox*P flanked exon II, either *in vivo* or *in vitro*, by Cre recombinase results in a deleted or c-*myb*^{d} allele.

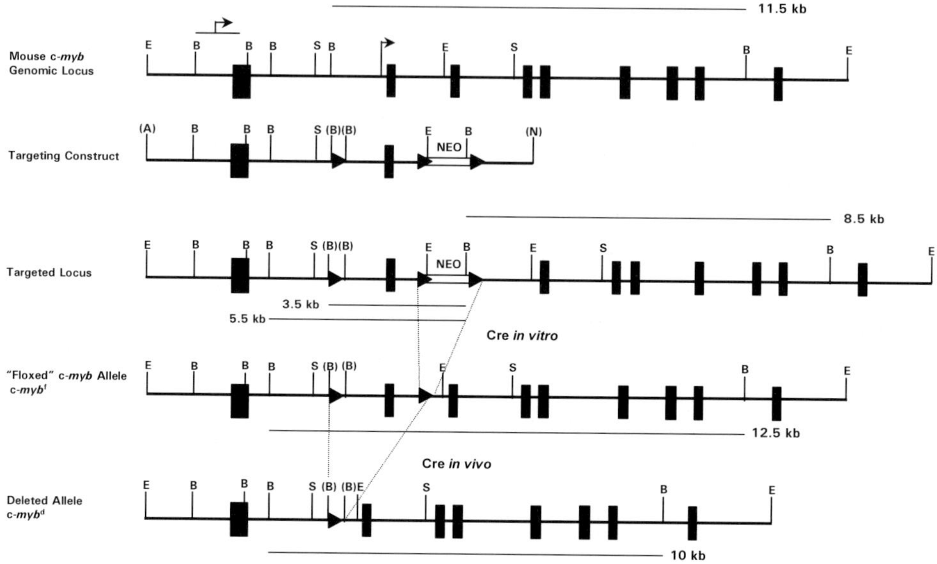

Figure 1
Strategy to create a conditionally targetted mouse *c-myb* allele.

Embryonic stem cells heterozygous for the c-myb^{f} allele were injected into C57BL/6 blastocysts and the resulting chimaeric mice crossed to C57BL/6 animals to establish mice carrying the c-myb^{f} allele in the germ line. Mice that are either homozygous or heterozygous for the floxed c-*myb* allele are born at a Mendelian ratio with no apparent defects in growth, development or fertility. To specifically inactivate the c-*myb* locus in B or T-lineage cells we have crossed c-myb^{f} mice to either CD19Cre (Rickert et al., 1995; Rickert et al., 1997) or lckCre (Gu et al., 1994; Lee et al., 2001) mice. CD19Cre mice carry Cre as an insertion in the first exon of the CD19 locus and specifically produce Cre in $CD19^{+}$ B-lineage cells. Thus, Cre is produced beginning in Fraction B pro-B cells (see Figure 2). In contrast, lckCre mice express Cre from a transgene using the *lck* proximal promoter. We have used two strains of lckCre mice that efficiently delete *lox*P targetted DNA at different stages of T cell development. One lckCre line (Lee et al., 2001) very efficiently deletes the c-myb^{f} allele beginning at the DN2 stage during T cell development (see Figure 2) while the other (Gu et al., 1994) deletes efficiently during the DP stage, which has allowed us to assess the role of c-*myb* during distinct stages of thymocyte development. We have also crossed mice carrying the floxed c-*myb* allele with hCMV-Cre mice (Schwenk et al., 1995) that provide ubiquitous expression of Cre to produce mice that carry a deleted c-*myb* allele (c-*myb*^{d}). Intercrosses between c-*myb*$^{f/d}$ or c-*myb*$^{w/d}$ mice have never produced c-*myb*$^{d/d}$ mice. However, timed pregnancies between c-*myb*$^{f/d}$ or c-*myb*$^{w/d}$ mice produce c-*myb*$^{d/d}$ embryos at appropriate Mendelian ratios that die by day 15 *post coitus* with severe anaemia as originally reported for traditional null c-*myb* mutations (Mucenski et al., 1991). Protein extracts produced from day 12 c-*myb*$^{d/d}$ foetal livers contain no detectable c-Myb protein while c-Myb is readily detected in livers from c-*myb*$^{w/d}$ or wild type day 12 embryos. Thus, deletion of exon II appears to result in a true null allele. Our preliminary analysis of these mice suggests an essential role for c-*myb* during B cell development at the pro-B to pre-B cell transition. Similarly, we have found that c-*myb* is important for transit through the CD4/CD8 double negative compartment, particularly transition from DN3 to DN4 and that c-Myb is required for efficient differentiation during the CD4/CD8 double positive stage of thymocyte development. In the following section we will discuss potential roles for c-Myb during B and T cell differentiation in the context of these mice and our initial observations.

3. LYMPHOCYTE DEVELOPMENT AND EXPRESSION OF C-MYB

Mature B and T cells are derived from pluripotent HSCs in the bone marrow. However, while B cells develop from HSC to membrane IgM (mIgM) bearing cells in the bone marrow environment, progenitor cells that give rise to T cells migrate from the bone marrow to the thymus where commitment to unipotential T cell development and differentiation to functional CD4 and CD8 single positive T cells takes place. There are striking parallels between B and T cell development (Figure 2). The early stages of both B and T cell development are devoted to the highly orchestrated series of gene rearrangement events, referred to as V(D)J recombination, that ultimately allow production of B or T cell antigen specific receptors, BCR and TCR respectively from variable (V), diversity (D) and joining (J) segments (Krangel, 2003; Tonegawa, 1983). Initially, productive rearrangement at the immunoglobulin heavy chain (B cells) or TCRβ (T cells) loci results in a proliferative burst that expands the number of cells that have successfully completed this process. Subsequently, these cells enter a quiescent phase during which a second set of rearrangement events takes place at the immunoglobulin light chain loci (B cells) or the TCRα locus (T cells). Productive rearrangement during this second round of recombination events results in pairing of the two sets of peptides that form the BCR or TCR and expression of the antigen specific receptor on the cell surface. Immature B and T lymphocytes then undergo the process of repertoire selection that allows testing to identify useful antigen specific receptors and negative selection, through several mechanisms, to remove cells with self-reactive receptors. Strikingly, c-Myb appears to be abundantly expressed from the earliest stages of B and T cell development till the point

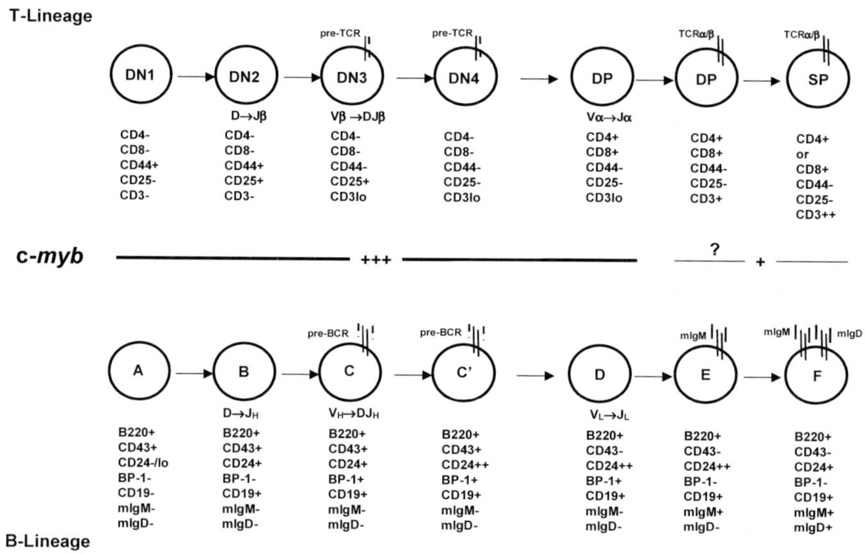

Figure 2

Expression of c-myb during B and T-lymphocyte development. The "?" represents uncertainty about when c-myb expression is down regulated in each lineage.

of repertoire selection when expression of c-Myb is down regulated to a low but detectable level, which is retained as antigen receptor bearing lymphocytes migrate from the primary lymphoid tissues to the periphery. During early lymphocyte development, the exons that encode the immunoglobulin and TCR variable regions are assembled from germ line V, D and J segments. The process of V(D)J recombination is initiated by the products of the lymphocyte specific recombination activating genes, RAG1 and RAG2, which form an endonuclease (Fugmann et al., 2000). The RAG recombinase introduces double-stranded DNA breaks between variable gene segments and flanking recombination signal sequences (RSS). Non-homologous end joining DNA repair proteins then ligate the double-stranded breaks to form contiguous V(D)J coding and recombination signal joints. While the RAG1 and RAG2 proteins are only expressed in developing lymphocytes, tissue specific expression does not explain why BCR encoding V-region gene segments undergo V(D)J recombination in developing B-lineage cells but not T-lineage cells and *vice versa*. In addition, rearrangement takes place in a specific order. For example, during T cell development, TCRβ rearrangement precedes TCRα rearrangement. Furthermore, $D \rightarrow J_\beta$ recombination precedes $V_\beta \rightarrow DJ_\beta$ recombination. The specificity and temporal regulation of V(D)J recombination is mediated by changes in higher order chromatin structure that control the accessibility of RSS to the RAG recombinase (Sleckman et al., 1996; Stanhope-Baker et al., 1996; Yancopoulos and Alt, 1985). Despite the interesting pattern of expression during lymphocyte development little is understood about the function of c-Myb during lymphocyte development or the activation of effector function mainly because of the previous lack of a tractable genetic model that allows the study of lymphocyte development in the absence of c-Myb.

3.1 Expression and Function of c-Myb in B Cells

Expression of *c-myb* during B cell development is not well characterised. Mouse tumours representing the pro-B and pre-B cell stages of development contain 10-fold more c-*myb* mRNA than tumours representing immature B cells, mature B cells or plasma cells, which suggested that *c-myb* might be differentially expressed during B-lymphocyte development (Bender and Kuehl, 1987). One report detected c-*myb* mRNA in common lymphoid progenitors (CLPs) but not in pro-B or pre-B cells (Akashi et al., 2000). However, we readily detect c-*myb* mRNA in electronically sorted bone marrow pro-B and pre-B cell populations (T. Bender, unpublished data), as do others (Dr. I-L Martensson, Cambridge, UK, personal communication) and it is likely that c-*myb* is expressed throughout B cell development. However, c-*myb* expression in Fraction D pre-B cells appears to be approximately five-fold greater than in $B220^+$ $CD43^-$ pro-B cells, suggesting that c-Myb may be particularly important at this stage or during transition to this stage during B cell development. Immature B cells (Fraction F) and splenic B cells also contain about five-fold less c-*myb* mRNA than pre-B cells. Expression of c-*myb* in pro-B and pre-B cell tumours was found to be constitutive but regulated during the cell cycle in immature and mature B cell lymphomas as well as plasmacytomas with maximum expression during S-phase (Catron et al., 1992; Isakson et al., 1991) and similar results were obtained in a study of c-*myb* expression in normal and leukaemic avian lymphocytes (Thompson et al., 1986). Human (Golay et al., 1991) and mouse (T. Bender, unpublished data) peripheral B-lymphocytes contain small amounts of c-*myb* mRNA and protein and expression is greatest in proliferating B cells. Treatment of human tonsillar B cells with a variety of mitogenic signals results in increased c-*myb* expression. However, signals that drive resting B cells into the G1 stage of the cell cycle are not sufficient to increase c-*myb* expression. For example, treatment of resting human B cells with anti-CD20 or anti-CD40 alone does not drive proliferation and did not induce c-*myb* expression. However, the mitogenic signal of anti-CD20 plus anti-CD40 did result in increased c-*myb* expression (Golay et al., 1991; Golay et al., 1992). Thus, expression of c-*myb* is detected throughout B cell development and in peripheral B cells. No information is currently available regarding the relative expression of c-*myb* in peripheral B-1, follicular and marginal zone B cell subsets and the role played by c-*myb* during B cell function remains entirely unstudied. The interesting changes in amounts of c-*myb* contained in cells at different stages of differentiation or in response to activation protocols may suggest that c-Myb plays different roles at different times during the life of B cells.

B cell development in the mouse takes place continuously throughout life, first in foetal liver during embryogenesis and then in the adult bone marrow. B cells are derived from pluripotent HSCs in the bone marrow and the most immature B cell progenitors are located near the bone marrow endosteum. Maturing B cells migrate toward the venous sinusoids as they progress through differentiation to the mIgM positive immature B cell stage. B cell development progresses from HSCs through sequentially more restricted progenitors, such as the CLP, and finally commitment to the B-lymphoid lineage (Melchers and Rolink, 1998). After commitment, developing B cells go through a series of successive developmental stages, first completing rearrangement of the immunoglobulin heavy chain locus, followed by a proliferative stage of clonal expansion, entrance into a quiescent state and finally rearranging the immunoglobulin light chain loci. Productive rearrangement of the kappa or lambda light chain loci and expression of a light chain protein leads to assembly and expression of IgM on the immature B cell surface (mIgM). The identity of cells from the CLP until DJ_H rearrangements are detected remains somewhat controversial (Hardy, 2003) but the stages of B cell development after this point are now well characterised in terms of successive expression of surface and intracellular markers and Ig gene rearrangements. Several transcription factors have been identified that are required for B cell development including E2A, early B cell factor (EBF) and Pax5 (Lin and Grosschedl, 1995; Nutt et al., 1997; Zhuang et al., 1994). While E2A and EBF are required for B cell development neither is sufficient to commit B-lineage progenitor cells to unipotential B cell development. In contrast, Pax5 is required for commitment to B cell development and is the only gene product known to mediate this function (Nutt et al., 1997). Pro-B cells from Pax5 deficient mice express B-lineage markers and begin rearrangement at the immunoglobulin heavy chain locus but fail to move beyond the point of $D \rightarrow J_H$ rearrangements and are capable of differentiating into other haemopoietic lineages (Nutt et al., 1997; Nutt et al., 1999; Rolink et al., 1999). It is interesting to speculate that c-Myb may play a role during the very early stages of B cell development at one or more steps between the CLP and later stage B-lineage cells since we did not detect B220$^+$ or CD19$^+$ B cells in the bone marrow of c-*myb*$^{-/-}$/*Rag1*$^{-/-}$ chimaeric mice although we did detect very early thymocyte progenitors. c-Myb function at stages of B cell development prior to Fraction B cannot be addressed using the CD19Cre mice since CD19 expression is first detected at this point. However, current inducible Cre producing strains of mice as well as new strains under production should allow this point to be addressed in the near future (Feil et al., 1997; Hayashi and McMahon, 2002; Kuhn et al., 1995).

Two main nomenclatures are currently used to describe B cell development in the bone marrow based on successive expression of cell surface markers and status of immunoglobulin gene rearrangement (Hardy et al., 1991; Melchers et al., 1995; Rolink et al., 1996). We use the scheme developed by Hardy and colleagues (Hardy et al., 1991) to follow B cell development (see Figure 2). The pro-B cell stage in this strategy is defined as B220$^+$ CD43$^+$ bone marrow cells and the process of V(D)J recombination is initiated in this subset. Developing pro-B cells (Fractions A-C') express c-Kit (Loffert et al., 1994), which has been reported to be a c-Myb target (Ratajczak et al., 1998). The role of c-Kit during B cell development is unclear. Monoclonal antibodies against c-Kit inhibit the proliferation of B cell progenitors, as well as other haemopoietic progenitors, during *in vitro* culture on stromal cells (Rolink et al., 1991b) yet the same antibodies result in enhanced B-lymphopoiesis *in vivo* (Ogawa et al., 1991). In addition, mice that carry a mutated c-*kit* locus (W^V) have apparently normal B-lymphopoiesis (Landreth et al., 1984). Not all pro-B cells are responsive to SCF and this population may expand in the absence of c-Kit (Kodama et al., 1992). Alternatively, c-Kit may simply not be essential for B-lymphocyte development or other cytokines or cellular interactions may compensate for loss of *c-kit*. However, it will be of interest to examine *c-kit* expression in B cell deficient B-lineage cells. In the Hardy strategy, pro-B cells are further divided into Fractions A, B, C and C' based on surface expression of two more surface markers, CD24 and BP-1. Cells in Fraction A are described as CD24 negative or low and this subset has been further divided to define very early precursors that appear to be committed to B-lineage development (Hardy, 2003; Li et al., 1996). Gene rearrangement at the immunoglobulin heavy chain locus is poorly detected, if at all, in

Fraction A cells though a subset contains germ line μ-transcripts that are believed to reflect changes in higher order chromatin structure that are required for accessibility of RSS to the V(D)J recombinase (Sleckman et al., 1996; Yancopoulos and Alt, 1985). Expression of the RAG1 and RAG2 as well as $D \rightarrow J_H$ rearrangements are first clearly detected in Fraction B and recombination at the immunoglobulin heavy chain locus is initiated beginning with D_H to J_H-segment rearrangement (Ehlich et al., 1993; Hardy et al., 1991; Li et al., 1993). In addition, lambda-5 and V-pre-B, components of the surrogate light chain (SLC), and Ig-α and Ig-β, signalling components of the BCR, are first detected at this stage (Hardy et al., 1991; Li et al., 1993). It is interesting to note that the RAG2 core promoter has been reported to be a target of c-Myb activity in B cells (Jin et al., 2002; Kishi et al., 2002). These reports were based on *in vitro* experiments but it remains a fascinating possibility that c-Myb activity may be required for expression of RAG2 and efficient V(D)J recombination during B cell development. After DJ_H rearrangement, changes in chromatin structure occur at the germ line V_H segments that are accompanied by the appearance of germ line V_H transcripts and $V_H \rightarrow DJ_H$ recombination is detected in Fraction C. Changes in chromatin structure appear to allow access of the V(D)J recombinase to the V_H recombination signal sequences and initiation of $V_H \rightarrow DJ_H$ rearrangement. The change from DJ_H to $V_H \rightarrow DJ_H$ recombination requires Pax-5 (Nutt et al., 1997; Urbanek et al., 1994) and IL-7 (Corcoran et al., 1996). Interestingly, changes in histone hyperacetylation occur in a stepwise fashion that correlates with recombination activity (Chowdhury and Sen, 2003; Hesslein et al., 2003). Initially, a domain that extends from the most 5' D-segment to the intergeneic region between the mu and delta heavy chain constant region coding sequences is hyperacetylated but V_H genes are not hyperacetylated. $D \rightarrow J_H$ recombination but not $V_H \rightarrow DJ_H$ recombination takes place in this context. After completion of $D \rightarrow J_H$ recombination, three independent domains in the V_H locus sequentially become hyperacetylated moving from D proximal to the most 5' V_H segments.

Productive (in frame) $V_H DJ_H$ rearrangement results in expression of intracellular mu heavy chain protein (c_μ) and c_μ is first detected in Fraction C' (Hardy et al., 1991). Once expressed, c_μ can pair with lambda-5 and Vpre-B and along with Ig-α and Ig-β is inserted in the cell membrane as the pre-B cell receptor (pre-BCR) (Melchers et al., 1999). It is not clear why c_μ is not detected in Fraction C since cells that contain a productively rearranged immunoglobulin heavy chain locus are present but it is likely that c_μ pairs with SLC components and are rapidly selected into Fraction C'. Assembly, of the pre-BCR serves as a major checkpoint during B cell development and provides signals that direct clonal expansion, allelic exclusion and differentiation to Fraction D (Martensson et al., 2002).

However, only about half of the H-chains produced by Fraction C' cells can form a pre-BCR (ten Boekel et al., 1997). Cells producing H-chains that cannot form a pre-BCR are not selected and fail to proliferate. Growth and differentiation of B cells beyond the pro-B cell stage *in vivo* is highly dependent on IL-7 *in vivo* (Peschon et al., 1994). Both pro-B and pre-B cells express IL-7R on their surface yet the role of IL-7 signalling in pre-BCR dependent proliferation is unclear. Rolink and colleagues (Rolink et al., 1991a) have reported that $CD19^{+}$ $c\text{-}Kit^{+}$ cells (roughly equivalent to Fraction B-C) can undergo two to five divisions in the absence of stromal cells or IL-7 (Rolink et al., 2000) while others have found these cells to be dependent on very low concentrations of IL-7 for proliferation (Marshall et al., 1998; Ray et al., 1998). It is likely that isolated pre-BCR bearing cells can undergo limited division that is enhanced by IL-7. In this context, it is interesting to note that the lambda-5 enhancer has been reported to be a target of c-Myb activity (Martensson et al., 2001). Thus, c-Myb may prove to play a significant role in regulating pro-B to pre-B cell transition by regulating expression of lambda-5. In this case, we may expect to find a strong but incomplete block to B cell differentiation, similar to that described for lambda-5 deficient mice in c-Myb deficient B cells (Kitamura et al., 1992). c-Myb expression is also associated with proliferation in developing B cells and a number of genes associated with proliferation have been reported to be c-Myb targets (Arsura and Sonenshein, 2001; Oh et al., 1999). The use of tissue specific c-*myb* deletion mutant mouse strains should allow us to assess the relative contribution of c-Myb to proliferation, survival and differentiation during B cell development.

Signalling through the pre-BCR leads to rapid down regulation of the recombination activating genes and transcription of genes encoding components of the SLC stops, limiting continued pre-BCR signalling and developing B cells enter a quiescent phase (Grawunder et al., 1995). Upon loss of pre-BCR expression the developing B cells enter Fraction D (small pre-B cells) and expression of the RAG proteins is re-established leading to immunoglobulin light chain gene rearrangement. Changes in higher order chromatin structure appear to allow ordered access of the recombinase complex to the kappa locus prior to the lambda locus (Engel et al., 1993). Again, it is interesting to note that the *Rag2* promoter is reported to be a c-Myb target in B cells and to speculate that c-Myb may be important for V(D)J recombination in pre-B cells (Jin et al., 2002; Kishi et al., 2002). In frame rearrangement at one of the light chain loci allows expression of cytoplasmic light chain. However, about 20% of small pre-B cells express cytoplasmic μ and kappa proteins but do not display mIgM, suggesting that a large number of kappa proteins produced cannot pair with the expressed heavy chain (Rolink et al., 2001; Yamagami et al., 1999b; Yamagami et al.,

1999a) or that further signals are required for display of mIgM (Henderson et al., 1992). In the first case, receptor editing in Fraction D offers the opportunity for rescue of pre-B cells that express a light chain that is unable to pair with the H-chain (Nemazee, 2000). In addition to potential regulation of the *Rag2* promoter, the relationship between c-Myb and survival makes it interesting to consider a role for c-Myb in pre-B cell survival, conceivably regulating a window during which receptor editing can occur. A potential result of limiting receptor editing might be to limit the prospect of rescuing these cells by ongoing immunoglobulin light chain gene rearrangement and possibly limit the diversity of the peripheral light chain V-region repertoire.

Display of mIgM, but not mIgD, on the surface of developing B cells defines the immature B cell stage (Fraction F). Immature B cells that encounter autoantigen fail to undergo maturation associated up regulation of mIgM, undergo developmental arrest and can induce receptor editing (Melamed et al., 1998). It should be stressed that nothing is known about when c-*myb* expression is down regulated during transition from the pre-B to the immature B cell stage of differentiation or if high level expression of c-*myb* can be re-induced in immature B cells, which poses an interesting proposition. If c-Myb is involved in regulating *Rag2* expression in B cells it is possible that it is involved regulating receptor editing. Similarly, if c-Myb is important for immature B cell survival it could limit a window during which receptor editing can take place. Receptor editing appears to be developmentally regulated because interaction with autoantigen in transitional B cells or mature B cells induces apoptosis or anergy (Tiegs et al., 1993). Mice produce about 2×10^7 immature B cells per day but about 90% percent of these are lost and much of this loss is likely due to interaction with autoantigen (Osmond, 1991). Immature B cells migrate to the spleen where they undergo further maturation through two (Loder et al., 1999) or three transitional stages (Allman et al., 2001) with a half life of about four days (Rolink et al., 1998) to become $mIgM^+$ $mIgD^+$ B cells. Signalling through the BCR appears to play a significant role in transition from the bone marrow because several mutations, including *syk* and an Ig-α mutant that lacks a cytoplasmic tail, appear to result in increased loss at this point in development (Torres et al., 1996; Turner et al., 1997). Though little is known about the expression of c-*myb* in peripheral B cell subsets, continued c-*myb* expression in peripheral B cells suggests a potential role for c-Myb in maintaining peripheral B cell subsets.

3.2 Expression and Function of c-Myb in T Cells

Expression of c-*myb* mRNA has been better characterised in T cells than in B cells but very little is known about the relevance of c-*myb* expression to either T cell development or mature T cell function. During ontogeny, c-*myb* mRNA has been detected in the thymus by *in situ* hybridisation as soon as T cell precursors migrate to the thymic anlage. As the developing thymus resolves into cortex and medullary regions c-*myb* expression is detected mainly in the cortex (Ess et al., 1999). One early study demonstrated abundant c-*myb* mRNA expression in PNA^+ thymocytes but not in the PNA^- population (Sica et al., 1992). Since the PNA^- population contains both DN and SP thymocytes it is likely that the small amount of c-*myb* mRNA detected in the PNA^- thymocytes was due to the DN component. We have recently examined c-*myb* mRNA expression in electronically sorted thymic CD4/CD8 subsets and find our results are in good agreement with earlier experiments in that c-*myb* mRNA is abundantly expressed in CD4/CD8 DN and DP subsets and is decreased greater than ten-fold in SP cells. Interestingly, we find that c-*myb* expression is decreased in post-positive selection DP thymocytes (T.P. Bender, unpublished data). Resting peripheral T cells contain small but detectable amounts of c-*myb* mRNA and protein. However, c-*myb* expression is greatly increased in small resting T cells and cloned T cell lines from both human and mouse after stimulation with mitogen or anti-TCR plus exogenous IL-2 (Reed et al., 1986; Reed et al., 1987; Shipp and Reinherz, 1987; Stern and Smith, 1986). Experiments using inhibitors of protein synthesis have demonstrated that expression of c-*myb* mRNA is on a direct signalling pathway from the IL-2R (Churilla et al., 1989). These experiments led to the notion that c-*myb* plays a significant role during T cell activation, which was initially supported by studies that used c-*myb* antisense oligonucleotides to inhibit proliferation of cloned T cell lines (Gewirtz et al., 1989). However, the oligonucleotides used in these experiments contained four consecutive guanidine residues and have been reported to inhibit proliferation non-specifically (Burgess et al., 1995; Villa et al., 1995). Importantly, stimulation of fresh T cells or cloned T cell lines via the TCR without a source of exogenous IL-2 is not sufficient to induce expression of c-*myb*, but does induce apoptosis. This strongly suggests that c-Myb may be more important for survival than proliferation in peripheral T cells and is consistent with studies using MEnT (Pauza, 1987; Reed et al., 1986; Reed et al., 1987; Taylor et al., 1996). However, any potential role for c-Myb during activation of mature T cells, peripheral T cell differentiation or function remains undefined. Mice that carry c-*myb* loci targetted with *lox*P sites in combination with inducible Cre producing strains should allow insight into the role of c-Myb in peripheral T cell function.

During ontogeny, thymocyte precursors that migrate to the developing thymus are committed to lymphoid development but remain multipotent for lymphoid lineages. These initial migrants undergo one or two rounds of proliferation and begin to express surface proteins associated with the T cell lineage. Two days after initially seeding the thymus developing T cells begin to undergo the process of gene rearrangements that are necessary for production of the T cell antigen receptor (TCR). There are two major families of T cells defined by the type of antigen specific receptor they produce. The predominate type carry α/β T cell receptors and the process of α/β T cell development is best understood. The second family is referred to as γ/δ T cells and development of these cells is less well characterised. Both types of receptors are heterodimers and are encoded by separate sets of genes (Fehling et al., 1999). During ontogeny, γ/δ T cells are the first to be produced and are the predominant type of T cell in the foetal thymus. Foetal thymic γ/δ T cells are produced in two major waves that home to different organs and are characterised by distinct patterns of V-gene usage. After birth, α/β T cells are by far the predominate type of T cell made in the thymus throughout adult life. Differentiation along the γ/δ T cell lineage is less well characterised but commitment to the γ/δ-lineage appears to take place at the DN stage and precludes differentiation along the α/β pathway. Transcriptional enhancers associated with both TCRγ and TCRδ loci are among the best characterised targets of c-Myb (Hernandez-Munain et al., 1996; Hernandez-Munain and Krangel, 1995; Hernandez-Munain and Krangel, 2002; Hsiang et al., 1995). While less is understood about the consequences of TCRγ enhancer regulation by c-Myb, experiments utilising a TCRδ minilocus in transgenic mice reported greatly suppressed V(D)J recombination when the c-Myb binding site was mutated (Hernandez-Munain et al., 1996). Whether c-Myb modulates the efficiency of V(D)J recombination at other TCR loci is currently being examined.

T cell development in the thymus has been defined in terms of CD4 and CD8 expression and the status of rearrangement at the TCR loci (see Figure 2). The earliest T cell precursor that migrates to the thymus lacks expression of the CD4 and CD8 co-receptors and this is referred to as the double negative (DN) stage of T cell development. Since we were able to detect very immature $CD44^{+/lo}$ $CD25^{-}$ thymocyte precursors in thymi of c-*myb*$^{-/-}$/*Rag1*$^{-/-}$ chimaeric mice this suggests that the production and migration of T cell precursors to the thymus may not be c-Myb-dependent, however, c-Myb may be required for transition to stages that are committed to unipotential T cell development or for efficient expansion or survival of these early precursors (Allen et al., 1999). The DN stage is further subdivided based on expression of the CD44 and CD25 surface markers (Godfrey et al., 1993; Godfrey and Zlotnik, 1993). The most immature DN thymocytes are $CD44^{+/lo}$ $CD25^{-}$ cells that retain the potential to produce T cells, B cells, NK cells and dendritic cells (Guidos et al., 1989a; Guidos et al., 1989b; Shortman and Wu, 1996). These cells commit to unipotential T cell development, become $CD44^{+}$ $CD25^{-}$ (DN1), further mature into $CD44^{+}$ $CD25^{+}$ (DN2) thymocytes and begin to rearrange the TCRβ locus (Godfrey et al., 1993). The process of TCRβ rearrangement continues into the $CD44^{-}$ $CD25^{+}$ DN3 stage. c-Myb has been implicated in the regulation of *Rag2* gene expression during T cell development via a regulatory region that appears to be distinct from that involved in c-Myb-mediated regulation of expression during B cell development (Wang et al., 2000). However, this region is more likely involved in the re-expression of *Rag2* during the CD4/CD8 DP stage than in DN T cells. It is interesting to note that TCRβ V-segment promoters often contain consensus c-Myb binding sequences and it is possible that c-Myb might be involved in regulating germ line Vβ-segment transcription that is associated with active Vβ→DJβ recombination. When a functional Vβ-segment is formed TCRβ is expressed on the surface in association with pre-Tα and the CD3 elements of the TCR complex (Aifantis et al., 1999; Saint-Ruf et al., 1994; von Boehmer et al., 1999). Signalling through the pre-TCR complex initiates the process of β-selection, which is analogous to signalling through the pre-BCR, and results in down regulation of the RAG proteins, proliferation, and differentiation to the DP stage (von Boehmer et al., 1999). Similar to components of the pre-BCR (see above), pre-TCRα regulatory sequences have been reported to be a target of c-Myb (Reizis and Leder, 2001). Thus, failure to move from DN3 to DN4 could reflect failure to form the pre-TCR. Studies using MEnT suggest that interfering with Myb activity does not suppress pre-TCRα expression but results in both decreased proliferation at the DN3/DN4 transition and expression of cyclins A2, D3 and B1 (Pearson and Weston, 2000). Our preliminary analysis of T cell development in c-*myb* deficient thymocytes suggests that a strong block to differentiation beyond the DN3 stage may lie upstream of β-selection and we do not find changes in expression of cyclins. Thus, c-Myb may be required at multiple points during the early stages of T cell development.

Upon transition to the DP stage developing T cells begin the process of TCRα gene rearrangement. Cells that make a functional TCRα rearrangement begin to express low levels of α/β TCR on the surface and, after further maturation (see below), express either CD4 or CD8 on the surface and are referred to as single positive (SP) thymocytes. After assembly and surface expression of the α/β TCR, maturing T cells become restricted for antigen recognition by the self major histocompatibility complex (MHC) through the process of positive and negative selection (Robey and Fowlkes, 1994). This process selects for T cells that can interact with self MHC molecules and eliminates cells that interact strongly with self peptides presented in the context of self MHC antigens. Repertoire selection begins at the DP stage when maturing T cells express both CD4 and CD8 and TCRα/β on the cell membrane. During positive selection, T cells expressing antigen receptors that do not interact with self MHC molecules die by apoptosis. Subsequently, T cells with strong affinity for self peptides in the context of self MHC antigen are eliminated by apoptosis. During this process, T cells that are restricted for antigen recognition by class I MHC molecules cease expression of CD4 and continue to express the CD8 co-receptor. Double positive T cells that are restricted by class II MHC complexes continue to express only the CD4 co-receptor. After further maturation in the thymus SP thymocytes migrate to the periphery. The abundant c-*myb* expression in DP thymocytes intriguingly places c-*myb* at the point during T cell development where repertoire selection takes place. Interestingly, among the best characterised genes that are thought to be targets of c-Myb, such as CD4 (Allen, III et al., 2001; Siu et al., 1992) and the adenosine deaminase (ADA) thymic control locus (Ess et al., 1995), are important for T cell development and function. At this point it is unclear what role, if any, c-Myb plays during repertoire selection but several possibilities can be envisioned. First, in the absence of c-Myb ADA may be poorly expressed if at all, and mice that lack ADA are deficient in DP and SP thymocytes (Apasov et al., 2001; Blackburn et al., 1998). Second, the association between c-Myb and survival makes it interesting to suggest that developing T cells may not be able to efficiently receive or process signals for positive selection (Badiani et al., 1994; Taylor et al., 1996). Third, if c-Myb is involved in regulating *Rag2* expression during the DP stage, $V_\alpha \rightarrow J_\alpha$ rearrangement may be suppressed in the absence of c-Myb. Fourth, during V(D)J rearrangement at the TCRα locus, rearrangements first take place into the Vα proximal J segments (Roth et al., 1991; Thompson et al., 1990). However, aberrant VαJα rearrangements can be rescued by further rearrangement events to downstream Jα segments (McGargill et al., 2000; Wang et al., 1998; Yannoutsos et al., 2001). Poor survival in the DP compartment might limit the window during which further rearrangements at the TCRα locus can take place (Guo et al., 2002). Finally, the CD4 locus has been reported to be a target of both positive and negative regulation by c-Myb (Allen et al., 2001; Siu et al., 1992). We have noted an inverted ratio of CD4 to CD8 SP cells in thymocytes that develop in the absence of c-Myb and it is interesting to speculate that differentiation of CD4 SP thymocytes may be suppressed in the absence of c-Myb. However, we do not detect any difference in the level of CD4 surface expression in DP or CD4 SP cells in c-Myb deficient thymocytes.

4. FUTURE PROSPECTS

The fascinating pattern of c-*myb* expression in lymphocytes has long suggested a significant role(s) for c-Myb during lymphocyte development and reports of putative targets of c-Myb activity have suggested that c-Myb may play diverse roles in mediating proliferation, survival and differentiation. However, the lack of a tractable genetic model has greatly impeded gaining insight into c-Myb function during lymphocyte development or in identifying or confirming physiologically relevant c-Myb targets. Tissue specific and conditional deletion at the c-*myb* locus provides a significant new model to understand c-Myb function during lymphocyte development and has already allowed a glimpse into these activities. Future experiments should allow parsing of the relative contribution of c-Myb to proliferation, survival and differentiation during lymphocyte development and mature lymphocyte effector function. These models will also allow insight into c-Myb function in other haemopoietic lineages as well as non-hematopoietic systems. Furthermore, the ability to delete c-*myb* at will, should allow the identification of novel, physiologically relevant targets of c-Myb activity as well as allow assignment of specific and potentially overlapping roles for the other Myb family members in the near future.

ACKNOWLEDGEMENTS

I thank Dr. K. Rajewsky for his generous encouragement, support and collaboration in producing and analysing the conditionally targetted c-*myb* mice as well as Christopher Kremer, Matthew Thomas and Amanda Duley, Manfred Kraus and Thorsten Buch for help and discussions. This work was supported in part by a United States Public Health Service grant (CA85842) to T.P.B.

REFERENCES

Aifantis, I., Feinberg, J., Fehling, H.J., Di Santo, J.P. and von Boehmer, H. (1999) Early T cell receptor beta gene expression is regulated by the pre-T cell receptor-CD3 complex. J Exp Med *190*, 141-144.

Akashi, K., Traver, D., Miyamoto, T. and Weissman, I.L. (2000) A clonogenic common myeloid progenitor that gives rise to all myeloid lineages. Nature. *404*, 193-197.

Allen, R.D., Bender, T.P. and Siu, G. (1999) c-Myb is essential for T cell development. Genes Dev *13*, 1073-1078.

Allen, R.D., Kim, H.K., Sarafova, S.D. and Siu, G. (2001) Negative regulation of CD4 gene expression by a HES-1-c-Myb complex. Mol Cell Biol *21*, 3071-3082.

Allen, R.D., III, Kim, H.K., Sarafova, S.D. and Siu, G. (2001) Negative regulation of CD4 gene expression by a HES-1-c-Myb complex. Mol. Cell. Biol. *21*, 3071-3082.

Allman, D., Lindsley, R.C., DeMuth, W., Rudd, K., Shinton, S.A. and Hardy, R.R. (2001) Resolution of three nonproliferative immature splenic B cell subsets reveals multiple selection points during peripheral B cell maturation. J Immunol *167*, 6834-6840.

Apasov, S.G., Blackburn, M.R., Kellems, R.E., Smith, P.T. and Sitkovsky, M.V. (2001) Adenosine deaminase deficiency increases thymic apoptosis and causes defective T cell receptor signaling. J Clin Invest *108*, 131-141.

Arsura, M. and Sonenshein, G.E. (2001) The roles of the c-*myc* and c-*myb* oncogenes in hematopoiesis and leukemogenesis. In Transcription Factors: Normal and Malignant Development of Blood Cells, K.Ravid and J.Licht, eds. (New York: Wiley-Liss), pp. 521-549.

Badiani, P., Corbella, P., Kioussis, D., Marvel, J. and Weston, K. (1994) Dominant interfering alleles define a role for c-Myb in T-cell development. Genes Dev *8*, 770-783.

Bender, T.P. and Kuehl, W.M. (1987) Differential expression of the c-myb proto-oncogene marks the pre-B cell/B cell junction in murine B lymphoid tumors. J Immunol *139*, 3822-3827.

Bender, T.P., Thompson, C.B. and Kuehl, W.M. (1987) Differential expression of c-myb mRNA in murine B lymphomas by a block to transcription elongation. Science *237*, 1473-1476.

Blackburn, M.R., Datta, S.K. and Kellems, R.E. (1998) Adenosine deaminase-deficient mice generated using a two-stage genetic engineering strategy exhibit a combined immunodeficiency. J Biol Chem *273*, 5093-5100.

Brown, K.E., Kindy, M.S. and Sonenshein, G.E. (1992) Expression of the c-myb proto-oncogene in bovine vascular smooth muscle cells. J Biol Chem *267*, 4625-4630.

Burgess, T.L., Fisher, E.F., Ross, S.L., Bready, J.V., Qian, Y.X., Bayewitch, L.A., Cohen, A.M., Herrera, C.J., Hu, S.S., Kramer, T.B. and et al (1995) The antiproliferative activity of c-myb and c-myc antisense oligonucleotides in smooth muscle cells is caused by a nonantisense mechanism. Proc Natl Acad Sci USA *92*, 4051-4055.

Catron, K.M., Purkerson, J.M., Isakson, P.C. and Bender, T.P. (1992) Constitutive versus cell cycle regulation of c-myb mRNA expression correlates with developmental stages in murine B lymphoid tumors. J Immunol *148*, 934-942.

Chowdhury, D. and Sen, R. (2003) Transient IL-7/IL-7R signaling provides a mechanism for feedback inhibition of immunoglobulin heavy chain gene rearrangements. Immunity *18*, 229-241.

Churilla, A.M., Braciale, T.J. and Braciale, V.L. (1989) Regulation of T lymphocyte proliferation. Interleukin 2-mediated induction of c-myb gene expression is dependent on T lymphocyte activation state. J Exp Med *170*, 105-121.

Corcoran, A.E., Smart, F.M., Cowling, R.J., Crompton, T., Owen, M.J. and Venkitaraman, A.R. (1996) The interleukin-7 receptor alpha chain transmits distinct signals for proliferation and differentiation during B lymphopoiesis. EMBO J *15*, 1924-1932.

Ehlich, A., Schaal, S., Gu, H., Kitamura, D., Muller, W. and Rajewsky, K. (1993) Immunoglobulin heavy and light chain genes rearrange independently at early stages of B cell development. Cell *72*, 695-704.

Engel, K., Plath, K. and Gaestel, M. (1993) The MAP kinase-activated protein kinase 2 contains a proline- rich SH3-binding domain. FEBS Lett *336*, 143-147.

Ess, K.C., Whitaker, T.A., Cost, G.J., Witte, D.P., Hutton, J.J. and Aronow, B.J. (1995) A central role for a single c-Myb binding site in a thymic locus control region. Mol Cell Biol *15*, 5707-5715.

Ess, K.C., Witte, D.P., Bascomb, C.P. and Aronow, B.J. (1999) Diverse developing mouse lineages exhibit high-level c-Myb expression in immature cells and loss of expression upon differentiation. Oncogene *18*, 1103-1111.

Fehling, H.J., Gilfillan, S. and Ceredig, R. (1999) Alpha beta/gamma delta lineage commitment in the thymus of normal and genetically manipulated mice. Adv Immunol *71*, 1-76.

Feil, R., Wagner, J., Metzger, D. and Chambon, P. (1997) Regulation of Cre recombinase activity by mutated estrogen receptor ligand-binding domains. Bioch Biophys Res Comm *237*, 752-757.

Foos, G., Grimm, S. and Klempnauer, K.H. (1994) The chicken A-myb protein is a transcriptional activator. Oncogene *9*, 2481-2488.

Fugmann, S.D., Lee, A.I., Shockett, P.E., Villey, I.J. and Schatz, D.G. (2000) The RAG proteins and V(D)J recombination: complexes, ends, and transposition. Ann Rev Immunol *18*, 495-527.

Gewirtz, A.M., Anfossi, G., Venturelli, D., Valpreda, S., Sims, R. and Calabretta, B. (1989) G1/S transition in normal human T-lymphocytes requires the nuclear protein encoded by c-myb. Science *245*, 180-183.

Godfrey, D.I., Kennedy, J., Suda, T. and Zlotnik, A. (1993) A developmental pathway involving four phenotypically and functionally distinct subsets of CD3-CD4-CD8- triple-negative adult mouse thymocytes defined by CD44 and CD25 expression. J Immunol *150*, 4244-4252.

Godfrey, D.I. and Zlotnik, A. (1993) Control points in early T-cell development. Immunol Today *14*, 547-553.

Golay, J., Capucci, A., Arsura, M., Castellano, M., Rizzo, V. and Introna, M. (1991) Expression of c-myb and B-myb, but not A-myb, correlates with proliferation in human hematopoietic cells. Blood *77*, 149-158.

Golay, J., Cusmano, G. and Introna, M. (1992) Independent regulation of c-myc, B-myb, and c-myb gene expression by inducers and inhibitors of proliferation in human B lymphocytes. J Immunol *149*, 300-308.

Graf, T. (1992) Myb: a transcriptional activator linking proliferation and differentiation in hematopoietic cells. Curr Opin Genet Dev *2*, 249-255.

Grawunder, U., Leu, T.M., Schatz, D.G., Werner, A., Rolink, A.G., Melchers, F. and Winkler, T.H. (1995) Down-regulation of RAG1 and RAG2 gene expression in preB cells after functional immunoglobulin heavy chain rearrangement. Immunity *3*, 601-608.

Gu, H., Marth, J.D., Orban, P.C., Mossmann, H. and Rajewsky, K. (1994) Deletion of a DNA polymerase beta gene segment in T cells using cell type-specific gene targeting. Science *265*, 103-106.

Guidos, C.J., Weissman, I.L. and Adkins, B. (1989a) Developmental potential of CD4-8- thymocytes. Peripheral progeny include mature CD4-8- T cells bearing alpha beta T cell receptor. J Immunol *142*, 3773-3780.

Guidos, C.J., Weissman, I.L. and Adkins, B. (1989b) Intrathymic maturation of murine T lymphocytes from CD8+ precursors. Proc Natl Acad Sci USA *86*, 7542-7546.

Guo, J., Hawwari, A., Li, H., Sun, Z., Mahanta, S.K., Littman, D.R., Krangel, M.S. and He, Y.W. (2002) Regulation of the TCRalpha repertoire by the survival window of CD4(+)CD8(+) thymocytes. Nat Immunol *3*, 469-476.

Hardy, R.R. (2003) B-cell commitment: deciding on the players. Curr Opin Immun 15(2):158-65.

Hardy, R.R., Carmack, C.E., Shinton, S.A., Kemp, J.D. and Hayakawa, K. (1991) Resolution and characterization of pro-B and pre-pro-B cell stages in normal mouse bone marrow. J Exp Med *173*, 1213-1225.

Hayashi, S. and McMahon, A.P. (2002) Efficient recombination in diverse tissues by a tamoxifen-inducible form of Cre: a tool for temporally regulated gene activation/inactivation in the mouse. Dev Biol *244*, 305-318.

Henderson, A.J., Narayanan, R., Collins, L. and Dorshkind, K. (1992) Status of kappa L chain gene rearrangements and c-kit and IL-7 receptor expression in stromal cell-dependent pre-B cells. J Immunol *149*, 1973-1979.

Hernandez-Munain, C. and Krangel, M.S. (1995) c-Myb and core-binding factor/PEBP2 display functional synergy but bind independently to adjacent sites in the T-cell receptor delta enhancer. Mol Cell Biol *15*, 3090-3099.

Hernandez-Munain, C. and Krangel, M.S. (2002) Distinct roles for c-Myb and core binding factor/polyoma enhancer-binding protein 2 in the assembly and function of a multiprotein complex on the TCR delta enhancer in vivo. J Immunol *169*, 4362-4369.

Hernandez-Munain, C., Lauzurica, P. and Krangel, M.S. (1996) Regulation of T cell receptor delta gene rearrangement by c-Myb. J Exp Med *183*, 289-293.

Hesslein, D.G., Pflugh, D.L., Chowdhury, D., Bothwell ,A.L., Sen, R. and Schatz, D.G. (2003) Pax5 is required for recombination of transcribed, acetylated, 5' IgH V gene segments. Genes Dev *17*, 37-42.

Hsiang, H.H., Goldman, J.P. and Raulet, D.H. (1995) The role of c-Myb or a related factor in regulating the T cell receptor gamma gene enhancer. J Immunol *154*, 5195-5204.

Isakson, P.C., Purkerson, J., Catron, K. and Bender, T.P. (1991) A novel method for synchronizing a B cell lymphoma. J Immunol Meth *145*, 137-142.

Jacobs, S.M., Gorse, K.M. and Westin, E.H. (1994) Identification of a second promoter in the human c-myb proto- oncogene. Oncogene *9*, 227-235.

Jin, Z.X., Kishi, H., Wei, X.C., Matsuda, T., Saito, S. and Muraguchi, A. (2002) Lymphoid Enhancer-Binding Factor-1 Binds and Activates the Recombination-Activating Gene-2 Promoter Together with c-Myb and Pax-5 in Immature B Cells. J Immunol *169*, 3783-3792.

Kishi, H., Jin,Z.X., Wei, X.C., Nagata, T., Matsuda, T., Saito, S. and Muraguchi, A. (2002) Cooperative binding of c-Myb and Pax-5 activates the RAG-2 promoter in immature B cells. Blood *99*, 576-583.

Kitamura, D., Kudo, A., Schaal, S., Muller, W., Melchers, F. and Rajewsky, K. (1992) A critical role of lambda 5 protein in B cell development. Cell *69*, 823-831.

Kodama, H., Nose, M., Yamaguchi, Y., Tsunoda, J., Suda, T., Nishikawa, S. and Nishikawa, S. (1992) In vitro proliferation of primitive hemopoietic stem cells supported by stromal cells: evidence for the presence of a mechanism(s) other than that involving c-kit receptor and its ligand. J Exp Med *176*, 351-361.

Krangel, M.S. (2003) Gene segment selection in V(D)J recombination: accessibility and beyond. Nat Immunol 4(7):624-30.

Kuhn, R., Schwenk, F., Aguet, M. and Rajewsky, K. (1995) Inducible gene targeting in mice. Science *269*, 1427-1429.

Landreth, K.S., Kincade, P.W., Lee, G. and Harrison, D.E. (1984) B lymphocyte precursors in embryonic and adult W anemic mice. J Immunol *132*, 2724-2729.

Lee, P.P., Fitzpatrick, D.R., Beard, C., Jessup, H.K., Lehar, S., Makar, K.W., Perez-Melgosa, M., Sweetser, M.T., Schlissel, M.S., Nguyen, S., Cherry, S.R., Tsai, J.H., Tucker, S.M., Weaver, W.M., Kelso, A., Jaenisch, R. and Wilson, C.B. (2001) A critical role for Dnmt1 and DNA methylation in T cell development, function, and survival. Immunity. *15*, 763-774.

Li, Y.S., Hayakawa, K. and Hardy, R.R. (1993) The regulated expression of B lineage associated genes during B cell differentiation in bone marrow and fetal liver. J Exp Med *178*, 951-960.

Li, Y.S., Wasserman, R., Hayakawa, K. and Hardy, R.R. (1996) Identification of the earliest B lineage stage in mouse bone marrow. Immunity *5*, 527-535.

Lin, H. and Grosschedl, R. (1995) Failure of B-cell differentiation in mice lacking the transcription factor EBF. Nature. *376*, 263-267.

Loder, F., Mutschler, B., Ray, R.J., Paige, C.J., Sideras, P., Torres, R., Lamers, M.C. and Carsetti, R. (1999) B cell development in the spleen takes place in discrete steps and is determined by the quality of B cell receptor-derived signals. J Exp Med *190*, 75-89.

Loffert, D., Schaal, S., Ehlich, A., Hardy, R.R., Zou, Y.R., Muller, W. and Rajewsky, K. (1994) Early B-cell development in the mouse: insights from mutations introduced by gene targeting. Immunol Rev *137*, 135-153.

Marshall, A.J., Fleming, H.E., Wu, G.E. and Paige, C.J. (1998) Modulation of the IL-7 dose-response threshold during pro-B cell differentiation is dependent on pre-B cell receptor expression. J Immunol *161*, 6038-6045.

Martensson, A., Xie, X.Q., Persson, C., Holm, M., Grundstrom, T. and Martensson, I.L. (2001) PEBP2 and c-myb sites crucial for lambda5 core enhancer activity in pre-B cells. Eur J Immunol *31*, 3165-3174.

Martensson, I.L., Rolink ,A., Melchers, F., Mundt, C., Licence, S. and Shimizu, T. (2002) The pre-B cell receptor and its role in proliferation and Ig heavy chain allelic exclusion. Sem Immunol *14*, 335-342.

McGargill, M.A., Derbinski, J.M. and Hogquist, K.A. (2000) Receptor editing in developing T cells. Nat Immunol *1*, 336-341.

Melamed, D., Benschop, R.J., Cambier, J.C. and Nemazee, D. (1998) Developmental regulation of B lymphocyte immune tolerance compartmentalizes clonal selection from receptor selection. Cell *92*, 173-182.

Melchers, F., Rolink, A., Grawunder, U., Winkler, T.H., Karasuyama, H., Ghia, P. and Andersson, J. (1995) Positive and negative selection events during B lymphopoiesis. Curr Opin Immunol *7*, 214-227.

Melchers, F. and Rolink, A.G. (1998) B-lymphocyet development and biology. In Fundamentl Immunology, W.E.Paul, ed. (Philadelphia: Lippincott-Raven Publishers), pp. 183-224.

Melchers, F., ten Boekel, E., Yamagami,T ., Andersson, J. and Rolink, A. (1999) The roles of preB and B cell receptors in the stepwise allelic exclusion of mouse IgH and L chain gene loci. Semin Immunol *11*, 307-317.

Mettus, R.V., Litvin, J., Wali, A., Toscani, A., Latham, K., Hatton, K. and Reddy, E.P. (1994) Murine A-myb: evidence for differential splicing and tissue-specific expression. Oncogene *9*, 3077-3086.

Mucenski, M.L., McLain, K., Kier, A.B., Swerdlow, S.H., Schreiner, C.M., Miller, T.A., Pietryga, D.W., Scott, W.J., Jr. and Potter, S.S. (1991) A functional c-myb gene is required for normal murine fetal hepatic hematopoiesis. Cell *65*, 677-689.

Nemazee, D. (2000) Receptor editing in B cells. Adv Immunol *74*, 89-126.

Nomura, N., Takahashi, M., Matsui M., Ishii, S., Date, T., Sasamoto, S. and Ishizaki, R. (1988) Isolation of human cDNA clones of myb-related genes, A-myb and B- myb. Nucl Acids Res *16*, 11075-11089.

Nutt, S.L., Heavey, B., Rolink, A.G. and Busslinger, M. (1999) Commitment to the B-lymphoid lineage depends on the transcription factor Pax5. Nature. *401*, 556-562.

Nutt, S.L., Urbanek, P., Rolink, A. and Busslinger, M. (1997) Essential functions of Pax5 (BSAP) in pro-B cell development: difference between fetal and adult B lymphopoiesis and reduced V-to-DJ recombination at the IgH locus. Genes Dev *11*, 476-491.

Ogawa, M., Matsuzaki, Y., Nishikawa, S., Hayashi, S., Kunisada, T., Sudo, T., Kina, T., Nakauchi, H. and Nishikawa, S. (1991) Expression and function of c-kit in hemopoietic progenitor cells. J Exp Med *174*, 63-71.

Oh, I.H., Reddy, E.P. (1999) The myb gene family in cell growth, differentiation and apoptosis. Oncogene *18*, 3017-3033.

Osmond, D.G. (1991) Proliferation kinetics and the lifespan of B cells in central and peripheral lymphoid organs. Curr Opin Immunol *3*, 179-185.

Pauza, C.D. (1987) Regulation of human T-lymphocyte gene expression by interleukin 2: immediate-response genes include the proto-oncogene c-myb. Mol Cell Biol *7*, 342-348.

Pearson, R. and Weston, K. (2000) c-Myb regulates the proliferation of immature thymocytes following beta-selection. EMBO J *19*, 6112-6120.

Peschon, J.J., Morrissey, P.J., Grabstein, K.H., Ramsdell, F.J., Maraskovsky, E., Gliniak, B.C., Park, L.S., Ziegler, S.F., Williams, D.E. and Ware, C.B. (1994) Early lymphocyte expansion is severely impaired in interleukin 7 receptor-deficient mice. J Exp Med *180*, 1955-1960.

Ratajczak, M.Z., Perrotti, D., Melotti, P., Powzaniuk, M., Calabretta, B., Onodera, Kregenow, D.A., Machalinski, B. and Gewirtz, A.M. (1998) Myb and Ets proteins are candidate regulators of c-kit expression in human hematopoietic cells. Blood *91*, 1934-1946.

Ray, R.J., Stoddart, A., Pennycook, J.L., Huner, H.O., Furlonger, C., Wu, G.E. and Paige, C.J. (1998) Stromal cell-independent maturation of IL-7-responsive pro-B cells. J Immunol *160*, 5886-5897.

Reed, J.C., Alpers, J.D., Nowell, P.C. and Hoover, R.G. (1986) Sequential expression of protooncogenes during lectin-stimulated mitogenesis of normal human lymphocytes. Proc Natl Acad Sci USA *83*, 3982-3986.

Reed, J.C., Alpers, J.D., Scherle, P.A., Hoover, R.G., Nowell, P.C. and Prystowsky, M.B. (1987) Proto-oncogene expression in cloned T lymphocytes: mitogens and growth factors induce different patterns of expression. Oncogene *1*, 223-228.

Reiss, K., Travali, S., Calabretta, B. and Baserga, R. (1991) Growth regulated expression of B-myb in fibroblasts and hematopoietic cells. J Cell Physiol *148*, 338-343.

Reizis, B. and Leder, P. (2001) The Upstream Enhancer Is Necessary and Sufficient for the Expression of the Pre-T Cell Receptor alpha Gene in Immature T Lymphocytes. J Exp Med *194*, 979-990.

Rickert, R.C., Rajewsky, K. and Roes, J. (1995) Impairment of T-cell-dependent B-cell responses and B-1 cell development in CD19-deficient mice. Nature *376*, 352-355.

Rickert, R.C., Roes, J. and Rajewsky, K. (1997) B lymphocyte-specific, Cre-mediated mutagenesis in mice. Nucl Acids Res *25*, 1317-1318.

Robey, E. and Fowlkes, B.J. (1994) Selective events in T cell development. Ann Rev Immunol *12*, 675-705.

Rolink, A., Kudo, A., Karasuyama, H., Kikuchi, Y. and Melchers, F. (1991a) Long-term proliferating early pre B cell lines and clones with the potential to develop to surface Ig-positive, mitogen reactive B cells in vitro and in vivo. EMBO J *10*, 327-336.

Rolink, A., Streb, M., Nishikawa, S. and Melchers, F. (1991b) The c-kit-encoded tyrosine kinase regulates the proliferation of early pre-B cells. Eur J Immunol *21*, 2609-2612.

Rolink, A., ten Boekel, E., Melchers, F., Fearon, D.T., Krop, I. and Andersson, J. (1996) A subpopulation of B220+ cells in murine bone marrow does not express CD19 and contains natural killer cell progenitors. J Exp Med *183*, 187-194.

Rolink, A.G., Andersson, J. and Melchers, F. (1998) Characterization of immature B cells by a novel monoclonal antibody, by turnover and by mitogen reactivity. Eur J Immunol *28*, 3738-3748.

Rolink, A.G., Nutt, S.L., Melchers, F. and Busslinger, M. (1999) Long-term in vivo reconstitution of T-cell development by Pax5-deficient B-cell progenitors. Nature. *401*, 603-606.

Rolink, A.G., Schaniel, C., Andersson, J. and Melchers, F. (2001) Selection events operating at various stages in B cell development. Curr Opin Immunol *13*, 202-207.

Rolink, A.G., Winkler, T., Melchers, F. and Andersson, J. (2000) Precursor B cell receptor-dependent B cell proliferation and differentiation does not require the bone marrow or fetal liver environment. J Exp Med *191*, 23-32.

Roth, M.E., Holman, P.O. and Kranz, D.M. (1991) Nonrandom use of J alpha gene segments. Influence of V alpha and J alpha gene location. J Immunol *147*, 1075-1081.

Saint-Ruf, C., Ungewiss, K., Groettrup ,M., Bruno, L., Fehling, H.J. and von Boehmer, H. (1994) Analysis and expression of a cloned pre-T cell receptor gene. Science *266*, 1208-1212.

Schwenk, F., Baron, U. and Rajewsky, K. (1995) A cre-transgenic mouse strain for the ubiquitous deletion of loxP-flanked gene segments including deletion in germ cells. Nucl Acids Res *23*, 5080-5081.

Shipp, M.A. and Reinherz, E.L. (1987) Differential expression of nuclear proto-oncogenes in T cells triggered with mitogenic and nonmitogenic T3 and T11 activation signals. J Immunol *139*, 2143-2148.

Shortman, K. and Wu, L. (1996) Early T lymphocyte progenitors. Ann Rev Immunol *14*, 29-47.

Sica, A., Tan, T.H., Rice, N., Kretzschmar, M., Ghosh, P. and Young, H.A. (1992) The c-rel protooncogene product c-Rel but not NF-kappa B binds to the intronic region of the human interferon-gamma gene at a site related to an interferon-stimulable response element. Proc Natl Acad Sci USA *89*, 1740-1744.

Siu, G., Wurster, A.L., Lipsick, J.S. and Hedrick, S.M. (1992) Expression of the CD4 gene requires a Myb transcription factor. Mol Cell Biol *12*, 1592-1604.

Sleckman, B.P., Gorman, J.R. and Alt, F.W. (1996) Accessibility control of antigen-receptor variable-region gene assembly: role of cis-acting elements. Ann Rev Immunol *14*, 459-481.

Stanhope-Baker, P., Hudson, K.M., Shaffer, A.L., Constantinescu, A. and Schlissel, M.S. (1996) Cell type-specific chromatin structure determines the targeting of V(D)J recombinase activity in vitro. Cell *85*, 887-897.

Stern, J.B. and Smith, K.A. (1986) Interleukin-2 induction of T-cell G1 progression and c-myb expression. Science *233*, 203-206.

Tanaka, Y., Patestos, N.P., Maekawa, T. and Ishii, S. (1999) B-myb is required for inner cell mass formation at an early stage of development. J Biol Chem *274*, 28067-28070.

Taylor, D., Badiani, P. and Weston, K. (1996) A dominant interfering Myb mutant causes apoptosis in T cells. Genes Dev *10*, 2732-2744.

ten Boekel, E., Melchers, F. and Rolink, A.G. (1997) Changes in the V(H) gene repertoire of developing precursor B lymphocytes in mouse bone marrow mediated by the pre-B cell receptor. Immunity *7*, 357-368.

Thompson, C.B., Challoner, P.B., Neiman, P.E. and Groudine, M. (1986) Expression of the c-myb proto-oncogene during cellular proliferation. Nature *319*, 374-380.

Thompson, S.D., Pelkonen, J. and Hurwitz, J.L. (1990) First T cell receptor alpha gene rearrangements during T cell ontogeny skew to the 5' region of the J alpha locus. J Immunol *145*, 2347-2352.

Tiegs, S.L., Russell, D.M. and Nemazee, D. (1993) Receptor editing in self-reactive bone marrow B cells. J Exp Med *177*, 1009-1020.

Tonegawa, S. (1983) Somatic generation of antibody diversity. Nature *302*, 575-581.

Torres R.M., Flaswinkel, H., Reth, M. and Rajewsky, K. (1996) Aberrant B cell development and immune response in mice with a compromised BCR complex. Science *272*, 1804-1808.

Torres, R.M. and Kuhn, R. (1997) Laboratory protocols for conditional gene targeting. Oxford University Press).

Toscani, A., Mettus, R.V., Coupland, R., Simpkins, H., Litvin, J., Orth, J., Hatton, K.S. and Reddy, E.P. (1997) Arrest of spermatogenesis and defective breast development in mice lacking A-myb. Nature *386*, 713-717.

Toth, C.R., Hostutler, R.F., Baldwin, A.S., Jr. and Bender, T.P. (1995) Members of the nuclear factor kB family transactivate the murine c-*myb* gene. J Biol Chem *270*, 7661-7671.

Trauth, K., Mutschler, B., Jenkins, N.A., Gilbert, D.J., Copeland, N.G. and Klempnauer, K.H. (1994) Mouse A-myb encodes a trans-activator and is expressed in mitotically active cells of the developing central nervous system, adult testis and B lymphocytes. EMBO J *13*, 5994-6005.

Turner, M., Gulbranson-Judge, A., Quinn, M.E., Walters, A.E., MacLennan, I.C. and Tybulewicz, V.L. (1997) Syk tyrosine kinase is required for the positive selection of immature B cells into the recirculating B cell pool. J Exp Med *186*, 2013-2021.

Urbanek, P., Wang, Z.Q., Fetka, I., Wagner, E.F. and Busslinger, M. (1994) Complete block of early B cell differentiation and altered patterning of the posterior midbrain in mice lacking Pax5/BSAP. Cell *79*, 901-912.

Villa, A.E., Guzman, L.A., Poptic, E.J., Labhasetwar, V., D'Souza, S., Farrell, C.L., Plow, E.F., Levy, R.J., DiCorleto, P.E. and Topol, E.J. (1995) Effects of antisense c-myb oligonucleotides on vascular smooth muscle cell proliferation and response to vessel wall injury. Circ Res *76*, 505-513.

von Boehmer, H., Aifantis, I., Feinberg, J., Lechner, O., Saint-Ruf, C., Walter, U., Buer, J. and Azogui, O. (1999) Pleiotropic changes controlled by the pre-T-cell receptor. Curr Opin Immunol *11*, 135-142.

Wang, F., Huang, C.Y. and Kanagawa, O. (1998) Rapid deletion of rearranged T cell antigen receptor (TCR) Valpha-Jalpha segment by secondary rearrangement in the thymus: role of continuous rearrangement of TCR alpha chain gene and positive selection in the T cell repertoire formation. Proc Natl Acad Sci USA *95* , 11834-11839.

Wang, Q.F., Lauring, J. and Schlissel, M.S. (2000) c-Myb binds to a sequence in the proximal region of the RAG-2 promoter and is essential for promoter activity in T-lineage cells. Mol Cell Biol *20*, 9203-9211.

Yamagami, T., ten Boekel,E., Andersson, J., Rolink, A. and Melchers, F. (1999a) Frequencies of multiple IgL chain gene rearrangements in single normal or kappa L chain-deficient B lineage cells. Immunity *11*, 317-327.

Yamagami, T., ten Boekel, E., Schaniel, C., Andersson, J., Rolink, A. and Melchers, F. (1999b) Four of five RAG-expressing JCkappa-/- small pre-BII cells have no L chain gene rearrangements: detection by high-efficiency single cell PCR. Immunity *11*, 309-316.

Yancopoulos, G.D. and Alt, F.W. (1985) Developmentally controlled and tissue-specific expression of unrearranged VH gene segments. Cell *40*, 271-281.

Yannoutsos, N., Wilson, P., Yu, W., Chen, H.T., Nussenzweig, A., Petrie, H. and Nussenzweig, M.C. (2001) The role of recombination activating gene (RAG) reinduction in thymocyte development in vivo. J Exp Med *194*, 471-480.

Zhuang, Y., Soriano, P. and Weintraub, H. (1994) The helix-loop-helix gene E2A is required for B cell formation. Cell *79*, 875-884.

Chapter 5

INVOLVEMENT OF C-MYB IN RED CELL AND MEGAKARYOCYTE DEVELOPMENT

Alexandros Vegiopoulos, Nikla R. Emambokus[1], Jon Frampton
Institute of Biomedical Research, Birmingham University Medical School, Edgbaston, Birmingham, B15 2TT, United Kingdom, [1]Harvard Medical School, 320 Longwood Avenue, Boston MA 02115, United States of America.

Abstract: The cell lineages that give rise to erythrocytes and platelets derive from a common progenitor. Several transcription factors are known to be involved in the control of differentiation along these two pathways, although it is unclear what transcriptional regulatory mechanisms operate during the commitment decision or to distinguish one lineage from the other. Historically, c-Myb has been thought to block erythroid differentiation at the stage of the committed precursor and to have no role in megakaryocytopoiesis. More recent data, especially that derived from novel engineered alleles of c-*myb*, indicate that c-Myb is important for the differentiation along both the erythroid and megakaryocytic lineages, and that the level of the protein may serve to control progression through differentiation and perhaps the commitment choice.

1. INTRODUCTION

Although red cells and platelets are very distinct cell types they have a common precursor during their development. This relationship is reflected in many of the transcriptional processes regulating their differentiation and the expression of key functional molecules. Of the mammalian *myb* genes only c-*myb* has been studied to any degree within this branch of haemopoiesis and this has largely involved work with model erythroid cell lines. This chapter will focus on the known and likely functions of c-Myb in the erythroid and megakaryocytic lineages and will discuss recent developments using mouse genetic tools and primary cell culture. In addition to describing the newer data that these systems have highlighted with respect to c-Myb function in erythroid and megakaryocytic cells we

J. Frampton (ed.), Myb Transcription Factors: Their Role in Growth, Differentiation and Disease, 107-131.

will also consider the prospects for their future use in defining the mechanisms of action of the factor.

2. OVERVIEW OF ERYTHROPOIESIS AND MEGAKARYOCYTOPOIESIS

Erythrocytes (red cells) and thrombocytes (platelets in mammals) carry out vital and very specialised functions in all vertebrates. Erythrocytes are responsible for the transport of oxygen and carbon dioxide between the lungs and the organs, while thrombocytes act in the repair of damaged blood vessels and in blood clotting. In mammals, both cell types lack a nucleus and have to be produced from immature dividing progenitor cells. Haemopoiesis is the process by which multipotential stem cells residing in specialised organs proliferate and differentiate to produce all blood cell types. The production of erythrocytes and megakaryocytes, the platelet-forming cells, from stem cells involves multiple intermediate precursor cell types. The haemopoietic stem cell (HSC) gives rise to so-called common lymphoid and myeloid progenitors (CLP and CMP, Akashi et al., 2000). The CMP is capable of giving rise to granulocytes and macrophage as well as erythrocytes and megakaryocytes. Accumulating evidence implies the existence of a megakaryocyte-erythroid progenitor (MEP), deriving from the CMP, which can give rise to cells of both the erythroid and megakaryocytic lineages (Debili et al., 1996; Vannucchi et al., 2000; Akashi et al., 2000).

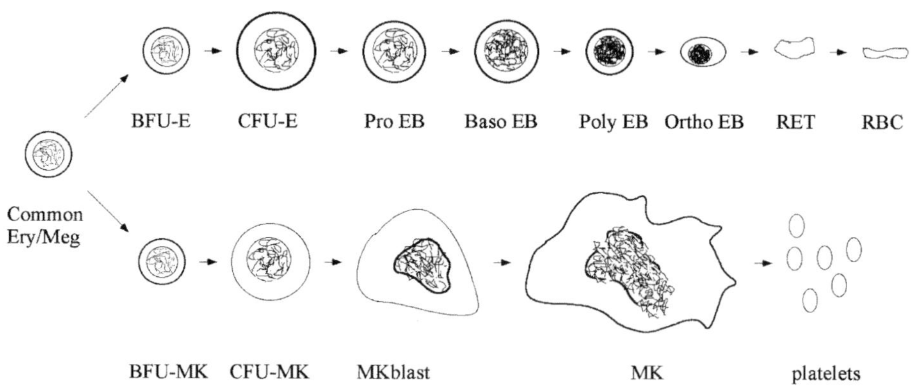

Figure 1

Cellular stages of erythroid and megakaryocytic differentiation. Pro EB: Proerythroblast. Baso EB: Basophilic erythroblast. Poly EB: Polychromatophilic erythroblast. Ortho EB: Orthochromatic erythroblast. RET: Reticulocyte. RBC: Erythrocyte. MKblast: Megakaryoblast. MK: Megakaryocyte.

Commitment of the MEP towards specific lineage differentiation can be defined through morphological and functional criteria (Figure 1). The most immature committed progenitor of the erythroid lineage is the burst-forming unit (BFU-E) that has a high proliferative potential. The direct progeny of the BFU-E, the colony-forming unit erythroid (CFU-E), is more differentiated and gives rise to proerythroblasts that in turn mature into erythroblasts. Erythroblasts undergo a series of terminal cell divisions in association with the maturation process, which is characterised by condensation of the nucleus and a reduction in size and results in the formation of enucleated reticulocytes that finally develop into functional erythrocytes. The terminal differentiation of erythrocytes is also marked by the intensive synthesis of several proteins including the oxygen carrying globins, carbonic anhydrase and specific cytoskeletal components. Similar to the erythroid lineage, the first committed megakaryocytic precursor is termed the BFU-MK and gives rise to CFU-MK that in turn differentiate to megakaryoblasts. Up to this latter stage, the megakaryocytic precursors expand by mitotic cell divisions. Thereafter, however, cellular division stops even though DNA replication continues, giving rise to polyploid cells that contain up to 64 times the normal amount of DNA in their multilobular nuclei (Levine, 1980). This process is called endoreplication and within haemopoiesis is a feature characteristic of the maturing megakaryocyte. Mature megakaryocytes are large in size and form cytoplasmic extensions, the proplatelets, which are released to the blood circulation as functional platelets. As megakaryocytic cells differentiate, expression of specific proteins contributing to platelet function increases. These specific proteins include molecules such as the GPIIb/GPIIIb (CD41/CD61) fibrinogen receptor and the GPIb/IX/V (CD42) receptor for the von Willebrand factor.

3. TRANSCRIPTIONAL REGULATION OF ERYTHROID AND MEGAKARYOCYTIC CELL DEVELOPMENT

The processes of commitment, proliferation and differentiation during haemopoiesis are coupled and precisely regulated at the molecular level. Haemopoietic cells process extrinsic signals from their microenvironment and integrate these into their intrinsic genetic program through modulation of transcription factor activity. These factors have complex expression patterns within the haemopoietic system and function in concert in multiprotein complexes that bind to gene regulatory elements to drive gene transcription. Experiments employing targetted gene disruption in mice have been primary in the demonstration of the importance of specific transcription factors in haemopoiesis. In terms of the regulation of the erythroid and megakaryocytic lineages, such gene knockout studies have indicated an essential role for several transcription factors at particular stages of differentiation. Although loss of some factors has an effect that is restricted to one or other lineage, the underlying similarity between many of the mechanisms regulating erythropoiesis and megakaryopoiesis is reflected in the fact that some transcriptional regulators have an essential function in both lineages. The results of experiments, including gene knockout studies, on many of the key transcription factors that have an influence on erythroid or megakaryocytic differentiation are summarised below.

Targetted disruption of the GATA-1 gene affects both the erythroid and megakaryocytic lineages. Embryos lacking the GATA-1 gene die around E10.5 due to severe anaemia (Fujiwara et al., 1996) caused by defective maturation and apoptosis of erythroblasts (Weiss and Orkin, 1995). GATA-1 acts by positively regulating the transcription of numerous erythroid genes including regulatory genes such as the erythropoietin receptor, the erythroid transcription factor EKLF and the anti-apoptotic factor Bcl-xL (Zon et al., 1991; Crossley et al., 1994; Gregory et al., 1999). A serendipitous lineage-selective knockout also revealed a role for GATA-1 in megakaryocytes in that cells lacking the factor hyperproliferate and show impaired endoreplication and maturation resulting in lower platelet numbers (Shivdasani et al., 1997; Vyas et al., 1999). One of interaction partners of GATA-1 is FOG-1, a multi-zinc-finger protein co-expressed with GATA-1 (Tsang et al., 1997). Targetted mutation of a residue in GATA-1 known to be crucial for the interaction between the two proteins demonstrated the dependence on FOG-1 for correct functioning of GATA-1 (Crispino et al., 1999). The dependence of GATA-1 on FOG-1 was also revealed in FOG-1$^{-/-}$ embryos, which also die of anaemia due to a block of erythroid development at the proerythroblast stage (Tsang et al., 1998). In contrast to the GATA-1 knockout, however, FOG-1$^{-/-}$ embryos exhibit a complete absence of megakaryocytopoiesis. This difference has recently been explained by showing that GATA-2, which is also partnered by FOG-1, can perform much the same role as GATA-1 in early megakaryopoiesis (Chang et al., 2002). The implication that GATA-1 plays a role in early parts of the megakaryocyte lineage, even if redundantly with GATA-2 to some extent, is consistent with experiments in which GATA-1 was able to induce erythroid/megakaryocytic or multilineage differentiation when ectopically expressed in committed myeloid cells (Kulessa et al., 1995; Heyworth et al., 2002). A role for GATA-1 in erythroid lineage specification is also suggested from the functional antagonism that is seen between GATA-1 and the Ets transcription factor PU.1 (Rekhtman et al., 1999). PU.1, which is required for myeloid and lymphoid development, interacts directly with GATA-1 and in so doing can inhibit erythroid differentiation.

A second transcription factor of central importance for erythroid development is the basic helix-loop-helix protein SCL. Although SCL is required for the formation of blood at the earliest embryonic stages of haemopoiesis (Robb et al., 1995; Shivdasani et al., 1995a) it also has been shown to act during erythroid differentiation. Over expression of SCL in human immature haemopoietic cells enhances the formation and differentiation of erythroid and megakaryocytic precursors (Elwood et al., 1998). Consistent with this, inducible targetted deletion of the SCL gene in adult mice led to a complete loss of functional BFU-E/CFU-E and BFU-MK/CFU-MK progenitors (Hall et al., 2003; Mikkola et al., 2003). Interestingly, DNA motifs recognised by SCL, the so-called E-boxes, are often located in the proximity of GATA-binding motifs in erythoid-specific *cis*-regulatory elements. Experiments performed by Wadman et al (1997) suggest the existence of a pentameric protein complex bound to such sites, including SCL and its heterodimeric partner E2A, GATA-1 and an LMO-2/Lbd1 dimer.

A transcription factor that has been implicated in erythroid differentiation primarily through its activating effect on globin gene expression is EKLF (erythroid Kruppel-like factor). EKLF is essential for proper globin gene expression in erythroid cells as knockout embryos die from a beta-thalassaemia-like anaemia (Nuez et al., 1995; Perkins et al., 1995). Although activation of transcription of adult-type β-globin genes might be the primary cause underlying the mutant phenotype, an involvement of EKLF in maturation and inhibition of proliferation was also demonstrated using cells immortalised from EKLF$^{-/-}$ embryos (Coghill et al., 2001).

The basic leucine zipper protein NF-E2 is involved in megakaryocytic differentiation but also seems to influence erythroid development. NF-E2 heterodimerizes with MafG or MafK, members of the small Maf protein family, and expression levels of these proteins are crucial for correct megakaryocyte differentiation (Motohashi et al., 2000). Consistent with this, the phenotypes of NF-E2$^{-/-}$ and mafG/mafK double knockout mice are very similar in that they lack platelets due to a failure of mature megakaryocytes to form and release proplatelets (Shivdasani and Orkin, 1995; Shivdasani et al., 1995b; Onodera et al., 2000). In addition, these mice display a mild anaemia that was attributed to a reduced response of red blood cells to oxidative stress (Chan et al., 2001).

The promoters of many megakaryocyte-expressed genes contain adjacent GATA and Ets binding elements that appear to function in the regulation of transcription. Among the known Ets factors, Fli-1 appears to be prominent in regulation of megakaryocytic differentiation. The consensus of the phenotypes of knockout mouse strains generated by different laboratories is that Fli-1 is necessary for normal megakaryocytic development and efficient platelet production, however, it is not clear which stages of megakaryopoiesis are affected (Hart et al., 2000; Spyropoulos et al., 2000; Kawada et al., 2001). A second conclusion drawn from these studies is the requirement for Fli-1 for the generation of normal numbers of erythroid progenitors. This is probably not surprising since Fli-1 over expression is associated with murine erythroleukaemia and can also enhance proliferation of primary erythroblasts while inhibiting their differentiation (Pereira et al., 1999; Lesault et al., 2002).

With the exception of SCL, most of the transcription factors presented above promote differentiation along the erythroid or megakaryocytic lineages. Two proteins that have been shown to be essential for proliferation of immature haemopoietic progenitors are the transcription regulators c-Myb and GATA-2. Generally, expression of both proteins peaks in dividing precursors of all haemopoietic lineages and is down regulated during differentiation. In fact, down-regulation seems to be a prerequisite for differentiation in several cell types. Within the erythro-megakaryocytic part of the haemopoietic hierarchy over-expression of GATA-2 can lead to an arrest of erythroid differentiation while increased megakaryocytic development was observed in one set of experiments implicating GATA-2 in the regulation of commitment to this lineage (Briegel et al., 1993; Ikonomi et al., 2000). Targetted disruption of GATA-2 causes an embryonic lethal phenotype with a general defect in haemopoiesis due to the inability of multipotential progenitors to expand efficiently and survive in response to cytokines although terminal erythroid differentiation appeared to be intact (Tsai and Orkin, 1997). Mice homozygous for a disrupted c-*myb* allele have normal numbers of primitive (yolk sac-derived) erythrocytes but mainly non-functional haemopoietic progenitors in the foetal liver and die around E15 due to severe anaemia (Mucenski et al., 1991; Sumner et al., 2000).

While there are some parallels in the knockout phenotypes of GATA-2 and c-*myb*, probably reflecting the requirement for these transcription factors in the expansion of immature haemopoietic progenitors, recent results indicate that c-Myb influences both commitment of bipotential erythro-megakaryocytic progenitors and the later stages of erythroid development.

4. THE INVOLVEMENT OF MYB PROTEINS IN ERYTHROID AND MEGAKARYOCYTIC CELLS

The expression pattern of A-Myb and the phenotype of A-*myb*$^{-/-}$ mice suggest that there is no function for this protein in the development of the erythroid or megakaryocytic lineages. In contrast, the wide expression of B-Myb in proliferating cells and its down regulation upon differentiation (Reiss et al., 1991; Sitzman et al., 1996) has certainly been seen in haemopoietic cells lines. Indeed, experiments employing antisense oligonucleotides indicate that B-Myb is essential for proliferation of certain myeloid and lymphoid cell lines (Arsura et al., 1992; Reiss et al., 1991). However, the failure of B-*myb*$^{-/-}$ embryos to develop beyond the blastocyst stage (Tanaka et al., 1999) does not allow any conclusions relevant either to haemopoiesis in general or more specifically to erythropoiesis and megakaryocytopoiesis. Similarly, c-Myb is abundantly expressed in immature blood cell types and is down regulated upon differentiation (Kastan et al, 1989). Examination of c-*myb* RNA levels in primary erythroid cells revealed a peak of mRNA levels in CFU-E and early erythroblast stages (Emilia et al., 1986; Valtieri et al., 1991), suggesting that c-Myb is relevant to the development of the erythroid lineage. Consistent with this, c-*myb* knockout embryos lack definitive nucleated erythrocytes (Mucenski et al., 1991), however, since almost all haemopoietic definitive lineages are affected by the targetted disruption of c-*myb*, it is reasonable to assume that this is merely a downstream consequence of a defect in immature multipotential progenitors. Indeed, analysis of *in vitro* differentiated c-*myb*$^{-/-}$ ES cells and a detailed examination of mutant embryos led to the conclusion that although definitive immature progenitors are generated in the absence of c-Myb they are not functional (Clarke et al., 2000; Sumner et al., 2000). Although there are a number of significant differences between primitive and definitive erythrocytes, the fact that primitive yolk sac erythropoiesis is normal in c-*myb*$^{-/-}$ embryos suggests that a transcriptional programme leading to the formation of functional red blood cells can, in principle, be executed in the absence of c-Myb function.

Unlike enucleated definitive erythrocytes, primitive nucleated red cells do not arise from a BFU-E progenitor and are generated independently of SCF and erythropoietin (Russell, 1979; Palis et al., 1999; Lee et al., 2001). An important role for c-Myb in BFU-E progenitors was suggested by the reduction in both the frequency and size of erythroid colonies when c-*myb* antisense oligonucleotides were applied to human bone marrow *in vitro* (Gewirtz and Calabretta, 1988; Caracciolo et al., 1990). Moreover, when purified erythroid progenitors were used, CFU-E progenitors were affected to a greater extent than BFU-Es and this effect also correlated with proliferation (Valtieri et al., 1991). Thus, it became apparent that c-Myb functions not only in BFU-E progenitors, but also in intermediate-late erythroid cells, at least with respect to proliferation. As discussed in the next section, the requirement for particularly high expression levels of c-Myb in certain stages of erythropoiesis suggests that the protein is most likely involved in several cellular processes.

Originally, it was concluded that megakaryocytopoiesis is not strongly affected by the disruption of the c-*myb* gene since megakaryocytes were a prominent cell type in the foetal liver of c-*myb*$^{-/-}$ embryos (Mucenski et al., 1991). A closer examination, however, revealed a reduced total number of megakaryocytes in comparison to the wild type foetal liver. c-*myb*$^{-/-}$ foetal liver megakaryocytes appeared to be fully mature and, although it still remains to be formally proven, they are likely to be producing functional platelets because the embryos exhibited no signs of hemorrhaging (Sumner et al., 2000). The implication of these observations is that the absence of c-Myb does not prevent terminal differentiation but still leaves open the possibility that earlier stages of megakaryocytic development are defective, at least in the extent of proliferation that is possible. Recent studies using a knockdown allele of c-*myb* described later in this chapter shed some light on this issue.

5. C-MYB FUNCTION IN INTERMEDIATE-LATE ERYTHROID PRECURSORS: ERYHTROLEUKAEMIC CELL MODELS

The difficulty in purifying large numbers of primary erythroid cells has hindered extensive experimentation on the role of c-Myb in this lineage, however, erythroleukaemic cell lines resembling pro-erythroblasts have provided a useful model. Murine erythroleukaemic (MEL) cell lines obtained from mice with leukaemia induced by the Friend virus can be differentiated by stimulation with chemical agents such as dimethyl sulphoxide (DMSO) or hexamethylene bisacetamide (HMBA). Numerous studies have assessed expression levels of c-*myb* RNA during induced differentiation of MEL cells and although there has been some controversy, some conclusions could be drawn (Kirsch et al., 1986; Ramsay et al., 1986). Thus, in the course of chemically induced differentiation, c-*myb* RNA levels often display a biphasic mode of regulation, in that a rapid decrease is followed by a second peak before a decline associated with terminal differentiation. In contrast, when erythropoietin was used to induce differentiation of responsive erythroid cell lines, a rapid and steady decline in c-*myb* mRNA was observed (Chern et al., 1991a). Although the difference between the simple decline and biphasic expression pattern is not understood, it is evident from over expression experiments that c-Myb down regulation is a prerequisite for induced differentiation to take place (Clarke et al., 1988; Tokodoro et al., 1988). Furthermore, using a c-Myb expression vector driven by a globin gene promoter in MEL cells it was demonstrated that in contrast to the late decline in c-*myb* expression, the early down regulation during commitment of the cells is not crucial for differentiation (McClinton et al., 1990; Danish et al., 1992). As expected from the over expression results, loss-of-function experimental strategies using either c-*myb* antisense oligonucleotides or a dominant-negative interfering Myb led to spontaneous differentiation of MEL cells to a degree similar to that seen using chemical inducers (Chern et al., 1991b; Chen et al., 2002).

In conclusion, the MEL cell model provides evidence of a role for c-Myb in the maintenance of an immature, proliferating state in intermediate erythroid precursors. However, due to the transformed state of erythroleukaemic cells it is difficult to investigate the pathways through which c-Myb is regulating proliferation versus differentiation and how these two processes might be connected. Furthermore, these cell models cannot be used to address the potential influence of c-Myb on specific lineage commitment of the bipotent erythro-megakaryocytic progenitors.

6. ONCOGENICALLY ACTIVATED C-MYB IS DIRECTLY INVOLVED IN ERYTHROID AND THROMBOCYTIC CELL TRANSFORMATION

The avian acute leukaemia virus E26 contains sequences encoding a fusion protein derived from the c-*myb* and c-*ets*-1 genes that is responsible for transformation and leukaemogenesis (Lipsick and Wang, 1999). The c-Myb-related sequences (v-Myb) were originally presumed to be involved in transformation of erythroid cells since E26 was thought to elicit an erythroleukaemia in chickens (Radke et al., 1982). However, it was subsequently discovered that cells transformed by E26 represent multipotential haemopoietic progenitors (Graf et al., 1992). Interestingly, selective inactivation or ablation of the two components of the fusion protein demonstrated that it is primarily the Ets sequence that accounts for inhibition of erythroid differentiation in transformed cells (Rossi et al., 1996) whereas the Myb portion controls entry into thrombocytic differentiation (Frampton et al., 1995). The latter findings are in line with the occurrence of differentiated megakaryocytes in c-*myb*$^{-/-}$ foetal liver and support the hypothesis that this is due to an inhibitory effect of c-Myb on megakaryocytic commitment or differentiation.

More recent results have provided evidence for the ability of v-Myb to transform erythroid cells under certain conditions and furthermore influence the differentiation phenotype of the target cell types. The AMV virus, like E26, encodes a v-Myb protein that is truncated and mutated version of c-

Myb and causes monoblastic leukaemia in the chicken (Lipsick and Wang, 1999). When early blastoderm cells were infected in the presence of basic fibroblast growth factor (bFGF) with AMV or a retrovirus encoding c-Myb, transformed erythroid cells could be obtained that could be differentiated by removal of the bFGF and addition of erythropoietin and insulin (Bartunek et al., 2002). In contrast to the MEL cell system, v-Myb or c-Myb did not block terminal differentiation of these erythroid cells. Thus, in co-operation with bFGF, v-Myb (and c-Myb) can establish or permit an erythroid phenotype in transformed cells, implying that c-Myb might have a function not only in inhibiting terminal differentiation, but also in directing normal erythroid differentiation.

Co-operation between v-Myb and a growth factor in the determination of erythroid differentiation has also been observed in our laboratory. A virus designated v-Myb^{ts}/EGFR, capable of expressing a temperature sensitive E26 v-Myb lacking the Ets sequences and the human EGF receptor (Khazaie et al., 1988) was found to transform bipotent erythroid/thrombocyte progenitors when used to infect 2 day old chick blastoderm cells in the presence of EGF at 37°C (JF, unpublished observations). Interestingly, these cells could be selectively committed to erythroid or thrombocytic differentiation by either removal of EGF from the culture or temperature-induced inactivation of v-Myb upon shift to 42°C (Figure 2).

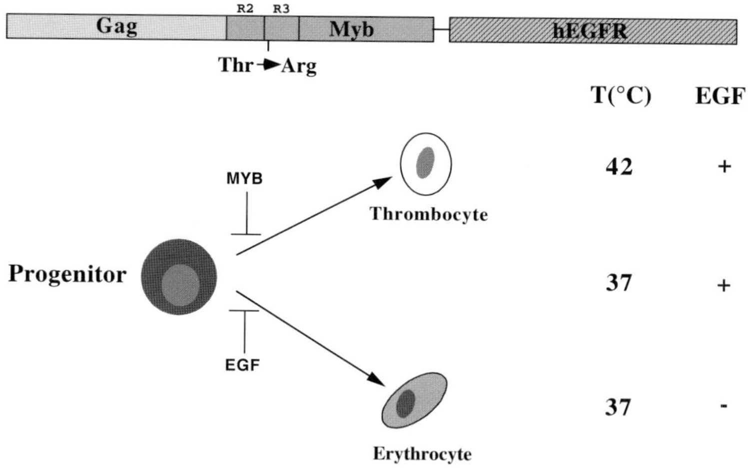

Figure 2

Bipotential erythro-megakaryocytic progenitor transformed by v-Myb^{ts}/EGFR. The upper part of the schematic shows the structure of the v-Myb^{ts}/EGFR sequences between the viral LTRs. The threonine to arginine mutation in Myb R3 confers temperature sensitivity to the DNA binding capacity of v-Myb. The lower part of the diagram summarises the conditions that maintain progenitor status or lead to commitment to the erythroid or thrombocytic lineages.

Differentiation along the erythroid or thrombocytic pathways was associated with expression of specific surface markers and RNAs (Figure 3A and data not shown). Intriguingly, analysis of this system showed that although the loss of v-Myb activity led to entry into thrombopoiesis, commitment to erythropoiesis upon withdrawl of EGF resulted in up regulation of expression of both c-*myb* and A-*myb* RNAs (Figure 3B).

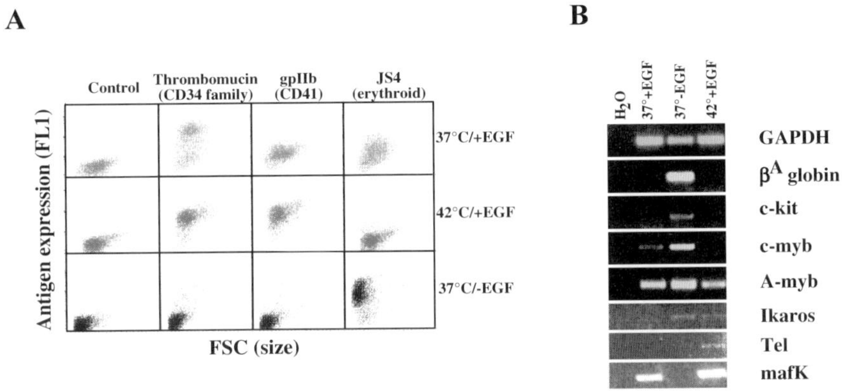

Figure 3

Phenotype of the v-Myb^{ts}/EGFR transformed bipotential progenitor and changes upon induced commitment. (A) Flow cytometric analysis of surface antigen expression. (B) RT-PCR analysis of gene expression.

7. COMMITMENT PROCESSES AND ERYTHROID DIFFERENTIATION DEPEND ON C-MYB LEVELS: LESSONS FROM A NOVEL ALLELE

Recently, in order to generate a c-*myb* allele that could be disrupted in a controlled fashion, the wild type gene has been modified by homologous recombination in mouse embryonic stem cells (Emambokus et al., 2003). Exons 3 to 6, encoding most of the DNA binding domain, were flanked by loxP sequences (Figure 4) thereby making possible *in vivo* excision by Cre recombinase either expressed from a tissue-specific transgene or selectively introduced into cells. For the purposes of gene targetting a neomycin-resistance (neo^R) cassette was incorporated in the modified locus. It was flanked by FRT sequences, which are a substrate for the yeast FLP recombinase, so that it could be subsequently removed in mice that express FLP.

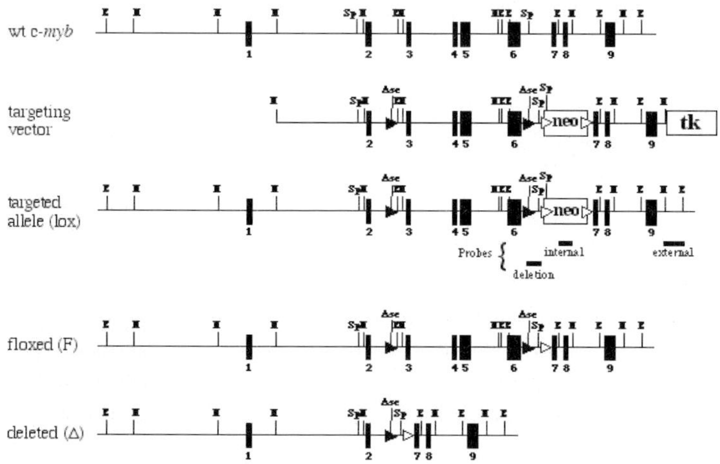

Figure 4

Generation of a conditional allele of *c-myb*. The targetting vector is shown together with the organisation of the wild type c-*myb* gene. A neomycin resistance cassette (neo) for positive selection and the herpes simplex virus-1 thymidine kinase gene (tk) for negative selection were introduced into intron 6 and just downstream of exon 9 respectively. The neo cassette was flanked by Flp recognition sites (open arrowheads). LoxP sites (filled arrowheads) were introduced into intron 2 and intron 6. The vertical black boxes represent exons. Relevant restriction endonuclease sites are indicated (E - EcoRI; H - HindIII and Sp - SpeI).

The loxP-modified, or “floxed”, c-*myb* allele (c-myb^{F}) containing no neo^{R} cassette could be bred to homozygosity and was found to be effectively deleted *in vivo* by Cre recombinase expressed from a transgene under the control of the Mx1 gene promoter. Cre expression is rapidly induced from the Mx1-Cre transgene by administration of type I interferon α/β or polyinosinic-polycytidylic acid (Kühn et al., 1995). Inducible deletion in c-$myb^{F/F}$:Mx1-Cre mice was efficient and led to a profound haemopoietic phenotype within 2-5 days (NE and JF, unpublished). Thus, as expected from the requirement for c-Myb in immature, cycling progenitors, these cells were diminished in the bone marrow. Similarly, erythroid progenitors were rapidly lost and were undetectable by flow cytometry at day 10 of Cre induction. In addition, the number of blood platelets increased initially and severely declined later, indicating aberrant megakaryocytic development. While these results confirmed the expected requirement for c-Myb for haemopoietic progenitor maintenance in the adult, they do not allow definitive conclusions about the processes affected by the disruption of c-Myb function.

Interestingly, embryos carrying a floxed c-*myb* allele retaining the neo^R cassette (c-myb^{loxP}) and a null-allele (c-myb^{-}) were anaemic and died around E15. Foetal livers of c-$myb^{-/loxP}$ embryos contained only 5% of the wild type levels of c-Myb protein, implying that the c-myb^{loxP} allele is expressed at a lower level compared to the wild type allele, and should therefore be considered to be a knockdown mutation. Similar to c-$myb^{-/-}$ embryos, the c-$myb^{-/loxP}$ embryos lacked definitive erythrocytes, although their foetal livers contained a high proportion of immature cells. The relative abundance of early erythroid precursors (pro-erythroblast stage) was increased, while later stages were reduced. Significantly, megakaryocytes and their precursors were present in a higher proportion compared to the wild type. When the functionality of haemopoietic progenitors was assessed in colony assays, no BFU-E or CFU-E colonies could be detected. Granulocyte-containing colonies were also absent, whereas macrophage (CFU-M) colony numbers were increased. Normal mixed multilineage colonies were replaced by a large number of colonies containing predominantly macrophages and megakaryocytes. c-$myb^{loxP/loxP}$ embryos were viable and displayed an intermediate phenotype confirming the dosage-dependent effect.

These recent results suggest that the normal development of certain stages in haemopoiesis is critically dependent on high levels of c-Myb protein. While low levels of c-Myb permit proliferation of immature progenitors, maintenance of normal numbers of late erythroid precursors depends on higher levels of the protein. Early proerythroblasts are present in the knockdown foetal liver, but they are not fully functional since they are unable to form CFU-E colonies in response to cytokines *in vitro*. The phenotype of the *in vitro* colonies that develop from knockdown foetal liver and the presence of increased megakaryocytic cells suggests that c-Myb might inhibit differentiation of multipotential progenitors along the macrophage and megakaryocytic lineages.

8. HOW CAN THE PRECISE ROLE OF C-MYB IN ERYTHROID AND MEGAKARYOCYTIC PRECURSORS BE DETERMINED?

In an effort to investigate the function of c-Myb in erythroid progenitors the current approach followed in our laboratory makes use of primary cell material from mice carrying floxed c-*myb* alleles. Cells with an immature progenitor phenotype can be purified from the foetal liver or the bone marrow by cell sorting, while a culture system described by Dolznig et al (2001) can be used to expand early proerythroblasts from foetal liver. The fate of cells after ablation of c-*myb* and their response to specific proliferation/differentiation signals can then be followed in detail. Since these primary cells can be terminally differentiated we can analyze c-Myb function in all intermediate-late stages of erythropoiesis. Similarly, in order to explore the question of a role of c-Myb in late megakaryocytic development, we can generate cultures of megakaryocytic cells that are able to undergo normal terminal differentiation. These systems are particularly suited to the identification of potential c-Myb target genes by mRNA profiling.

The molecular mechanism of c-Myb action is still not well understood in spite of the isolation of a number of candidate target genes (see Chapters 13 and 14). This lack of understanding is particularly relevant to the erythroid and megakaryocytic cell lineages including the bipotent erythro-megakaryocytic erythroid progenitor. Are there clues about the likely mechanisms of c-Myb action based on the existing knowledge of both c-Myb function in other haemopoietic cells and the cellular regulatory processes known to characterise erythropoiesis and megakaryocytopoiesis?

From a signalling point of view in erythroid cells, it is reasonable to place c-Myb downstream of the cytokine receptors c-Kit and EpoR as a mediator of their growth and differentiation-promoting effects. This suggestion is clearly speculative and simplistic, especially since multiple microenvironmental signals are expected to regulate erythropoiesis *in vivo*. As described earlier, SCF/c-Kit signalling promotes proliferation and survival in early erythroid progenitors and contributes to the maintenance of an immature state. These effects have clearly been attributed to c-Myb as well. Interestingly, the biological function of c-Kit is also important in melanocyte development and spermatogenesis (Silvers, 1979; Rossi et al., 2000), two processes in which v-Myb and A-Myb, respectively, have been implicated as regulators of proliferation and differentiation (Bell and Frampton, 1999; Oh and Reddy, 1999). The situation in megakaryocytic progenitors is even less well defined, although a number of signalling pathways from cytokine receptors could impinge on c-Myb. Analysis of growth factor responsiveness of cells from genetically modified mice as described above should provide a means to test possible links with c-Myb function.

Which basic cellular processes, namely proliferation, differentiation and cell death might be linked directly to the function of c-Myb (and possibly B-Myb) in erythroid and megakaryocytic cells? That c-Myb might have effects on several processes, probably involving multiple target genes, is suggested by observations on the inhibition of c-Myb activity in haemopoietic cell lines. Depending on the exact system, such approaches resulted in a cessation of growth, accumulation of cells in G1/S phase, and increased apoptosis and differentiation (Gewirtz et al., 1989; White and Weston, 2000).

Several proliferation- or cell cycle-related genes have been proposed as c-Myb target genes, including cyclin A, cdc2, DNA polymerase alpha and c-*myc* (Oh and Reddy, 1999; Muller et al., 1999). Interestingly, the phenotype of mice homozygous for a knockout allele of DNA ligase I, an enzyme with a key role in the joining of short replication intermediates during DNA replication, closely resembles the phenotype of c-$myb^{-/-}$ embryos, implying that this gene could be regulated by c-Myb (Bentley et al., 1996). c-*myb* antisense experiments in a promyelocytic cell line indicate that c-Myb may regulate expression of the receptor of the growth factor IGF-1, which is an important stimulus to proliferation and survival of late erythroid cells (Reiss et al., 1992; Muta et al., 1994). c-Kit has also been proposed to be a c-Myb target gene (Hogg et al., 1997; Ratajczak et al., 1998). Interestingly, it was reported last year that the *Drosophila* Myb protein (DMyb) is involved directly in DNA replication (Beall et al., 2002). DMyb mediates DNA replication through specific binding to sequences comprising enhancers of replication. In line with these findings, DMyb has been shown to be essential for normal DNA replication in S-phase as well as for G2/M progression and mitosis (Katzen et al., 1998; Manak et al., 2002). DMyb is most closely related to vertebrate B-Myb, and thus the possibility of a direct role for B-Myb, and perhaps also c-Myb, in DNA replication will have to be considered in future experiments.

Megakaryocyte progenitors do not undergo the extensive expansion that characterises the erythroid lineage. Instead, endoreplication is a central event in megakaryocytic development and accordingly cell cycle regulation displays specific features allowing alternate S phases and abortive mitoses. Cyclin E and possibly cyclin A are among the proteins that have been shown to mediate this process (García et al., 2000). As B-Myb is positively regulated by phosphorylation involving these two proteins together with cyclin-dependent kinases, there is the possibility that it is a mediator of endoreplication (Sala et al., 1997; Saville and Watson, 1998). Interestingly, research on DMyb has suggested that it has a role as a suppresser of endoreplication in certain cell types (Katzen et al., 1998; Fitzpatrick et al., 2002).

Myb proteins have been implicated in the regulation of apoptosis. Hence, v-Myb and c-Myb have been shown to promote survival by up-regulating the expression of *bcl*-2 in myelomonocytic and T cells, respectively (Frampton et al., 1996; Taylor et al., 1996). A critical anti-apoptotic pathway triggered by erythropoietin in late erythroid precursors involves the Bcl-x_L protein (Gregory et al., 1999; Motoyama et al., 1999) and although different transcription factors, including GATA-1 and Stat-5 (Gregory et al., 1999; Socolovsky et al., 1999), have been shown to up-regulate the *bcl*-x_L gene in erythroid cells, a possible role for c-Myb remains to be explored. However, our experiments using conditional deletion of c-*myb* and examination of haemopoietic tissues in c-*myb* knockdown embryos gave no indication of increased apoptosis in either the erythroid or megakaryocytic lineages (NE and JF, unpublished observations).

The aberrant development of intermediate erythroid progenitors in c-*myb* knockdown embryos could indicate a requirement for c-Myb in specific differentiation processes, some of which might include necessary cell divisions associated with maturation. The importance of cell division for terminal erythroid differentiation has been demonstrated. Thus, interruption of DNA synthesis by specific drugs or vitamin deficiencies leads to apoptosis in S phase and incomplete maturation of erythrocytes (Koury et al., 1997; Koury et al., 2000). Although over-expression of c-Myb in immature cells has been shown to act as an inhibitor of differentiation in erythroid cells, certain aspects of research presented above raise the possibility that it could also have a function in the induction of erythroid-specific gene expression. Using the controllable dominant negative Myb derivative MEnT in MEL cells Chen and Bender (2001) have identified several target gene candidates including the carbonic anhydrase I isozyme

A final consideration in terms of the direct mechanistic effects of c-Myb in erythroid and megakaryocytic cells is the possibility that it is involved in protein complexes that can have altered regulatory properties dependent upon the balance of component factors. Several proteins have been identified as interaction partners of c-Myb and a functional relevance is beginning to emerge for some of the interactions (Ness, 1999; see also Chapter 12). One might expect, therefore, that a subset of processes mediated by c-Myb in erythropoiesis and megakaryocytopoiesis does not involve direct transcriptional regulation of c-Myb target genes *per se*. The ubiquitous and highly related molecules CBP and p300, which act as cofactors in association with transcription factors (Blobel, 2000), bind to c-Myb and enhance its transactivating capacity (Dai et al., 1996; Tomita et al.,

2000). Their co-activating potential has been attributed mainly to their intrinsic histone acetyltransferase activity, however, they are also able to acetylate transcription factor proteins, including c-Myb. Strikingly, when the interaction between c-Myb and p300 was abolished *in vivo* through targetted mutation of the c-Myb binding sites in p300, a phenotype with some aspects similar to the c-*myb* knockdown phenotype emerged (Kasper et al., 2002). Mice heterozygous for both the p300 mutation and a c-*myb* null allele showed elevated blood platelet numbers and perturbations in the balance of differentiating megakaryocytes. Since CBP and p300 have been shown to mediate synergistic as well as antagonistic interactions between transcription factors, a role as integrators of the transcription factor network has been suggested. Hence, GATA-1 and c-Myb antagonise for binding to CBP and cross-inhibit their transactivation potential (Takahashi et al., 2000). Although it is unclear whether this antagonism is functional, it is tempting to speculate that antagonism for limiting molecules of CBP/p300 is contributing to the balance between proliferation and differentiation of erythroid and megakaryocytic progenitors.

9. CONCLUDING REMARKS

To date, our understanding of the role of c-Myb in the erythroid and megakaryocytic lineages indicates that a number of distinct stages are under its control and that this includes both proliferation- and differentiation-related events (Figure 5). The involvement of c-Myb, a proliferation-promoting factor, in the differentiation of certain lineages makes it an interesting molecule in that it seems to act as a link between lineage-specific factors and general regulators of proliferation and survival. Many questions remain, but at least the tools and systems are now available to help resolve them, especially in the case of c-Myb. Specific deletion of the c-*myb* gene at precise points along the differentiation pathways of both the erythroid and megakaryocytic lineages should be achieved in the near future. This, in combination with microarray screening approaches, should also facilitate more rapid identification of the target genes of c-Myb. The impending generation of a similar conditional allele of the B-*myb* gene (P. García and JF, unpublished) will likewise open up possibilities for investigating whether B-Myb is solely associated with proliferation control or whether it also has specific differentiation related effects. Lastly, the finding of A-*myb* expression in avian erythroid cells suggests that further investigation into its expression in mammalian systems is worthwhile.

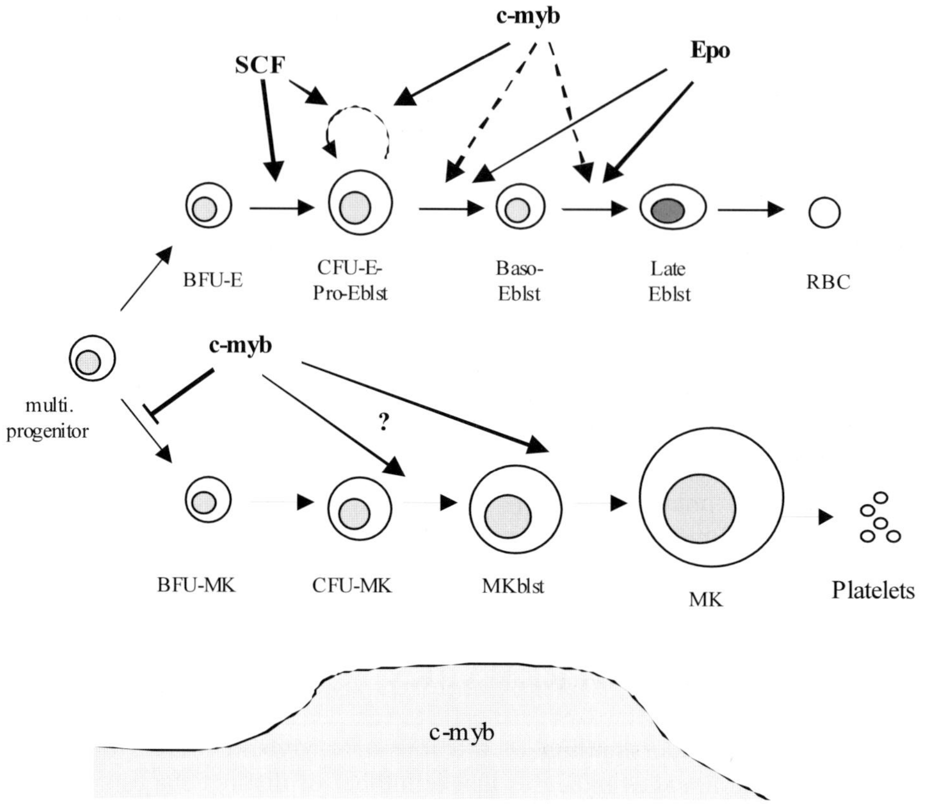

Figure 5

c-Myb function in erythropoiesis and megakaryocytopoiesis. Dashed arrows indicate conclusions supported by recent experiments in c-*myb* knockdown mice. Processes dependent on SCF and erythropoietin (Epo) are indicated. The curve represents c-*myb* expression levels.

ACKNOWLEDGEMENTS

JF and AV are supported by the Wellcome Trust.

REFERENCES

Akashi, K., Traver, D., Miyamoto, T. and Weissman, I.L. (2000) A clonogenic common myeloid progenitor that gives rise to all myeloid lineages. Nature *404*, 193-197.

Alexander, W.S., Roberts, A.W., Nicola, N.A., Li, R. and Metcalf, D. (1996) Deficiencies in progenitor cells of multiple haemopoietic lineages and defective megakaryocytopoiesis in mice lacking the thrombopoietic receptor c-Mpl. Blood *87*, 2162-2170.

Arsura, M., Introna, M., Passerini, F., Mantovani, A. and Golay, J. (1992) B-myb antisense oligonucleotides inhibit proliferation of human haemopoietic cell lines. Blood *79*, 2708-2716.

Aurigemma, R.E., Blair, D.G. and Ruscetti, S.K. (1992) Transactivation of erythroid transcription factor GATA-1 by a myb-ets-containing retrovirus. J Virol *66*, 3056-3061.

Bartunek, P., Pajer, P., Karafiat, V., Blendinger, G., Dvorak, M. and Zenke, M. (2002) bFGF signaling and v-Myb cooperate in sustained growth of primitive erythroid progenitors. Oncogene *21*, 400-410.

Beall, E.L., Manak, J.R., Zhou, S., Bell, M., Lipsick, J.S. and Botchan, M.R. (2002) Role for a Drosophila Myb-containing protein complex in site-specific DNA replication. Nature *420* 833-837.

Bell, M.V. and Frampton, J. (1999) v-Myb can transform and regulate the differentiation of melanocyte precursors. Oncogene *18*, 7226-7233.

Bentley, D., Selfridge, J., Millar, J.K., Samuel, K., Hole, N., Ansell, J.D. and Melton, D.W. (1996) DNA ligase I is required for foetal liver erythropoiesis but is not essential for mammalian cell viability. Nat Genet 13, 489-491.

Blobel, G.A. (2000) CREB-binding protein and p300: molecular integrators of haemopoietic transcription. Blood *95*, 745-755.

Briegel, K., Lim, K.C., Plank, C., Beug, H., Engel, J.D. and Zenke, M. (1993) Ectopic expression of a conditional GATA-2/estrogen receptor chimera arrests erythroid differentiation in a hormone-dependent manner. Genes Dev *7*, 1097-1109.

Cantor, A.B. and Orkin, S.H. (2002) Transcriptional regulation of erythropoiesis: an affair involving multiple partners. Oncogene *21*, 3368-3376.

Caracciolo, D., Venturelli, D., Valtieri, M., Peschle, C., Gewirtz, A.M. and Calabretta, B. (1990) Stage-related proliferative activity determines c-myb functional requirements during normal human haemopoiesis. J Clin Invest *85*, 55-61.

Chan, J.Y., Kwong, M., Lo, M., Emerson, R. and Kuypers, F.A. (2001) Reduced oxidative-stress response in red blood cells from p45NFE2-deficient mice. Blood *97*, 2151-2158.

Chang, A.N., Cantor, A.B., Fujiwara, Y., Lodish, M.B., Droho, S., Crispino, J.D., Orkin, S.H. (2002) GATA-factor dependence of the multitype zinc-finger protein FOG-1 for its essential role in megakaryopoiesis. Proc Natl Acad Sci USA *99*, 9237-9242.

Chen, J. and Bender, T.P. (2001) A novel system to identify Myb target promoters in friend murine erythroleukemia cells. Blood Cells Mol Dis *27*, 429-436.

Chen, J., Kremer, C.S. and Bender, T.P. (2002) A Myb dependent pathway maintains Friend murine erythroleukemia cells in an immature and proliferating state. Oncogene *21*, 1859-1869.

Chern, Y., Spangler, R., Choi, H.S. and Sytkowski, A.J. (1991a) Erythropoietin activates the receptor in both Rauscher and Friend murine erythroleukemia cells. J Biol Chem *266*, 2009-2012.

Chern, Y.J., O'Hara, C. and Sytkowski, A. J. (1991b) Induction of hemoglobin synthesis by downregulation of MYB protein with an antisense oligodeoxynucleotide. Blood *78*, 991-996.

Clarke, D., Vegiopoulos, A., Crawford, A., Mucenski, M., Bonifer, C. and Frampton, J. (2000) In vitro differentiation of c-myb(-/-) ES cells reveals that the colony forming capacity of unilineage macrophage precursors and myeloid progenitor commitment are c-Myb independent. Oncogene *19*, 3343-3351.

Clarke, M.F., Kukowska-Latallo, J.F., Westin, E., Smith, M. and Prochownik, E. V. (1988) Constitutive expression of a c-myb cDNA blocks Friend murine erythroleukemia cell differentiation. Mol Cell Biol *8*, 884-892.

Coghill, E., Eccleston, S., Fox, V., Cerruti, L., Brown, C., Cunningham, J., Jane, S. and Perkins, A. (2001) Erythroid Kruppel-like factor (EKLF) coordinates erythroid cell proliferation and hemoglobinization in cell lines derived from EKLF null mice. Blood *97*, 1861-1868.

Crispino, J.D., Lodish, M.B., MacKay, J.P. and Orkin, S.H. (1999) Use of altered specificity mutants to probe a specific protein-protein interaction in differentiation: the GATA-1:FOG complex. Mol Cell *3*, 219-228.

Crossley, M., Tsang, A.P., Bieker, J.J. and Orkin, S.H. (1994) Regulation of the erythroid Kruppel-like factor (EKLF) gene promoter by the erythroid transcription factor GATA-1. J Biol Chem *269*, 15440-15444.

Dai, P., Akimaru, H., Tanaka, Y., Hou, D.X., Yasukawa, T., Kanei-Ishii, C., Takahashi, T. and Ishii, S. (1996) CBP as a transcriptional coactivator of c-Myb. Genes Dev *10*, 528-540.

Danish, R., el-Awar, O., Weber, B.L., Langmore, J., Turka, L.A., Ryan, J.J. and Clarke, M.F. (1992) c-myb effects on kinetic events during MEL cell differentiation. Oncogene *7*, 901-907.

Debili, N., Coulombel, L., Croisille, L., Katz, A., Guichard, J., Breton-Gorius, J. and Vainchenker, W. (1996) Characterization of a bipotent erythro-megakaryocytic progenitor in human bone marrow. Blood *88*, 1284-1296.

Dolznig, H., Boulme, F., Stangl, K., Deiner, E.M., Mikulits, W., Beug, H. and Mullner, E.W. (2001) Establishment of normal, terminally differentiating mouse erythroid progenitors: molecular characterization by cDNA arrays. FASEB J *15*, 1442-1444.

Elwood, N.J., Zogos, H., Pereira, D.S., Dick, J.E. and Begley, C.G. (1998) Enhanced megakaryocyte and erythroid development from normal human CD34(+) cells: consequence of enforced expression of SCL. Blood *91*, 3756-3765.

Emambokus, N.R., Vegiopoulos, A., Harman, B., Jenkinson, E.J., Anderson, G. and Frampton, J. (2003) Progression through key stages of the hematopoietic hierarchy is dependent on distinct threshold levels of c-Myb. EMBO J *22*, 4478-4488.

Emilia, G., Donelli, A., Ferrari, S., Torelli, U., Selleri, L., Zucchini, P., Moretti, L., Venturelli, D., Ceccherelli, G. and Torelli, G. (1986) Cellular levels of mRNA from c-myc, c-myb and c-fes onc-genes in normal myeloid and erythroid precursors of human bone marrow: an in situ hybridization study. Br J Haematol *62*, 287-292.

Fitzpatrick, C.A., Sharkov, N.V., Ramsay, G. and Katzen, A.L. (2002) Drosophila myb exerts opposing effects on S phase, promoting proliferation and suppressing endoreduplication. Development *129*, 4497-4507.

Frampton, J., McNagny, K., Sieweke, M., Philip, A., Smith, G. and Graf, T. (1995) v-Myb DNA binding is required to block thrombocytic differentiation of Myb-Ets-transformed multipotent haematopoietic progenitors. EMBO J *14*, 2866-2875.

Frampton, J., Ramqvist, T. and Graf, T. (1996) v-Myb of E26 leukemia virus up-regulates bcl-2 and suppresses apoptosis in myeloid cells. Genes Dev *10*, 2720-2731.

Fujiwara, Y., Browne, C.P. Cunniff, K., Goff, S.C. and Orkin, S.H. (1996) Arrested development of embryonic red cell precursors in mouse embryos lacking transcription factor GATA-1. Proc Natl Acad Sci USA *93*, 12355-12358.

García, P., Frampton, J., Ballester, A. and Calés, C. (2000) Ectopic expression of cyclin E allows non-endomitotic megakaryoblastic K562 cells to establish re-replication cycles. Oncogene *19*, 1820-1833.

Gewirtz, A.M. and Calabretta, B. (1988) A c-myb antisense oligodeoxynucleotide inhibits normal human haemopoiesis in vitro. Science *242*, 1303-1306.

Gewirtz, A.M., Anfossi, G., Venturelli, D., Valpreda, S., Sims, R. and Calabretta, B. (1989) G1/S transition in normal human T-lymphocytes requires the nuclear protein encoded by c-myb. Science *245*, 180-183.

Graf, T., McNagny, K., Brady, G. and Frampton, J. (1992) Chicken "erythroid" cells transformed by the Gag-Myb-Ets-encoding E26 leukemia virus are multipotent. Cell *70*, 201-213.

Gregory, T., Yu, C., Ma, A., Orkin, S.H., Blobel, G.A. and Weiss, M.J.(1999) GATA-1 and erythropoietin cooperate to promote erythroid cell survival by regulating bcl-xL expression. Blood *94*, 87-96.

Hall, M.A., Curtis, D.J., Metcalf, D., Elefanty, A.G., Sourris, K., Robb, L., Gothert, J.R., Jane, S.M. and Begley, C.G. (2003) The critical regulator of embryonic haemopoiesis, SCL, is vital in the adult for megakaryopoiesis, erythropoiesis, and lineage choice in CFU-S12. Proc Natl Acad Sci USA *100*, 992-997.

Hart, A., Melet, F., Grossfeld, P., Chien, K., Jones, C., Tunnacliffe, A., Favier, R. and Bernstein, A. (2000) Fli-1 is required for murine vascular and megakaryocytic development and is hemizygously deleted in patients with thrombocytopenia. Immunity *13*, 167-177.

Heyworth, C., Pearson, S., May, G. and Enver, T.(2002) Transcription factor-mediated lineage switching reveals plasticity in primary committed progenitor cells. EMBO J *21*, 3770-3781.

Hogg, A., Schirm, S., Nakagoshi, H., Bartley, P., Ishii, S., Bishop, J.M. and Gonda, T.J. (1997) Inactivation of a c-Myb/estrogen receptor fusion protein in transformed primary cells leads to granulocyte/macrophage differentiation and down regulation of c-kit but not c-myc or cdc2. Oncogene *15*, 2885-2898.

Ikonomi, P., Rivera, C.E., Riordan, M., Washington, G., Schechter, A.N. and Noguchi, C.T. (2000) Overexpression of GATA-2 inhibits erythroid and promotes megakaryocyte differentiation. Exp Hematol *28*, 1423-1431.

Karafiat, V., Dvorakova, M., Pajer, P., Kralova, J., Horejsi, Z., Cermak, V., Bartunek, P., Zenke, M. and Dvorak, M. (2001) The leucine zipper region of Myb oncoprotein regulates the commitment of haemopoietic progenitors. Blood *98*, 3668-3676.

Kasper, L.H., Boussouar, F., Ney, P.A., Jackson, C.W., Rehg, J., van Deursen, J.M. and Brindle, P.K. (2002) A transcription-factor-binding surface of coactivator p300 is required for haematopoiesis. Nature *419*, 738-743.

Katzen, A.L., Jackson, J., Harmon, B.P., Fung, S.M., Ramsay, G. and Bishop, J.M. (1998) Drosophila myb is required for the G2/M transition and maintenance of diploidy. Genes Dev *12*, 831-843.

Kawada, H., Ito, T., Pharr, P.N., Spyropoulos, D.D., Watson, D.K. and Ogawa, M. (2001) Defective megakaryopoiesis and abnormal erythroid development in Fli-1 gene-targetted mice. Int J Hematol *73*, 463-468.

Khazaie, K., Dull, T.J., Graf, T., Schlessinger, J., Ullrich, A., Beug, H., Vennstrom, B. (1988) Truncation of the human EGF receptor leads to differential transforming potentials in primary avian fibroblasts and erythroblasts. EMBO J. *7*, 3061-3071.

Kirsch, I.R., Bertness, V., Silver, J. and Hollis, G.F. (1986) Regulated expression of the c-myb and c-myc oncogenes during erythroid differentiation. J Cell Biochem *32*, 11-21.

Koury, M.J., Horne, D.W., Brown, Z.A., Pietenpol, J.A., Blount, B.C., Ames, B.N., Hard, R. and Koury, S.T. (1997) Apoptosis of late-stage erythroblasts in megaloblastic anaemia: association with DNA damage and macrocyte production. Blood *89*, 4617-4623.

Koury, M.J., Price, J.O. and Hicks, G.G. (2000) Apoptosis in megaloblastic anaemia occurs during DNA synthesis by a p53-independent, nucleoside-reversible mechanism. Blood *96*, 3249-3255.

Kuhn, R., Schwenk, F., Aguet, M. and Rajewsky, K. (1995) Inducible gene targeting in mice. Science *269*, 1427-1429.

Kulessa, H., Frampton, J. and Graf, T. (1995) GATA-1 reprograms avian myelomonocytic cell lines into eosinophils, thromboblasts, and erythroblasts. Genes Dev *9*, 1250-1262.

Lee, R., Kertesz, N., Joseph, S.B., Jegalian, A. and Wu, H. (2001) Erythropoietin (Epo) and EpoR expression and 2 waves of erythropoiesis. Blood *98*, 1408-1415.

Lesault, I., Quang, C.T., Frampton, J. and Ghysdael, J. (2002) Direct regulation of BCL-2 by FLI-1 is involved in the survival of FLI-1-transformed erythroblasts. EMBO J *21*, 694-703.

Levine, R.F. (1980) Isolation and characterization of normal human megakaryocytes. Br. Journal Haematology *45*, 487.

Lipsick, J.S. and Wang, D.M. (1999) Transformation by v-Myb. Oncogene *18*, 3047-3055.

Manak, J.R., Mitiku, N. and Lipsick, J.S. (2002) Mutation of the Drosophila homologue of the Myb protooncogene causes genomic instability. Proc Natl Acad Sci USA *99*, 7438-7443.

McClinton, D., Stafford, J., Brents, L., Bender, T.P. and Kuehl, W.M. (1990) Differentiation of mouse erythroleukemia cells is blocked by late up-regulation of a c-myb transgene. Mol Cell Biol *10*, 705-710.

Mikkola, H.K., Klintman, J., Yang, H., Hock, H., Schlaeger, T.M., Fujiwara, Y. and Orkin, S.H. (2003) Haematopoietic stem cells retain long-term repopulating activity and multipotency in the absence of stem-cell leukaemia SCL/tal-1 gene. Nature *421*, 547-551.

Motohashi, H., Katsuoka, F., Shavit, J.A., Engel, J.D. and Yamamoto, M. (2000) Positive or negative MARE-dependent transcriptional regulation is determined by the abundance of small Maf proteins. Cell *103*, 865-875.

Motoyama, N., Kimura, T., Takahashi, T., Watanabe, T. and Nakano, T. (1999) bcl-x prevents apoptotic cell death of both primitive and definitive erythrocytes at the end of maturation. J Exp Med *189*, 1691-1698.

Mucenski, M.L., McLain, K., Kier, A.B., Swerdlow, S.H., Schreiner, C.M., Miller, T.A., Pietryga, D.W., Scott, W.J., Jr. and Potter, S.S. (1991) A functional c-myb gene is required for normal murine foetal hepatic haemopoiesis. Cell *65*, 677-689.

Muller, C., Yang, R., Idos, G., Tidow, N., Diederichs, S., Koch, O.M., Verbeek, W., Bender, T.P. and Koeffler, H.P. (1999) c-myb transactivates the human cyclin A1 promoter and induces cyclin A1 gene expression. Blood *94*, 4255-4262.

Murphy, G.J. and Leavitt, A.D. (1999) A model for studying megakaryocyte development and biology. Proc Natl Acad Sci USA *96*, 3065-3070.

Muta, K., Krantz, S.B., Bondurant, M.C. and Wickrema, A. (1994) Distinct roles of erythropoietin, insulin-like growth factor I, and stem cell factor in the development of erythroid progenitor cells. J Clin Invest *94*, 34-43.

Muta, K., Krantz, S.B., Bondurant, M.C. and Dai, C.H. (1995) Stem cell factor retards differentiation of normal human erythroid progenitor cells while stimulating proliferation. Blood *86*, 572-580.

Ness, S.A. (1999) Myb binding proteins: regulators and cohorts in transformation. Oncogene *18*, 3039-3046.

Nuez, B., Michalovich, D., Bygrave, A., Ploemacher, R. and Grosveld, F. (1995) Defective haematopoiesis in foetal liver resulting from inactivation of the EKLF gene. Nature *375*, 316-318.

Oh, I.H. and Reddy, E.P. (1999) The myb gene family in cell growth, differentiation and apoptosis. Oncogene *18*, 3017-3033.

Onodera, K., Shavit, J.A., Motohashi, H., Yamamoto, M. and Engel, J.D. (2000) Perinatal synthetic lethality and haemopoietic defects in compound mafG::mafK mutant mice. EMBO J *19*, 1335-1345.

Palis, J., Robertson, S., Kennedy, M., Wall, C. and Keller, G. (1999) Development of erythroid and myeloid progenitors in the yolk sac and embryo proper of the mouse. Development *126*, 5073-5084.

Pereira, R., Quang, C.T., Lesault, I., Dolznig, H., Beug, H. and Ghysdael, J. (1999) FLI-1 inhibits differentiation and induces proliferation of primary erythroblasts. Oncogene *18*, 1597-1608.

Perkins, A.C., Sharpe, A.H. and Orkin, S. H. (1995) Lethal beta-thalassaemia in mice lacking the erythroid CACCC-transcription factor EKLF. Nature *375*, 318-322.

Radke, K., Beug, H., Kornfeld, S. and Graf, T. (1982) Transformation of both erythroid and myeloid cells by E26, an avian leukemia virus that contains the myb gene. Cell *31*, 643-653.

Ramsay, R.G., Ikeda, K., Rifkind, R.A. and Marks, P.A. (1986) Changes in gene expression associated with induced differentiation of erythroleukemia: protooncogenes, globin genes, and cell division. Proc Natl Acad Sci USA *83*, 6849-6853.

Ratajczak, M.Z., Perrotti, D., Melotti, P., Powzaniuk, M., Calabretta, B., Onodera, K., Kregenow, D.A., Machalinski, B. and Gewirtz, A.M. (1998) Myb and ets proteins are candidate regulators of c-kit expression in human haemopoietic cells. Blood *91*, 1934-1946.

Reiss, K., Travali, S., Calabretta, B. and Baserga, R. (1991) Growth regulated expression of B-myb in fibroblasts and haemopoietic cells. J Cell Physiol *148*, 338-343.

Reiss, K., Porcu, P., Sell, C., Pietrzkowski, Z. and Baserga, R. (1992) The insulin-like growth factor 1 receptor is required for the proliferation of hemopoietic cells. Oncogene *7*, 2243-2248.

Rekhtman, N., Radparvar, F., Evans, T. and Skoultchi, A.I. (1999) Direct interaction of haemopoietic transcription factors PU.1 and GATA-1: functional antagonism in erythroid cells. Genes Dev *13*, 1398-1411.

Robb, L., Lyons, I., Li, R., Hartley, L., Kontgen, F., Harvey, R.P., Metcalf, D. and Begley, C.G. (1995) Absence of yolk sac haemopoiesis from mice with a targetted disruption of the scl gene. Proc Natl Acad Sci USA *92*, 7075-7079.

Rossi, F., McNagny, K.M., Logie, C., Stewart, A.F. and Graf, T. (1996) Excision of Ets by an inducible site-specific recombinase causes differentiation of Myb-Ets-transformed haemopoietic progenitors. Curr Biol *6*, 866-872.

Rossi, P., Sette, C., Dolci, S. and Geremia, R. (2000) Role of c-kit in mammalian spermatogenesis. J Endocrinol Invest *23*, 609-615.

Russell, E.S. (1979) Hereditary anaemias of the mouse: a review for geneticists. Adv Genet *20*, 357-459.

Sala, A., Casella, I., Grasso, L., Bellon, T., Reed, J.C., Miyashita, T. and Peschle, C. (1996) Apoptotic response to oncogenic stimuli: cooperative and antagonistic interactions between c-myb and the growth suppressor p53. Cancer Res *56*, 1991-1996.

Sala, A., Kundu, M., Casella, I., Engelhard, A., Calabretta, B., Grasso, L., Paggi, M.G., Giordano, A., Watson, R.J., Khalili, K. and Peschle, C. (1997) Activation of human B-MYB by cyclins. Proc Natl Acad Sci USA *94*, 532-536.

Saville, M.K. and Watson, R.J. (1998) The cell-cycle regulated transcription factor B-Myb is phosphorylated by cyclin A/Cdk2 at sites that enhance its transactivation properties. Oncogene *17*, 2679-2689.

Shivdasani, R.A. and Orkin, S.H. (1995) Erythropoiesis and globin gene expression in mice lacking the transcription factor NF-E2. Proc Natl Acad Sci USA *92*, 8690-8694.

Shivdasani, R.A., Mayer, E.L. and Orkin, S.H. (1995a) Absence of blood formation in mice lacking the T-cell leukaemia oncoprotein tal-1/SCL. Nature *373*, 432-434.

Shivdasani, R.A., Rosenblatt, M.F., Zucker-Franklin, D., Jackson, C.W., Hunt, P., Saris, C.J. and Orkin, S.H. (1995b) Transcription factor NF-E2 is required for platelet formation independent of the actions of thrombopoietin/MGDF in megakaryocyte development. Cell *81*, 695-704.

Shivdasani, R.A., Fujiwara, Y., McDevitt, M.A. and Orkin, S.H. (1997) A lineage-selective knockout establishes the critical role of transcription factor GATA-1 in megakaryocyte growth and platelet development. EMBO J *16*, 3965-3973.

Silvers,W.K. (1979) The Coat Colors of Mice. Springer Verlag, New York, NY.

Sitzmann, J., Noben-Trauth, K. and Klempnauer, K.H. (1995) Expression of mouse c-myb during embryonic development. Oncogene *11*, 2273-2279.

Sitzmann, J., Noben-Trauth, K., Kamano, H. and Klempnauer, K.H. (1996) Expression of B-Myb during mouse embryogenesis. Oncogene *12*, 1889-1894.

Socolovsky, M., Fallon, A.E., Wang, S., Brugnara, C., Lodish, H.F. (1999) Foetal anaemia and apoptosis of red cell progenitors in Stat5a-/-5b-/- mice: a direct role for Stat5 in Bcl-X(L) induction. Cell *98*, 181-191.

Spyropoulos, D.D., Pharr, P.N., Lavenburg, K.R., Jackers, P., Papas, T.S., Ogawa, M. and Watson, D.K. (2000) Hemorrhage, impaired haemopoiesis, and lethality in mouse embryos carrying a targetted disruption of the Fli1 transcription factor. Mol Cell Biol *20*, 5643-5652.

Sumner, R., Crawford, A., Mucenski, M. and Frampton, J. (2000) Initiation of adult myelopoiesis can occur in the absence of c-Myb whereas subsequent development is strictly dependent on the transcription factor. Oncogene *19*, 3335-3342.

Takahashi, T., Suwabe, N., Dai, P., Yamamoto, M., Ishii, S. and Nakano, T. (2000) Inhibitory interaction of c-Myb and GATA-1 via transcriptional co-activator CBP. Oncogene *19*, 134-140.

Tanaka, Y., Patestos, N.P., Maekawa, T. and Ishii, S. (1999) B-myb is required for inner cell mass formation at an early stage of development. J Biol Chem *274*, 28067-28070.

Tanikawa, J., Ichikawa-Iwata, E., Kanei-Ishii, C., Nakai, A., Matsuzawa, S., Reed, J.C. and Ishii, S. (2000) p53 suppresses the c-Myb-induced activation of heat shock transcription factor 3. J Biol Chem *275*, 15578-15585.

Taylor, D., Badiani, P. and Weston, K. (1996) A dominant interfering Myb mutant causes apoptosis in T cells. Genes Dev *10*, 2732-2744.

Thompson, C.B., Challoner, P.B., Neiman, P.E. and Groudine, M.(1986) Expression of the c-myb proto-oncogene during cellular proliferation. Nature *319*, 374-380.

Todokoro, K., Watson, R.J., Higo, H., Amanuma, H., Kuramochi, S., Yanagisawa, H. and Ikawa, Y. (1988) Down-regulation of c-myb gene expression is a prerequisite for erythropoietin-induced erythroid differentiation. Proc Natl Acad Sci USA *85*, 8900-8904.

Tomita, A., Towatari, M., Tsuzuki, S., Hayakawa, F., Kosugi, H., Tamai, K., Miyazaki, T., Kinoshita, T. and Saito, H. (2000) c-Myb acetylation at the carboxyl-terminal conserved domain by transcriptional co-activator p300. Oncogene *19*, 444-451.

Tsai, F.Y. and Orkin, S.H. (1997) Transcription factor GATA-2 is required for proliferation/survival of early haemopoietic cells and mast cell formation, but not for erythroid and myeloid terminal differentiation. Blood *89*, 3636-3643.

Tsang, A.P., Visvader, J.E., Turner, C.A., Fujiwara, Y., Yu, C., Weiss, M. J., Crossley, M. and Orkin, S.H. (1997) FOG, a multitype zinc finger protein, acts as a cofactor for transcription factor GATA-1 in erythroid and megakaryocytic differentiation. Cell *90*, 109-119.

Tsang, A.P., Fujiwara, Y., Hom, D.B. and Orkin, S.H. (1998) Failure of megakaryopoiesis and arrested erythropoiesis in mice lacking the GATA-1 transcriptional cofactor FOG. Genes Dev *12*, 1176-1188.

Vannucchi, A.M., Paoletti, F., Linari, S., Cellai, C., Caporale, R., Ferrini, P.R., Sanchez, M., Migliaccio, G. and Migliaccio, A.R. (2000) Identification and characterization of a bipotent (erythroid and megakaryocytic) cell precursor from the spleen of phenylhydrazine-treated mice. Blood *95*, 2559-2568.

Valtieri, M., Venturelli, D., Care, A., Fossati, C., Pelosi, E., Labbaye, C., Mattia, G., Gewirtz, A.M., Calabretta, B. and Peschle, C. (1991) Antisense myb inhibition of purified erythroid progenitors in development and differentiation is linked to cycling activity and expression of DNA polymerase alpha. Blood *77*, 1181-1190.

Vyas, P., Ault, K., Jackson, C.W., Orkin, S.H. and Shivdasani, R.A. (1999) Consequences of GATA-1 deficiency in megakaryocytes and platelets. Blood *93*, 2867-2875.

Wadman, I.A., Osada, H., Grutz, G.G., Agulnick, A.D., Westphal, H., Forster, A. and Rabbitts, T.H. (1997) The LIM-only protein Lmo2 is a bridging molecule assembling an erythroid, DNA-binding complex which includes the TAL1, E47, GATA-1 and Ldb1/NLI proteins. EMBO J *16*, 3145-3157.

Weiss, M.J. and Orkin, S.H. (1995) Transcription factor GATA-1 permits survival and maturation of erythroid precursors by preventing apoptosis. Proc Natl Acad Sci USA *92*, 9623-9627.

White, J.R. and Weston, K. (2000) Myb is required for self-renewal in a model system of early haemopoiesis. Oncogene *19*, 1196-1205.

Wu, H., X. Liu, Jaenisch, R. and Lodish, H.F. (1995) Generation of committed erythroid BFU-E and CFU-E progenitors does not require erythropoietin or the erythropoietin receptor. Cell *83*, 59-67.

Whyatt, D. and Grosveld, F. (2002) Cell-nonautonomous function of the retinoblastoma tumour suppressor protein: new interpretations of old phenotypes. EMBO Rep *3*, 130-135.

Ziebold, U., Bartsch, O., Marais, R., Ferrari, S. and Klempnauer, K.H. (1997) Phosphorylation and activation of B-Myb by cyclin A-Cdk2. Curr Biol *7*, 253-260.

Zon, L.I., Youssoufian, H., Mather, C., Lodish, H.F. and Orkin, S.H. (1991) Activation of the erythropoietin receptor promoter by transcription factor GATA-1. Proc Natl Acad Sci U S A *88*, 10638-10641.

Chapter 6

C-MYB AS A KEY PLAYER IN THE CONTROL OF MYELOID CELL DIFFERENTIATION

Sandrine Sarrazin and Michael H. Sieweke
Centre d'Immunologie de Marseille Luminy, Campus de Luminy, Case 906, 13288 Marseille Cedex09, France.

Abstract: c-Myb is required for definitive haemopoiesis and is expressed in immature proliferating haemopoietic cells. This includes the myeloid compartment, where c-Myb is strongly expressed in progenitor cells and is down regulated as they mature. c-Myb activity inhibits differentiation, promotes cellular proliferation and survival of myeloid progenitor cells. We review the known target genes and molecular mechanisms contributing to these phenomena. Beyond influencing myeloid differentiation through the control of progenitor cell proliferation and survival, c-Myb may also play a direct role in lineage choice. Leukaemic retroviruses containing different mutant versions of c-Myb transform different types of myeloid cells. These naturally selected mutants appear to represent distinct 'frozen' activity states of c-Myb that differ in their ability to associate with cofactors and to activate specific target genes, with important consequences for the differentiation choice between the granulocytic and monocytic lineages.

1. INTRODUCTION

The role of c-Myb in myeloid differentiation appears to involve both the control of myeloid progenitor proliferation and the ability to select between the gene expression programs of the granulocytic and monocytic lineages in response to extracellular signals. We will review both of these aspects, drawing from gene inactivation, gain of function and molecular studies in different cell culture and *in vivo* systems. Much of our knowledge about the biological functions of c-Myb, especially in myeloid cells, is derived from the study of the v-Myb proteins encoded by the avian E26 and AMV leukaemic retroviruses or experimentally generated mammalian equivalents. This is in part grounded in the historical development of the field and in part due to the stronger biological effects of the activated viral proteins that lack

J. Frampton (ed.), Myb Transcription Factors: Their Role in Growth, Differentiation and Disease, 133-144.

negative regulatory mechanisms. Experience has shown that observations made with the v-Myb proteins are mostly transferable to c-Myb and have taught us important lessons about its function. Therefore we will refer to the v-Myb proteins where they have provided useful insights relevant to c-Myb function in normal myeloid differentiation.

2. CONTROL OF MYELOID PROGENITOR PROLIFERATION AND SURVIVAL BY C-MYB

2.1 Expression Profile of c-Myb in Myeloid Differentiation

In contrast to other members of the Myb family of transcription factors, c-Myb is predominantly expressed in the haemopoietic system. Besides its expression in the erythroid, thrombocytic and lymphoid lineages, c-Myb is also expressed in immature myeloid cells (Duprey and Boettiger, 1985; Bjerregaard et al., 2003). The observation that retrovirally transduced or experimentally created oncogenic versions of Myb transformed myeloid cells and induced myeloid leukaemias in experimental animals led to the assumption that the normal cellular protein c-Myb might be involved in maintaining the proliferative state of myeloid progenitors. Indeed, high expression levels of c-Myb correlate with cellular proliferation (Gonda and Metcalf, 1984; Thompson et al., 1986; Studzinski and Brelvi, 1987) and are dramatically down-regulated during differentiation of myeloid progenitors to monocytes, both in murine WEHI-3B (Gonda and Metcalf, 1984) or human HL-60 myeloid cell lines (Studzinski and Brelvi, 1987) and primary chicken yolk sac progenitors (Duprey and Boettiger, 1985). During granulocytic differentiation of the murine EML progenitor cell line, c-Myb levels remain high at the myeloblast and promyelocyte stage and successively decrease as cells differentiate to mature granulocytes (Du et al., 2002). These observations in experimental systems correlate well with c-*myb* mRNA and protein levels found in haemopoietic cells from normal human bone marrow, where c-Myb is detected in $CD34^+$ progenitors (Kastan et al., 1989), myeloblasts and promyelocytes but not in mature granulocytes (Bjerregaard et al., 2003). The functional significance of this expression profile is indicated both by gene inactivation and over-expression studies.

2.2 Gene Inactivation Studies

c-Myb deficient mice are defective for definitive haemopoiesis in the foetal liver (Mucenski et al., 1991), the para-aortic splanchnopleural (P-Sp) and the aorta-gonad-mesonephros (AGM) region (Mukouyama et al., 1999), but still capable of primitive yolk sac haemopoiesis (Mucenski et al., 1991). The defect is cell autonomous, since *c-myb*$^{-/-}$ cells also fail to contribute to foetal or adult definitive haemopoiesis in chimaeric mice (Sumner et al., 2000). Consistant with this, *in vitro* differentiation of *c-myb*$^{-/-}$ ES cells gives rise to primitive uni-lineage precursors but is defective for definitive haemopoiesis (Clarke et al., 2000). On the other hand careful chimaera analysis also revealed that c-Myb deficiency does not completely abrogate the development of CD34^{+}/c-Kit^{+} definitive progenitors but rather prevents their expansion (Sumner et al., 2000). Consistent with the residual presence of immature progenitors in *c-myb*$^{-/-}$ embryos the infection of cultivated AGM cells with a c-Myb expressing retrovirus can rescue multi lineage haemopoiesis (Mukouyama et al., 1999). Together these studies indicate that c-Myb is required to maintain the proliferation and expansion of early multipotent haemopoietic progenitors. Gene inactivation in knockout mice can only reveal the earliest essential function of a gene, but experiments with *c-myb* antisense oligonuclotides indicate that c-Myb may have a similar function in later myeloid restricted GM-progenitors. Thus, the abrogation of *c-myb* using antisense oligonuclotides inhibited the growth of several myeloid cell lines blocked at different stages of myeloid differentiation (Anfossi et al., 1989) or primary myeloid leukemia cell samples (Calabretta et al., 1991) and resulted in a decrease in both size and number of CFU-GM colonies developing from normal human bone marrow mono-nucleated cells (Gewirtz and Calabretta, 1988).

2.3 Over-expression Studies

Strong support for the notion that c-Myb maintains myeloid progenitor proliferation comes from the observations made using constitutively active alleles. The hypothesis was originally put forward after the identification of mutated versions of c-Myb encoded by avian leukaemia viruses that transform myeloid blood cells (Beug et al., 1979; Roussel et al., 1979; Graf et al., 1981). Indeed truncated versions of c-Myb in which the negative regulatory domains have been deleted, hence in this respect resembling the viral versions, lead to the expansion of myeloblasts (Metz et al., 1991; Metz and Graf, 1991; Grasser et al., 1991) or GM progenitors (Gonda et al., 1989a; Gonda et al., 1989b; Gonda et al., 1993) that can give rise to myeloid cell lines in the chick and mouse systems, respectively. Consistent with this, the constitutive expression of truncated c-Myb versions leads to a block of IL-6- or LIF-induced monocytic differentiation in the mouse M1 cell line (Hoffman-Liebermann and Liebermann, 1991; Selvakumaran et al., 1992) and to continued proliferation and inhibition of late stage maturation in G-CSF-induced granulocytic differentiation of the Il-3-dependent 32DCl3 cell line (Patel et al., 1993; Bies et al., 1995; Patel et al., 1996; Kumar et al., 2003). The fact that over-expression of full length c-Myb can also block differentiation in these cell lines (Selvakumaran et al., 1992; Bies et al., 1995) and enhances proliferation of primary haemopoietic cells (Gonda et al., 1989b), suggests that the biological effect of the truncated proteins is not a *de novo* acquired function but represents the effect of continuous constitutive activity of the wild type protein, which during normal myelopoiesis must be tightly regulated. In addition to controlling proliferation, activated Myb can also protect immature myeloid cells from apoptosis (Frampton et al., 1996).

2.4 Relevant Targets of c-Myb Activity in Myeloid Progenitor Proliferation and Survival

Several potential target genes have been identified that may contribute to the enhanced proliferation and life-span observed in cells with increased Myb activity. Thus, it has been shown that c-Myb positively regulates transcriptional activation of genes involved in cell cycle control such as *cdc2, cyclin A1* and *topoisomerase II*α (Ku et al., 1993; Muller et al., 1999; Brandt et al., 1997) or in signal cascades transmitting proliferative stimuli such as *myeloblastin*, *c-kit*, *gbx-2* and *c-myc* (Hogg et al., 1997; Kowenz-Leutz et al., 1997; Lutz et al., 2001; Schmidt et al., 2000). In addition, c-Myb represses transcription of the tumour suppressor gene *ink4b*, a cyclin-dependent kinase inhibitor that is up-regulated during myeloid differentiation and promotes growth arrest (Wolff et al., 2001). Finally, Myb may also promote myeloid progenitor survival through activation of the anti-apoptotic *bcl-2* gene (Frampton et al., 1996; Schmidt et al., 2000). However, it is not always clear whether all potential targets are also controlled by c-Myb *in vivo* (Hogg et al., 1997), and it remains to be determined, which of them are most relevant for its biological function(s). Recently it has been shown that *Drosophila* Myb has essential non-transcriptional functions in S-phase (Manak et al., 2002) and that it is a critical component of a protein complex mediating site-specific DNA replication (Beall et al., 2002). *Drosophila* Myb is most closely related to B-Myb, from which A-Myb and c-Myb appear to have arisen later in evolution by gene duplications that also involved the acquisition of a transcriptional activation domain (Ganter and Lipsick, 1999; Simon et al., 2002). Whether A-Myb and c-Myb have kept a direct role in DNA replication beyond their function as transcriptional activators that may contribute to their role in cellular proliferation is an interesting question that merits future investigation.

3. THE ROLE OF C-MYB IN MYELOID DIFFERENTIATION DECISIONS

Since proliferation and differentiation are linked processes it is theoretically possible that the block of differentiation observed upon over-expression of activated Myb proteins is simply an indirect consequence of increased proliferation. However, there is evidence to indicate that c-Myb directly affects myeloid differentiation and lineage choice independently of proliferation.

3.1 Inhibiton of Differentation

Early on it had been observed that avian leukaemia viruses that can transform myeloid cells arrested cells at a different stages of differentiation depending on the oncogene (Beug et al., 1979). This suggested that beyond inducing proliferation they also actively influenced the differentiation status of the cell. This hypothesis was confirmed when it was shown that the introduction of v-Myb into v-Myc transformed proliferating macrophages changed their phenotype to more immature myeloblasts (Ness et al., 1987). Consistent with this, c-Myb is down-regulated in myeloid cell lines upon induction of terminal differentiation but not upon growth inhibition without differentiation (Hoffman-Liebermann and Liebermann, 1991). Furthermore, it has also been shown in two different inducible systems that v-Myb can induce the retro-differentiation of macrophages. Thus, myeloblasts transformed by a temperature-sensitive mutant version of v-Myb differentiate into resting macrophage 4-6 days after v-Myb inactivation by temperature shift. A back-shift to the permissive temperature induces them to lose macrophage characteristics and gradually reacquire an immature phenotype (Beug et al., 1984; Beug et al., 1987). Similarly, estrogen induction of a v-Myb-estrogen receptor fusion protein in the macrophage cell-line HD11 induced loss of macrophage characteristics and restored an immature proliferating progenitor phenotype in an estrogen-dependent manner (Burk and Klempnauer, 1991). Together, these results indicate that Myb activity is not only important for maintaining myeloid progenitor proliferation but also actively inhibits terminal maturation of myelo-monocytic cells.

3.2 Lineage Choice Between Granulocyte and Monocyte Pathways

c-Myb-mediated inhibition of myeloid differentiation also appears to be a crucial checkpoint for the lineage choice between the monocytic and granulocytic pathways. Whereas c-Myb expression appears not to be compatible with monocytic differentiation, as outlined above, it is expressed in the immature proliferating stages of the granulocyte pathway (Du et al., 2002; Bjerregaard et al., 2003). Furthermore it appears to be required for granulocytic but not monocytic progenitor proliferation (Ferrari et al., 1990). Down-regulation of c-Myb during the normal monocytic differentiation process may be assured by the Maf proteins MafB and c-Maf, which induce monocytic differentiation in myeloid progenitors (Kelly et al., 2000; Hedge et al., 1999), are strongly expressed in the monocytic lineage (Sieweke et al., 1996; Eichmann et al., 1997) and inhibit c-Myb transactivation activity (Hedge et al., 1998).

Insight into the mechanisms that control Myb activity at the branch point between granulocytic and monocytic pathways has originally come from the study of the E26 and AMV chicken retroviruses, which harbour different *v-myb* genes. Whereas E26 derived v-Myb or truncated c-Myb transform cells resembling myeloblasts in the chicken (Metz et al., 1991; Metz and Graf, 1991; Grasser et al., 1991) or GM progenitors in the mouse system (Gonda et al., 1989a; Gonda et al., 1989b; Gonda et al., 1993), cells transformed by AMV v-Myb, which harbours several point mutations with respect to c-Myb, have a monoblast phenotype. This lineage specificity is also reflected in the target genes activated by the different Myb versions.

E26 v-Myb and c-Myb, but not AMV v-Myb, synergise with the b-Zip factor C/EBPβ to activate the *mim-1* gene (Ness et al., 1993; Burk et al., 1993), which is expressed in the granulocytic lineage but not in monocytes (Ness et al., 1989; Ness et al., 1993). The regulation of the *mim-1* promoter appears to be representative for the promoters of several myeloid genes expressed in the granulocytic lineage and in myeloid progenitors, such as *lysozyme* (Ness et al., 1993), *tom-1a* (Burk et al., 1997), *myeloid peroxidase* (Britos-Bray and Friedman, 1997), *neutrophil elastase* (Oelgeschlager et al., 1996; Verbeek et al., 1999), and *myeloblastin* (Lutz et al., 2001), where E26 v-Myb or c-Myb have also been found to cooperate with different members of the C/EBP family (Burk et al., 1993; Mink et al., 1996; Oelgeschlager et al., 1996; Nuchprayoon et al., 1997; Verbeek et al., 1999; Lutz et al., 2001), which at least in the case of C/EBPβ (Mink et al., 1996; Tahirov et al., 2002) and C/EBPε (Verbeek et al., 1999) appears to involve direct protein interactions. This cooperation is consistent with the critical role of C/EBPs in granulocytic differentiation. Thus C/EBPα or C/EBPβ enhance granulocytic differentiation when expressed in progenitor cells (Radomska et al., 1998; Duprez et al., 2003) and C/EBPα and C/EBPε deficient mice have defects at different stages of granulocyte differentiation (Yamanaka et al., 1997; Zhang et al., 1997).

Conversely AMV v-Myb, but not E26 v-Myb or c-Myb, can activate transcription of the gene encoding GBX-2, a homeo-box transcription factor, which in turn can transactivate the cMGF gene (Kowenz-Leutz et al., 1997), an avian myeloid growth factor related to Il-6 and G-CSF (Leutz et al., 1989) that is responsible for factor independent growth of AMV transformed cells (Metz et al, 1991). Induction of GBX-2 appears to be critical for the monocytic phenotype of AMV v-Myb transformed cells, since co-expression of GBX-2 in myeloblasts transformed by E26 v-Myb confers a monoblast phenotype (Kowenz-Leutz et al., 1997).

It appears that the mutations occurring in AMV v-Myb mimic the input from signalling cascades that control c-Myb activity and change its target gene specificity. Thus a constitutive activated RAS signalling cascade confers the ability on c-Myb to activate GBX-2 (Kowenz-Leutz et al., 1997). This cooperation of c-Myb with RAS signalling in myeloid progenitor cells appears to be also conserved in the mammalian system. Murine foetal liver cells transformed by a truncated, activated c-Myb construct are dependent on GM-CSF for colony formation in semisolid medium, proliferation and survival (Gonda et al., 1993; Donovan et al., 2002). By contrast, under conditions of constitutively activated RAS signalling, such as in a knockout for NF1, a GTPase activation protein (GAP) that negatively regulates $p21^{ras}$, Myb transformed cells survive, proliferate, and form colonies in the absence of GM-CSF, partly because of autocrine production of GM-CSF (Donovan et al., 2002). This is reminiscent of the cooperation of c-Myb with RAS signalling in the chicken system, but it is unknown whether GBX-2 or related Myb target genes are involved in GM-CSF transactivation.

In summary it appears that extracellular signals can alter c-Myb activity and divert a progenitor cell from the granulocytic into the monocytic pathway. The molecular basis for this is unclear but the mutations present in AMV v-Myb may provide some clues.

3.3 Molecular Basis for Changes of Myb Activity in Lineage Choice

The chicken viruses harbouring *v-myb* genes have been very useful for the dissection of c-Myb function. As outlined above, the naturally selected mutations in AMV v-Myb appear to represent a "frozen" alternative activity state of c-Myb in the normal differentiation process. Three mutations of AMV v-Myb are important to confer a monoblast phenotype, activate the *gbx-2* gene and lose *mim-1* activation (Ness et al., 1989; Introna et al., 1990; Kowenz-Leutz et al., 1997). They exchange three hydrophobic residues to polar ones on the outer surface of R2 repeat of the c-Myb DNA binding domain and thus eliminate a hydrophobic patch. All three changes are necessary, since single back mutations to the wild type sequence result in a shift of v-Myb transformed cells towards the granulocytic lineage (Introna et al., 1990). So what is the molecular basis for such a dramatic effect of a few point mutations on cellular phenotype? Both structural and biochemical studies have demonstrated that the hydrophobic patch on R2 represents a protein interaction surface for cofactors that is abolished by the AMV mutations (Tahirov et al., 2002); Leverson et al., 1998). The fact that c/EBPβ can interact with the c-Myb repeat R2 via a leucine zipper extension, even at a distance on nonadjacent c-Myb and c/EBP binding sites via DNA looping, provides an intriguing structural explanation for the cooperativity of the factors in transactivation of myeloid genes expressed in the granulocytic lineage (Tahirov et al., 2002). The R2 mutations present in AMV v-Myb abolish the C/EBPβ interaction and thus explain why AMV v-Myb cannot transactivate *mim-1* or similarly controlled genes. Whether loss of C/EBPβ interaction is sufficient for Myb proteins to transactivate the GBX-2 gene is an interesting question, but since the Myb DNA binding domain can interact with several proteins (Ganter et al., 1998; Leverson et al., 1998; Ying et al., 2000; Tahirov et al., 2002), it is also quite possible that other cofactor interactions with Myb proteins may be important for differentiation along the monocytic lineage and that the cell fate decision between the granulocytic and monocytic lineage critically involves a cofactor exchange on the R2 surface. In the case of AMV v-Myb, differential cofactor interaction is achieved via point mutations affecting the interaction surface. How increased RAS signalling may achieve this exchange in the case of c-Myb is unknown. None of the AMV mutations themselves are direct phosphorylation sites but it is conceivable that RAS triggered kinase signalling phosphorylates c-Myb or its partner molecules and causes a conformational change that promotes cofactor exchange. In any case, c-Myb and its partner molecules appear to provide an intriguing example of how changes in transcription factor complex composition can critically influence lineage choice in the haemopoietic system (Sieweke and Graf, 1998).

REFERENCES

Anfossi, G., Gewirtz, A.M. and Calabretta, B. (1989) An oligomer complementary to c-myb-encoded mRNA inhibits proliferation of human myeloid leukemia cell lines. Proc Natl Acad Sci USA *86*, 3379-3383.

Beall, E.L., Manak, J.R., Zhou, S., Bell, M., Lipsick, J.S. and Botchan, M.R. (2002) Role for a Drosophila Myb-containing protein complex in site-specific DNA replication. Nature *420*, 833-837.

Beug, H., von Kirchbach, A., Doderlein, G., Conscience, J.F. and Graf, T. (1979) Chicken haemopoietic cells transformed by seven strains of defective avian leukemia viruses display three distinct phenotypes of differentiation. Cell *18*, 375-390.

Beug, H., Leutz, A., Kahn, P. and Graf, T. (1984) Ts mutants of E26 leukemia virus allow transformed myeloblasts, but not erythroblasts or fibroblasts, to differentiate at the nonpermissive temperature. Cell *39*, 579-588.

Beug, H., Blundell, P. and Graf, T. (1987) Reversibility of differentiation and proliferative capacity in avian myelomonocytic cells transformed by tsE26 leukemia virus. Genes Dev. *1*, 277-286.

Bies, J., Mukhopadhyaya, R., Pierce, J. and Wolff, L. (1995) Only late, nonmitotic stages of granulocyte differentiation in 32Dcl3 cells are blocked by ectopic expression of murine c-myb and its truncated forms. Cell Growth Differ *6*, 59-68.

Bjerregaard, M.D., Jurlander, J., Klausen, P., Borregaard, N. and Cowland, J.B. (2003) The in vivo profile of transcription factors during neutrophil differentiation in human bone marrow. Blood *101*, 4322-4332.

Brandt, T.L., Fraser, D.J., Leal, S., Halandras, P.M., Kroll, A.R. and Kroll, D.J. (1997) c-Myb trans-activates the human DNA topoisomerase IIalpha gene promoter. *J Biol Chem* **272**, 6278-6284.

Britos-Bray, M. and Friedman, A.D. (1997) Core binding factor cannot synergistically activate the myeloperoxidase proximal enhancer in immature myeloid cells without c-Myb. Mol Cell Biol *17*, 5127-5135.

Burk, O. and Klempnauer, K.H. (1991) Estrogen-dependent alterations in differentiation state of myeloid cells caused by a v-myb/estrogen receptor fusion protein. EMBO J *10*, 3713-3719.

Burk, O., Mink, S., Ringwald, M. and Klempnauer, K.H. (1993) Synergistic activation of the chicken mim-1 gene by v-myb and C/EBP transcription factors. EMBO J *12*, 2027-2038.

Burk, O., Worpenberg, S., Haenig, B. and Klempnauer, K.H. (1997) tom-1, a novel v-Myb target gene expressed in AMV- and E26-transformed myelomonocytic cells. EMBO J *16*, 1371-1380.

Calabretta, B., Sims, R.B., Valtieri, M., Caracciolo, D., Szczylik, C., Venturelli, D., Ratajczak, M., Beran, M. and Gewirtz, A.M. (1991) Normal and leukemic haemopoietic cells manifest differential sensitivity to inhibitory effects of c-myb antisense oligodeoxynucleotides: an in vitro study relevant to bone marrow purging. Proc Natl Acad Sci USA *88*, 2351-2355.

Clarke, D., Vegiopoulos, A., Crawford, A., Mucenski, M., Bonifer, C. and Frampton, J. (2000) In vitro differentiation of c-myb(-/-) ES cells reveals that the colony forming capacity of unilineage macrophage precursors and myeloid progenitor commitment are c-Myb independent. Oncogene *19*, 3343-3351.

Donovan, S., See, W., Bonifas, J., Stokoe, D. and Shannon, K.M. (2002) Hyperactivation of protein kinase B and ERK have discrete effects on survival, proliferation, and cytokine expression in Nf1-deficient myeloid cells. Cancer Cell *2*, 507-514.

Du Y, Campbell, J.L., Nalbant, D., Youn, H., Bass, A.C., Cobos, E., Tsai, S., Keller, J.R. and Williams, S.C. (2002) Mapping gene expression patterns during myeloid differentiation using the EML haemopoietic progenitor cell line. Exp Hematol *30*, 649-658.

Duprey, S.P. and Boettiger, D. (1985) Developmental regulation of c-*myb* in normal myeloid progenitor cells. Proc. Natl. Acad. Sci. USA *82*, 6937-6941.

Duprez, E., Wagner, K., Koch, H. and Tenen, D.G. (2003) C/EBPbeta: a major PML/RARA responsive gene in retinoic acid-induced differentiation of APL cells. EMBO J, in press.

Eichmann, A., Grapin-Botton, A., Kelly, L., Graf, T., Le Douarin, N.M. and Sieweke, M. (1997) The expression pattern of the mafB/kr gene in birds and mice reveals that the kreisler phenotype does not represent a null mutant. Mech Dev *65*, 111-122.

Ferrari, S., Donelli, A., Manfredini, R., Sarti, M., Roncaglia, R., Tagliafico, E., Rossi, E., Torelli, G. and Torelli, U. (1990) Differential effects of c-myb and c-fes antisense oligodeoxynucleotides on granulocytic differentiation of human myeloid leukemia HL60 cells. Cell Growth Differ *1*, 543-548.

Frampton, J., Ramqvist, T. and Graf, T. (1996) v-Myb of E26 leukemia virus up-regulates bcl-2 and suppresses apoptosis in myeloid cells. Genes Dev *10*, 2720-2731.

Ganter, B., Fu, S. and Lipsick, J.S. (1998) D-type cyclins repress transcriptional activation by the v-Myb but not the c-Myb DNA-binding domain. EMBO J *17*, 255-268.

Ganter, B. and Lipsick, J.S. (1999) Myb and oncogenesis. Adv Cancer Res *76*, 21-60.

Gewirtz, A.M. and Calabretta, B. (1988) A c-myb antisense oligodeoxynucleotide inhibits normal human haemopoiesis in vitro. Science *242*, 1303-1306.

Gonda, T.J. and Metcalf, D. (1984) Expression of myb, myc and fos proto-oncogenes during the differentiation of a murine myeloid leukaemia. Nature *310*, 249-251.

Gonda, T.J., Ramsay, R.G. and Johnson, G.R. (1989a) Murine myeloid cell lines derived by in vitro infection with recombinant c-myb retroviruses express myb from rearranged vector proviruses. EMBO J *8*, 1767-1775.

Gonda, T.J., Buckmaster, C. and Ramsay, R.G. (1989b) Activation of c-myb by carboxy-terminal truncation: relationship to transformation of murine haemopoietic cells in vitro. EMBO J *8*, 1777-1783.

Gonda, T.J., Macmillan, E.M., Townsend, P.V. and Hapel, A.J. (1993) Differentiation state and responses to haemopoietic growth factors of murine myeloid cells transformed by myb. Blood *82*, 2813-2822.

Graf, T., von Kirchbach, A. and Beug, H. (1981) Characterization of the haemopoietic targetcells of AEV, MC29 and AMV avian acute leukemia viruses. Exp. Cell Res. *131*, 331-343.

Grasser, F.A., Graf, T. and Lipsick, J.S. (1991) Protein truncation is required for the activation of the c-myb proto-oncogene. Mol Cell Biol *11*, 3987-3996.

Hedge, S.P., Kumar, A., Kurschner, C. and Shapiro, L.H. (1998) c-Maf interacts with c-Myb to regulate transcription of an early myeloid gene during differentiation. Mol Cell Biol *18*, 2729-2737.

Hegde, S.P., Zhao, J., Ashmun, R.A. and Shapiro, L.H. (1999) c-Maf induces monocytic differentiation and apoptosis in bipotent myeloid progenitors. Blood 94, 1578-1589.

Hoffman-Liebermann, B. and Liebermann, D.A. (1991) Suppression of c-myc and c-myb is tightly linked to terminal differentiation induced by IL6 or LIF and not growth inhibition in myeloid leukemia cells. Oncogene *6*, 903-909.

Hogg, A., Schirm, S., Nakagoshi, H., Bartley, P., Ishii, S., Bishop, J.M. and Gonda, T.J. (1997) Inactivation of a c-Myb/estrogen receptor fusion protein in transformed primary cells leads to granulocyte/macrophage differentiation and down regulation of c-kit but not c-myc or cdc2. Oncogene *15*, 2885-2898.

Introna, M., Golay, J., Frampton, J., Nakano, T., Ness, S.A. and Graf, T. (1990) Mutations in v-*myb* alter the differentiation of myelomonocytic cells transformed by the oncogene. Cell *63*, 1287-1297.

Kastan, M.B., Stone, K.D. and Civin, C.I. (1989) Nuclear oncoprotein expression as a function of lineage, differentiation stage, and proliferative status of normal human haemopoietic cells. Blood *74*, 1517-1524.

Kelly, L.M., Englmeier, U., Lafon, I., Sieweke, M.H. and Graf, T. (2000) MafB is an inducer of monocytic differentiation. EMBO J *19*, 1987-1997.

Kowenz-Leutz, E., Herr, P., Niss, K. and Leutz, A. (1997) The homeobox gene GBX2, a target of the myb oncogene, mediates autocrine growth and monocyte differentiation. Cell *91*, 185-195.

Ku, D.H., Wen, S.C., Engelhard, A., Nicolaides, N.C., Lipson, K.E., Marino, T.A. and Calabretta, B. (1993) c-myb transactivates cdc2 expression via Myb binding sites in the 5'-flanking region of the human cdc2 gene. J Biol Chem *268*, 2255-2259.

Kumar, A., Lee, C.M. and Reddy, E.P. (2003) c-Myc is essential but not sufficient for c-Myb-mediated block of granulocytic differentiation. J Biol Chem *278*, 11480-11488.

Leutz, A., Damm, K., Sterneck, E., Kowenz, E., Ness, S., Frank, R., Gausepohl, H., Pan, Y.C., Smart, J., Hayman, M., et al. (1989) Molecular cloning of the chicken myelomonocytic growth factor. EMBO J. *8*, 175-181.

Leverson, J.D. and Ness, S.A. (1998) Point mutations in v-Myb disrupt a cyclophilin-catalyzed negative regulatory mechanism. Mol Cell *1*, 203-211.

Lutz, P.G., Houzel-Charavel, A., Moog-Lutz, C. and Cayre, Y.E. (2001) Myeloblastin is an Myb target gene: mechanisms of regulation in myeloid leukemia cells growth-arrested by retinoic acid. Blood *97*, 2449-2456.

Manak, J.R., Mitiku, N. and Lipsick, J.S. (2002) Mutation of the Drosophila homologue of the Myb protooncogene causes genomic instability. Proc Natl Acad Sci USA *99*, 7438-7443.

Metz, T. and Graf, T. (1991) v-myb and v-ets transform chicken erythroid cells and cooperate both in trans and in cis to induce distinct differentiation phenotypes. Genes Dev *5*, 369-380.

Metz, T., Graf, T. and Leutz, A. (1991) Activation of cMGF expression is a critical step in avian myeloid leukemogenesis. EMBO J *10*, 837-844.

Mink, S., Kerber, U. and Klempnauer, K.H. (1996) Interaction of C/EBPbeta and v-Myb is required for synergistic activation of the mim-1 gene. Mol Cell Biol *16*, 1316-1325.

Mucenski, M.L., McLain, K., Kier, A.B., Swerdlow, S.H., Schreiner, C.M., Miller, T.A., Pietryga, D.W., Scott, W.J. Jr. and Potter, S.S. (1991) A functional c-myb gene is required for normal murine fetal hepatic haemopoiesis. Cell *65*, 677-689.

Mukouyama, Y., Chiba, N., Mucenski, M.L., Satake, M., Miyajima, A., Hara, T. and Watanabe, T. (1999) Haemopoietic cells in cultures of the murine embryonic aorta-gonad-mesonephros region are induced by c-Myb. Curr Biol *9*, 833-836.

Muller, C., Yang, R., Idos, G., Tidow, N., Diederichs, S., Koch, O.M., Verbeek, W., Bender, T.P. and Koeffler, H.P. (1999) c-myb transactivates the human cyclin A1 promoter and induces cyclin A1 gene expression. Blood *94*, 4255-4262.

Ness, S., Beug, H. and Graf, T. (1987) v-*myb* dominance over v-*myc* in doulby transformed chick myelomonocytic cells. Cell *51*, 41-50.

Ness, S.A., Marknell, Å. and Graf, T. (1989) The v-*myb* oncogene product binds to and activates the promyelocyte-specific *mim*-1 gene. Cell *59*, 1115-1125.

Ness, S.A., Kowenz-Leutz, E., Casini, T., Graf, T. and Leutz, A. (1993) Myb and NF-M: combinatorial activators of myeloid genes in heterologous cell types. Genes Dev *7*, 749-759.

Nuchprayoon, I., Simkevich, C.P., Luo, M., Friedman, A.D. and Rosmarin, A.G. (1997) GABP cooperates with c-Myb and C/EBP to activate the neutrophil elastase promoter. Blood *89*, 4546-4554.

Oelgeschlager, M., Nuchprayoon, I., Luscher, B. and Friedman, A.D. (1996) C/EBP, c-Myb, and PU.1 cooperate to regulate the neutrophil elastase promoter. Mol Cell Biol *16*, 4717-4725.

Patel, G., Kreider, B., Rovera, G. and Reddy, E.P. (1993) v-myb blocks granulocyte colony-stimulating factor-induced myeloid cell differentiation but not proliferation. Mol Cell Biol *13*, 2269-2276.

Patel, G., Tantravahi, R., Oh, I.H. and Reddy, E.P. (1996) Transcriptional activation potential of normal and tumor-associated myb isoforms does not correlate with their ability to block

GCSF-induced terminal differentiation of murine myeloid precursor cells. Oncogene *13*, 1197-1208.

Radomska, H.S., Huettner, C.S., Zhang, P., Cheng, T., Scadden, D.T. and Tenen, D.G. (1998) CCAAT/enhancer binding protein alpha is a regulatory switch sufficient for induction of granulocytic development from bipotential myeloid progenitors. Mol Cell Biol *18*, 4301-4314.

Roussel, M., Saule, S., Lagrou, C., Rommens, C., Beug, H., Graf, T. and Stehelin, D. (1979) Three new types of viral oncogene of cellular origin specific for haematopoietic cell transformation. Nature *281*, 452-455.

Schmidt, M., Nazarov, V., Stevens, L., Watson, R. and Wolff, L. (2000) Regulation of the resident chromosomal copy of c-myc by c-Myb is involved in myeloid leukemogenesis. Mol Cell Biol *20*, 1970-1981.

Selvakumaran, M., Liebermann, D.A. and Hoffman-Liebermann, B. (1992) Deregulated c-myb disrupts interleukin-6- or leukemia inhibitory factor-induced myeloid differentiation prior to c-myc: role in leukemogenesis. Mol Cell Biol *12*, 2493-2500.

Sieweke, M.H., Tekotte, H., Frampton, J. and Graf, T. (1996) MafB is an interaction partner and repressor of Ets-1 that inhibits erythroid differentiation. Cell 85, 49-60.

Sieweke, M.H. and Graf, T. (1998) A transcription factor party during blood cell differentiation. Curr Opin Genet Dev *8*, 545-551.

Simon, A.L., Stone, E.A. and Sidow, A. (2002) Inference of functional regions in proteins by quantification of evolutionary constraints. Proc Natl Acad Sci USA *99*, 2912-2917.

Studzinski, G.P. and Brelvi, Z.S. (1987) Changes in proto-oncogene expression associated with reversal of macrophage-like differentiation of HL 60 cells. J Natl Cancer Inst *79*, 67-76.

Sumner, R., Crawford, A., Mucenski, M. and Frampton, J. (2000) Initiation of adult myelopoiesis can occur in the absence of c-Myb whereas subsequent development is strictly dependent on the transcription factor. Oncogene *19*, 3335-3342.

Tahirov, T.H., Sato, K., Ichikawa-Iwata, E., Sasaki, M., Inoue-Bungo, T., Shiina, M., Kimura, K., Takata, S., Fujikawa, A., Morii, H., Kumasaka, T., Yamamoto, M., Ishii, S. and Ogata, K. (2002) Mechanism of c-Myb-C/EBP beta cooperation from separated sites on a promoter. Cell *108*, 57-70.

Thompson, C.B., Challoner, P.B., Neiman, P.E. & Groudine, M. (1986) Expression of the c-myb proto-oncogene during cellular proliferation. Nature *319*, 374-380.

Verbeek, W., Gombart, A.F., Chumakov, A.M., Muller, C., Friedman, A.D. and Koeffler, H.P. (1999) C/EBPepsilon directly interacts with the DNA binding domain of c-myb and cooperatively activates transcription of myeloid promoters. Blood *93*, 3327-3337.

Wolff, L., Schmidt, M., Koller, R., Haviernik, P., Watson, R., Bies, J. and Maciag, K. (2001) Three genes with different functions in transformation are regulated by c-Myb in myeloid cells. Blood Cells Mol Dis *27*, 483-488.

Yamanaka, R., Barlow, C., Lekstrom-Himes, J., Castilla, L.H., Liu, P.P., Eckhaus, M., Decker, T., Wynshaw-Boris, A. and Xanthopoulos, K.G. (1997) Impaired granulopoiesis, myelodysplasia, and early lethality in CCAAT/enhancer binding protein epsilon-deficient mice. Proc Natl Acad Sci USA *94*, 13187-13192.

Ying, G.G., Proost, P., van Damme, J., Bruschi, M., Introna, M. and Golay, J. (2000) Nucleolin, a novel partner for the Myb transcription factor family that regulates their activity. J Biol Chem *275*, 4152-4158.

Zhang, D.E., Zhang, P., Wang, N.D., Hetherington, C.J., Darlington, G.J. and Tenen, D.G. (1997) Absence of granulocyte colony-stimulating factor signaling and neutrophil development in CCAAT enhancer binding protein alpha- deficient mice. Proc Natl Acad Sci USA *94*, 569-574.

Chapter 7

DOES C-MYB HAVE A ROLE IN HAEMOPOIETIC STEM CELLS AND MULTILINEAGE PROGENITORS?

[1]Nikla R. Emambokus and Jon Frampton
Institute of Biomedical Research, Birmingham University Medical School, Edgbaston, Birmingham, B15 2TT, United Kingdom, [1]Harvard Medical School, 320 Longwood Avenue, Boston MA 02115, United States of America.

Abstract: The c-Myb transcription factor is widely expressed throughout the haemopoietic hierarchy, including in stem cells and multilineage progenitors. Although the formation of such immature haemopoietic cells during development is possible in the absence of c-Myb, the function of progenitors, at least, is strongly dependent on the factor. However, it is not clear whether c-Myb has a role in stem cells or what function it performs in progenitor cells. The recent creation of novel alleles of c-*myb* that are either expressed at lower levels or can be conditionally inactivated is now enabling a closer examination of the involvement of c-Myb in these immature stages of haemopoiesis.

1. INTRODUCTION

In normal haemopoietic cells, c-Myb expression is high in more immature cells and is generally downregulated upon terminal differentiation (Kastan et al., 1989). The importance of c-Myb in the haemopoietic system is vividly demonstrated by the death of c-*myb* knockout mice due to their failure to develop foetal haemopoiesis (Mucenski et al., 1991). Previous chapters have considered the specific role of c-Myb in individual haemopoietic lineages and clearly a picture is emerging concerning the processes and genes that it regulates. In contrast, our understanding of whether c-Myb is important in more immature blood cells, especially in haemopoietic stem cells (HSCs), is minimal or totally lacking. This chapter will summarise what is known about c-Myb in these early stages of haemopoiesis, but will be largely a perspective on what we might expect to find through further investigation and how this might be achieved.

J. Frampton (ed.), Myb Transcription Factors: Their Role in Growth, Differentiation and Disease, 145-161.

2. OVERVIEW OF HAEMOPOIETIC STEM CELLS AND PROGENITORS

2.1 The Haemopoietic Hierarchy

The haemopoietic system is a paradigm for the complex processes that come together to ensure that cells with distinct differentiated phenotypes are produced at the correct time and in appropriate numbers throughout development and during adult life. Each and every day during the life of an adult human a staggering 10^{12} blood cells can be produced. Each of the blood cell types arises from precursors that can expand before maturing along a single differentiation pathway. These committed unilineage precursors derive from progenitor cells that have the potential both to proliferate and to enter into more than one pathway of differentiation. Further up this hierarchy, the progenitors have less restricted potential and at the top sit a small number of stem cells (HSCs) that are able to keep supplying the system without themselves becoming depleted.

Stem cells and progenitors within the haemopoietic hierarchy have been defined by a combination of functional bioassays and phenotypic characterisation. Different bioassays are required for a full assessment of cells. Precursors and multilineage progenitors can be defined from the morphology of *in vitro* colonies grown in semi-solid media containing lineage-specific cytokines (Moore and Metcalf, 1970). Multilineage myeloid progenitors are also often assessed using the *in vivo* spleen colony forming (CFU-S) assay originally described by Till and McCulloch (1961). Stem cells are assayed by their ability to reconstitute lethally irradiated animals. Phenotypic characterisation and purification of stem cells and progenitors is largely achieved by use of fluorescence activated cell sorting (FACS) (reviewed in Pohlmann et al., 2001). Mouse HSCs are contained within a population of cells representing 0.05% of bone marrow cells that do not express specific lineage markers, express low levels of Thy1.1 and are positive for Sca-1 and c-Kit (Thy1.1^{lo}Sca-1^{+}c-Kit^{+}Lin^{-}). This population can be further divided into three populations: long-term HSCs (LT-HSCs), short-term HSCs (ST-HSCs), and multipotent progenitor (MPP) cells. The LT-HSC population represents about 0.005% to 0.01% of bone marrow cells and has extensive self-renewal capacity providing long-term reconstituting ability. ST-HSCs have limited self-renewal capacity and are able to contribute to haemopoiesis for 6 to 8 weeks. MPP cells cannot self-renew and can reconstitute bone marrow for no more than 4 weeks. Retention of dyes such as rhodamine 123 (Rho) by cells can be used to subdivide HSC populations further by FACS. Hence, LT-HSCs exhibit the highest rates of efflux of the dye stain most weakly (Rho^{lo}). More recently, the use of multiple combinations of surface antigens has permitted the isolation of most of the multilineage progenitor cells, including the common lymphoid progenitor (CLP), the common myeloid progenitor (CMP), the granulocyte-macrophage progenitor (GMP) and the megakaryocyte-erythroid progenitor (MEP) (Akashi et al., 2000).

The hierarchical structure of haemopoiesis has the advantage of allowing a rapid response to the demands for differentiated cells through expansion of progenitors and precursors. In order to maintain the hierarchy each step has to be tightly regulated. Progenitor expansion and commitment to differentiation are obvious points of control, but the most challenging demand is the maintenance of only a small number of stem cells. Each HSC that commits towards a progenitor fate must be replaced by a single, identical cell (self-renewal). If this replacement is not achieved then the stem cell complement will eventually be depleted, whereas if there is an excess of self renewal this could result in over expansion potentially leading to leukaemia (Dick, 2003).

2.2 Establishing the Haemopoietic System During Development

In mammalian development, the first wave of haemopoiesis is initiated in the developing yolk sac. This phase of so called "primitive" haemopoiesis is characterised by the rapid production of nucleated erythrocytes as the predominant haemopoietic cell. Primitive haemopoiesis is short-lived, lasting until about mid gestation, and is thereafter replaced by adult or "definitive" haemopoiesis. This phase is characterised by the generation of all of the mature blood cell types present in the adult. The developmental origin of HSCs has been rather controversial, and it is still unclear whether all blood lineages arise from the yolk sac or from the embryo proper. Nevertheless, studies on the mouse embryo have given us a picture of the sequence of events (see review by Dzierzak et al., 1998). HSCs can be directly assayed from E11.5 onwards, prior to this cells from explanted tissues must be cultured before HSCs can be determined by engraftment. Broadly speaking, progenitors for myeloid lineages appear in the yolk sac at E7-7.5 while multipotential progenitors with the capacity to give rise to both myeloid and lymphoid cells are generated in the para-aortic splanchnopleura (PAS) at E7.5 and can be found in the yolk sac at E8.5 when the circulation is established. Beginning at E9 CFU-S can be detected in the aorta-gonad-mesonephros (AGM) region, and to a lesser extent the yolk sac, and subsequently in the foetal liver. HSCs that are capable of repopulating lethally irradiated adults appear in cultured AGM derived at E10 and are thereafter found in the yolk sac and later in the foetal liver. In the adult,

HSCs reside in the bone marrow (Galloway and Zon, 2003). HSCs probably arise from the endothelial region of embryonic blood vessels, consistent with the long held view that endothelial and haemopoietic cells are derived from a common precursor, the haemangioblast (Sabin, 1917). Although current dogma supports the idea that the HSCs from the foetal liver populate the adult bone marrow, this has not yet been conclusively proven. In fact there are suggestions that this is not the case as marked foetal liver cells are not later traced in the adult bone marrow (Emambokus and Frampton, 2003). Also, Dieterlen-Lievre and colleagues recently showed that both the embryonic portion of the placenta, in both mice and men, have higher clonogenic multilineage potential than the embryonic foetal liver and hypothesise that the placenta might in fact be the source of the adult bone marrow haemopoietic populations (Alvarez-Silva et al., 2003).

2.3 Transcriptional Regulation of Haemopoietic Stem Cells and Progenitors

The regulation of self-renewal and commitment of stem cells and progenitors requires the interpretation by the cell of extrinsic signals in the form of cytokines and microenvironmental influences. Part, at least, of such an interpretation must involve the establishment of a programme of transcription factor activity. Indeed, identification of expressed RNAs and the results of gene ablation or over expression studies have suggested specific functions for certain transcription factors in HSCs and haemopoietic progenitors. Although it is beyond the scope of this chapter to review all of these, some selected examples will be considered because of the parallels or differences that might be seen with respect to the potential role of c-Myb in these cells.

2.3.1 Gene expression

A number of studies have sought to define a molecular signature for haemopoietic progenitors and stem cells, and these have demonstrated the expression of a vast array of transcriptional regulators, many of which to date have no defined targets or function. Earlier studies relied on the purification of specific populations of progenitors or stem cells and an analysis of transcriptional regulators that were suspected to be of importance (for example, Orlic et al., 1995; Zinovyeva et al., 2000; Akashi et al., 2000). The study by Akashi and colleagues was particularly detailed in that extensive use of differences in surface antigen expression enabled fractionation of cells not only into HSC and progenitors but allowed division of the latter into specific multi- and bipotential subtypes (see section 2.1

above). Recently screening of large collections of cDNA clones or microarray chips has produced more comprehensive and less biased profiles of gene expression. Phillips et al (2000) created a cDNA library from foetal liver enriched for HSCs by sorting for $Lin^-Sca^+Kit^+AA4.1^+$ cells and identified almost 10% of the 2119 sequenced genes as transcription factors. Similarly, Park et al (2002) used FACS to purify cells (in this case $Lin^- Thy1.1^{lo}Sca^+Kit^+$) for cDNA library construction. Using retention of the dye Rhodamine 123 they were additionally able to distinguish HSCs from multilineage progenitors and identified transcriptional regulators expressed in one or other or both of these populations. The most comprehensive analysis so far is probably that described by Ivanova et al (2002) who used highly subtracted cDNAs generated from purified HSCs, progenitors and committed cell populations from foetal liver and bone marrow to screen Affymetrix oligonucleotide arrays. Their screen encompassed approximately 80% of the genes expressed (about 14% of these were identified as transcription factors) and was able to distinguish transcripts present in mouse compared to human foetal liver HSCs ($Lin^-Sca^+Kit^+AA4.1^+$ and $Lin^-CD34^+CD38^-$, respectively), in foetal liver compared to adult bone marrow HSCs ($Lin^-Sca^+Kit^+AA4.1^+$ and $Lin^-Sca^+Kit^+$, respectively), and in long-term compared to short-term bone marrow HSCs ($Lin^-Sca^+Kit^+Rho^{lo}$ and $Lin^-Sca^+Kit^+ Rho^{hi}$, respectively).

A relatively small set of the transcription factors that are known from such studies to be expressed in HSCs or progenitors have been subjected to detailed investigations that shed light on their likely role in these cells. Most of the relevant findings to date have been the somewhat serendipitous result of ablation or over expression of genes originally identified because of their involvement in committed haemopoietic cells.

2.3.2 Gene ablation

Inactivation of gene function by targetted homologous recombination in ES cells has been used to assess the function of several transcriptional regulators relevant to haemopoiesis (Cantor and Orkin, 2002). In several cases, homozygosity for the inactivated allele leads to embryonic lethality that has been interpreted as a failure to generate haemopoietic stem cells or progenitors during development. In such cases, it has not been possible to conclude whether the particular transcription factor performed a similar crucial function in adult haemopoiesis, although recent application of conditional Cre-LoxP gene deletion technology (Rajewsky et al., 1996) has made this a tractable issue.

Inactivation of the SCL gene, encoding a basic helix-loop-helix protein that is expressed in immature haemopoietic cells and a number of committed cell types, revealed that it is essential for the formation of blood at the earliest embryonic stages of haemopoiesis (Robb et al., 1995; Shivdasani et al., 1995). Interestingly, the conditional deletion of SCL in the adult did not affect bone marrow HSC function (Mikkola et al., 2003). Embryonic lethality at about E11 was also seen in mice homozygous for a null allele of the AML/Runx-1 gene, although primitive haemopoiesis was still apparent and the major defect seemed to be one of loss of the capacity to produce definitive HSCs in the AGM (Okuda et al., 1996). Ablation of several other transcription factor genes, such as that encoding GATA-2 (Tsai et al., 1994) and Pbx-1 (DiMartino et al., 2001), also results in a lethal failure of haemopoiesis during embryogenesis, although the appearance of defective immature cells implies that these regulators are required for the maintenance, but not the initiation, of definitive haemopoiesis.

Deletion of some transcription factor genes has been found to have more subtle effects in that mice homozygous for the targetted allele are viable, but close investigation of HSC function does reveal a degree of deficiency. For example, bone marrow HSCs derived from STAT-5a/b$^{-/-}$ mice have cell autonomous defects in competitive long-term repopulating activity at least partly resulting from a reduced potential for expansion (Bradley et al., 2002).

Negative regulation of gene expression by transcription factors is likely to play a significant role in many aspects of haemopoiesis, and the tight control of HSC self renewal versus commitment is clearly a case in point. For example, homozygosity for a null allele of the *Polycomb group* gene *bmi-1* causes a progressive failure of the haemopoietic system (van der Lugt et al., 1994) which has more recently been attributed to a role for Bmi-1 in HSC maintenance through restriction of its proliferation and commitment (Lessard and Sauvageau, 2003; Park et al., 2003).

Many transcription factors are part of families of related genes and as a consequence exhibit some functional redundancy that can effectively mask the effect of a single gene ablation. Such a situation has been observed in relation to gene function in HSCs in the case of the HoxB3 and HoxB4 genes. Mice deficient in both genes have defects in haemopoiesis that have been traced to impaired proliferative and repopulating capacity of HSCs (Bjornsson et al., 2003).

2.3.3 Over expression studies

Another approach to assessing a specific role for a transcription factor in haemopoietic stem cell or progenitor function involves over expression. An obvious advantage over gene ablation studies is that the likelihood of embryonic lethality is much less, although this strategy has not been undertaken extensively. Nevertheless, a number of recent publications have addressed the role of HoxB4 and Pbx-1 through over expression and have generated conclusions that fit well with those derived from the gene ablation studies on these genes already discussed above (section 2.3.2). Hence, introduction of a retrovirus expressing HoxB4 into bone marrow cells cultured *ex vivo* resulted in a massive expansion of HSCs that retained full repopulating capacity (Antonchuk et al., 2002). This enhancing effect of HoxB4 was found to be further enhanced if the expression of Pbx-1 was simultaneously reduced (Krosl et al., 2003). A similar study involving HoxB4 over expression in pre-circulation yolk sac cells by retroviral infection or in haemopoietic precursors generated by *in vitro* differentiation of embryonic stem (ES) cells containing an inducible transgene resulted in the production of definitive HSCs capable of repopulating irradiated animals (Kyba et al., 2002).

3. EXPRESSION OF C-MYB IN HAEMOPOIETIC STEM CELLS AND PROGENITORS

It has long been known that c-Myb is highly expressed in immature haemopoietic progenitors (Kastan et al., 1989). c-*myb* RNA can be detected at sites of definitive haemopoietic stem cell emergence in the PAS (Labastie et al., 1998) and is expressed at the very onset of definitive haemopoietic activity in the mouse yolk sac (Palis et al., 1999). The expression of c-*myb* together with that of AML-1/Runx-1 is among the first molecular features to distinguish the emerging embryonic haemopoietic cells from the haemogenic endothelium. Other transcription factors, such as SCL and GATA-2, have a broader pattern of expression including haemogenic and non-haemogenic endothelium.

Several of the studies on gene expression in immature haemopoietic cells revealed the presence of c-*myb* RNA in c-Kit^{+} progenitors and HSCs (Orlic et al., 1995; Zinovyeva et al., 2000; Akashi et al., 2000). The analysis performed by Zinovyeva et al (2000) went further by suggesting that only short-term HSCs express c-*myb*. Of the three large scale cDNA library or oligonucleotide microarray screening efforts described above c-*myb* expression was reported in stem cells by Ivanova et al., (2002). Again, short-term HSCs were defined as c-*myb* expressing, although the analysis did also detect c-*myb* RNA in LT-HSCs. Not unexpectedly, expression was also seen in mutlilineage and committed progenitors.

4. C-MYB HAS AN ESSENTIAL ROLE IN HAEMOPOIETIC DEVELOPMENT

4.1 Definitive Haemopoiesis is Initiated but cannot Progress in c-myb-/- Embryos

Given the pattern of c-*myb* RNA expression in haemopoietic cells, it came as no great surprise that mice homozygous for an inactivated allele of c-*myb* died *in utero* at about E15 due to a failure to develop foetal liver haemopoiesis (Mucenski et al., 1991). However, in spite of being first characterised over a decade ago, it is still not clear why haemopoiesis fails in c-*myb*$^{-/-}$ embryos. Yolk sac haemopoiesis is apparently unaffected by the absence of c-Myb since primitive nucleated erythrocytes were present (Mucenski et al., 1991) and were produced during *in vitro* differentiation of c-*myb*$^{-/-}$ ES cells (Clarke et al., 2000). Apart from primitive erythrocytes, the only mature cells present in the c-*myb*$^{-/-}$ foetal liver were megakaryocytes and macrophages. Several questions are raised from these basic observations. Firstly, is commitment to definitive haemopoiesis occurring in the absence of c-Myb? Secondly, if definitive haemopoiesis is present are the non-erythroid cells the result of aberrant differentiation of progenitors? Finally, why do c-*myb*$^{-/-}$ embryos survive until E15, when mice homozygous for gene knockouts that also fail to establish definitive haemopoiesis die earlier?

The first two questions have been at least partly answered by subsequent studies utilising the c-*myb* knockout allele. A more detailed examination of the foetal liver from c-*myb*$^{-/-}$ embryos and analysis of cells arising during the *in vitro* differentiation of c-*myb*$^{-/-}$ ES cells revealed that cells with the phenotypic characteristics of progenitors were produced in the absence of c-Myb. Hence, cells co-expressing the progenitor antigen CD34 and the pan-haemopoietic marker CD45 were detected in the c-*myb*$^{-/-}$ foetal liver (Sumner et al., 2000) and amongst the cells derived from c-*myb*$^{-/-}$ ES cells (Clarke et al., 2000). However, these cells did not possess functional activity as progenitors when assayed for colony formation *in vitro*. The study by Sumner et al (2000) also showed that c-*myb*$^{-/-}$ ES cells introduced into wild type blastocysts became part of a chimaeric animal contributing widely to non-haemopoietic tissues although the presence of c-*myb*$^{-/-}$ haemopoietic cells was very restricted. In this latter study no contribution of c-*myb*$^{-/-}$ cells to adult haemopoietic cells could be detected but examination of the foetal liver of E11-12 chimaeras revealed the presence of c-*myb*$^{-/-}$ progenitor like cells that subsequently failed to expand or were lost by terminal differentiation. A related study involving chimaeric animals suggested that some c-*myb*$^{-/-}$ haemopoietic progenitors could be detected in the adult. Allen et al (1999) were able to identify a small number of very immature c-*myb*$^{-/-}$ thymocytes in the adult thymus of chimaeras between c-*myb*$^{-/-}$ ES cells and *rag-1*$^{-/-}$ host blastocysts. The highly selective environment used in this latter study is the likely explanation for the existence of detectable c-*myb*$^{-/-}$ haemopoietic cells nevertheless, like the study by Sumner et al (2000), their presence is indicative of the generation of definitive haemopoietic cells. Whether these cells are the descendants of HSCs originating in the AGM or from some later haemogenic endothelium is not known. Certainly, haemogenic sites within the AGM appear to be functioning in c-*myb*$^{-/-}$ embryos since haemopoietic cells could be detected emerging from the endothelial layer with the same frequency in the wild type and knockout (NE and JF, unpublished).

Regarding the survival of c-*myb*$^{-/-}$ embryos to E15, it is somewhat surprising that mice homozygous for null alleles of the erythropoietin receptor die at an earlier point around E13 (Lin et al., 1996), when they too are defective for definitive erythropoiesis. Likewise, why does the absence of definitive haemopoiesis caused by ablation of AML/Runx-1 or GATA-2 (Okuda et al., 1996; Tsai et al., 1994) result in embryonic lethality at about E11? Several possibilities can be suggested, but there is currently no substantive evidence in support of any of them. Firstly, there might be increased or prolonged production of primitive erythroid cells in c-*myb*$^{-/-}$ embryos. This has not been described although a detailed comparative enumeration has not been performed. Intriguingly, it has been suggested that Myb can programme primitive erythroid cells towards a definitive progenitor phenotype (McNagny and Graf, 2003). This conclusion was based on the apparent primitive erythroid nature of the targets for the progenitor transforming Myb-Ets encoding avian leukaemia virus E26. A second possibility is that in c-*myb*$^{-/-}$ embryos there is a continued presence of other haemopoietic cells that ameliorate the effects of the absence of definitive red cells. One candidate might be megakaryocytes and the platelets they yield, indeed their persistence in c-*myb*$^{-/-}$ embryos has been noted (Mucenski et al., 1991; Sumner et al., 2000). Finally, and perhaps most likely, is the possibility that the inactivation of genes leading to embryonic lethality earlier than seen for c-*myb*$^{-/-}$ embryos results in additional uncharacterised defects either in primitive erythropoiesis or in essential non-haemopoietic cells.

4.2 c-Myb is Required for the Maintenance of Adult Haemopoiesis

To investigate the function of c-Myb in adult haemopoiesis we have generated a conditional loxP-modified ("floxed") allele of c-*myb*

(Emambokus et al., 2003). We introduced loxP sites upstream of exon 3 and downstream of exon 6 and a neo^R selection cassette flanked by Flp recombinase recognition sites (FRT) was cloned into intron 6. Removal of the neo^R cassette by crossing with animals expressing Flp recombinase mice yielded the c-myb^F allele.

To bring about conditional deletion we crossed c-myb^F to the Mx-Cre transgenic line (Kühn et al, 1995). Transcription from the Mx promoter, and hence expression of Cre, can be stimulated to high levels in most tissues by type I interferon (IFNα/β), effective production of IFN being elicited by injection of double-stranded RNA (polyinosine-polycytosine, pIpC). Three month old c-$myb^{F/+}$ and c-$myb^{F/F}$ mice carrying the Mx-Cre transgene were injected with 250 μg polydI:dC, repeat injections being performed on day 2 and day 4. By day 2, deletion was already achieved in the majority of cells in the bone marrow of c-$myb^{F/F}$ mice. Although overall cellularity was little affected by day 5 the bone marrow exhibited a dramatic decrease in the number of c-Kit^+ progenitor cells in c-$myb^{F/F}$ compared to c-$myb^{F/+}$ mice. By day 10, the downstream effects of loss of progenitor cells were apparent in that the bone marrow of c-$myb^{F/F}$ mice was hypocellular and peripheral blood counts showed signs of anaemia and thrombocytopenia (NE and JF, unpublished).

5. REDUCED LEVELS OF C-MYB LEAD TO ABERRANT PROGENITOR BEHAVIOUR

As discussed above, the analysis of c-$myb^{-/-}$ foetal liver has suggested that definitive progenitor expansion or differentiation fate are influenced by c-Myb, while induced deletion of the floxed c-*myb* allele in adults also hints at a function for c-Myb in "gating" entry into differentiation, at least along some lineages. Additional indications concerning the importance of c-Myb in regulating the balance of expansion versus commitment of MPPs has recently come from examination of embryos and adults that express reduced levels of c-Myb (Emambokus et al., 2003). During the generation of the floxed c-myb^F allele it was found that the intermediate allele containing the neo^R selection cassette (c-myb^{loxP}) is expressed at a much lower level than normal (5 to 10% of wild type). Through crossing with mice carrying the null allele (c-myb^-) it was possible to generate embryos containing either one (c-$myb^{loxP/-}$) or two (c-$myb^{loxP/loxP}$) of this "knockdown" allele. Like c-*myb* null embryos, those containing only a single knockdown allele died *in utero* at E15 whereas c-$myb^{loxP/loxP}$ mice reached adulthood. Closer examination of the foetal livers of c-$myb^{loxP/-}$ embryos revealed an absence of definitive erythroid cells, but they could be distinguished from c-$myb^{-/-}$ embryos by the presence of functional progenitor cells. However, these progenitors were abnormal in that the majority were multipotential but showing an emphasis on differentiation towards the macrophage and megakaryocyte lineages. The c-*myb*$^{loxP/-}$ progenitors differentiated *in vitro* earlier than their wild type counterparts and exhibited a rapid proliferative outburst of terminally differentiated cells. Although showing a less severe and non-lethal phenotype, homozygous c-*myb*$^{loxP/loxP}$ embryos also contained progenitors with aberrant differentiation potential.

Taken together, the results from mice containing the null, conditional and knockdown alleles of c-*myb* imply that MPPs, and by implication HSCs preceding them, can be generated in the absence of c-Myb. However, normal numbers and behaviour of MPPs requires c-Myb. Low levels of c-Myb seem to be all that is required for the production of normal numbers of MPPs, but higher levels are necessary for the maintenance of progenitor numbers and the correct commitment to differentiation.

6. WHAT MIGHT BE THE ROLE OF C-MYB IN HAEMOPOIETIC STEM CELLS AND PROGENITORS?

Two key issues concerning the role of c-Myb in HSCs/MPPs arise out of the discussion above. Firstly, even though it is not obligatory for HSC formation, does c-Myb nevertheless play a role in the HSC? Secondly, what does the transient presence of normal numbers of aberrant MPPs in c-*myb*$^{loxP/-}$ embryos indicate? Does the aberrant differentiation of MPPs in the presence of only a low level of c-Myb lead to their depletion because of limitations either in the potential of the MPP for expansion or in their generation from more immature HSCs?

6.1 Potential Functions of c-Myb in HSCs/MPPs

c-Myb is expressed in ST-HSCs, although if it is present in LT-HSCs it must be at lower levels (see section 3). Since c-Myb has been implicated in proliferation, differentiation and apoptosis, which aspects of these processes that requires c-Myb function might be expected to play a part in HSCs and MPPs? Several studies have shown the importance of cyclin-dependent kinase inhibitors in the maintenance of stem cell kinetics (Cheng et al., 2000a; Lessard and Sauvageau, 2003; Park et al., 2003) and in progenitor expansion (Cheng et al., 2000b). Although c-Myb has been suggested as a possible regulator of the cell cycle through effects on cyclins or cyclin-dependent kinases, the evidence is at best sketchy. A more credible link to proliferative potential is the possibility that c-Myb regulates expression of c-Kit, the receptor for stem cell factor (Hogg et al., 1997; Ratajczak et al., 1998). Interestingly, the expression profiling of stem cells and progenitors by Ivanova et al (2002) revealed an exact parallel between c-*myb* and c-*kit* RNAs. It is also feasible that c-Myb could perform a function similar to that of HoxB4 as discussed above.

Apoptosis has been postulated as an important factor in the regulation of HSCs and progenitors. For example, transgenic over expression of the anti-apoptotic protein Bcl-2 resulted in an increased number of bone marrow HSCs that had an increased potential in competitive repopulation assays (Domen et al., 2000). The *bcl*-2 gene has been shown to be a target for positive regulation by Myb in a variety of committed haemopoietic cells (Frampton et al., 1996; Taylor et al., 1996) and could therefore be a relevant target of c-Myb in more immature cells.

The majority of characterised c-Myb targets are lineage-specific differentiation-associated genes (see Chapters 13 and 14). As suggested by our observations on aberrant lineage specification from progenitors in the presence of low levels of the protein, it is also possible that c-Myb plays a controlling role on the commitment to differentiation (see Section 5). The apparently uncontrolled differentiation seen in the presence of reduced levels of c-Myb is consistent with results from model haemopoietic cell systems showing that c-Myb blocks differentiation and maintains an immature state (reveiwed in Oh and Reddy, 1999). Whether such a controlling influence by c-Myb on the commitment of differentiation from MPPs also applies in the more immature HSCs remains to be seen. There is also no present indication what target genes might be relevant to this activity of c-Myb.

6.2 The Importance of the Level of c-Myb Expression

It is becoming recognised that the absolute level of a specific transcription factor is an important component of the mechanism of lineage specification from progenitors. One of the first examples of this came from experiments demonstrating that the level of ectopic expression of GATA-1 in chicken myelomonocytic cells determined the phenotype of the "reprogrammed" cells (Kulessa et al 1994). More recently, DeKoter and Singh (2000) showed that graded expression of PU.1 regulates B-cell and macrophage development.

There are probably many ways in which variations in the level of a specific transcription factor might influence gene regulation differentially. Most likely, it is the consequence of distinct combinatorial interactions and the formation of multiprotein complexes. Such complexes may acquire the ability to regulate specific genes. c-Myb has a number of known partners with which it cooperates to regulate lineage specific genes, for example C/EBP on myelomonocytic gene promoters and HES-1 on the CD4 gene (Ness et al., 1993; Allen et al., 2001). Alternatively, complex formation may have an indirect effect by "titrating out" particular factors. In this way, the interaction between PU.1 and GATA-1 is a major determinant of erythroid versus myelomonocytic differentiation (Rekhtman et al., 1999; Nerlov et al., 2000). In a variant of this idea, c-Myb has been shown to be in "competition" with GATA-1 for formation of a complex with CBP (Takahashi et al., 2000). This interaction between c-Myb and CBP may well underlie several of the observations seen with respect to the erythroid and megakaryocyte lineages. Hence, Kaspar et al (2002) recently showed that mutations in the protein interaction surface of the p300 co-activator have a profound effect on haemopoiesis, and that some of this effect may be mediated through an altered interaction with c-Myb. How various factors interact in multi-protein complexes involving c-Myb will be an important focus in the coming years, and the role of c-Myb in HSCs and progenitors will undoubtedly involve some of these interactions.

7. FUTURE PROSPECTS

Although suspected of being of importance in haemopoietic progenitors for many years, the role of c-Myb in these cells is only just now beginning to be investigated. The development of genetic tools for the manipulation of c-Myb activity as well as sophisticated technologies for the analysis of the complete pattern of gene expression should yield important insights in the coming few years. As discussed above, a key issue is whether c-Myb has an influence on any of the definable HSC subsets. The profound effect of the absence of c-Myb on the development of haemopoietic cells from progenitors has made it impossible to assay HSC numbers or potential in haemopoietic tissues derived from the c-*myb* knock out. However, it might now be feasible to perform reconstitution assays using purified HSCs derived from the foetal livers or adult bone marrow of mice harbouring the knock down allele. Indications of a role for c-Myb in HSCs could also become apparent from an appropriate analysis of haemopoietic cells sorted from mice over-expressing c-Myb in immature cells.

Another area that may be worthy of investigation is the role that c-Myb could play in stromal cells. In the various niches in which HSCs reside, often in a quiescent state, it is possible that some of the component stromal cells have functions that are c-Myb-dependent. There is little direct evidence for this although Rob Ramsay and colleagues recently reported that foetal liver stromal cells express c-Myb that seems to have a positive effect on SCF expression (Sicurella et al., 2001).

ACKNOWLEDGEMENTS

JF is supported by a Wellcome Trust Senior Research Fellowship.

REFERENCES

Akashi, K., Traver, D., Miyamoto, T. and Weissman, I.L. (2000) A clonogenic myeloid progenitor that gives rise to all myeloid lineages. Nature *404*, 193-197.

Allen, R.D. 3rd, Bender, T.P. and Siu, G. (1999) c-Myb is essential for early T cell development. Genes Dev *13*, 1073-1078.

Allen, R.D., Kim, H.K., Sarafova, S.D. and Siu, G. (2001) Negative regulation of CD4 gene expression by a HES-1-c-Myb complex. Mol Cell Biol *21*, 3071-3082.

Alvarez-Silva, M., Belo-Diabangouaya, P., Salaun, J. and Dieterlen-Lievre, F. (2003) Mouse placenta is a major hematopoietic organ. Development, in press.

Antonchuk, J., Sauvageau, G. and Humphries, R.K. (2002) HOXB4-induced expansion of adult hematopoietic stem cells ex vivo. Cell *109*, 39-45.

Bjornsson, J.M., Larsson, N., Brun, A.C., Magnusson, M., Andersson, E., Lundstrom, P., Larsson, J., Repetowska, E., Ehinger, M., Humphries, R.K. and Karlsson, S. (2003) Reduced proliferative capacity of hematopoietic stem cells deficient in hoxb3 and hoxb4. Mol Cell Biol *23*, 3872-3883.

Bradley, H.L., Hawley, T.S. and Bunting, K.D. (2002) Cell intrinsic defects in cytokine responsiveness of STAT5-deficient hematopoietic stem cells. *Blood* 100, 3983-3989.

Cai, Z., de Bruijn, M., Ma, X., Dortland, B., Luteijn, T., Downing, R.J. and Dzierzak, E. (2000) Haploinsufficiency of AML1 affects the temporal and spatial generation of haemopoietic stem cells in the mouse embryo. Immunity *4*, 423-431.

Cheng, T., Rodrigues, N., Shen, H., Yang, Y., Dombkowski, D., Sykes, M. and Scadden, D.T. (2000a) Hematopoietic stem cell quiescence maintained by $p21^{cip/waf1}$. Science *287*, 1804-1808.

Cheng, T., Rodrigues, N., Dombkowski, D., Stier, S. and Scadden, D.T. (2000b) Stem cell repopulation efficiency but not pool size is governed by p27kip1. Nat Med *6*, 1235-1240.

Clarke, D., Vegiopoulos, A., Crawford, A., Mucenski, M., Bonifer, C. and Frampton, J. (2000) *In vitro* differentiation of c-$myb^{-/-}$ ES cells reveals that the colony forming capacity of unilineage macrophage precursors and myeloid progenitor commitment are c-Myb independent. Oncogene *19*, 3343-3351.

DeKoter, R.P. and Singh, H. (2000) Regulation of B lymphocyte and macrophage development by graded expression of PU.1. Science *288*, 1439-1441.

Dick, J.E. (2003) Self-renewal writ in blood. Nature *423*, 231-233.

DiMartino, J.F., Selleri, L., Traver, D., Firpo, M.T., Rhee, J., Warnke, R., O'Gorman, S., Weissman, I.L. and Cleary, M.L. (2001) The Hox cofactor proto-oncogene Pbx1 is required for maintenance of definitive hematopoiesis in the fetal liver. Blood *98*, 618-626.

Domen, J., Cheshier, S.H. and Weissman, I.L. (2000) The role of apoptosis in the regulation of hematopoietic stem cells: Overexpression of Bcl-2 increases both their number and repopulation potential. J Exp Med *191*, 253-264.

Dzierzak, E., Medvinsky, A. and de Bruijn, M. (1998) Qualitative and quantitative aspects of haematopoietic cell development in the mammalian embryo. Immunol Today *5*, 228-236.

Emambokus, N.R. and Frampton, J. (2003) The glycoprotein IIb molecule is expressed on early murine hematopoietic progenitors and regulates their numbers in sites of hematopoiesis. Immunity *19*, 33-45.

Emambokus, N.R., Vegiopoulos, A., Harman, B., Jenkinson, E.J., Anderson, G. and Frampton, J. (2003) Progression through key stages of the hematopoietic hierarchy is dependent on distinct threshold levels of c-Myb. EMBO J *22*, 4478-4488.

Frampton, J., Ramqvist, T. and Graf, T. (1996) v-Myb of E26 leukemia virus up-regulates *bcl*-2 and suppresses apoptosis in myeloid cells. Genes Dev *10*, 2720-2731.

Galloway, J.L. and Zon, L.I. (2003) Ontogeny of haemopoiesis: examining the emergence of haemopoietic cells in the vertebrate embryo. Curr Top Dev Biol *53*, 139-158.

Hogg, A., Schirm, S., Nakagoshi, H., Bartley, P., Ishii, S., Bishop, J.M. and Gonda, T.J. (1997) Inactivation of a c-Myb/estrogen receptor fusion protein in transformed primary cells leads to granulocyte/macrophage differentiation and down regulation of c-*kit* but not c-*myc* or *cdc2*. Oncogene *15*, 2885-2898.

Ivanova, N.B., Dimos, J.T., Schaniel, C., Hackney, J.A., Moore, K.A. and Lemischka, I.R. (2002) A stem cell molecular signature. Science *298*, 601-604.

Kastan, M.B., Slamon, D.J. and Civin, C.I. (1989) Expression of protooncogene c-myb in normal human haemopoietic cells. Blood *73*, 1444-1451.

Kasper, L.H., Boussouar, F., Ney, P.A., Jackson, C.W., Rehg, J., van Deursen, J.M. and Brindle, P.K. (2002) A transcription-factor-binding surface of coactivator p300 is required for haematopoiesis. Nature *419*, 738-743.

Krosl, J., Beslu, N., Mayotte, N., Humphries, R.K. and Sauvageau, G. (2003) The competitive nature of HOXB4-transduced HSC is limited by PBX1: The generation of ultra-competitive stem cells retaining full differentiation potential. Immunity *18*, 561-571.

Kühn, R., Schwenk, F., Aguet, M. and Rajewsky, K. (1995) Inducible gene targeting in mice. Science *269*, 1427-1429.

Kulessa, H., Frampton, J. and Graf, T. (1995) GATA-1 reprograms avian myelomonocytic cell lines into eosinophils, thromboblasts and erythroblasts. Genes & Dev *9*, 1250-1262.

Kyba, M., Perlingeiro, R.C.R. and Dalet, G.Q. (2002) HoxB4 confers definitive lymphoid-myeloid engraftment potential on embryonic stem cell and yolk sac hematopoietic progenitors. Cell *109*, 29-37.

Labastie, M.C., Cortes, F., Romeo, P.H., Dulac, C. and Peault, B. (1998) Molecular identity of haemopoietic precursor cells emerging in the human embryo. Blood *92*, 3624-3635.

Lessard, J. and Sauvageau, G. (2003) *Bmi-1* determines the proliferative capacity of normal and leukaemic stem cells. Nature *423*, 255-260.

Lin, C.S., Lim, S.K., D'Agati, V. and Costantini, F. (1996) Differential effects of an erythropoietin receptor gene disruption on primitive and definitive erythropoiesis. Genes Dev. *10*, 154-164.

McNagny, K.M. and Graf, T. (2003) E26 leukaemia virus converts primitive erythroid cells into cycling multilineage progenitors. Blood *101*, 1103-1110.

Mikkola,H.K.A., Klintman,J., Yang,H., Hock,H., Schlaeger,T.M., Fujiwara,Y. and Orkin,S.H. (2003) Hematopoietic stem cells retain long-term repopulating activity and multipotency in the absence of stem-cell leukaemia SCL/*tal-1* gene. Nature *421*, 547-551.

Moore, M.A. and Metcalf, D. (1970) Ontogeny of the haemopoietic system: yolk sac origin of in vivo and in vitro colony forming cells in the developing mouse embryo. Br J Haematol. *18*, 279-296.

Mucenski, M.L., McLain, K., Kier, A.B., Swerdlow, S.H., Schreiner, C.M., Miller, T.A., Pietryga, D.W., Scott, W.J. Jr and Potter, S.S. (1991) A functional c-myb gene is required for normal murine foetal hepatic haemopoiesis. Cell *65*, 677-689.

Ness, S.A., Kowenz-Leutz, E., Casini, T., Graf, T. and Leutz, A. (1993) Myb and NF-M: combinatorial activators of myeloid genes in heterologous cell types. Genes Dev *7*, 749-759.

Nerlov, C., Querfurth, E., Kulessa, K. and Graf, T. (2000) GATA-1 interacts with the myeloid PU.1 transcription factor and represses PU.1-dependent transcription. Blood *95*, 2543-2551.

Oh, I.H. and Reddy, E.P. (1999) The myb gene family in cell growth, differentiation and apoptosis. Oncogene *18*, 3017-3033.

Okuda, T., van Deursen, J., Hiebert, S.W., Grosveld, G. and Downing, J.R. (1996) AML1, the target of multiple chromosomal translocations in human leukaemia, is essential for normal foetal liver haemopoiesis. Cell *84*, 321-330.

Orlic, D., Anderson, S., Biesecker, L.G., Sorrentino, B.P. and Bodine, D.M. (1995) Pluripotent haemopoietic stem cells contain high levels of mRNA for c-kit, GATA-2, p45 NF-E2, and c-myb and low levels or no mRNA for c-fms and the receptors for granulocyte colony-stimulating factor and interleukins 5 and 7. Proc Natl Acad Sci USA *92*, 4601-4605.

Palis, J., Robertson, S., Kennedy, M., Wall, C. and Keller, G. (1999) Development of erythroid and myeloid progenitors in the yolk sac and embryo proper of the mouse. Development *126*, 5073-5084.

Park, I., He, Y., Lin, F., Laerum, O.D., Tian, Q., Bumgarner, R., Klug, C.A., Li, K., Kuhr, C., Doyle, M.J., Xie, T., Schummer, M., Sun, Y., Goldsmith, A., Clarke, M.F., Weissman, I.L., Hood, L. and Li, L. (2002) Differential gene expression of adult murine hematopoietic stem cells. Blood *99*, 488-498.

Park, I., Qian, D., Kiel, M., Becker, M.W., Pihalja, M., Weissman, I.L., Morrison, S.J. and Clarke, M.F. (2003) Bmi-1 is required for maintenance of adult self-renewing hematopoietic stem cells. Nature *423*, 302-305.

Phillips, R.L., Ernst, R.E., Brunk, B., Ivanova, N., Mahan, M.A., Deanehan, J.K., Moore, K.A., Overton, G.C. and Lemischka, I.R. (2000) The genetic program of hematopoietic stem cells. Science *288*, 1635-1640.

Pohlmann, S.J., Slayton, W.B. and Spangrude, G.J. (2001) Stem cell populations: purification and behaviour. In: Zon, L.I. ed. Hematopoiesis - A Developmental Approach. Oxford, United Kingdom: Oxford University Press, 35-47.

Rajewsky, K., Gu, H., Kuhn, R., Betz, U.A., Muller, W., Roes, J. and Schwenk, F. (1996) Conditional gene targeting. J Clin Invest *98*, 600-603.

Ratajczak, M.Z., Pernotti, D., Melotti, P., Powzaniuk, M., Calabretta, B., Onodera, K., Kregenow, D.A., Machalinski, B. and Gewirtz, A.M. (1998) Myb and *ets* proteins are candidate regulators of c-*kit* expression in human haemopoietic cells. Blood *91*, 1934-1946.

Rektman, N., Radparvar, F., Evans, T. and Skoultchi, A.I. (1999) Direct interaction of haemopoietic transcription factors PU.1 and GATA-1: functional antagonism in erythroid cells. Genes Dev *13*, 1398-1411.

Robb, L., Lyons, I., Li, R., Hartley, L., Kontgen, F., Harvey, R.P., Metcalf, D. and Begley, C.G. (1995) Absence of yolk sac hematopoiesis from mice with a targeted disruption of the scl gene. Proc Natl Acad Sci USA *92*, 7075-7079.

Sabin, F.R. (1917) Origin and development of the primitive vessels of the chick and of the pig. Contrib Embryol Carnegie Inst *226*, 61-124.

Shivdasani, R.A., Mayer, E.L. and Orkin, S.H. (1995) Absence of blood formation in mice lacking the T-cell leukaemia oncoprotein tal-1/SCL. Nature *373*, 432-434.

Sicurella, C., Freeman, R., Micallef, S., Mucenski, M.L., Bertoncello, I. and Ramsay, R.G. (2001) Defective stem cell factor expression in c-myb null fetal liver stroma. Blood Cells Mol Dis *27*, 470-478.

Sumner, R., Crawford, A., Mucenski, M. and Frampton, J. (2000) Initiation of adult myelopoiesis can occur in the absence of c-Myb whereas subsequent development is strictly dependent on the transcription factor. Oncogene *19*, 3335-3342.

Takahashi, T., Suwabe, N., Dai, P., Yamamoto, M., Ishii, S. and Nakano, T. (2000) Inhibitory interaction of c-Myb and GATA-1 via transcriptional co-activator CBP. Oncogene *19*, 134-140.

Taylor, D., Badiani, P. and Weston, K. (1996) A dominant negative interfering Myb mutant causes apoptosis in T cells. Genes Dev *10*, 2732-2744.

Till, J. E. and McCulloch, E. A. (1961) A direct measurement of the radiation sensitivity of normal mouse bone marrow cells Radiat Res. *14*, 213-22.

Tsai, F.Y., Keller, G., Kuo, F.C., Weiss, M., Chen, J., Rosenblatt, M., Alt, F.W. and Orkin, S.H. (1994) An early haematopoietic defect in mice lacking the transcription factor GATA-2. Nature *371*, 221-226.

van der Lugt, N.M., Domen, J., Linders, K., van Roon, M., Robanus-Maandag, E., te Riele, H., van der Valk, M., Deschamps, J., Sofroniew, M., van Lohuizen, M., et al. (1994) Posterior trnsformation, neurological abnormalities, and severe hematopoietic defects in mice with a targeted deletion of the bmi-1 proto-oncogene. Genes Dev *8*, 757-759.

Zinovyeva, M.V., Zijlmans, J.M.J.M., Fibbe, W.E., Visser, J.W.M. and Belyavsky, A.V. (2000) Analysis of gene expression in subpopulations of murine hematopoietic stem and progenitor cells. Exp Hematol *28*, 318-334.

Chapter 8

A-MYB IN DEVELOPMENT AND CANCER

Ramana V. Tantravahi, Stacey J. Baker and E. Premkumar Reddy
Fels Institute for Cancer Research and Molecular Biology, Temple University School of Medicine, 3307 N. Broad Street, Philadelphia, PA 19140, United States of America.

Abstract: The mammalian *myb* gene family consists of a set of three genes, *c-myb*, *A-myb*, and *B-myb*. Of these, *c-myb* is the most extensively studied. The three *myb* genes encode transcription factors that bind DNA in a sequence specific manner and regulate complex cellular processes, such as proliferation, differentiation and histogenesis. Myb proteins play a central role in the maintenance of the differentiation state of cells; thus implicating deregulated *myb* gene expression or Myb protein function in establishment or maintenance of the neoplastic state. *Myb* gene sequences are conserved through evolution. While the three *myb* genes appear to code for proteins that bind to similar DNA sequences, each of these proteins exhibits a characteristic pattern of expression and intrinsic biochemical activity. This review describes the structure, function and regulation of the A-*myb* gene and its protein product and compares the properties of A-Myb with the more extensively studied c-Myb protein.

1. INTRODUCTION

Myb gene sequences were first isolated from the avian acute transforming retrovirus avian myeloblastosis virus (AMV) in 1941 (Hall et al., 1941). Further studies by Ivanov in 1964 (Ivanov et al., 1964) led to the discovery of a different virus isolate named avian erythroblastosis virus E26. Purified AMV and E26 particles are remarkable in their ability to rapidly induce myeloblastic and erythroblastic leukaemia in infected birds with very high efficiency.

Molecular biological methods, including cDNA cloning and DNA sequencing, led to the observation that a common transforming gene sequence named *v-myb* exists in both virus isolates. Subsequent observations of the DNA from uninfected birds revealed that the *myb* transforming gene sequences were present in the genomes of host organisms,

J. Frampton (ed.), Myb Transcription Factors: Their Role in Growth, Differentiation and Disease, 163-179.

thus indicating that a capture of these sequences (retroviral transduction) had occurred during the genesis of both acute transforming retroviruses.

2. MYB RELATED GENES

2.1 A-*myb* Gene Structure and Expression

Examination of the cellular *myb* gene (*c-myb*) began after the cloning of the *c-myb* cDNA from chicken, mouse and human cells by various groups (Gonda et al., 1985; Rosson and Reddy, 1986; Bender and Kuehl, 1986; Majell, et al., 1986; Katzen et al., 1985). *Myb* gene sequences are conserved highly in nearly every metazoan species thus far examined (see the chapter by Davidson et al.).

In 1988, Nomura and colleagues used cDNA libraries produced from cultured human tumour cell lines to isolate *myb*-related sequences (Nomura et al., 1988). From these screening studies, two novel genes, *A-myb* and *B-myb* were isolated. Sequence analysis of the two open reading frames revealed sequence capable of encoding proteins of similar size to the human *c-myb* gene product. Figure 1 is a schematic depiction of the Myb proteins encoded by the three *myb* genes. Of these, it is now well established that the c-*myb* gene encodes two proteins of 75 and 89 kDa, due to alternative splicing (Dudek and Reddy, 1989; Dasgupta and Reddy, 1989; Shen-Ong, et al., 1989). The p89 isoform, which is less abundant (constituting approximately 1-10% of the total Myb protein), is encoded by a mRNA which contains an additional exon termed 9A. It is interesting to note that both the A-Myb and B-Myb proteins contain exon 9A sequences and thus are more homologous to the p89 isoform of c-Myb (reviewed in Oh and Reddy, 1999). Each protein possesses the characteristic tripartite DNA binding motif, central transactivation domain, and C-terminal negative regulatory domain.

The DNA binding domain of the Myb proteins is present at their amino termini. Consisting of three 50 amino acid repeat sequences termed R1, R2 and R3, the DNA binding domain is the most highly conserved stretch of sequence found in all MYB protein isoforms, all of which seem to bind to bind to PyAACG/TG *in vitro* (Biedenkapp et al., 1988). Two important features have been noted within these tandem repeats. First, there is a periodic occurrence of tryptophans (Anton and Frampton, 1988); each of the three repeats has three tryptophans which are separated by 18 or 19 amino acid residues and this feature is conserved between mouse, human, chicken, and *Drosophila* c-Myb, corn c1 and yeast Bas1 as well as A-Myb and B-Myb (reviewed in Lipsick, 1996). The tryptophan repeat of the DNA binding domain is a characteristic feature that defines all Myb family proteins and mutational analysis of c-Myb has shown that alterations within this structure produce profound effects on the ability of the mutant proteins to bind to DNA (Saikumar et al., 1990, Frampton et al., 1991). While extensive mutational analysis has not been performed with A-Myb, the high degree of sequence homology and similar binding specificities between c-Myb and A-Myb suggest that these tryptophan repeats and the basic amino acids that lie adjacent to the tryptophan residues which mediate the contact with DNA play a very similar role in the two proteins.

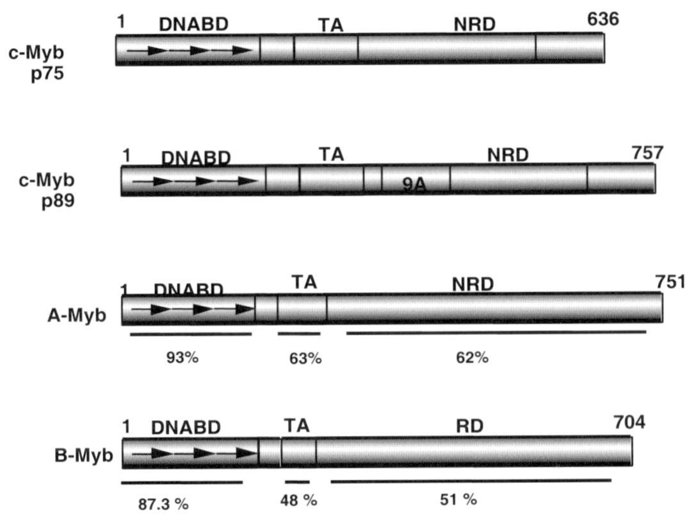

Figure 1

Structural comparison of Myb gene family products. Schematic structure of A-, B- and c-Myb proteins is represented. The c-*myb* gene encodes two proteins, one with exon 9A sequences (p89 c-Myb) and one without (p75-c-Myb). The numbers below are percentage homology to c-Myb. DNAB, DNA binding domain; TA, Transactivating domain, NRD, Negative regulatory domain, RD, Regulatory domain.

The transactivation domain of A-Myb and c-Myb resides in a central portion of the molecule, and measures between 52 and 85 amino acids in length. The transactivation domain of A-Myb is the least well characterised portion of the molecule. Its overall charge is negative, reminiscent of such domains in other transcription factors. Even in the case of c-Myb, mutational analysis of the transactivation domain has yet to reveal a specific stretch of sequence responsible for activity (Sakura et al., 1989; Weston and Bishop, 1989; Ibanez and Lipsick, 1990; Lane et al., 1990; Kalkbrenner et al., 1990). However, deletion of this domain was found to result in complete loss of the transactivation potential suggesting an essential role for this domain (Golay et al., 1994; Takashi et al., 1995).

The negative regulatory domain of c-Myb was first identified through characterisation of transforming retroviruses that lacked this domain. c-*myb* and A-*myb* cDNA clones lacking these carboxy-terminal sequences demonstrate consistently higher levels of transactivation in reporter gene studies suggesting that the C-terminal domain of c-Myb and A-Myb code for their negative regulatory domain (Golay et al., 1994; Takashi et al., 1995; Oh and Reddy, 1997; Trauth et al., 1994). It has been presumed that the negative regulatory domain functions as a docking site for *trans* regulators and indeed, a protein has been identified which binds to this domain of c-Myb (Sakura et al., 1989; Tavner et al., 1998), although its mechanism of action has yet to be ascertained clearly. This domain has also been postulated to be a phosphorylation target of Cyclin/CDK complexes that regulate A-Myb activity during cell cycle progression (Ziebold and Klempnauer, 1997).

Clearly, the domain of highest homology is the DNA binding domain, and indeed, this high level of homology is reflected in the ability of all three Myb proteins to bind to the same consensus DNA binding sequence. Nevertheless, the *A-myb* and *B-myb* genes encode proteins that are distinct from c-Myb. The expression patterns of the various *myb* family members differs. It has been well established, for example, that *c-myb* expression is predominantly restricted to the immature cells of haemopoietic lineages. On the other hand, *B-myb* transcripts can be detected in nearly every tissue. The expression pattern of *A-myb* reveals a great deal about the role of this protein in embryonic and adult development (reviewed in Oh and Reddy, 1999).

2.2 A-*myb* Expression

Using *in situ* hybridisation and Northern blot analysis, a number of investigators have determined that *A-myb* expression is regulated in a tissue specific manner as well as developmentally (Golay et al., 1994; Mettus et al., 1994; Trauth et al., 1994, Latham et al., 1996; Sleeman, 1993; Toscani et al., 1997). *A-myb* is expressed predominantly in male germ cells in all species examined. *A-myb* transcripts are also expressed in ovarian and brain tissues, as well as in the B-lymphocytes present in the germinal centers of the spleen. In adult male mice, *A-myb* expression is most readily detected in the germ cells, as is shown in Figure 2. During development, *A-myb* expression begins to increase at post-natal day 10, during which time, primary spermatocytes appear. In adult male mice, *A-myb* expression is highest in spermatagonia and in primary spermatocytes. Expression of *A-myb* decreases as meiosis proceeds, and spermatids are formed. This expression pattern reveals a role for *A-myb* in progression through the first meiotic prophase in spermatogenesis.

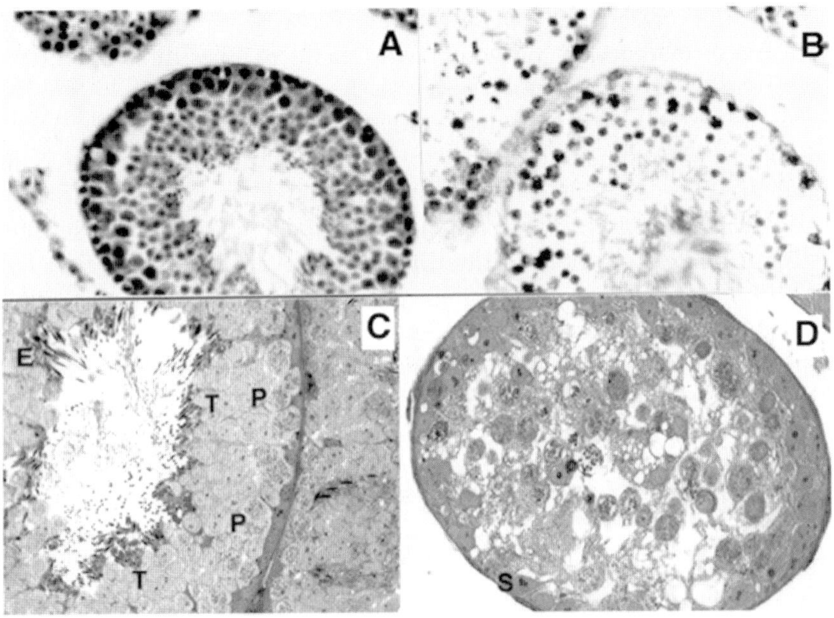

Figure 2
Expression pattern of A-*myb* and its role in the development of testis. (A) Haematoxylin and eosin-stained sections of adult mouse testis. (B) Adjacent section after *in situ* hybridisation to the anti-sense A-*myb* cRNA probe. (C and D) Sections of seminiferous tubules from either wild type or A-*myb*$^{-/-}$ mice. P, primary spermatocytes at pachytene; T, round spermatids. (see colour section p. xix)

A-myb expression is also detectable in the mammary ductal epithelia of pregnant and nursing adult females. This expression pattern was revealed in *in situ* hybridisation studies of mammary tissue from virgin, pregnant and nursing female mice (Toscani et al., 1997). Expression of *A-myb* correlates precisely with the morphological changes observed in differentiating mammary ductal epithelia. In addition, *A-myb* expression is also observed in some cultured B-lymphocytes and in the germinal centers of the spleen leading to the suggestion that this gene may play an important role in the development of germinal centers.

3. A-MYB FUNCTION IN VIVO

The role of A-Myb *in vivo* was tested directly through the generation of mutant mice nullizygous at the A-*myb* locus (Toscani et al., 1997). Over 400 intercrosses between mice heterozygous for a disrupted A-*myb* locus were performed leading to progeny with each of the three predicted genotypes (A-$myb^{+/+}$, A-$myb^{+/-}$ and A-$myb^{-/-}$).

Mice heterozygous for a disrupted *A-myb* allele (A-$myb^{+/-}$) are unremarkable in appearance compared to normal littermates. The mice are fertile, and are thus capable of producing nullizygous A-$myb^{-/-}$ progeny. Overall, A-$myb^{-/-}$ mice are notable for their small size. At birth, these mice are indistinguishable from normal littermates, however, during the first few weeks of life they lag behind in their growth. A-$myb^{-/-}$ pups are small, wrinkled and have a hunched posture. As they reach adulthood, some of the more easily detectable differences become less obvious. A-$myb^{-/-}$ females reach 90% of the size and weight of their normal littermates, while A-$myb^{-/-}$ males reach 70%. More rigorous inspection of the nullizygous animals revealed deficits in fertility, and in testis and mammary gland development (Toscani et al., 1997).

3.1 Phenotype of A-$myb^{-/-}$ Males

Mating behavior of male A-$myb^{-/-}$ mice is normal. However, wild type females mated with A-$myb^{-/-}$ males never became pregnant. Examination of the testis from A-$myb^{-/-}$ males revealed a complete absence of spermatozoa while testis from both A-$myb^{+/+}$ and A-$myb^{+/-}$ animals contained sperm counts in excess of 1 x 10^7 spermatazoa per testis. These differences were not attributable to serum testosterone levels, as animals of all three genotypes contained no significant variations (Toscani et al., 1997).

Histological examination of the testis from A-$myb^{-/-}$ males revealed qualitative differences in the appearance of the seminiferous tubules present, although the number of tubules appeared to be similar. A number of differences can be observed upon gross examination of the histological sections in Figure 2. Development of sperm proceeds from the periphery of the tubule to the center of the lumen. As the sprematagonia undergo meiosis, the daughter cells travel toward the lumen. Normal differentiation and luminal progression are observed in the seminiferous tubules of a normal mouse testis (Figure 2C). At the center of the lumen, mature sperm can be observed.

In A-$myb^{-/-}$ males, the morphology of the seminiferous tubules is abnormal. Primary spermatocytes appear to be degenerating; yet the less developed pre-leptonene spermatocytes and spermatagonia appear normal

(Figure 2D). Although spermatoagonial mitosis was observed, meiosis of primary spermatocytes appeared to be arrested at the pachytene stage. Arrest of meiosis at this stage resulted in significant and measurable levels of apoptosis of spermatocytes, vaculoization of Sertoli cell cytoplasm, and loss of formation of mature spermatocytes, resulting in the observed infertility of A-*myb*$^{-/-}$ males.

3.2 Phenotype of A-*myb*$^{-/-}$ Females

Unlike A-*myb*$^{-/-}$ males, A-*myb*$^{-/-}$ females show no pathology with respect to their germ cells. A-*myb*$^{-/-}$ females have normal ovaries, and when mated with wild type or A-*myb*$^{+/-}$ males, became pregnant and produced litters of normal size. Upon delivery of their pups, however, A-*myb*$^{-/-}$ females demonstrated an inability to nurse their pups. Examination of A-*myb*$^{-/-}$ mammary tissue revealed developmental deficits associated with proliferation and differentiation of the mammary ductal epithelia.

In normal females, *A-myb* expression is observed predominantly in mammary ductal epithelia. Development of the female murine mammary gland begins in embryogenesis and terminates in adulthood. The gland itself responds to hormonal and environmental cues throughout the life cycle, and its various developmental states are divided into distinct stages (reviewed in Topper and Freeman, 1980; Howlett and Bissell, 1993).

3.2.1 The embryonic stage

This stage of murine breast development begins at day 11 of gestation, at which time the epidermal tissue on each side of the ventral midline gives rise to the mammary ridge. Those cells which collect within the ridge ultimately form the mammary buds which, in the final days of gestation (days 16-21), proliferate and elongate into the mammary cord; it is the opening at one end of the cord which forms the nipple while the other end develops intricate branching to form mammary ducts. In male mice, androgen receptors are induced on the fibroblastic mesenchyme (by the mammary epithelium) around day 12 of embryogenesis (Kratochwil, 1986). Testosterone then stimulates a condensation of fibroblasts around the epithelial rudiments that ultimately leads to glandular disintegration.

3.2.2 The adolescent stage

Sexual maturation, which occurs in the mouse from weeks 4 to 6, is a period during which the ductal system of the mammary gland is actively proliferating and branching within the mammary fat pad stroma. The key event during this developmental process is the formation of terminal end buds at the tips of the ducts. The buds of the developing mammary gland are the sites at which cues for growth and development are received from both ovarian hormones and growth factors generated by fat pad adipocytes. The ductal elongation seen in puberty is directly stimulated by oestrogen but is refractory to progesterone treatment (reviewed in Topper and Freeman, 1980; Haslam, 1988; Howlett and Bissell, 1993). Once mice reach sexual maturity and their breast tissue becomes fully developed, the mammary epithelial cells acquire progesterone receptors and no longer proliferate in response to oestrogens alone. Indeed, they now require both oestrogen and progesterone for their proliferation (Haslam, 1988; Wang et al., 1990).

3.2.3 Pregnancy

Pregnancy allows for further development of mouse mammary cells that, by the end of the adolescent stage, are left both undifferentiated and quiescent. During the second half of pregnancy, the mitogenic and differentiative effects of oestrogen and progesterone are evidenced by the formation of additional ductal networks and the emergence of specialised alveolar cells at the tips of the ducts with the capacity to synthesize and secrete milk (reviewed in Topper and Freeman, 1980; Neville and Daniel, 1987; Howlett and Bissell, 1993; Gilbert, 1984). At mid pregnancy, only a small percentage of mammary cells contain the organelles that are necessary to synthesize the proteins that are required during lactation. Prior to lactation, the induction of rough endoplasmic reticulum by glucocorticoids enables the cells to synthesise prolactin. At birth, prolactin is secreted which induces transcription as well as stabilisation of the casein message (reviewed in Gilbert, 1984). Following the delivery of pups, further development of the breast tissue, especially the alveolar structures is greatly enhanced by the suckling action of the pups, which provides hormonal cues for further development of the mammary gland and allows it to produce milk. Caseins, with ß-casein being the most abundant, comprise the majority of milk proteins.

3.2.4 Involution

Weaning of the offspring leads to a decreased capacity for both milk synthesis and secretion. The alveolar cells which mostly comprise the breast during lactation undergo apoptosis and are resorbed and are replaced by adipocyte stroma in preparation for subsequent pregnancies (Hurley, 1989).

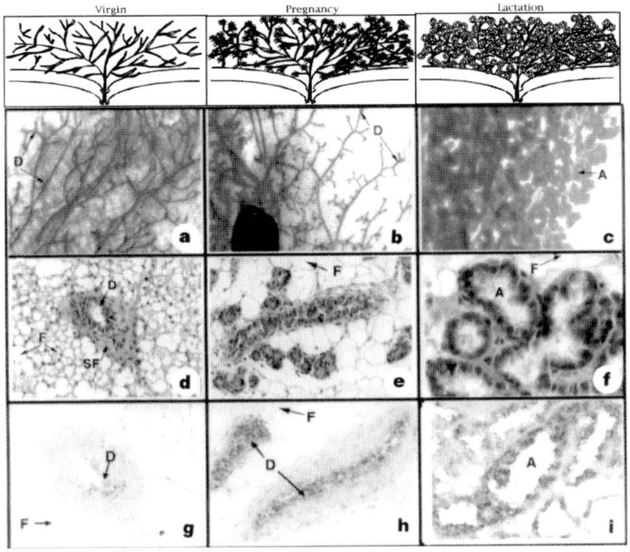

Figure 3

Expression of A-*myb* in mouse mammary tissue. The top panel shows a schematic representation of ductal branching in virgin, preganat and lactating mammary gland. (A-C) Whole mount preparations of mammary glands derived from a nulliparous, 10-day preganant and a lactating mouse two days after delivery. (D-F) Sections of the same tissues stained with haemotoxylin and eosin. (G-I) *In situ* hybridisation pattern of the breast sections with A-*myb* specific probe. A, alveoli; D, ductal epithelial cells; F, adipocytes; SF, fibroblasts.
(see colour section p. xix)

The morphological changes associated with the developmental stages described above can be observed in Figure 3. The rudimentary branching pattern of mammary ductal epithelia present in the nulliparous animal (Figure 3A) becomes more convoluted during pregnancy (Figure 3B), and still more convoluted at the beginning of lactation (Figure 3C). Ductal structures become more and more prevalent, to the exclusion of adipocytes (Figures 3D-F).

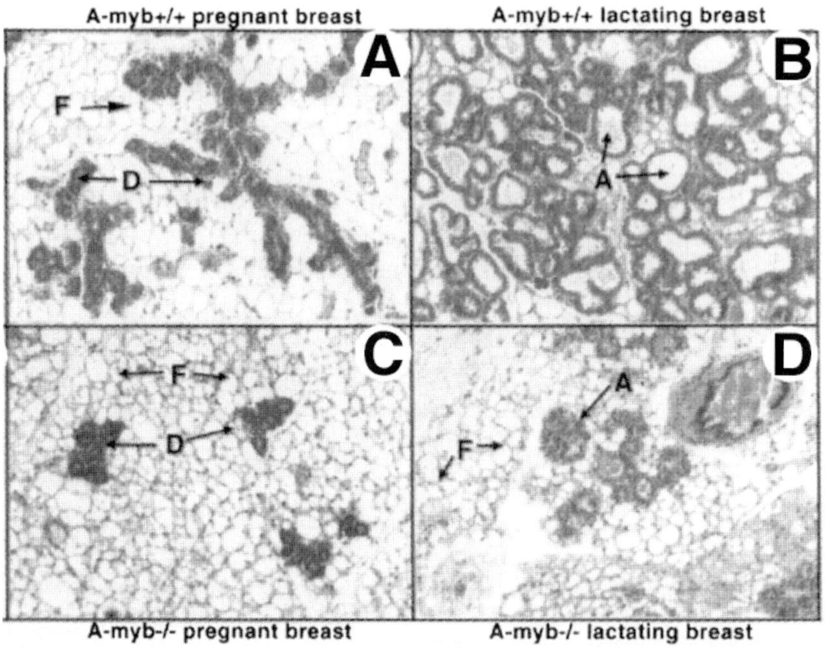

Figure 4
Defective breast development in mice lacking A-Myb. Mammary glands derived from 10 day pregnant and lactating (2 days after delivery of pups) wild-type and A-Myb-/- mice were used for histopathological analysis. Note the reduced proliferation of ductal cells and incompletely formed alveolar structure in A-Myb -/- mice, which leads to a failure of the A-Myb-/- mice to lactate. A, alveoli;D, ductal epithelial cells; F, adipocytes. (see colour section p. xx)

In situ hybridisation of sections of mammary tissue at these various stages reveals little or no *A-myb* expression in the mammary tissue of virgin mice. Expression of *A-myb* increases dramatically during the periods of cell division that accompany pregnancy, resulting in ductal branching and development of alveolar structures. This increase in expression is cell type specific, as *A-myb* expression is confined to the ductal epithelium. No hybridisation is observable in the surrounding fibroblasts and adipocytes (Figure 3).

The inability of A-$myb^{-/-}$ females to nurse their offspring can be traced to the poorly developed mammary glands. Figure 4 shows histological sections of mammary tissue from A-$myb^{+/+}$ and A-$myb^{-/-}$females that are pregnant (Figure 4A and C) and lactating (Figure 4B and D). Both nulliparous animals were found to have similar mammary gland structures consisting primarily of adipose tissue, fibroblasts, and some rudimentary ductal epithelium. The pregnant mammary gland of the wild type animal (Figure

4A) begins to develop an extensive network of ductal epithelia, while the corresponding A-*myb*$^{-/-}$ mammary gland remains composed primarily of adipose tissue, with only marginal increases in ductal epithelia (Figure 4C).

In the lactating mammary gland of the wild type mouse, nearly all of the adipose tissue is replaced with a highly differentiated network of ductal cells containing milk-producing alveolar structures (Figure 4B: alveloi are marked "A"). Even after delivery of offspring, the A-*myb*$^{-/-}$ mammary glands contain a sizeable percentage of adipose tissue, only a rudimentary ductal network, and few, if any alveolar structures (Figure 4D). Loss of A-*myb* sequences correlates with inability to produce functional mammary gland structures, and thus, inability to nurse the young.

3.3 Effect of the Absence of A-Myb on Germinal Centre Function

A-*myb* expression in human germinal centers has been sublocalised to the dark zone resident centroblast population (Vora et al., 2001). In addition, A-*myb* expression was found to be characteristic of certain subsets of mature B cell neoplasias such as Burkitt's lymphoma, sIg^{+} B-cell acute lymphocytic leukaemia, and subsets of chronic lymphocytic leukaemia. Based on these observations, it has been proposed that A-Myb plays a critical role in the regulation of the germinal center reaction, including the promotion of high-rate B cell proliferation and antibody V gene somatic hypermutation. An examination of the germinal center response driven by T cell-dependent antigen immunisation and the associated process of antibody V gene somatic hypermutation and heavy chain class switching in A-*myb*$^{-/-}$ mice was found to be overtly normal. Nonetheless, these mice displayed mild splenic white pulp hypoplasia and blunted primary serum antibody response, suggesting that although A-Myb is not directly involved in the regulation of the memory B cell response, it may play a role in enhancing peripheral B cell survival or proliferative capacity.

4. MECHANISMS OF A-MYB ACTION

The results of nucleotide and amino acid sequence homology studies suggested that A-*myb*, like *c-myb*, encodes a DNA binding transcription factor (Oh and Reddy, 1997; Oh and Reddy, 1999). Assessment of the transactivation function of A-Myb has been performed in cultured cell lines derived from both myeloid and mammary epithelial cells. An expression plasmid encoding murine A-Myb was used to confirm the ability of the protein to transactivate *c*-Myb-responsive promoter sequences. A-Myb, like *c*-Myb, was also shown to cooperate with the product of the *ets-2* gene in activation of the *mim-1* promoter, as well as a synthetic Myb-responsive element linked to the Herpes Simplex Virus Thymidine Kinase promoter. Mutant A-Myb isoforms produced from 3' truncated A-*myb* cDNA clones demonstrated greater transactivation activity than their wild type counterparts, while retaining the ability to cooperate with Ets-2 (Dudek et al., 1992; Golay et al., 1994).

4.1 A-Myb Transactivation and Granulocytic Differentiation

Structure/function analysis of Myb proteins has been performed in the cultured myeloid precursor line, 32Dcl3 (Valteri et al., 1987). 32Dcl3 cells are derived from normal mouse bone marrow and are dependent upon interleukin-3 (IL-3) for survival and maintenance of the undifferentiated (myeloblastic) state. Replacement of IL-3 with granulocyte colony stimulating factor (G-CSF) results in induction of terminal differentiation into neutrophilic granulocytes. Terminal differentiation of 32Dcl3 cells is accompanied by significant decreases in *c-myb* mRNA and protein levels. Constitutive expression of a *v-myb* transgene in 32D cells renders the cells refractory to the G-CSF-induced differentiation signal (Patel et al., 1993).

Because A-*myb* and *c-myb* encode transcription factors capable of binding the 5'YAACKG3' sequence, Oh and Reddy tested the ability of A-Myb to block G-CSF-induced differentiation of 32Dcl3 cells (Oh and Reddy, 1997). Surprisingly, 32D/A-Myb cell lines failed to overcome the differentiation-inducing effects of G-CSF while constitutive expression of c-Myb could readily induce such a block to G-CSF-induced differentiation. In addition, expression of c-Myb in these cells was found to increase the proliferative potential of this cell line in G-CSF, while the expression of A-Myb resulted in a complete block in the ability of this cell line to proliferate in the presence of G-CSF. However, transfection of a C-terminal truncation mutant of A-Myb that lacks the negative regulatory domain was found to render 32Dcl3 cells refractory to G-CSF-induced terminal differentiation. These results suggest that the c-Myb and A-Myb proteins do not exhibit identical biological function in spite of their extensive sequence homology and their ability to bind to very similar DNA sequences. They further suggest that the nature of their targets is dictated by the co-factors with which they interact in a given cellular environment. However, the observation that a truncated mutant of A-Myb can bring about the transformation of 32Dcl3 cells suggests that A-Myb can act an oncogene, when the negative regulation is lost.

4.2 A-Myb and Cell Cycle Regulation

Studies with synchronised Swiss 3T3 cells and serum-starved bovine vascular smooth muscle cells (SMC) have revealed that A-*myb* mRNA expression reaches maximum levels during the G_1 and early S-phases of the cell cycle (Ziebold and Klempnauer, 1997; Marhamati et al., 1997). Nuclear run-on assays have shown that this increase, at lease in the case of SMCs, is due to an increase in the rate of transcription. Co-expression of c-Myc with A-Myb, but not c-Myb or B-Myb was shown to promote progression into S-phase, suggesting that not only is A-Myb differentially expressed during the cell cycle, it has the ability to regulate the cell cycle (Marhamati et al., 1997).

5. A-MYB AND CANCER

Early studies performed with the *A-myb* cDNA established an expression pattern noticeably distinct from either *c-myb*, or *B-myb*. Nevertheless, *A-myb* and *c-myb* were found to be co-expressed in the lymphoid compartment. *c-myb* expression is limited to the precursors of all lymphocytes, with expression abating upon terminal differentiation. Deregulated expression of *A-myb* has been associated for some time with lymphocytic leukaemia. In 1996, Golay et al surveyed mRNA and protein expression in a variety of lymphoid leukaemias from both the B and T cell lineages (Golay et al., 1996). *A-myb* expression was observed in most Burkitt's lymphoma cell lines, but was less apparent in Non-Hodgkin's lymphoma, Epstein-Barr virus transformed lymphoblasts, or myelomas. Leukaemic B cell acute lymphocytic leukaemias that expressed surface immunoglobulin (sIg^+ B-ALL) also expressed high levels of *A-myb*. Promyelocytic leukaemias did not express *A-myb*, and a small percentage (25%) of chronic lymphocytic leukaemias (CLL) expressed *A-myb*.

The molecular mechanism underlying deregulated *A-myb*-associated B lymphocytic leukaemia has been studied using a variety of approaches. Cultured cell lines and transgenic animals have been used to determine the changes in patterns of gene expression associated with *A-myb* overexpression. Transgenic technology has been used to produce B-lymphocytic leukaemic animal models. In 1997, DeRocco et al inserted an *A-myb* cDNA into the germ line of mice (DeRocco et al., 1997). The resulting mouse line expressed *A-myb* in a variety of tissues, but developed pathology only in the B-lymphoid compartment after a several month-long latency period. Gross pathological examination revealed hyperplasia of the spleen and lymph nodes. Increased rates of DNA synthesis were observed in splenocytes of these transgenic animals in comparison with controls, suggesting a role for A-Myb in splenic B cell proliferative response to mitogen.

A-Myb-associated leukaemias are observed only in B-lymphocytes. Ying et al examined the basis for this B-cell tropism in 1997 (Ying et al., 1997). In this study, A-Myb transcriptional transactivation activity was studied in both B and T-lymphocyte cell lines. These studies revealed that the A-Myb protein could transactivate a reporter gene containing *cis* acting regulatory sequences derived from the *c-myc* promoter, or from the native *myc* promoter itself. In this study, B-lymphocytes were shown to express a tissue-specific transcriptional co-activator that binds specifically to the A-Myb DNA binding domain.

Upregulation of *c-myc* mRNA has been shown to be vital for induction of transformation (Arsura et al., 2000). Treatment of WEHI 231 and CH33B cell lymphomas with anti IgM leads to growth arrest and apoptosis. Ectopic expression of A-Myb in these leukaemic lines leads to relief from IgM-mediated growth arrest and apoptosis via increased expression of *c-myc.* Treatment of A-Myb-expressing B cell leukaemias with antisense *c-myc* oligonucleotides restores the potency of the IgM-mediated apoptotic response. Upregulation of anti-apoptotic genes, such as *bcl-2,* has been shown to occur in follicular lymphomas via chromosomal translocation (Heckman et al., 2000). The translocated *bcl-2* locus contains a binding site for a Cdx homeodomain transcription factor. A-Myb has been shown to interact with the Cdx protein to upregulate *bcl-2.*

6. CONCLUSIONS AND FUTURE DIRECTIONS

While the function of A-Myb in cell growth, differentiation and development is just being unraveled, its role as a transforming protein still remains to be established. In addition to its over-expression in Burkitt's lymphomas and other B cell leukaemias, our recent studies show that A-Myb is also over-expressed in several human breast cancers, suggesting the possibility that it may play a critical role in these tumours. The observation that A-Myb is required for the proper development of breast epithelium further lends support to this notion. Recent development of mouse models for B cell leukaemias and breast cancer and the availability of mice with genetic modification in the A-*myb* locus should allow us to precisely define the role of A-Myb in these cancers in the near future.

REFERENCES

Anton, I.A. and Frampton, J. (1988) Tryptophans in Myb proteins. Nature *336,* 719.

Arsura, M., Hofmann, C.S., Golay, J., Introna, M. and Sonenshein, G.E. (2000) A-Myb rescues murine B cell lymphomas from Igm-receptor-mediated apoptosis through c-myc transcriptional regulation. Blood *96,* 1013-1020

Bender, T.P. and Kuehl, W.M. (1986) Murine Myb protooncogene: cDNA sequence and evidence for 5' heterogeneity. Proc. Nat. Acad. Sci. USA *83,* 3204-3208

Biedenkapp, H., Borgmeyer, U., Sippel, A.E. and Klempnauer,K.-H. (1988) Viral Myb oncogene encodes a sequence-specific DNA-binding activity. Nature *335,* 835-837.

Dasgupta, P. and Reddy, E.P. (1989) Identification of alternatively spliced transcripts for human c-Myb: Molecular cloning and sequence analysis of human exon 9A sequences. Oncogene *4,* 1419-1423.

DeRocco, S.E., Iozzo, R., Ma, X., Schwarting, R., Peterson, D. and Calabretta, B. (1997) Ectopic expression of A-Myb in transgenic mice causes follicular hyperplasia and enhanced B lymphocyte proliferation. Proc. Nat. Acad. Sci. USA *94,* 3240-3244.

Dudek, H. and Reddy, E.P. (1989) Identification of two translational products of c-Myb. Oncogene *4,* 1061-1066.

Dudek,, H., Tantravahi, R.V., Rao, V.N., Reddy, E.S., and Reddy, E.P. (1992) Myb and Ets proteins cooperate in transcriptional activation of the mim-1 promoter. Proc. Nat. Acad. Sci. USA *89,* 1291-1295.

Frampton, J., Gibson, T.J., Ness, S.A., Doderlein, G. and Graf, T. (1991) Proposed structure for the DNA-binding domain of the Myb oncoprotein based on novel building and mutational analysis. Protein Engineering *4,* 891-901.

Gilbert, S.F. (1994) Cell interactions at a distance: Hormones as mediators of development. In: Developmental Biology. Sinauer Associates, Inc., Sutherland Massachusettes, 716-753.

Golay, J., Loffarelli, L., Luppi, M., Castellano, M. and Introna, M. (1994) The human A-Myb protein is a strong activator of transcription. Oncogene *9,* 2469-2479.

Golay, J., Luppi, M., Songia, S., Palvarini, C., Lombardi, L., Aiello, A., Delia, D., Lam, K., Crawford, D.H., Biondi, A., Barbui, T., Rambaldi, A. and Introna, M. (1996) Expression of A-Myb but not c-Myb and B-Myb, is restricted to Burkitt's lymphoma, sIg+ B-acute lymphocytic leukemia, and a subset of chronic lymphocytic leukemias Blood *87,* 1900-1911.

Gonda, T.J., Gough, N.M., Dunn, A.R. and deBlaquire, J. (1985) Nucleotide sequence of cDNA clones of the murine Myb protooncogene. EMBO J. *4,* 2003-2008.

Hall, W.J., Bean C. W. and Pollard, M. (1941) Transmission of fowl leukosis through chick embryos and adult chicks. Am. J. Vet. Res. *2,* 272-279.

Haslam, S.Z. (1988) Progesterone effects on DNA synthesis in normal mouse mammary glands. Endocrinology *122,* 464-470.

Heckman, C.A., Mehew, J.W., Ying, G.G., Introna, M., Golay, J. and Boxer, L.M. (2000) A-Myb up-regulates Bcl-2 through a Cdx binding site in t(14;18) lymphoma cells. J. Biol. Chem. *275,* 6499-6508.

Howlett, A.R. and Bissell, M.J. (1993) The influence of tissue microenvironment (Stroma and Extracellular Matrix) on the development and function of mammary epithelium. Epith. Cell Biol. *2,* 79-89.

Hurley, W.L. (1989) Mammary gland fuvnction during involution. J. Dairy Sci. *72,* 1637-1646.

Ibanez, C.E. and Lipsick, J.S. (1990) trans activation of gene expression by v-Myb. Mol. Cell. Biol. *10,* 2285-2293.

Ivanov, X., Mladnov, Z., Nedyalkov, S. *et al*, (1964) Experimental investigations into avian leukosis V. Transmission, hematology and morphology of avian myelocytomatosis. Bull. Inst. Pathol. Comp. Animaux Domest. (Sophia) *10,* 5-38.

Kalkbrenner, F., Guehmann, S. and Moelling, K. (1990) Transcriptional activation by human c-Myb and v-Myb genes. Oncogene *5,* 657-661.

Katzen, A.L., Kornberg, T.B. and Bishop, J.M. (1985) Isolation of the protooncogene c-Myb from D. melonagaster. Cell *41,* 449-456

Kratochwil, K. (1986). The importance of epithelial-stromal interaction in mammary gland development. In: Rich, M.M., Hager, J.C., and J. Taylor-Papadimitriou (eds) Breast cancer origins, detection and treatment. Martinus Nijoff, Boston, 1-12.

Lane, T., Ibanez, C., Garcia, A., Graf, T. and Lipsick, J.S. (1990) Transformation by v-Myb correlates with trans-activation of gene expression. Mol. Cell. Biol. *10,* 2591-2598.

Latham, K.E., Litvin, J., Orth, J.M., Patel, B., Mettus, R. and Reddy, E.P. (1996) Temporal patterns of A-Myb and B-Myb gene expression during testis development. Oncogene *13,* 1161-1168.

Lipsick, J.S. (1996) One billion years of Myb. Oncogene *13,* 223-235.

Marhamati, D.J., Bellas, R.E., Arsura, M., Kypreos, K.E. and Sonenshein, G.E. (1997) A-Myb is expressed in bovine vascular smooth muscle cells during the late G1-to-S phase transition and cooperates with c-myc to mediate progression to S phase. Mol. Cell. Biol. *17,* 2448-2457.

Majello, B., Kenyon, L.C. and Dalla-Favera, R. (1986) Human c-Myb protooncogene: nucleotide sequence of cDNA and organization of the genomic locus. Proc. Natl. Acad. Sci. USA *83,* 9636-9640.

Mettus, R.V, Litvin, J., Wali, A., Toscani, A., Latham, K., Hatton, K. and Reddy, E.P. (1994). Murine A-Myb: evidence for differential splicing and tissue-specific expression. Oncogene *9,* 3077-3086.

Neville, M.C. and Daniel, C.W. (eds) (1987) The mammary gland: development, regulation and function. Plenum Press, New York.

Nomura, N., Takahashi, M., Matsui, M., Ishii, S., Date, T., Sasamoto, S. and Ishizaki, R. (1988) Isolation of human cDNA clones of Myb-related genes, A-Myb and B-Myb. Nucl. Acids Res. *16,* 11075-11080.

Oh, I-H. and Reddy, E.P. (1997) Murine A-Myb gene encodes a transcription factor, which cooperates with Ets-2 and exhibits distinctive biochemical and biological activities from c-Myb. J. Biol. Chem. *272,* 21432-21443.

Oh, I-H. and Reddy, E.P. (1999) The Myb gene family in cell growth, differentiation and apoptosis. Oncogene *18,* 3017-3033.

Patel, G., Kreider, B., Rovera, G. and Reddy, E.P. (1993). v-Myb blocks granulocyte colony-stimulating factor-induced myeloid cell differentiation but not proliferation. Mol. Cell. Biol. *13,* 2269-2276.

Rosson, D. and Reddy, E.P. (1986) Nucleotide sequence of chicken c-Myb complementary DNA and implications for Myb oncogene activation. Nature *319,* 604-606.

Saikumar, P., Murali, R. and Reddy, E.P. (1990) *Role of Tryptophan repeats and flanking amino acids in Myb-DNA interactions*. Proc. Nat. Acad. Sci. USA *87,* 8452-8456.

Sakura, H., Kanei-Ishii, C., Nagase, T., Nakagoshi, H., Gonda, T.J. and Ishii, S. (1989) Delineation of three functional domains of the transcriptional activator encoded by the c-Myb proto-oncogene. Proc. Nat. Acad. Sci. USA *86,* 5758-5762.

Shen-Ong, G.L., Lüscher, B. and Eisenman, R.N. (1989) A second c-Myb protein is translated from an alternatively spliced mRNA expressed from normal and 5'-disrupted Myb loci. Mol. Cell. Biol. *9,* 5456-5463.

Sleeman, J.P. (1993) Xenopus A-Myb is expressed during early spermatogenesis. Oncogene *8,* 1931-1941.

Tavner, F.J., Simpson, R., Tashiro, S., Favier, D., Jenkins, N.A., Gilbert, D.J., Copeland, N.G., Macmillan, E.M., Lutwyche, J., Keough, R.A., Ishii, S. and Gonda, T.J. (1998) Molecular cloning reveals that the p160 Myb-binding protein is a novel, predominantly nucleolar protein which may play a role in transactivation by Myb. Mol. Cell. Biol. *18,* 989-1002.

Takahashi, T., Nakagoshi, H., Sarai, A., Nomura, N., Yamamoto, T. and Ishii, S. (1995) Human A-Myb gene encodes a transcriptional activator containing the negative regulatory domains. FEBS Lett. *358,* 89-96.

Topper, Y.J. and Freeman, C.S. (1980). Multiple hormone interactions in the developmental biology of the mammary gland. Physiol. Rev. *60,* 1049-1106.

Toscani, A., Mettus, R.V., Coupland, R., Simpkins, H., Litvin, J., Orth, J., Hatton, K.S. and Reddy, E.P. (1997). Arrest of spermatogenesis and defective breast development in mice lacking A-Myb. Nature *386,* 713-717.

Trauth K., Mutschler B., Jenkins N.A., Gilbert D.J., Copeland N.G. and Klempnauer K-H. (1994) Mouse A-Myb encodes a trans-activator and is expressed in mitotically active cells of the developing central nervous system, adult testis and B lymphocytes. EMBO J. *13,* 5994-6005.

Valtieri, M., Tweardy, D.J., Caracciolo, D., Johnson, K., Mavilio, F., Altmann, S., Santoli, D. and Rovera, G. (1987) Cytokine-dependent granulocytic differentiation. Regulation of proliferative and differentiative responses in a murine progenitor cell line. J. Immunol. *138,* 3829-3835.

Vora, K.A., Lentz, V.M., Monsell, W., Rao, S.P., Mettus, R., Toscani, A., Reddy, E.P. and Manser, T. (2001) The T cell-dependent B cell immune response and germinal center reaction are intact in A-Myb-deficient mice. J. Immunol. *166,* 3226-3230.

Wang, S., Counterman, L.J. and Haslam, S.Z. (1990) Progesterone action in normal mouse mammary gland. Endocrinology *127,* 2183-2189.

Weston, K. and Bishop, J.M. (1989) Transactivation by the v-Myb oncogene and is cellular progenitor, c-Myb. Cell *58,* 85-93.

Ying, G.G., Arsura, M., Introna, M. and Golay, J. (1997) The DNA binding domain of the A-MYB transcription factor is responsible for its B cell-specificity and binds to a B cell 110 kDa nuclear protein. J. Biol. Chem. *272,* 24921-24926.

Ziebold, U. and Klempnauer, K-H. (1997) Linking Myb to the cell cycle: cyclin-dependent phosphorylation and regulation of A-Myb activity. Oncogene *15,* 1011-1019.

Chapter 9

B-MYB: A HIGHLY REGULATED MEMBER OF THE MYB TRANSCRIPTION FACTOR FAMILY

Robert J. Watson
Ludwig Institute for Cancer Research and Department of Virology, Faculty of Medicine, Imperial College London, Norfolk Place, London W2 1PG, United Kingdom.

Abstract: Expression of the *B-myb* transcription factor gene is regulated at two major levels during the mammalian cell cycle. Transcriptional regulation by an E2F-dependent mechanism directs maximal expression levels of B-Myb protein to late G1/S, while phosphorylation of B-Myb by Cyclin A/Cdk2 at the G1/S transition and during S phase enhances its transactivation properties. *B-myb* is an essential gene for early embryonic development, and the timing of its regulation strongly suggests that its most critical functions are required during S phase. In addition to its presumptive role in regulating gene expression, recent evidence also suggests that B-Myb displays additional non-transcriptional activities, for example through its binding to the p107 retinoblastoma-related protein. This chapter reviews how B-Myb activity is regulated by hyperphophorylation during the cell cycle and addresses how this may contribute to cell growth control.

1. INTRODUCTION

Uniquely amongst the mammalian *myb* genes, *B-myb* is expressed in all cell lineages. Expression of *B-myb* is not ubiquitous, however, as transcription is very tightly linked to the proliferation status of the cell. Indeed, *B-myb* is a classical E2F-regulated gene, and as such its transcription is induced to maximal levels at the G1/S boundary during the cell cycle (reviewed by Fiona Tavner). Furthermore, activity of the B-Myb protein is regulated by Cyclin A2/Cdk2 kinase-mediated phosphorylation, which itself is maximally active at the G1/S transition and throughout S phase. It is evident, therefore, that transcriptional and post-translational controls combine to restrict active B-Myb to the late G1 and S phases of the cell

J. Frampton (ed.), Myb Transcription Factors: Their Role in Growth, Differentiation and Disease, 181-199.

cycle, strongly suggesting that it is at these stages where B-Myb's function is most critical.

A number of lines of investigation have provided circumstantial evidence to indict B-Myb as an essential regulator of the cell cycle. Perhaps the most incriminating is the finding that genetic knockout of B-Myb in mice resulted in embryonic lethality at a very early (E4.5-E6.5) stage of embryogenesis (Tanaka et al., 1999). Although it is unclear whether death resulted from effects on cell proliferation, the fact that the inner cell mass failed to outgrow when *B-myb*$^{-/-}$ blastocycts were cultured *in vitro* suggests, at the least, an essential role in growth of these primitive pluripotent cells. A role for B-Myb in cell proliferation is also supported by studies showing that *B-myb* antisense oligonucleotides inhibit proliferation of myeloid, lymphoid, glioblastoma, fibroblast and neuroblastoma cell lines (Arsura et al., 1992; Lin et al., 1994; Raschellá et al., 1995; Sala and Calabretta, 1992). In contrast, constitutive *B-myb* expression allows BALB/c 3T3 fibroblasts to grow in low serum conditions (Sala and Calabretta, 1992) and prevents cell cycle arrest in IL6-induced differentiation of M1 myeloid leukemia cells (Bies et al., 1996).

This chapter will review how B-Myb protein function is regulated during the cell cycle, and will address how this activity may contribute to control of cell growth.

2. PHOSPHORYLATION OF B-MYB BY CYCLIN-DEPENDENT KINASES

As expected from regulation of the mRNA, re-entry of quiescent Swiss 3T3 fibroblasts into the cell cycle was found to be accompanied by a marked increase in B-Myb protein in late G1, which reached a maximal level in S phase (Robinson et al., 1996). Unexpectedly, it was also observed that a distinct B-Myb form with lower mobility on SDS-PAGE gels appeared precisely at the G1/S junction, and this form persisted throughout S phase. Notably, this lower mobility B-Myb form could not be detected in G0/early G1 in fibroblasts which over-expressed an ectopic *B-myb* gene (Robinson et al., 1996). It was concluded from this study that B-Myb is subject to specific modification by an S phase-active kinase.

2.1 Identification of Kinases Phosphorylating B-Myb

All the evidence gathered to date indicates that Cyclin A2/Cdk2 is the primary enzyme responsible for B-Myb phosphorylation in S phase. Initial experiments in which baculovirus vectors encoding Cyclin A2/Cdk2, Cyclin

E1/Cdk2 or Cyclin D1/Cdk4 were co-infected into Sf9 cells with a B-Myb virus, showed that only Cyclin A2/Cdk2 induced a mobility shift on SDS-PAGE consistent with the distinct S phase form of B-Myb (Robinson et al., 1996). Similarly, co-transfection of various Cyclins with B-Myb into either primate COS-7 cells or human Saos-2 cells showed that Cyclin A2 was able to induce the characteristic mobility shift (Sala et al., 1997; Ziebold et al., 1997), whereas Cyclin D1 and Cyclin B1 had no effect. In these experiments Cyclin E1 had little or no apparent activity on B-Myb (Sala et al., 1997; Ziebold et al., 1997). In some respects it is curious that B-Myb is preferentially phosphorylated by Cyclin A2/Cdk2 rather than Cyclin E1/Cdk2, since these enzymes have similar substrate consensus sequence requirements (Holmes and Solomon, 1996). Indeed, when B-Myb and Cyclin E1/Cdk2 were brought together on protein G-agarose beads as a co-immunoprecipitate, B-Myb was efficiently phosphorylated *in vitro* (Johnson et al., 1999). We have been unable to show direct interactions between Cyclin A2 and B-Myb, although the related Cyclin A1 (which is expressed in restricted cell lineages) is able both to bind and mediate phosphorylation of B-Myb (Müller-Tidow et al., 2001). Therefore, the general preference for Cyclin A2 can not be accounted for by its propensity to target B-Myb physically. Surprisingly, we have found that Cyclin E1 is able to bind B-Myb (M. Joaquin and RJW, unpublished data), and the significance of this interaction deserves further investigation.

2.2 Identification of Cdk2 Phosphorylation Sites on B-Myb

Further evidence that Cyclin A2/Cdk2 is the authentic enzyme responsible for B-Myb modification came from phosphopeptide mapping. Initially it was shown that similar spots were obtained on 2-D tryptic phosphopeptide maps when comparing bacterially expressed B-Myb phosphorylated *in vitro* with Cyclin A2/Cdk2 and B-Myb protein immunoprecipitated from Swiss 3T3 fibroblasts in S phase (Ziebold et al., 1997). Subsequently, a number of the key Cyclin A2/Cdk2 phosphorylation sites in B-Myb were identified (Bartsch et al., 1999; Johnson et al., 2002; Johnson et al., 1999; Saville and Watson, 1998). These studies showed that the major sites detected in mouse B-Myb on 2D tryptic phosphopeptide maps (T447, T490, T497, T524 and S581) conform either to the preferred Cyclin A2/Cdk2 consensus site (S/T-P-X-K/R) or near-optimal sites (Bartsch et al., 1999; Saville and Watson, 1998). Automated peptide radiosequencing of tryptic phosphopeptides [^{32}P]phosphate-labelled *in vivo* identified a further ten B-Myb phosphorylation sites (Johnson et al., 2002; Johnson et al., 1999), all of which contained the consensus sequence S/T-P consistent with phosphorylation by Cdks. It is therefore apparent that the majority (if not all) of the 22 S/T-P sites present in B-Myb may be modified *in vivo*. A summary of the phosphorylation sites is shown in Figure 1.

It is notable that virtually all these sites cluster in the central part of the B-Myb protein encompassing the Transactivation Domain and the Conserved Region (Figure 1). All in all, it is evident that Cyclin A2/Cdk2 (potentially supplemented by Cyclin E1/Cdk2 and Cyclin A1/Cdk2) is the major kinase that modifies B-Myb *in vivo*. However, it is notable that at least one phosphorylation site (S581) was still modified in cells in which Cdk2 activity was inhibited with a dominant-negative protein, Cdk2DN (Saville and Watson, 1998), suggesting that B-Myb can additionally be modified by other proline-directed kinases, at least at certain sites.

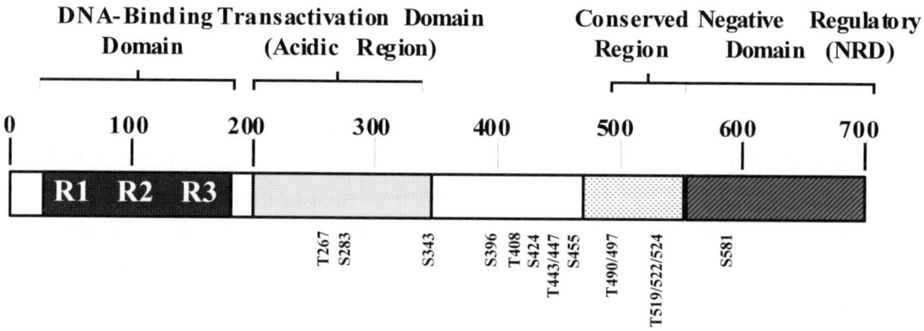

Figure 1

Cyclin A/Cdk2 phosphorylation sites in B-Myb. The locations of domains within the 704 amino acid mouse B-Myb protein are represented schematically. Indicated below are the positions of threonine and serine residues which have been shown experimentally to be phosphorylated *in vivo*, and in most instances to be substrates for cyclin A2/Cdk2 *in vitro* (see text for details).

2.3 Regulation of B-Myb Function by Phosphorylation

Phosphorylation of B-Myb by Cyclin A2/Cdk2 has been shown to have two important functions: (1) it greatly enhances its transcriptional activation properties and (2) it marks the activated protein for degradation.

B-Myb was initially regarded in some quarters as a repressive member of the Myb transcription factor family, as it was found to have little or no transactivation activity itself but was able to inhibit the activity of co-expressed c-Myb (Foos et al., 1992; Watson et al., 1993). It became evident, however, that B-Myb is actually a highly repressed protein which can be activated either by artificial removal of a C-terminal Negative Regulatory Domain (NRD) or by Cyclin A/Cdk2-mediated phosphorylation (Ansieau et al., 1997; Lane et al., 1997; Sala et al., 1997; Ziebold et al., 1997). The obvious conclusion to be drawn from these observations is that phosphorylation overcomes the inhibitory effect of the NRD. Evidence accrued to date is broadly consistent with this interpretation, but implies that this is an over-simplification.

2.3.1 The negative regulatory domain (NRD)

The NRD has been defined by its function rather than by any consideration of protein domain structure, and it is debatable whether it is really a discrete entity or rather reflects a number of activities determined by the B-Myb C-terminus. Deletion of just 29 amino acids from the C-terminus of B-Myb was found to increase B-Myb's transactivation activity quite markedly when measured on a promoter containing three strong Myb-binding sites (MBS) (Bessa et al., 2001b). In a series of C-terminal deletion mutants, maximal transcriptional activity was obtained with B-Myb+561, which lacks the C-terminal 143 amino acids (Bessa et al., 2001b). Of the 15 B-Myb phosphorylation sites identified (Johnson et al., 2002), only one (S581) maps to the region deleted in this mutant. Although a point mutation of S581 did have some effect, the mutant was still quite responsive to enhancement by Cyclin A2 (Saville and Watson, 1998). It is probable, therefore, that the NRD is not the sole target of Cyclin A-mediated phosphorylation, rather additional modification at other sites combine to negate the activity of this region. Consistent with this notion, small interstitial deletions within the transactivation domain were unexpectedly found to increase B-Myb transactivation activity (Joaquin et al., 2002), suggesting that these mutations affected function of the NRD. The picture that emerges from a number of studies is that B-Myb adopts a transcriptionally inactive configuration that is disrupted to a greater or lesser extent by deletions. Phosphorylation is also able to affect the repressed B-Myb state, and it appears that what is most important in this respect is attaining a hyperphosphorylated state by modifying the protein at multiple sites rather than targetting key single sites.

2.3.2 Nuclear localisation and DNA-binding

Phosphorylation of B-Myb potentially could modify its function in a number of ways. Analysis of one possible mechanism was prompted by the observation that S581 lies within a bipartite nuclear localisation sequence (NLS), analogous to a cdc2 phosphorylation site in yeast SW15 which regulates cell localisation during the cell cycle (Takemoto et al., 1994). However, mutation of S581 to alanine was found to have no effect on nuclear localisation of B-Myb, even when a second more N-terminal NLS was removed from the B-Myb protein (Takemoto et al., 1994). Indeed, we observed that B-Myb remains resolutely nuclear at all stages of the cell cycle, even in G0/G1 when it is clearly not hyperphosphorylated (Robinson et al., 1996). Therefore, enhancement of B-Myb activity by Cyclin A/Cdk2 can not be explained by a nuclear localisation mechanism.

We have also addressed whether B-Myb DNA-binding is affected by phosphorylation (Bessa et al., 2001b). In contrast to C-terminally truncated mutants, full-length B-Myb binds poorly *in vitro* to oligonucleotide probes containing a single MBS, and the only bandshifts seen in these electrophoretic mobility shift assays correspond to proteolytically degraded B-Myb (Bessa et al., 2001b). This suggested that the DNA-binding domain was occluded by the presence of the NRD and raised the possibility that ablation of NRD function by phosphorylation may unmask latent DNA-binding activity. In fact, we found that hyperphosphorylation of B-Myb did not result in conspicuous DNA-binding by full-length B-Myb in this assay. In contrast to oligonucleotide probes, we were able to detect binding of full-length B-Myb to a short DNA fragment probe containing three MBS, however, no increase in DNA-binding to this probe was evident when B-Myb was hyperphophorylated (Bessa et al., 2001b). Additionally, co-transfection with Cdk2DN, which effectively inhibits transactivation activity of B-Myb, had no obvious effect on DNA-binding activity.

Using a different binding assay with an immobilised MBS, it was found that mutation of certain phosphorylation sites (most notably S581) actually increased B-Myb DNA-binding activity, and it was concluded from this study that phosphorylation at these sites therefore inhibited DNA-binding activity (Johnson et al., 1999). The DNA-binding activity of phosphorylated B-Myb was not directly tested in this assay, however, and an alternative explanation for the altered DNA-binding activity of these mutants is that the tightly repressive state of B-Myb was mildly disrupted by the mutation. However, neither interpretation of these results properly explains why transactivation activity exhibited by the S581A mutant is diminished (Johnson et al., 1999; Saville and Watson, 1998). Nonetheless, it can be concluded from these studies that there is no evidence favouring the notion that hyperphosphorylation of B-Myb enhances DNA-binding..

2.3.3 Interactions of B-Myb with co-activators

Another potential means through which the activity of hyperphosphorylated B-Myb could be enhanced is by regulating its interactions with co-activators. In common with other Myb proteins, B-Myb interacts both functionally and physically with the transcriptional co-activators, CREB-binding protein (CBP) and the related p300 protein (Bessa et al., 2001b; Johnson et al., 2002; Li and McDonnell, 2002). CBP and p300 act as scaffolding proteins that connect sequence-specific transcription factors to the basal transcriptional machinery. Additionally, they have intrinsic histone acetyltransferase (HAT) activity, and therefore may be involved in chromatin remodelling through acetylation of nucleosomal histones as well as direct acetylation of transcription factors with which they associate (reviewed in Chan and La Thangue, 2001). We found that the ability of CBP to co-activate B-Myb was inhibited by the Cdk2DN protein, while conversely Cyclin A2 synergised with CBP to enhance B-Myb transactivation (Bessa et al., 2001b). In view of these findings, it was surprising to find that neither Cdk2DN nor Cyclin A2 affected binding of B-Myb to CBP in cotransfected cells (Bessa et al., 2001b). Similarly, Johnson and colleagues found that binding to p300 was unaffected by mutation of the 15 known Cyclin A/Cdk2 phosphorylation sites in B-Myb (Johnson et al., 2002). This published evidence is therefore inconsistent with a simple model in which CBP/p1300 specifically interact with and co-activates hyperphosphorylated B-Myb. It is notable that the B-Myb+561 deletion mutant, which is constitutively hyperactive and unresponsive to phosphorylation, was found not to synergise with CBP (Bessa et al., 2001a). Although experimental approaches have yet to address this issue, this finding suggests that a critical outcome following interaction with CBP/p300 is increased phosphorylation of B-Myb, resulting in its transcriptional enhancement. Potentially this may arise through conformational changes in B-Myb which expose critical phosphorylation sites. It is now known that p300, and putatively CBP, is able to acetylate B-Myb (Johnson et al., 2002), presumably at one or more of 4 conserved lysine residues which are acetylated by these protein in c-Myb (Sano and Ishii, 2001; Tomita et al., 2000). To speculate further, acetylation of B-Myb may induce conformational changes that facilitate phosphorylation of the protein by Cyclin A/Cdk2. Further work in this area is required to test these ideas.

The poly(ADP-ribose) polymerase (PARP) protein has also been shown to co-activate with B-Myb (Cervellera and Sala, 2000). This activity results from a direct physical interaction between these two proteins mediated through the B-Myb DNA-binding domain, but is not dependent upon the poly-ADP ribosylation activity of PARP. Notably, PARP and Cyclin A2 were found to act synergistically to drive B-Myb transcriptional activity, while a B-Myb mutant containing substitutions in 10 Cyclin A/Cdk2 phosphorylation sites (B-Myb10Mut) failed to respond to PARP (Santilli et al., 2001). Although these data initially suggested that PARP may interact specifically with phosphorylated B-Myb, binding studies showed that binding to PARP was unaffected in B-Myb10Mut. Rather, current evidence strongly suggests that synergism between PARP and Cyclin A2 is due to the ability of PARP to facilitate phosphorylation of B-Myb by Cyclin A2/Cdk2 (Santilli et al., 2001). Cyclin A2 was unable to enhance B-Myb activity in $PARP^{-/-}$ fibroblasts, suggesting that PARP may be a physiological facilitator of B-Myb phosphorylation. It would therefore be of interest to determine whether PARP bridges the interaction between Cyclin A2/Cdk2 and B-Myb, or rather whether it induces conformational changes in B-Myb which expose the sites to be modified.

Very recently, B-Myb has been reported to interact with two additional co-activators, a major component of TFIID, TAFII250 (Bartusel and Klempnauer, 2003), and the zinc finger protein ZPR8 (Seong et al., 2003). B-Myb transcriptional activity is substantially dependent upon active TAFII250 in hamster ts13 cells, in which this co-activator is temperature-sensitive. ts13 cells display a G1 block and undergo apoptosis when TAFII250 is inactivated at the restrictive temperature, suggesting a functional link with cell cycle regulators such as B-Myb, however, simply over-expressing B-Myb did not rescue temperature-sensitivity (Bartusel and Klempnauer, 2003). It has yet to be established whether interactions with either TAFII250 or ZPR8 are influenced by the phosphorylation status of B-Myb. In conclusion, evidence to date does not favour the notion that B-Myb hyperphosphorylation affects its interactions with co-activators.

2.3.4 Interactions of B-Myb with co-repressors

A further way, in principle, in which the activity of hyperphosphorylated B-Myb could be enhanced is by precluding interactions with negative regulators. B-Myb has been reported to interact with two classes of co-repressor proteins, N-CoR/SMRT and BS69 (Li and McDonnell, 2002; Masselink et al., 2001), which inhibit its transcriptional activity. The closely related N-CoR and SMRT proteins are components of multiprotein complexes containing histone deacetylases (HDAC), which bind and repress many different types of transcription factors, including nuclear hormone receptors (Jepsen and Rosenfeld, 2002). The B-Myb C-terminus was shown to interact with the nuclear receptor binding domain of N-CoR. Transcriptional repression of B-Myb resulting from this interaction could be overcome by co-expressing the unliganded thyroid hormone nuclear receptor, an effect that presumably resulted from competition for the binding site on N-CoR (Li and McDonnell, 2002). Repression of B-Myb activity by N-CoR could also be overcome with trichostatin A (TSA), an inhibitor of histone deacetylase (HDAC) activity, which suggests that recruitment of HDAC to an N-CoR/B-Myb or SMRT/B-Myb complex may account for the inhibitory effect on B-Myb. Dominant-negative derivatives of N-CoR and SMRT synergised with B-Myb in a transcriptional assay (Li and McDonnell,

2002), suggesting that ectopically-expressed B-Myb activity is repressed by endogenous levels of N-CoR/SMRT present in the transfected cells. Thus N-CoR/SMRT may be general regulators of B-Myb activity, as these proteins are widely expressed in different cell lineages. Could repression of B-Myb activity by N-CoR/SMRT explain the sensitivity of B-Myb to inhibition mediated by the NRD and responsiveness to Cyclin A/Cdk2-mediated hyperphosphorylation? In favour of this notion, it was found by Li and McDonnell (2002) that several non-overlapping C-terminal deletions of B-Myb which resulted in hyperactivation of B-Myb transcriptional properties also resulted in loss of binding of B-Myb to N-CoR/SMRT. As these deletions overcome the effect of the NRD, but do not necessarily map to this domain, this suggests that the NRD is necessary for B-Myb to adopt a conformation that enables it to bind the transcriptional inhibitory N-CoR/SMRT proteins. Parenthetically, our own studies suggest a similar scenario for the interaction of B-Myb with the p107 pocket protein (Joaquin et al., 2002). Significantly, the association with N-CoR was reduced when B-Myb was modified by Cyclin A/Cdk2 (Li and McDonnell, 2002), suggesting that this modification disrupted the potentially repressive conformation dictated by the NRD. Li and McDonnell suggest a model in which disruption of N-CoR/SMRT binding to B-Myb upon Cyclin A/Cdk2-mediated phosphorylation enable association with the CBP/p300 co-activators (Li and McDonnell, 2002). As a result of the latter interaction, acetylase activity provided by CBP/p300, now unfettered by HDAC activity associated with N-CoR/SMRT, may modify histones and other regulatory proteins (possibly including B-Myb itself) and stimulate transcription. Although this is an interesting model to explain activation of B-Myb by Cyclin A/Cdk2, further work is required to prove its validity. One major sticking point is that other researchers find that C-terminally truncated B-Myb mutants actually bind N-CoR as efficiently as wt B-Myb (Masselink et al., 2001). Masselink et al. (2001) also report that interactions with N-CoR and B-Myb are unaffected by Cyclin A co-expression. They suggest instead that physical interactions with the BS69 protein, which was first identified as a protein that binds and inhibits the adenovirus 5 E1A oncogene (Hateboer et al., 1995), contributes to B-Myb's transcriptional repression activity. In summary, conflicting data regarding the B-Myb-N-CoR/SMRT intraction do not allow definitive conclusions to be drawn about its role in mediating the response of B-Myb to hyperphosphorylation. However, this is clearly a very interesting possibility that needs further investigation.

2.4 Proteasomal Degradation of Hyperphosphorylated B-Myb

Another means to control protein activity is at the level of its stability. Indeed, many regulators of the cell cycle and transcription are short-lived proteins regulated by protein degradation through the ubiquitin-proteasome pathway (Desterro et al., 2000; Elledge and Harper, 1998). It is apparent from several studies that B-Myb protein levels were substantially reduced when Cyclin A was co-expressed and conversely were stabilised when Cyclin/Cdk2 phosphorylation sites were mutated or when kinase activity was inhibited with the Cdk2DN protein (Bessa et al., 2001b; Charrase et al., 2000; Johnson et al., 1999; Saville and Watson, 1998). Detailed study of this phenomenon showed that B-Myb protein half-life was reduced from 2.7 hours to 50 minutes after Cyclin A2 expression, while this effect was abolished by proteasome inhibitors such as lactacystin (Charrase et al., 2000). This suggested that hyperphosphorylation may mark B-Myb for ubiquitination and subsequent proteasomal degradation. Consistent with this notion, the reduced stability of wild type B-Myb in Cyclin A2 co-transfected cells was correlated with a marked increase in polyubiquitination of B-Myb (Charrase et al., 2000). Notably, Cyclin A2 had little or no effect upon the stability of a C-terminally truncated protein (B-$Myb_{1\text{-}508}$), moreover, this mutant was scarcely modified by ubiquitin. This result with B-$Myb_{1\text{-}508}$ suggests that lysine residues modified by ubiquitin are localised to C-terminal B-Myb sequences, or alternatively, the C-terminus may be recognised by a ubiquitin ligase. In respect of the latter suggestion, it may be significant that degradation of B-Myb was accelerated in cells co-transfected with the F-box protein $p45^{Skp2}$, which is a component of the SCF complex that functions as an E3 ubiquitin ligase (Charrase et al., 2000). Although it was shown that $p45^{Skp2}$ physically interacted with wt B-Myb (Charrase et al., 2000), it was not reported whether this association was affected by C-terminal deletions.

Ubiquitination may play an important role in limiting Cyclin A/Cdk2-modified B-Myb to S phase, thereby ensuring that transcriptionally hyperactivated protein is not carried through to subsequent stages of the cell cycle. More speculatively, this process may be an end component of a system that activates B-Myb as a direct consequence of the hyperphosphorylation step. Thus, $p45^{Skp2}$ has been shown to complex with Cyclin A2/Cdk2 (Zhang et al., 1995), and it is therefore conceivable that its binding to B-Myb is coupled with hyperphosphorylation of B-Myb. Interestingly, $p45^{Skp2}$ has recently been shown to be an essential co-activator of the c-Myc transcription factor (Kim et al., 2003; von der Lehr et al., 2003). If $p45^{Skp2}$ can similarly co-activate B-Myb, this may explain how hyperphosphorylation of B-Myb is linked both with enhancement of its transcriptional activity and reduction of its stability through ubiquitination. It is certain that future work will investigate this possibility.

3. REGULATION OF GENE EXPRESSION BY B-MYB DURING THE CELL CYCLE

In principle, an obvious requirement for B-Myb in the cell cycle is the transcriptional regulation of genes necessary for cell proliferation, particularly at the G1/S transition and during S phase where B-Myb activity is expected to be at its greatest. Despite several years of work on this topic, there remains an almost total void in our knowledge of B-Myb target genes that give an adequate explanation for why its activity is so tightly regulated during the cell cycle.

Direct evidence that B-Myb transcriptional activity is required for cell cycling is itself limited, but the best indication for such a role has come from the use of inducible dominant-interfering Myb proteins. Such proteins are expected to bind to MBS in Myb target genes and block transcription by the endogenous Myb proteins. Mouse embryonal stem (ES) cells expressing a dominant-interfering protein consisting of the c-Myb DNA-binding domain linked to an Engrailed transcriptional repression module regulated by a linked estrogen receptor hormone binding domain, showed a defect in G1/S transition and an apparently independent reduction in cell adhesion (Iwai et al., 2001). It was presumed that these effects resulted from inhibition of B-Myb transcriptional activity, since this is the only member of the Myb family which was found to be expressed in these primitive progenitor cells (Iwai et al., 2001). These data are consistent with the early embryonic lethal phenotype observed in B-*myb*$^{-/-}$ mice (Tanaka et al., 1999). Definitive evidence that the dominant-interfering protein works in the way specified will require the identification and characterisation of genes whose expression is inhibited as a result of its expression in ES cells.

B-Myb has been reported to activate gene expression from a number of different promoters, including c-*myc*, topoisomerase IIα, Cyclin A1, ApoJ/clusterin, *bcl*-2 and DNA polymerase α (reviewed in Joaquin and Watson, 2003; Sala and Watson, 1999). In cases where studied, as for example *bcl*-2 (Grassilli et al., 1999), Cyclin A2 co-expression has been shown to enhance transactivation of the gene promoter by B-Myb. As it is expected that B-Myb hyperphosphorylation would be restricted to late G1/S phase, authentic B-Myb target genes would themselves most likely demonstrate cyclical expression during the cell cycle. To date, however, no cell-cycle-regulated gene has been shown to depend on B-Myb for its temporal regulation. The identification of B-Myb target genes with the expected properties is of key importance in determining how the various regulatory mechanisms outlined in this chapter control B-Myb's activity in the cell cycle.

4. INTERACTIONS OF B-MYB WITH CELL-CYCLE REGULATORY PROTEINS

B-Myb has been reported to bind to several cell cycle regulatory proteins, including Cyclin D1 (Horstmann et al., 2000), Cyclin A1 (Müller-Tidow et al., 2001) and p107 (Bessa et al., 2001a; Joaquin et al., 2002; Sala et al., 1996b). These interactions may have significance in regulating B-Myb's transcriptional activity during the cell cycle, moreover, they may signify other non-transcriptional activities of B-Myb which affect cell proliferation. In regard to this point, it is notable that a dual role as transcriptional activator and component of a DNA replication complex has been proposed for the *Drosophila* Myb protein, DMyb, which appears to be more closely related to B-Myb than to either c-Myb or A-Myb (Simon et al., 2002). In support of an auxiliary role for B-Myb, it has been reported that the ability of B-Myb to overcome certain inhibitory effects on the cell cycle does not depend upon its transcriptional activity. Thus, ectopically expressed B-Myb was found to bypass a proliferation block induced by p53/p21$^{Waf1/Cip1}$ in a human glioblastoma cell line (Lin et al., 1994), and this activity could also be conferred by a transcriptionally defective B-Myb mutant. Similarly, B-Myb can overcome a G1 cell cycle block imposed by p107 (Sala et al., 1996b), a member of the retinoblastoma pocket protein family, and we have shown that this activity is also independent of B-Myb's transcriptional activity (Joaquin et al., 2002).

4.1 Effects on B-Myb Transcriptional Activity

Binding to Cyclin D1 does not appear to result in phosphorylation of B-Myb, and this interaction does not require the Cdk4/Cdk6 enzymatic subunits (Horstmann et al., 2000). In marked contrast to Cyclin A1 (Müller-Tidow et al., 2001), Cyclin D1 binding actually inhibits B-Myb transcriptional activity, indicating that it is a negative regulator. As Cyclin D1 binds a region of B-Myb that contains its transactivation domain, inhibition may result from masking the activation region. It has been suggested that the interaction with Cyclin D1 may hold B-Myb in a transcriptionally inactive complex during G1 until such time in late G1/S when Cyclin D1 is proteolytically degraded or B-Myb is activated by Cyclin

A/Cdk2 (Horstmann et al., 2000). The opposing effects of these two cyclins could enable fine-tuning of B-Myb activity during the cell cycle. In support of this hypothesis, it has recently been reported that B-Myb transcriptional activity is enhanced during the early stages of neural differentiation, and this coincided with reduced association with Cyclin D1 as levels of this protein declined (Cesi et al., 2002).

The p107 and Cdk9 proteins have also been reported to inhibit B-Myb's transcriptional activity (Sala et al., 1996b). It is unclear whether inhibition by p107 reflects a direct interaction between these proteins, or is an indirect effect resulting from its ability to inhibit Cyclin A/Cdk2 activity and thus prevent hyperphosphorylation of B-Myb. Indeed, we have found that a C-terminally truncated p107 mutant, which has retained the ability to bind B-Myb but lost Cyclin A2 binding, does not inhibit B-Myb transcriptional activity (MJ and RJW, unpublished data). This would suggest that if p107 does inhibit B-Myb activity directly, this does not depend upon physical masking of the transactivation domain as described for Cyclin D1. Like B-*myb*, transcription of p107 is also induced at the G1/S transition through E2F regulation. Levels of B-Myb and p107 proteins are therefore co-regulated during the cell cycle, and it seems inherently unlikely that p107 would normally function to inhibit B-Myb at a stage where it is most transcriptionally active. We have found no evidence that B-Myb hyperphosphorylation influences its interaction with p107 (M. Bessa and RJW, unpublished data). The Cdk9 protein is the enzymatic subunit associated with Cyclin T that is responsible for phosphorylation and activation of RNA polymerase II. Inhibition of B-Myb does not require Cdk9 enzymatic activity (De Falco et al., 2000), and in this respect its activity appears to be functionally similar to Cyclin D1. Cdk9 levels are not cell cycle-regulated and it is not obvious what the physiological relevance of the interaction with B-Myb could be.

Expression of the Cyclin A1 gene is restricted to a very few tissues, namely, testis, haemopoietic precursors and several myeloid leukaemic cell lines (Yang et al., 1997; Yang et al., 1999). Cyclin A1 interacts with the B-Myb C-terminus *in vitro* and these proteins appear to be associated *in vivo* in leukaemic cells. In contrast to Cyclin D1 and Cdk9, Cyclin A1/Cdk2 induces B-Myb activity by phosphorylating key threonine and serine residues (Müller-Tidow et al., 2001). Despite the great similarity between Cyclin A1 and Cyclin A2, no association between B-Myb and Cyclin A2 has been detected either *in vitro* or *in vivo*, suggesting that the B-Myb-Cyclin A1 association may provide a tissue-specific function.

4.2 Non-Transcriptional Regulatory Properties of B-Myb

The ability of B-Myb to overcome cell cycle arrest mediated by p53/p21$^{Waf1/Cip1}$ appears to be an indirect effect, and we have been unable to find evidence that these proteins physically interact (M. Joaquin and RJW, unpublished data). In contrast, bypass of a p107-mediated cell cycle arrest correlates precisely with the capacity of B-Myb to bind to p107 (Joaquin et al., 2002). Thus, transcriptionally hyperactivated B-Myb mutants lacking the C-terminal NRD were unable to bind p107 in *in vivo* interaction assays and had no effect on a p107-mediated G1 block. In contrast, a transcriptionally inert mutant (B-MybΔ205-243) that retained p107 binding was able to bypass the G1 block (Joaquin et al., 2002).

In certain cells, such as the Saos-2 osteosarcoma cell line, p107 causes G1 arrest through two mechanisms that are specified by distinct protein domains: a large pocket domain that binds E2F and an extensive N-terminal domain that binds and inhibits Cyclin/Cdk2 complexes (Zhu et al., 1995; Zhu et al., 1993). We mapped B-Myb binding to an N-terminal region of p107 between amino acids 10-486 (Figure 2) which, significantly, overlaps with a Cyclin A/E binding motif of p107 (Joaquin et al., 2002). This finding suggested that effects on p107-mediated cell cycle arrest relate to the ability of B-Myb to overcome the Cyclin/Cdk2 inhibitory activity of p107. In support of this notion, we demonstrated that co-expression of B-Myb overcame inhibition of Cyclin E/Cdk2 by p107 in transfected Saos-2 cells. Moreover, binding of Cyclin A2 and B-Myb to p107 was found to be mutually exclusive, indicating that the formation of p107/B-Myb complexes would preclude inhibitory interactions of p107 with Cyclin A2/Cdk2. Cyclin E/Cdk2 and Cyclin A2/Cdk2 complexes play important roles in cell replication both at the G1/S transition and during S phase (Krude et al., 1997). It may be proposed, therefore, that B-Myb functions to counteract inhibition of Cyclin/Cdk2 activity by *de novo* synthesized p107 at the critical G1/S boundary. The requirement for this interaction may be cell type-dependent, and would be influenced by the relative abundances of B-Myb and p107. Notably, B-Myb is highly expressed in ES cells, and a large proportion of p107 can be depleted by immunoprecipitation from ES cell extracts with a B-Myb antibody (M. Joaquin and RJW, unpublished data). Possibly the interaction with p107 accounts, at least in part, for the critical function of B-Myb in ES cells.

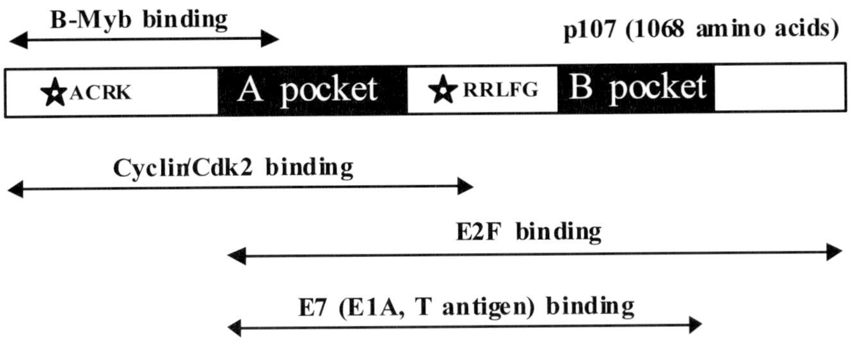

Figure 2
The B-Myb binding region on p107 overlaps with the Cyclin/Cdk2 binding domain. The domain structure of the 1068 amino acid p107 protein is represented schematically. The positions of short amino acid motifs in the p107 N-terminal and spacer regions required for binding and inhibiting Cyclin/Cdk2 activity are indicated by stars. Arrows represent p107 domains required for binding the proteins indicated. Note that the minimal region of p107 required for binding B-Myb has yet to be defined.

5. CONCLUSIONS AND PERSPECTIVES

The B-*myb* gene is subject to two levels of control during the cell cycle: at the transcriptional level through the action of the E2F and Rb family of proteins and at the post-translational level through the action of late G1/S phase-specific kinases, in particular Cyclin A/Cdk2. The first mechanism ensures that B-*myb* is transcribed only in proliferating cells, and directs maximal synthesis of B-Myb protein to late G1/S in cycling cells. The second mechanism ensures that B-Myb's transcriptional activity in cycling cells, where B-Myb protein is present throughout the cell cycle (albeit at fluctuating levels), is enhanced specifically in late G1/S. The fact that these coupled regulatory modes exist strongly suggests that B-Myb plays a significant role in the cell cycle during late G1/S. The precise nature of this role remains to be defined.

Experimental evidence points to B-Myb acting during the cell cycle both to regulate expression of genes required for cell cycling and by making direct interactions with cell cycle regulators such as p107. It may be significant that the related *Drosophila* Myb protein (DmMyb) has been shown to have a dual role in the cell cycle, both in transcriptional regulation of the cyclin B gene and as part of a DNA replication complex (Beall et al., 2002; Okada et al., 2002). Studies of DmMyb in the fly may therefore help us understand what B-Myb does in the mammalian cell cycle. In this respect, it is interesting that DmMyb mutants display defects in progression through both S and M phases of the cell cycle and accumulate chromosomal abnormalities (Fung et al., 2002; Katzen et al., 1998; Manak et al., 2002). Conversely, over-expressed DMyb induces cell cycle progression through S and M phases and suppresses endoreduplication (Fitzpatrick et al., 2002). Concordant with the effects of DmMyb, over-expression of B-*myb* has been shown to promote progression of at least certain cell into S phase, in particular when it is co-expressed with Cyclin A2 (Lane et al., 1997; Sala et al., 1996a). It is unclear at present whether this results from the induction of target gene transcription, direct effects upon cell cycle regulators or indeed a combination of both. Advances in technology, in particular the development of gene ablation using conditional gene knockouts based on the Cre/loxP system and RNA interference, should enable rapid progress to be made on understanding B-Myb's role in the cell cycle.

REFERENCES

Ansieau, S., Kowentz-Leutz, E., Dechend, R. and Leutz, A. (1997) B-Myb, a repressed *trans*-activating protein. J. Mol. Med. *75*, 815-819.

Arsura, M., Introna, M., Passerini, F., Mantovani, A. and Golay, J. (1992) B-myb antisense oligonucleotides inhibit proliferation of human hematopoietic cell lines. Blood *79*, 2708-2716.

Bartsch, O., Horstmann, S., Toprak, K., Klempnauer, K.-H. and Ferrari, S. (1999) Identification of cyclin A/Cdk2 phosphorylation sites in B-Myb. Eur. J. Biochem. *260*, 384-391.

Bartusel, T. and Klempnauer, K.-H. (2003) Transactivation mediated by B-Myb is dependent on TAF(II)250. Oncogene *22*, 2932-2941.

Beall, E.L., Manak, J.R., Zhou, S., Bell, M., Lipsick, J.S. and Botchan, M.R. (2002) Role for a *Drosophila* Myb-containing protein complex in site-specific DNA replication. Nature *420*, 833-837.

Bessa, M., Joaquin, M., Tavner, F., Saville, M.K. and Watson, R.J. (2001a) Regulation of the cell cycle by b-myb. Blood Cells Mol. Dis. *27*, 416-421.

Bessa, M., Saville, M.K. and Watson, R. J. (2001b) Inhibition of cyclin A/CDK2 phosphorylation impairs B-Myb transactivation function without affecting interactions with DNA or the CBP coactivator. Oncogene *20*, 3376-3386.

Bies, J., Hoffman, B., Amanullah, A., Giese, T. and Wolff, L. (1996) B-Myb prevents growth arrest associated with terminal differentiation of monocytic cells. Oncogene *12*, 355-363.

Cervellera, M.N. and Sala, A. (2000) Poly(ADP-ribose) polymerase is a B-MYB coactivator. J. Biol. Chem. *275*, 10692-10696.

Cesi, V., Tanno, B., Vitali, R., Mancini, C., Giuffrida, M.L., Calabretta, B. and Raschella, G. (2002) Cyclin D1-dependent regulation of B-myb activity in early stages of neuroblastoma differentiation. Cell Death Differ. *9*, 1232-1239.

Chan, H.M. and La Thangue, N.B. (2001) p300/CBP proteins: HATs for transcriptional bridges and scaffolds. J. Cell Sci. *114*, 2363-2373.

Charrase, S., Carena, I., Brondani, Klempnauer, K.-H. and Ferrari, S. (2000) Degradation of B-Myb by ubiquitin-mediated proteolysis: involvement of the Cdc34-$SCF^{p45Skp2}$ pathway. Oncogene *19*, 2986-2995.

De Falco, G., Bagella, L., Claudio, P.P., De Luca, A., Fu, Y., Calabretta, B., Sala, A. and Giordano, A. (2000) Physical interaction between CDK9 and B-Myb results in suppression of B-Myb gene autoregulation. Oncogene *19*, 373-379.

Desterro, J.M.P., Rodriguez, M.S. and Hay, R.T. (2000) Regulation of transcription factors by protein degradation. Cell Mol. Life Sci. *57*, 1207-1219.

Elledge, S.J. and Harper, J.W. (1998) The role of protein stability in the cell cycle and cancer. Biochim. Biophys. Acta. *1377*, M61-M70.

Fitzpatrick, C.A., Sharkov, N.V., Ramsay, G. and Katzen, A.L. (2002) *Drosophila* myb exerts opposing effects on S phase, promoting proliferation and suppressing endoreduplication. Development *129*, 4497-4507.

Foos, G., Grimm, S. and Klempnauer, K.-H. (1992) Functional antagonism between members of the myb family: B-Myb inhibits v-myb-induced gene activation. EMBO J. *11*, 4619-4629.

Fung, S.M., Ramsay, G. and Katzen, A.L. (2002) Mutations in *Drosophila* myb lead to centrosome amplification and genomic instability. Development *129*, 347-359.

Grassilli, E., Salomoni, P., Perrotti, D., Franceschi, C. and Calabretta, B. (1999) Resistance to apoptosis in CTLL-2 cells overexpressing B-Myb is associated with B-Myb-dependent bcl-2 induction. Cancer Res. *59*, 2451-2456.

Hateboer, H., Gennissen, Y.F., Ramos, R.M., Kerkhoven, V., Sonntag-Buck, H.G., Stunnenberg, R. and Bernards, R. (1995) BS69, a novel adenovirus E1A-associated protein that inhibits E1A transactivation. EMBO J. *14*, 3159-3169.

Holmes, J.K. and Solomon, M.J. (1996) A predictive scale for evaluating cyclin-dependent kinase substrates. J. Biol. Chem. *271*, 25240-25246.

Horstmann, S., Ferrari, S. and Klempnauer, K.-H. (2000) Regulation of B-Myb activity by cyclin D1. Oncogene *19*, 298-306.

Iwai, N., Kitajima, K., Sakai, K., Kimura, T. and Nakano, T. (2001) Alteration of cell adhesion and cell cycle properties of ES cells by an inducible dominant interfering Myb mutant. Oncogene *20*, 1425-1434.

Jepsen, K. and Rosenfeld, M.G. (2002) Biological Roles and mechanistic actions of co-repressors complexes. J. Cell Sci. *115*, 689-698.

Joaquin, M., Bessa, M., Saville, M.K. and Watson, R.J. (2002) B-Myb overcomes a p107-mediated cell proliferation block by interacting with an N-terminal domain of p107. Oncogene, *21*, 7923-7932.

Joaquin, M. and Watson, R.J. (2003) Cell cycle regulation by the B-Myb transcription factor. Cell Mol. Life Sci. *in press*.

Johnson, L.R., Johnson, T.K., Desler, M., Luster, T.A., Nowling, T., Lewis, R.E. and Rizzino, A. (2002) Effects of B-Myb on gene transcription: phosphorylation-dependent activity ans acetylation by p300. J. Biol. Chem. *277*, 4088-4097.

Johnson, T.K., Schweppe, R.E., Septer, J. and Lewis, R.E. (1999) Phosphorylation of B-Myb regulates its transactivation potential and DNA binding. J. Biol. Chem. *274*, 36741-36749.

Katzen, A.L., Jackson, J., Harmon, B.P., Fung, S.-M., Ramsay, G. and Bishop, J.M. (1998) *Drosophila* myb is required for the G2/M transition and maintenance of diploidy. Genes Dev. *12*, 831-843.

Kim, S.Y., Herbst, A., Tworkowski, K.A., Salghetti, S.E. and Tansey, W.P. (2003) Skp2 regulates myc protein stability and activity. Mol. Cell *11*, 1177-1188.

Krude, T., Jackman, M., Pines, J. and Laskey, R.A. (1997) Cyclin/Cdk-dependent initiation of DNA replication in a human cell-free system. Cell *88*, 109-119.

Lane, S., Farlie, P. and Watson, R. (1997) B-Myb function can be markedly enhanced by cyclin A-dependent kinase and protein truncation. Oncogene *14*, 2445-2453.

Li, X. and McDonnell, D.P. (2002) The transcription factor B-Myb is maintained in an inhibited state in target cells through its interaction with the nuclear corepressors N-CoR and SMRT. Mol. Cell. Biol. *22*, 3663-3673.

Lin, D., Fiscella, M., O'Connor, P.M., Jackman, J., Chen, M., Luo, L.L., Sala, A., Travali, S., Appella, E. and Mercer, W.E. (1994) Constitutive expression of B-*myb* can bypass p53-induced Waf1/Cip1-mediated G_1 arrest. Proc. Natl. Acad. Sci. USA *91*, 10079-10083.

Manak, J.R., Mitiku, N. and Lipsick, J.S. (2002) Mutation of the *Drosophila* homologue of the Myb protooncogene causes genomic instability. Proc. Natl. Acad. Sci. U S A *99*, 7438-7443.

Masselink, H., Vastenhouw, N. and Bernards, R. (2001) B-myb rescues ras-induced premature senescence, which requires its transactivation domain. Cancer Lett. *171*, 87-101.

Müller-Tidow, C., Wang, W., Idos, G.E., Diederichs, S., Yang, R., Readhead, C., Berdel, W. E., Serve, H., Saville, M., Watson, R. and Koeffler, H.P. (2001) Cyclin A1 directly interacts with B-Myb and cyclin A1/cdk2 phosphorylate B-Myb at functionally important serine and threonine residues: tissue-specific regulation of B-Myb function. Blood *97*, 2091-2097.

Okada, M., Akimaru, H., Hou, D.X., Takahashi, T. and Ishii, S. (2002) Myb controls G(2)/M progression by inducing cyclin B expression in the *Drosophila* eye imaginal disc. EMBO J. *21*, 675-684.

Raschellá, G., Negroni, A., Sala, A., Pucci, S., Romeo, A. and Calabretta, B. (1995) Requirement of B-Myb function for survival and diferentiative potential of human neuroblastoma cells. J. Biol. Chem. *270*, 8540-8545.

Robinson, C., Light, Y., Groves, R., Mann, D., Marais, R. and Watson, R. (1996) Cell-cycle regulation of B-Myb protein expression: specific phosphorylation during the S phase of the cell cycle. Oncogene *12*, 1855-1864.

Sala, A. and Calabretta, B. (1992) Regulation of BALB/c 3T3 fibroblast proliferation by B-myb is accompanied by selective activation of cdc2 and cyclin D1 expression. Proc. Natl. Acad. Sci. USA *89*, 10415-10419.

Sala, A., Casella, I., Bellon, T., Calabretta, B., Watson, R.J. and Peschle, C. (1996a) B-myb Promotes S Phase and Is a Downstrean Target of the Negative Regulator p107 in Human Cells. J. Biol. Chem. *271*, 9363-9367.

Sala, A., De Luca, A., Giordano, A. and Peschle, C. (1996b) The Retinoblastoma Family Member p107 Binds to B-Myb and Supresses Its Autoregulatory Activity. J. Biol. Chem. *271*, 28738-28740.

Sala, A., Kundu, M., Casella, I., Engelhard, A., Calabretta, B., Grasso, L., Paggi, M.G., Giordano, A., Watson, R.J., Khalili, K. and Peschle, C. (1997) Activation of human B-MYB by cyclins. Proc. Natl. Acad. Sci. USA *94*, 532-536.

Sala, A. and Watson, R. (1999) B-Myb protein in cellular proliferation, transcription control, and cancer: latest developments. J. Cell Physiol. *179*, 245-250.

Sano, Y. and Ishii, S. (2001) Increased affinity of c-Myb for CREB-binding protein (CBP) after CBP-induced acetylation. J. Biol. Chem. *276*, 3674-3682.

Santilli, G., Cervellera, M.N., Johnson, T.K., Lewis, R.E., Iacobelli, S. and Sala, A. (2001) PARP co-activates B-MYB through enhanced phosphorylation at cyclin/cdk2 sites. Oncogene *20*, 8167-8174.

Saville, M.K. and Watson, R.J. (1998) The cell-cycle regulated transcription factor B-Myb is phosphorylated by cyclin A/Cdk2 at sites that enhance its transactivation properties. Oncogene *17*, 2679-2689.

Seong, H.A., Kim, K.T. and Ha, H. (2003) Enhancement of B-myb transcriptional activity by ZPR9, a novel zinc finger protein. J. Biol. Chem. *6*, 6.

Simon, A.L., Stone, E.A. and Sidow, A. (2002) Inference of functional regions in proteins by quantification of evolutionary constraints. Proc. Natl. Acad. Sci. USA *99*, 2912-2917.

Takemoto, Y., Tashiro, S., Handa, H. and Ishii, S. (1994) Multiple nuclear localization signals of the B-*myb* gene product. FEBS Lett. *350*, 55-60.

Tanaka, Y., Patestos, N.P., Maekawa, T. and Ishii, S. (1999) B-myb Is Required for Inner Cell Mass Formation at an Early Stage of Development. J. Biol. Chem. *274*, 28067-28070.

Tomita, A., Towatari, M., Tsuzuki, S., Hayakawa, F., Kosugi, H., Tamai, K., Miyazaki, T., Kinoshita, T. and Saito, H. (2000) c-Myb acetylation at the carboxyl-terminal conserved domain by transcriptional co-activator p300. Oncogene *19*, 444-451.

von der Lehr, N., Johansson, S., Wu, S., Bahram, F., Castell, A., Cetinkaya, C., Hydbring, P., Weidung, I., Nakayama, K., Nakayama, K.I., *et al.* (2003) The F-Box Protein Skp2 Participates in c-Myc Proteosomal Degradation and Acts as a Cofactor for c-Myc-Regulated Transcription. Mol. Cell *11*, 1189-1200.

Watson, R.J., Robinson, C. and Lam, E. W.-F. (1993) Transcription regulation by murine B-*myb* is distinct from that by c-*myb*. Nucl. Acids Res. *21*, 267-272.

Yang, R., Morosetti, R. and Koeffler, H.P. (1997) Characterization of a second human cyclin A that is highly expressed in testis and in several leukemia cell lines. Cancer Res. *57*, 913-920.

Yang, R., Nakamaki, T., Lubbert, M., Said, J., Sakashita, A., Freyaldenhoven, B.S., Spira, S., Huynh, V., Müller, C. and Koeffler, H.P. (1999) Cyclin A1 is expressed in blasts of leukemic patients and during hematopoiesis. Blood *93*, 2067-2074.

Zhang, H., Kobayashi, R., Galaktionov, K. and Beach, D. (1995) p19Skp1 and p45Skp2 are essential elements of the cyclin A-CDK2 S phase kinase. Cell *82*, 915-925.

Zhu, L., Harlow, E. and Dynlacht, D. (1995) p107 uses a $p21^{CIP1}$-related domain to bind cyclin/cdk2 and regulate interactions with E2F. Genes Dev. *9*, 1740-1752.

Zhu, L., van den Heuvel, S., Helin, K., Fattaey, A., Ewen, M., Livingston, D., Dyson, N. and Harlow, E. (1993) Inhibition of cell proliferation by p107, a relative of the retinoblastoma protein. Genes Dev. *7*, 1111-1125.

Ziebold, U., Bartsch, O., Marais, R., Ferrari, S. and Klempnauer, K.-H. (1997) Phosphorylation and activation of B-Myb by cyclin A-Cdk2. Curr. Biol. *7*, 253-260.

Chapter 10

REGULATION OF MAMMALIAN MYB GENE EXPRESSION

Fiona J. Tavner
Ludwig Institute for Cancer Research and Department of Virology, Faculty of Medicine, Imperial College London, Norfolk Place, London W2 1PG, United Kingdom.

Abstract: Although the precise molecular mechanisms that govern mammalian *myb* (c-*myb*, B-*myb* (*MybL2*) and A-*myb* (*MybL1*)) gene expression are yet to be resolved, a collective understanding is beginning to emerge. At present, it is evident that distinct regulatory factors and mechanisms control expression of the mammalian *myb* genes, and this is presumably reflected in the defined expression patterns of individual family members. A review of the current state of knowledge pertaining to the molecular regulation of mammalian *myb* gene expression, and including an historical perspective, is presented within this chapter.

1. INTRODUCTION

The mammalian *myb* genes (c-*myb*, B-*myb* (*MybL2*) and A-*myb* (*MybL1*)) display distinct spatial and temporal patterns of expression in adult tissues and during murine embryogenesis. However, *myb* family members are also found co-expressed in certain tissues. It is apparent that the molecular mechanisms controlling the expression of these genes are distinct and particularly in the case of c-*myb*, complex. On the whole, the development of an understanding of the regulatory factors controlling *myb* expression has progressed relatively slowly since the cloning of these genes. Though, the B-*myb* promoter has recently emerged as a model to study cell cycle-mediated regulatory mechanisms of gene expression. This chapter will focus on the molecular entities and mechanisms involved in the control of *myb* gene expression and identify areas to be further progressed.

J. Frampton (ed.), Myb Transcription Factors: Their Role in Growth, Differentiation and Disease, 201-221.

2. EXPRESSION OF MYB GENES

2.1 In Adult Tissues

In mammalian adult tissues, both c-*myb* and A-*myb* display restricted patterns of expression. Comparatively high levels of c-*myb* expression are found predominantly in immature haemopoietic cells of the myeloid, lymphoid and erythroid lineages. In contrast, terminally differentiated cells of these lineages do not exhibit detectable c-*myb* expression. In resting, mature T and B lymphocytes however, c-*myb* expression is re-established upon induction of proliferation (Catron et al., 1992; Stern and Smith, 1986; Torelli et al., 1985). Expression of c-*myb* persists until later stages of haemopoietic cell maturation, directly preceding terminal differentiation. Critically, in order for haemopoietic cell maturation to proceed to terminal differentiation, down-regulation of c-*myb* expression is an essential prerequisite. Constitutive expression of c-*myb* in later stages of cellular maturation blocks the ability of cells to terminally differentiate and thus maintains cells in a proliferative state (McClinton et al., 1990; McMahon et al., 1988; Todokoro et al., 1988; Yanagisawa et al., 1991). Overexpression of c-*myb* is commonly found among haemopoietic neoplasias (Slamon et al., 1986; Slamon et al., 1984; Wolff, 1996). Expression of c-*myb* is also detected outside of the haemopoietic system, namely in colonic epithelia and in vascular smooth muscle cells (Brown et al., 1992; Ess et al., 1999; Thompson and Ramsay, 1995). Elevated expression is seen in human pre/malignant colonic epithelia and in vascular smooth muscle cells after arterial injury (Gunn et al., 1997; Ramsay et al., 1992). Interestingly, immunohistochemical staining of murine colonic crypts indicated that c-*myb* expression is found in some differentiated, non-proliferating cell types, revealing that expression is not solely correlated with proliferation in the colon (Rosenthal et al., 1996).

The expression of A-*myb* is similarly restricted, with greatest levels found in mature B lymphocytes and spermatocytes (Golay et al., 1991; Trauth et al., 1994). Expression of A-*myb* is also detectable to a lesser extent in brain, heart and lung tissues (Trauth et al., 1994). A substantial increase in A-*myb* expression is evident in murine female breast ductal epithelia during pregnancy and is coincident with cellular proliferation (Toscani et al., 1997). Certain haemopoietic neoplasias, in particular Burkitt's lymphomas and some chronic lymphocytic leukaemias display elevated A-*myb* expression (Golay et al., 1996). Unlike c-*myb*, neoplasia associated expression of A-*myb* is restricted to a sub-group of maturation-specific B cells (Golay et al., 1996).

In contrast to c-*myb* and A-*myb*, expression of B-*myb* is ubiquitous among established cell lines studied and is correlated with cellular proliferation (Arsura et al., 1994; Lam et al., 1992; Nomura et al., 1988). Expression of B-*myb* is induced upon mitogenic stimulation, but furthermore displays periodic expression during the cell cycle (Golay et al., 1991; Lam et al., 1992). Therefore, in contrast to c-Myb and A-Myb, it is assumed that B-Myb functions in a more general sense by association with cell cycle activities. In view of the fact that B-*myb* expression is ubiquitously associated with cellular proliferation, co-expression with at least one other *myb* family member occurs in certain cell types. This suggests that each Myb protein has a specific role when co-expressed within the same cell. Indeed, this is manifest by recent gene expression profiling (Rushton et al., 2003).

An interesting situation exists during spermatogenesis in the adult mouse, where B-*myb* expression is not seen in spermatagonia actively undergoing mitosis (Sitzmann et al., 1996). Rather, expression is seen in more mature, meiotically active sperm cells and is lacking in terminally differentiated, non-proliferative cells, as expected (Sitzmann et al., 1996). Hence, in this developmental system, the expression of B-*myb* is not entirely associated with cellular proliferation. It is important to note that A-*myb* is also expressed in the testis, specifically in spermatagonia and spermocytes, and may therefore replace the function of B-Myb in these cells (Trauth et al., 1994). In addition, co-expression of both A-*myb* and B-*myb* within the testis indicates that B-Myb may have an important function in meiotically dividing cells, which cannot be substituted by A-Myb. Interestingly, an additional, larger B-*myb* transcript is detected in testis tissue, the significance of which currently remains unknown (Sitzmann et al., 1996).

2.2 During Murine Embryogenesis

The expression patterns of *myb* family members during murine embryogenesis reflect their generalised spatial distribution in adult tissues. Thus, both c-*myb* and A-*myb* show restricted expression, whereas that of B-*myb* is widespread among proliferating cells and additionally detected in extra-embryonic tissues, including the yolk sac (Sitzmann et al., 1996). The developing central nervous system (CNS) displays prevalent B-*myb* expression, which is extinguished as cells progressively lose their proliferative capacity upon terminal differentiation (Sitzmann et al., 1996). Expression of A-*myb* is also detected in the developing CNS, but within distinct regions of the neural tube, brain, eye and olfactory epithelium containing actively dividing neuronal precursor cells (Trauth et al., 1994). The developing urogenital ridge also displays A-*myb* expression, specifically within the genital ridge (Trauth et al., 1994). Expression of c-*myb* is similarly restricted during murine embryogenesis and is found at sites of haemopoiesis, within the foetal liver and the cortex of the developing thymus (Ess et al., 1999; Sitzmann et al., 1995). The yolk sac and in particular, the blood islands, do not express c-*myb* indicating no functional requirement for c-Myb during early haemopoiesis (Mucenski et al., 1991; Sitzmann et al., 1995). A number of non-haemopoietic tissues also display c-*myb* expression, these being nasal, tracheal and bronchial epithelia of the respiratory tract, gastrointestinal tract epithelia, hair follicles and toothbuds (Ess et al., 1999; Sitzmann et al., 1995). During murine embryogenesis, c-*myb* expression rarely overlaps with that of A-*myb*. However, co-expression of both c-*myb* and A-*myb* does occur in the developing eye, within the neural retina (Sitzmann et al., 1995).

An important distinction in the temporal expression of the *myb* genes is also evident. During early embryogenesis (before day E10), B-*myb* is the only family member to be expressed (Sitzmann et al., 1996). Consistent with this observation, embryonic development does not proceed beyond formation of the inner cell mass (ICM) in mice homozygous for B-*myb* null alleles (B-*myb* 'knockout' mice), with early embryonic death occurring between days E4.5-6.5 (Tanaka et al., 1999). Failure of ICM formation is attributed to defective cellular proliferation and demonstrates a critical requirement for B-Myb function during early murine embryogenesis.

In keeping with their distinct and restricted expression, c-*myb* and A-*myb* knockout mice have very specific biological phenotypes. Mice lacking functional c-Myb are embryonic lethal (by day E15) due to abnormal foetal liver haemopoiesis and in particular severe anaemia (Mucenski et al., 1991). Mice lacking functional A-Myb survive embryonic development, but display defective adult phenotypes. Male mice do not produce mature spermatozoa and are therefore sterile (Toscani et al., 1997). In females, breast morphogenesis during pregnancy is severely compromised because of lack of mammary epithelial proliferation (Toscani et al., 1997). Consequently, females lacking A-Myb are unable to suckle their newborn pups.

3. REGULATION OF C-MYB EXPRESSION

The regulation of c-*myb* expression is complex and currently with little definition of the *cis*- and *trans*-acting regulatory components. In haemopoietic cells, down-regulation of c-*myb* expression is an essential prerequisite in order for terminal differentiation to occur. Regulation of transcription initiation does not appear to be the major mechanism by which c-*myb* expression is controlled. Initiation of transcription within the 5' promoter of c-*myb*, which does not contain a TATA box, is constitutive and promiscuous, with transcripts displaying 5' heterogeneity as a consequence of being initiated from multiple cap sites (Bender and Kuehl, 1986; Watson et al., 1987). Interestingly, this heterogeneity is more prevalent in cells expressing high levels of c-*myb* (generally more immature haemopoietic cells) than in cells with low levels of expression, which show restricted cap site usage (Watson et al., 1987). Instability of c-*myb* mRNA does not appear to account for the reduction of c-*myb* expression in mature haemopoietic cells, though c-*myb* transcripts contain an AU-rich sequence within their 3' untranslated region, which in other genes does contribute to mRNA instability (Watson, 1988a).

3.1 The Transcriptional Arrest Site

Examination of murine and human c-*myb* transcripts by nuclear run-on analyses indicated that changes in the level of gene expression during differentiation of haemopoietic cells are predominantly attributed to a mechanism of transcriptional arrest within the first intron (Bender et al., 1987; Castellano et al., 1992; Watson, 1988a; Watson, 1988b). Thus, in cells that do not produce a mature full-length c-*myb* transcript, RNA polymerase II-mediated transcription is initiated and continues to proceed, but is subsequently arrested at a site located in a central region of the first intron (Watson, 1988a; Watson, 1988b). The site of arrest within the murine c-*myb* gene has been more precisely mapped to a location approximately 1.7kb downstream of the 5' boundary of the first intron by RNase protection analysis in *Xenopus* oocytes (Yuan, 2000). Alignment of murine and human intron 1 sequences identified a region with high sequence similarity encompassing the site of transcriptional arrest, indicating potentially conserved regulatory elements (Toth et al., 1995; Yuan, 2000). It is apparent that sequences outside of the conserved region in intron 1 also contribute to the transcriptional arrest in certain cell types (eg fibroblasts) (Yuan, 2000). Continuation of transcriptional elongation through the arrest site appears to be regulated by sequences also residing within the first intron and located in a flanking region upstream of the arrest site itself (Yuan, 2000). It is therefore proposed that intronic sequences play an important role in directing cell type-specific expression of c-*myb* by regulation of transcriptional elongation.

3.2 Hypersensitivity Sites

Several DNase I hypersensitivity sites have been mapped within the first intron and 5' flanking region of the murine c-*myb* gene (Bender et al., 1987;

Reddy and Reddy, 1989). In particular, one hypersensitivity site (IV) maps to the region within the first intron in which transcriptional arrest occurs (Bender et al., 1987). Site IV displays a differential sensitivity to DNase I in accordance with the expression of c-*myb*. Thus, in the 70Z/3B pre-B cell lymphoma cell line highly expressing c-*myb*, site IV was sensitive to DNase I digestion, but not in A20.2J B cell lymphoma cells that express comparatively little c-*myb* (Bender et al., 1987). Site IV may therefore indicate the presence of a DNA-binding protein complex or higher order chromatin structure within the region of transcriptional arrest, which may furthermore be responsible for imposing the actual arrest mechanism.

3.3 Binding Proteins

3.3.1 Associated with Intron 1

In attempts to identify regulators of c-*myb* expression, a multitude of DNA-binding complexes has been detected at various sequences residing within the first intron. Electrophoretic mobility shift assays (EMSAs) have generally been employed in order to detect complexes that display differential binding activity correlated with c-*myb* expression. Binding complexes that are associated with intron 1 sequences when expression is downregulated are putative negative regulators, which may play a role in the transcriptional arrest mechanism. A recently reported activity, ABF (Attenuator Binding Factor), binds to a 15 bp sequence within the first intron of murine c-*myb* near the transcriptional arrest site, in cells with downregulated c-*myb* expression (Perkel et al., 2002). The DNA-binding activity of ABF increased concurrently with reduction in c-*myb* expression as murine erythroleukaemia (MEL) cells were induced to differentiate upon treatment with DMSO (Perkel et al., 2002). ABF contains an unidentified 64 kDa DNA-binding protein that awaits further investigation. Reporter assays indicated that deletion of the ABF binding site did not relieve transcriptional arrest imposed by larger DNA fragments containing the arrest site (Perkel et al., 2002). Additionally, the ABF site alone was unable to induce transcriptional arrest, indicating that sequences spanning a wider region than the singular site defined in this study are required. In the human c-*myb* gene, interferon regulatory factors (IRF) 1 and 2 bind to a site within the region of transcriptional arrest (Manzella et al., 2000). Overexpression of IRF-1, but not IRF-2, suppressed reporter activity driven by the c-*myb* 5' flanking/exon 1 and intron 1 sequences encompassing the IRF binding site (Manzella et al., 2000). In addition, c-*myb* expression failed to be downregulated in HL60 cells induced to differentiate with phorbol ester (TPA) when antisense IRF-1 was expressed (Manzella et al., 2000). Therefore, IRF-1 is implicated in the negative regulation of c-*myb* expression and possibly in the context of transcriptional arrest.

Members of the NF-κB family are found associated with DNA elements located within the first intron. Two NF-κB binding elements flank the arrest site and were firstly shown to bind NF-κB/Rel members in mature haemopoietic cell lines and thus correlated with reduced c-*myb* expression (Toth et al., 1995). Paradoxically, co-expression of NF-κB family members in murine thymoma EL4 cells resulted in activation of a reporter construct containing the c-*myb* 5' flanking/exon 1 and intron 1 sequences (Toth et al., 1995). Similarly, NF-κB p50 and RelB bind to the 3' NF-κB binding element within intron 1 concurrent with persistent, increased c-*myb* expression when hexamethylene bisacetamide (HMBA)-induced MEL cell differentiation was blocked by cAMP analogues (Suhasini et al., 1997). Furthermore, stable co-expression of NF-κB p50 and RelB in MEL cells prevented down-regulation of c-*myb* expression in the presence of HMBA, and consequently blocked erythroid differentiation (Suhasini and Pilz, 1999). Therefore, NF-κB binding activity is correlated with positive regulation of c-*myb* expression and by virtue of the location of cognate binding sites, may function by acting upon the mechanism of transcriptional arrest. Other potential activators of c-*myb* expression that bind within the vicinity of the transcriptional arrest site have also been described, but their identity and activity remains uncharacterised (Dooley et al., 1996; Reddy and Reddy, 1989).

3.3.2 Associated with the 5' flanking region

As well as intronic sequences being important in the regulation of c-*myb* expression, the 5' flanking region (promoter) also contributes to cell type-specific expression. It is apparent that distinct factors positively regulate the expression of c-*myb* in different haemopoietic lineages (Sullivan et al., 1997). In the Molt-4 T cell line, two Ets-like sites (5' and 3') have been identified by deletion analysis of the human c-*myb* 5' flanking region that contribute to activation, but to differing degrees (Sullivan et al., 1997). The 3' site binds c-Ets-1 *in vitro* and contributes significantly more to activation than the 5' site (Sullivan et al., 1997). The 5' site does not appear to bind Ets factors, however a 67 kDa protein of unknown identity was detected *in vitro* (Sullivan et al., 1997). The activating regions identified in the Molt-4 cells do not contribute to activation of c-*myb* expression in cells of other haemopoietic lineages, namely the DHL-9 B cell and K562 myeloid cell lines (Sullivan et al., 1997). Instead, another site within the c-*myb* 5' flanking sequence positively regulates expression within these cell lines and binds an unidentified protein of 50.5 kDa (Sullivan et al., 1997).

Factors ascribed a negative regulatory role in c-*myb* expression have also been detected at sequences located in the 5' flanking region. The Wilm's tumour product (WT1) is implicated in repression of c-*myb* expression in lymphoid cells. Although overexpression of WT1 in T and B cell lines can specifically repress the c-*myb* promoter, it appears that WT1 associates with different sites in each cell type, indicating differential site function in a lineage-specific context (McCann et al., 1995). The presence of three Myb-binding sites within the 5' flanking region has prompted investigations of autoregulation of expression. These sites were first shown to act positively in fibroblasts (Nicolaides et al., 1991). A subsequent study of these sites in haemopoietic cell lines indicates a negative regulatory role for sites I and II in T cells (Guerra et al., 1995).

3.3.3 Associated with expression in mature lymphoid cells

The expression of c-*myb* is re-established upon induction of mature T and B cell proliferation, and therefore presents a situation where activation of gene expression is required. In addition, the expression of c-*myb* appears to be cell cycle regulated in proliferating mature lymphoid cells, which is in contrast to the constitutive expression seen in immature cells (Catron et al., 1992; Thompson et al., 1986). It is not entirely understood what regulatory factors govern the induction of c-*myb* expression in proliferating mature lymphoid cells, however the presence of a conserved E2F site within the 5' promoter (flanking region) may be a critical determinant in this context.

The E2F site, together with a conserved NF-κB site in close proximity, have been implicated in the induction of c-*myb* expression in activated T cells in response to IL-2 stimulation (Lauder et al., 2001). In particular, IL-2-mediated induction of c-*myb* is transduced specifically via the phosphoinositide 3-kinase (PI3K) and protein kinase B (PKB) signalling pathway (Lauder et al., 2001). Reporter assays carried out in NIH3T3 fibroblasts indicated that activated PI3K and PKB not only activated the c-*myb* promoter, but also relieved transcriptional arrest imposed by the 5' flanking/exon 1 and intron 1 sequences. It remains to be determined whether PI3K signalling converges specifically upon the mechanism of transcriptional arrest. Mutation of the E2F site resulted in a markedly greater inhibition of promoter activity than mutation of the NF-κB site (Lauder et al., 2001). Mutation of both sites almost completely abolished promoter activity, indicating a functional requirement for both sites, albeit with differing contributions to overall activity. The E2F site binds *in vitro* translated E2F1/DP1 and in addition, several complexes were detected in activated T cell extracts that bind to the E2F site *in vitro* (Lauder et al., 2001). The composition of these complexes is yet to be identified. The NF-κB site binds purified p65, but not p50 homodimers *in vitro*. However, p65/p50 heterodimers were able to bind to this site. A complex detected in activated T cell extracts binds to the NF-κB site *in vitro*, with a small proportion of this complex exhibiting a mobility shift in the presence of c-Rel and p50 antibodies (Lauder et al., 2001). Hence E2F and NF-κB family members can bind *in vitro* to their cognate binding sequences within the c-*myb* promoter, but further characterisation of complexes binding to the E2F and NF-κB sites will be required to assess any functional contribution towards induction of c-*myb* expression in activated T cells.

In another study that focused specifically on the c-*myb* E2F site, certain E2Fs and pocket proteins were detected in complexes binding *in vitro* to the E2F site in lymphoblastoid X50-7 nuclear extracts (Campanero et al., 1999). The Sp1 transcription factor was also found to bind to the E2F site *in vitro*, but with reduced affinity compared to a typical Sp1-binding element (Campanero et al., 1999). In addition to the E2F/pocket protein complexes detected in X50-7 extracts, a distinct complex termed E2Fmyb-sp binds to a site that overlaps the E2F site (Campanero et al., 1999). Reporter assays indicated that both the E2F and overlapping E2Fmyb-sp sites contribute to activation of c-*myb* expression in asynchronous cells and mutation of these sites impairs activation of the c-*myb* promoter in G_1 phase, in NIH3T3 fibroblasts synchronised by serum deprivation/stimulation (Campanero et al., 1999). Collectively, these studies reveal that the c-*myb* E2F site operates in a positive regulatory role. It is noteworthy that expression of c-*myb* is considerably upregulated in response to conditional activation of E2F1, 2 and 3 (Müller et al., 2001).

A site has been identified in the 5' flanking region of the human c-*myb* gene that mediates an increase in expression upon activation of Jurkat T cells and which displays occupancy only in activated T cells (Phan et al., 1996). This site binds an unidentified complex termed CMAT (c-*myb* in activated T cells) that shows DNA-binding kinetics consistent with induction of c-*myb* expression (Phan et al., 1996). The functional significance of CMAT awaits further investigation.

3.4 Summary

It is clear that the regulation of c-*myb* expression is complicated, involving multiple control elements dispersed throughout a wide region of the gene. The mechanism of transcriptional arrest remains unknown and may be mediated by physical impediment of the transcription machinery by specific DNA-binding protein complexes or nucleosomes, or by properties of the sequence itself such as RNA secondary structure. Examination of the human c-*myb* gene revealed a region within the vicinity of the transcriptional arrest site that conforms to the requirements for potential formation of a RNA stem-loop structure and possible interference with transcription (Thompson et al., 1997). It is also necessary to elucidate how the transcriptional arrest is relieved and how particular activators interact to mediate expression of c-*myb* in a cell type-specific manner.

4. CELL CYCLE REGULATION OF B-MYB EXPRESSION

It is now well established that the expression of B-*myb* is cell cycle regulated. Earlier experiments identified that expression of B-*myb* is not detected in quiescent cells, but is significantly induced in late G_1 upon mitogenic stimulation, with maximal expression observed in S phase (Golay et al., 1991; Lam et al., 1992). Furthermore, arrest of cycling NIH3T3 fibroblasts with nocodazole and subsequent release from the G_2/M block, demonstrated periodic fluctuation of B-*myb* expression during the cell cycle (Lam et al., 1992). Transcription is initiated from multiple sites within the murine B-*myb* promoter, but in contrast to the regulation of c-*myb* expression, changes in B-*myb* expression are attributed to regulation of transcription initiation (Lam and Watson, 1993). Both murine and human B-*myb* 5' promoter regions confer cell cycle regulation of reporter activity in transfected, serum deprived/stimulated NIH3T3 cells, consistent with the kinetics of endogenous B-*myb* expression (Lam et al., 1995; Lam and Watson, 1993). Deletion analysis of the murine promoter indicated that sequences upstream of the transcription initiation sites are responsible for inherent basal promoter activity (Lam and Watson, 1993).

Sequence comparison of the human and murine promoter regions revealed extensive conservation throughout the transcription initiation and 5' untranslated regions, GC-richness and absence of a TATA box (Lam et al., 1995; Lam and Watson, 1993). A significantly lesser degree of sequence similarity is observed upstream of the transcription initiation region, although short intermittent tracts of conserved sequence exist (Lam et al., 1995). Further inspection of a region of the murine B-*myb* promoter conferring cell cycle regulation, localised by deletion analysis, identified a single E2F site that is entirely conserved in the human B-*myb* promoter (Lam et al., 1995; Lam and Watson, 1993).

4.1 The E2F Site

The conserved E2F site is located within the region of transcription initiation (Lam et al., 1995). Mutation of this site in the murine and human B-*myb* promoter resulted in loss of cell cycle regulation of reporter activity in fibroblasts, with constitutive high expression in quiescent and cycling cells (Lam et al., 1995; Lam and Watson, 1993). This defined a critical role for the E2F site in directing the induction of expression to a discrete stage (G_1/S) of the cell cycle and indicated that the B-*myb* E2F site is implicated in the repression of gene expression, an unfamiliar context at the time.

It has transpired in recent years that E2F sites additionally function in a negative manner to repress gene expression, as well as functioning to activate expression at particular stages of the cell cycle. The composition and activity of the repressor complex controlling B-*myb* expression via the E2F site has provided a basis for continuous investigation of this promoter. Cells rendered quiescent by serum deprivation have generally provided the environment in which to study potential repressors of B-*myb* expression. In quiescent (G_0) NIH3T3 cells, complexes that bind to the B-*myb* E2F site *in vitro* are composed of E2F4 and the pRb-related pocket proteins, p130 and p107 (Bennett et al., 1996; Lam et al., 1994). Furthermore, this G_0/G_1 binding complex was modified by the addition of cyclin E/cdk2 and later, cyclin A/cdk2, in parallel with derepression of B-*myb* (Bennett et al., 1996; Lam and Watson, 1993). It seems likely that key regulators of G_1/S progression may facilitate derepression of B-*myb in vivo*.

Evidence supporting the involvement of pocket protein complexes in the repression of B-*myb* expression, and specifically the interaction of p130 and p107 rather than pRb, has emanated from several studies. Co-expression of the human papillomavirus (HPV) E7 or adenovirus E1A oncoproteins, which interact with certain members of the pocket protein family and disrupt their interaction with E2F, derepress expression of B-*myb* (Lam et al., 1994; Lam and Watson, 1993). In particular, a specificity for p107 involvement in repressor function was demonstrated by expression of mutated E7 proteins unable to bind pRb, but which maintain the ability to interact with p107 and presumably also with the closely related p130 (Lam et al., 1994). In murine embryo fibroblasts (MEFs) derived from mice lacking functional p107 and p130 proteins (p107/p130 knockout mice: *p107*$^{-/-}$;*p130*$^{-/-}$), B-*myb* is among a number of genes whose cell cycle expression is deregulated to varying degrees (Hurford et al., 1997). Significant derepression of B-*myb* is visible in G_0/G_1 in *p107*$^{-/-}$;*p130*$^{-/-}$ MEFs with expression further elevated upon serum stimulation of these cells (Hurford et al., 1997). No deregulation of B-*myb* expression was evident in MEFs derived from pRb knockout mice or from single knockouts of p107 or p130, demonstrating a functional redundancy for p107 and p130 activity in B-*myb* repression (Hurford et al., 1997). The consequential lack of both p107 and p130 impinges upon the function of the B-*myb* E2F site, as revealed by reporter analyses in MEFs. Firstly, mutation of the B-*myb* E2F site gave rise to substantially less derepression in G_0 *p107*$^{-/-}$;*p130*$^{-/-}$ MEFs, than in control MEFs (Catchpole et al., 2002; Hurford et al., 1997). Secondly, re-introduction of p107 or p130 into G_0 *p107*$^{-/-}$;*p130*$^{-/-}$ MEFs reinstated repression of a reporter containing the wild type B-*myb* promoter, but not when the E2F site was mutated (Catchpole et al., 2002; Hurford et al., 1997). In contrast, overexpression of pRb was unable to efficiently repress the B-*myb* promoter (Catchpole et al., 2002).

Demonstration of an association of p107 and p130 with the endogenous B-*myb* promoter *in vivo* has been facilitated by chromatin immunoprecipitation (ChIP) analyses, which have also implicated the activity of other factors in the regulation of B-*myb* expression. In quiescent cells, the B-*myb* promoter is specifically occupied *in vivo* by p130/E2F4 and p107/E2F4 complexes (Takahashi et al., 2000; Wells et al., 2000). Association of pRb with the B-*myb* promoter was not evident, although this protein is associated with other E2F-regulated promoters (Wells et al., 2000). Importantly, the necessary requirement for a functional E2F site in order to recruit a p130/E2F4 repressor complex in G_0 cells has been shown by performing ChIP analyses of stable integrated B-*myb* promoter transgenes in NIH3T3 fibroblasts (Rayman et al., 2002). In the absence of a wild type E2F site, p130/E2F4 and p107/E2F4 repressor complexes were unable to be recruited to the B-*myb* promoter *in vivo*.

The influence of chromatin structure within the endogenous B-*myb* promoter may serve as an important regulatory factor in the imposition of repression. This mechanism is suggested by the *in vivo* association of the HDAC1 and mSin3B chromatin modifying factors with the B-*myb* promoter in quiescent cells, and the onset of histone acetylation coincident with B-*myb* derepression (Rayman et al., 2002; Takahashi et al., 2000). The association of both HDAC1 and mSin3B with the B-*myb* promoter is dependent upon an intact E2F site and the presence of p107 or p130, but not pRb (Rayman et al., 2002). This implies that chromatin modifiers are recruited to the endogenous B-*myb* promoter via p130/E2F and p107/E2F DNA-binding repressor complexes.

Members of the E2F family predominate in either a negative or positive role in regulating gene expression. Thus, E2F family members have been sub-grouped based on 'activating' or 'repressive' functions, with E2F4 and E2F5 associated with repression, and E2F1-3 associated with activation (reviewed in Trimarchi and Lees, 2001). In contrast to other E2F-regulated genes, E2F is not associated with the B-*myb* promoter in S phase even though 'activating' E2Fs are most prevalent at this stage (Takahashi et al., 2000; Wells et al., 2000). This absence is supported by an earlier examination of B-*myb* E2F site occupancy *in vivo*. Genomic footprinting of the E2F site in NIH3T3 fibroblasts demonstrated occupation in G_0 cells and loss of occupancy concurrent with induction of B-*myb* expression in late G_1 (Zwicker et al., 1996). Prior to loss of associated E2F in S phase, 'activating' E2Fs (E2F1 and E2F3) are found associated with the B-*myb* promoter in late G_1 in T98G glioblastoma cells stimulated to re-enter the cell cycle from quiescence (Takahashi et al., 2000). It is unclear whether these E2Fs bind specifically to the E2F site, or elsewhere within the B-*myb* promoter. Interestingly, cell cycle induction of B-*myb* expression is severely impaired in MEFs derived from E2F3 knockout mice, implying that E2F3 is required to activate B-*myb* expression in late G_1 (Humbert et al., 2000).

Recent genomic footprinting of the B-*myb* E2F site in *p107*$^{-/-}$;*p130*$^{-/-}$ MEFs has revealed that this site is unoccupied in G_0 despite the presence of Rb/E2F complexes and further reinforces the point of pocket protein specificity for the B-*myb* promoter *in vivo* (Catchpole et al., 2002). However, Rb/E2F complexes do bind to the B-*myb* E2F site in EMSAs, revealing a lack of stable association of Rb/E2F complexes with the B-*myb* promoter *in vivo* and furthermore, a lack of pocket protein specificity for the E2F site *in vitro* (Catchpole et al., 2002). This discrepancy indicates that other factors associated with or inherent to the B-*myb* promoter operate in a physiological context, but are not recapitulated in an *in vitro* setting. The absence of *in vivo* E2F association with the B-*myb* promoter in S phase suggests that the context of the B-*myb* E2F site within the promoter is important in determining specific interactions with this site. In this respect, the identification of an adjacent site that influences the activity of the E2F site has contributed significantly to an understanding of how the B-*myb* E2F site operates in the context of this promoter.

4.2 The Downstream Repression Site

Mutation of sequences flanking the B-*myb* E2F site identified an adjacent downstream site, termed the DRS (Downstream Repression Site), which contributes to repression of the B-*myb* promoter in G_0 (Bennett et al., 1996). Mutation of both the E2F site and DRS revealed that these sites function in a cooperative manner to enforce maximal repression of B-*myb* in G_0 (Bennett et al., 1996; Catchpole et al., 2002). Thorough mutagenesis of the DRS has recently led to delineation of a consensus sequence containing a critical core, based on the ability of this site to co-repress reporter activity in G_0 (Catchpole et al., 2002). Interestingly, mutation of core nucleotides within the DRS resulted in substantially greater promoter activity in S phase, but not when the E2F site was additionally mutated (Catchpole et al., 2002). In the absence of a functional DRS, increased promoter activity in S phase may be indicative of the association of 'activating' E2Fs with the E2F site and thus a restrictive role for the wild type DRS *in vivo*.

Bipartite regulatory elements have also been identified in several other cell cycle regulated genes. These contiguous elements are termed the cell cycle-dependent element (CDE) and cell cycle genes homology region (CHR), and are found in the promoter regions of the *cdc25C*, *cyclin A*, *cdc2* and *cyclin B2* genes (Lange-zu Dohna et al., 2000; Zwicker et al., 1995). Like B-*myb*, these genes are repressed in G_0 and derepressed in G_1/S, but maximal gene expression occurs later in the cell cycle than that of B-*myb* (Lange-zu Dohna et al., 2000; Lucibello et al., 1997; Zwicker et al., 1995). Not only do CDE/CHR elements have the same spatial arrangement as the B-*myb* E2F site/DRS, but the CHR elements conform to the consensus sequence determined for DRS, and the CDEs contain GC-rich cores similar to E2F sites. Taken together, these similarities suggest that common factors may interact with these regulatory sites to repress gene expression. However, it is evident from studies of the *cdc25C* and B-*myb* promoters that their cell cycle regulatory elements have distinct functional properties. Substitution of the *cdc25C* CHR with the B-*myb* DRS deregulated the *cdc25C* promoter in G_0 (Liu et al., 1996). Conversely, the *cdc25C* CHR can function in the context of the B-*myb* E2F site to repress reporter activity in G_0, and furthermore, to a better extent than that observed with the wild type B-*myb* promoter (Catchpole et al., 2002). Therefore, the *cdc25C* CHR has specific attributes in the context of the CDE that cannot be substituted by the B-*myb* DRS. In addition, the *cdc25C* CDE weakly binds E2F *in vitro* and displays a differential nucleotide requirement for repression with respect to the B-*myb* E2F site (Bennett et al., 1996; Zwicker et al., 1995).

The relative positions of the E2F site and DRS are important for their functional cooperativity to enforce maximal repression. Introduction of an additional 2 or 4 nucleotides between these sites resulted in depression in G_0 with increasing effect, respectively (Catchpole et al., 2002). This spatial relationship may be indicative of a functional interaction of factors binding to each of these sites. It is therefore significant that the *in vivo* binding of p130/E2F4 and p107/E2F4 complexes to a stable integrated B-*myb* promoter was severely compromised by mutation of the DRS, and that these pocket proteins require an intact DRS for repression of the B-*myb* promoter in G_0 (Catchpole et al., 2002). However, mutation of the DRS does not affect the ability of repressor complexes to bind to the B-*myb* E2F site *in vitro*, and again highlights a discrepancy between *in vitro* and *in vivo* binding analyses (Bennett et al., 1996). The identification of DRS-binding proteins currently remains elusive, despite previous indications that DRS-specific interactions had been detected (Catchpole et al., 2002; Liu et al., 1996; Saville and Watson, 1998).

4.3 Activators

Several putative Sp1-binding sites are located upstream of the B-*myb* E2F site/DRS and CDE/CHR elements of certain cell cycle regulated genes (Reviewed in Zwicker and Muller, 1997). Characterisation of these sites within the B-*myb* promoter is currently lacking, but genomic footprinting indicated possible constitutive occupation in G_0 and during the cell cycle (Zwicker et al., 1996). This implies that repressor complexes recruited by the downstream E2F site/DRS inhibit transcriptional activator activity, possibly by direct interaction. The presence of transactivators may serve to increase expression of B-*myb* beyond the level reached by derepression alone and in the absence of 'activating' E2F.

Interestingly, B-Myb was shown to be able to transactivate its own promoter through upstream Sp1-binding sites, rather than Myb-binding sites (Sala et al., 1999). Therefore B-Myb may co-operate with Sp1 to transactivate B-*myb* expression, but the precise mechanism remains unknown at present.

4.4 Summary

Significant progress has been made in understanding how B-*myb* expression is regulated during the cell cycle. It is also noteworthy that the cell cycle regulation of B-*myb* expression currently defines a unique example of an E2F-regulated gene. However, the precise mechanism of repression is yet to be determined and will be facilitated by the further identification of proteins that comprise the repressor complex, in particular DRS-binding proteins. The implication that the DNA architecture of the B-*myb* promoter may be an important regulatory factor in constituting repression represents another area for future investigation. Characterisation of the upstream activators is currently lacking. Further information concerning these activators will aid elucidation of the mechanism of repression. The identification and characterisation of the DRS demonstrates that the context of an E2F site within a promoter region is an important determinant governing specific regulatory protein interaction.

5. REGULATION OF A-MYB EXPRESSION

Characterisation of the A-*myb* promoter has only recently been described (Facchinetti et al., 2000). Three functionally important regions have been recognised within the human A-*myb* 5' promoter region by performing deletion analyses. These comprise the minimal promoter region, positive upstream and negative downstream regulatory regions, but do not appear to confer tissue-specific expression. Alignment of the human and murine promoter sequences has identified a conserved CCAAT box and two Sp1 consensus sites within the minimal promoter region. In addition, the promoter region is GC-rich and no TATA box was identified. Multiple transcription initiation sites have been mapped to a region immediately downstream of the CCAAT box in epithelial and B cells. Therefore, the A-*myb* promoter shows similarities with both the B-*myb* and c-*myb* promoter regions.

In vitro binding analyses revealed that the A-*myb* CCAAT box binds NF-Y with an affinity similar to that for a canonical Y-box sequence and that both Sp1 sites bind complexes containing Sp1 from the non-Hodgkin lymphoma-derived BJAB cell line (Facchinetti et al., 2000). Mutation of each of these sites (CCAAT, Sp-I and Sp-II) resulted in reduction of reporter activity to varying extents. Mutation of both Sp1 sites only marginally decreased reporter activity more than mutation of the stronger Sp1-I site alone. Mutation of all three sites reduced promoter activity to 20% and therefore indicates that these sites collectively contribute to the majority of promoter activity (Facchinetti et al., 2000).

The expression of A-*myb* is also subject to cell cycle regulation, with induction of expression occurring at the G1/S transition and maximal expression in S phase (Golay et al., 1998; Marhamati et al., 1997; Ziebold and Klempnauer, 1997). Reporter constructs containing regions of the A-*myb* promoter showed induction of activity in S phase upon serum stimulation of quiescent NIH3T3 fibroblasts (Facchinetti et al., 2000). Furthermore, reporter activity failed to be induced in S phase upon mutation of the CCAAT box and Sp1 sites (Facchinetti et al., 2000). It must be considered that these sites collectively contribute to the majority of promoter activity and will require further rigorous analysis in order to ascertain any cell cycle regulatory function.

5.1 Summary

Investigation of the regulatory factors controlling the expression of A-*myb* has only recently begun and thus presents an area for further expansion, particularly with respect to the cell cycle mode of regulation. Tissue-specific regulatory elements have yet to be identified and may require isolation of control regions in addition to those already identified.

6. CONCLUSIONS

It is evident that distinct mechanisms and factors control expression of the mammalian *myb* genes. However, the 5' promoter regions of these genes exhibit features in common. In particular, they are in general GC-rich, lack a TATA box and initiate transcription from multiple sites. In addition, all *myb* genes show evidence of being cell cycle regulated, with induction of expression near the G1/S transition. This implies a requirement for Myb function of sorts in S phase, but the nature of this role is yet to be exposed and will be facilitated by the identification of target genes and interacting proteins.

Determination of the regulatory factors and mechanisms controlling *myb* gene expression is imperative in understanding how deregulated expression of these genes is manifest in neoplastic cells. Importantly, this may aid in identifying any role that deregulated *myb* expression may have in tumourigenesis.

ACKNOWLEDGEMENTS

I thank Roger Watson for critical reading of the manuscript and to all members of the laboratory for useful discussion. FJT is supported by the Biotechnology and Biological Sciences Research Council (BBSRC) of the United Kingdom.

REFERENCES

Arsura, M., Luchetti, M.M., Erba, E., Golay, J., Rambaldi, A. and Introna, M. (1994) Dissociation between p93B-myb and p75c-myb expression during the proliferation and differentiation of human myeloid cell lines. Blood *83*, 1778-1790.

Bender, T.P. and Kuehl, W.M. (1986) Murine myb protooncogene mRNA: cDNA sequence and evidence for 5' heterogeneity. Proc Natl Acad Sci U S A *83*, 3204-3208.

Bender, T.P., Thompson, C.B. and Kuehl, W.M. (1987) Differential expression of c-myb mRNA in murine B lymphomas by a block to transcription elongation. Science *237*, 1473-1476.

Bennett, J.D., Farlie, P.G. and Watson, R.J. (1996) E2F binding is required but not sufficient for repression of B-myb transcription in quiescent fibroblasts. Oncogene *13*, 1073-1082.

Brown, K.E., Kindy, M.S. and Sonenshein, G.E. (1992) Expression of the c-myb proto-oncogene in bovine vascular smooth muscle cells. J Biol Chem *267*, 4625-4630.

Campanero, M.R., Armstrong, M. and Flemington, E. (1999) Distinct cellular factors regulate the c-myb promoter through its E2F element. Mol Cell Biol *19*, 8442-8450.

Castellano, M., Golay, J., Mantovani, A. and Introna, M. (1992) Detection of a transcriptional block in the first intron of the human c-myb gene. Int J Clin Lab Res *22*, 159-164.

Catchpole, S., Tavner, F., Le Cam, L., Sardet, C. and Watson, R.J. (2002) A B-myb promoter corepressor site facilitates in vivo occupation of the adjacent E2F site by p107 x E2F and p130 x E2F complexes. J Biol Chem *277*, 39015-39024.

Catron, K.M., Purkerson, J.M., Isakson, P.C. and Bender, T.P. (1992) Constitutive versus cell cycle regulation of c-myb mRNA expression correlates with developmental stages in murine B lymphoid tumors. J Immunol *148*, 934-942.

Dooley, S., Seib, T., Welter, C. and Blin, N. (1996) c-myb intron I protein binding and association with transcriptional activity in leukemic cells. Leuk Res *20*, 429-439.

Ess, K.C., Witte, D.P., Bascomb, C.P. and Aronow, B.J. (1999) Diverse developing mouse lineages exhibit high-level c-Myb expression in immature cells and loss of expression upon differentiation. Oncogene *18*, 1103-1111.

Facchinetti, V., Lopa, R., Spreafico, F., Bolognese, F., Mantovani, R., Tavner, F., Watson, R., Introna, M., and Golay, J. (2000) Isolation and characterization of the human A-myb promoter: regulation by NF-Y and Sp1. Oncogene *19*, 3931-3940.

Golay, J., Broccoli, V., Lamorte, G., Bifulco, C., Parravicini, C., Pizzey, A., Thomas, N.S., Delia, D., Ferrauti, P., Vitolo, D. and Introna, M. (1998) The A-Myb transcription factor is a marker of centroblasts in vivo. J Immunol *160*, 2786-2793.

Golay, J., Capucci, A., Arsura, M., Castellano, M., Rizzo, V. and Introna, M. (1991) Expression of c-myb and B-myb, but not A-myb, correlates with proliferation in human hematopoietic cells. Blood *77*, 149-158.

Golay, J., Luppi, M., Songia, S., Palvarini, C., Lombardi, L., Aiello, A., Delia, D., Lam, K., Crawford, D.H., Biondi, A., Barbui, T., Rambaldi, A. and Introna, M. (1996) Expression of A-myb, but not c-myb and B-myb, is restricted to Burkitt's lymphoma, sIg+ B-acute lymphoblastic leukemia, and a subset of chronic lymphocytic leukemias. Blood *87*, 1900-1911.

Guerra, J., Withers, D.A. and Boxer, L.M. (1995) Myb binding sites mediate negative regulation of c-myb expression in T-cell lines. Blood *86*, 1873-1880.

Gunn, J., Holt, C.M., Francis, S.E., Shepherd, L., Grohmann, M., Newman, C.M., Crossman, D.C. and Cumberland, D.C. (1997) The effect of oligonucleotides to c-myb on vascular smooth muscle cell proliferation and neointima formation after porcine coronary angioplasty. Circ Res *80*, 520-531.

Humbert, P.O., Verona, R., Trimarchi, J.M., Rogers, C., Dandapani, S. and Lees, J.A. (2000) E2f3 is critical for normal cellular proliferation. Genes Dev *14*, 690-703.

Hurford, R.K., Cobrinik, D., Lee, M.H. and Dyson, N. (1997) pRB and p107/p130 are required for the regulated expression of different sets of E2F responsive genes. Genes Dev *11*, 1447-1463.

Lam, E.W.-F., Robinson, C. and Watson, R.J. (1992) Characterization and cell cycle-regulated expression of mouse B-myb. Oncogene *7*, 1885-1890.

Lam, E.W.-F., Bennett, J. and Watson, R.J. (1995) Cell-cycle regulation of human B-*myb* transcription. Gene *160*, 277-281.

Lam, E.W.-F., Morris, J.D.H., Davies, R., Crook, T., Watson, R.J. and Vousden, K.H. (1994) HPV16 E7 oncoprotein deregulates B-*myb* expression: correlation with targeting of p107/E2F complexes. EMBO J *13*, 871-878.

Lam, E.W.-F. and Watson, R.J. (1993) An E2F-binding site mediates cell-cycle regulated repression of mouse B-*myb* transcription. EMBO J *12*, 2705-2713.

Lange-zu Dohna, C., Brandeis, M., Berr, F., Mossner, J. and Engeland, K. (2000) A CDE/CHR tandem element regulates cell cycle-dependent repression of cyclin B2 transcription. FEBS Lett *484*, 77-81.

Lauder, A., Castellanos, A. and Weston, K. (2001) c-myb transcription is activated by protein kinase B (PKB) following interleukin 2 stimulation of T cells and is required for PKB-mediated protection from apoptosis. Mol Cell Biol *21*, 5797-5805.

Liu, N., Lucibello, F.C., Zwicker, J., Engeland, K. and Müller, R. (1996) Cell cycle-regulated repression of B-myb transcription: cooperation of an E2F site with a contiguous corepressor element. Nucl Acids Res *24*, 2905-2910.

Lucibello, F.C., Liu, N., Zwicker, J., Gross, C. and Müller, R. (1997) The differential binding of E2F and CDF repressor complexes contributes to the timing of cell cycle-regulated transcription. Nucl Acids Res *25*, 4921-4925.

Manzella, L., Gualdi, R., Perrotti, D., Nicolaides, N.C., Girlando, G., Giuffrida, M.A., Messina, A. and Calabretta, B. (2000) The interferon regulatory factors 1 and 2 bind to a segment of the human c-myb first intron: possible role in the regulation of c-myb expression. Exp Cell Res *256*, 248-256.

Marhamati, D.J., Bellas, R.E., Arsura, M., Kypreos, K.E. and Sonenshein, G.E. (1997) A-myb is expressed in bovine vascular smooth muscle cells during the late G1-to-S phase transition and cooperates with c-myc to mediate progression to S phase. Mol Cell Biol *17*, 2448-2457.

McCann, S., Sullivan, J., Guerra, J., Arcinas, M. and Boxer, L.M. (1995) Repression of the c-myb gene by WT1 protein in T and B cell lines. J Biol Chem *270*, 23785-23789.

McClinton, D., Stafford, J., Brents, L., Bender, T.P. and Kuehl, W.M. (1990) Differentiation of mouse erythroleukemia cells is blocked by late up-regulation of a c-myb transgene. Mol Cell Biol *10*, 705-710.

McMahon, J., Howe, K.M. and Watson, R.J. (1988) The induction of Friend erythroleukaemia differentiation is markedly affected by expression of a transfected c-*myb* cDNA. Oncogene *3*, 717-720.

Mucenski, M.L., McLain, K., Kier, A.B., Swerdlow, S.H., Schreiner, C.M., Miller, T.A., Pietryga, D.W., Scott, W.J. and Potter, S.S. (1991) A functional c-myb gene is required for normal murine fetal hepatic hematopoiesis. Cell *65*, 677-689.

Müller, H., Bracken, A.P., Vernell, R., Moroni, M.C., Christians, F., Grassilli, E., Prosperini, E., Vigo, E., Oliner, J.D. and Helin, K. (2001) E2Fs regulate the expression of genes involved in differentiation, development, proliferation, and apoptosis. Genes Dev *15*, 267-285.

Nicolaides, N.C., Gualdi, R., Casadevall, C., Manzella, L. and Calabretta, B. (1991) Positive autoregulation of c-myb expression via Myb binding sites in the 5' flanking region of the human c-myb gene. Mol Cell Biol *11*, 6166-6176.

Nomura, N., Takahashi, M., Matsui, M., Ishii, S., Date, T., Sasamoto, S. and Ishizaki, R. (1988) Nucl Acids Res *16*, 11075-11089.

Perkel, J.M., Simon, M.C. and Rao, A. (2002) Identification of a c-myb attenuator-binding factor. Leuk Res *26*, 179-190.

Phan, S.C., Feeley, B., Withers, D. and Boxer, L.M. (1996) Identification of an inducible regulator of c-myb expression during T-cell activation. Mol Cell Biol *16*, 2387-2393.

Ramsay, R.G., Thompson, M.A., Hayman, J.A., Reid, G., Gonda, T.J. and Whitehead, R.H. (1992) Myb expression is higher in malignant human colonic carcinoma and premalignant adenomatous polyps than in normal mucosa. Cell Growth Differ *3*, 723-730.

Rayman, J.B., Takahashi, Y., Dannenberg, J.H., Catchpole, S., Watson, R., te Riele, H. and Dynlacht, B.D. (2002) E2F mediates cell cycle-dependent transcriptional repression *in vivo* by recruitment of specific co-repressor complex. Genes Dev *16*, 933-947.

Reddy, C.D. and Reddy, E.P. (1989) Differential binding of nuclear factors to the intron 1 sequences containing the transcriptional pause site correlates with c-myb expression. Proc Natl Acad Sci U S A *86*, 7326-7330.

Rosenthal, M.A., Thompson, M.A., Ellis, S., Whitehead, R.H. and Ramsay, R.G. (1996) Colonic expression of c-myb is initiated in utero and continues throughout adult life. Cell Growth Differ *7*, 961-967.

Rushton, J.J., Davis, L.M., Lei, W., Mo, X., Leutz, A. and Ness, S.A. (2003) Distinct changes in gene expression induced by A-Myb, B-Myb and c-Myb proteins. Oncogene *22*, 308-313.

Sala, A., Saitta, B., De Luca, P., Cervellera, M.N., Casella, I., Lewis, R.E., Watson, R., and Peschle, C. (1999) B-MYB transactivates its own promoter through SP1-binding sites. Oncogene *18*, 1333-1339.

Saville, M.K. and Watson, R.J. (1998) B-Myb: a key regulator of the cell cycle. Advances Cancer Res *72*, 109-140.

Sitzmann, J., Noben-Trauth, K., Kamano, H. and Klempnauer, K.-H. (1996) Expression of B-Myb during mouse embryogenesis. Oncogene *12*, 1889-1894.

Sitzmann, J., Noben-Trauth, K. and Klempnauer, K.H. (1995) Expression of mouse c-myb during embryonic development. Oncogene *11*, 2273-2279.

Slamon, D.J., Boone, T.C., Murdock, D.C., Keith, D.E., Press, M.F., Larson, R.A. and Souza, L.M. (1986) Studies of the human c-myb gene and its product in human acute leukemias. Science *233*, 347-351.

Slamon, D.J., deKernion, J.B., Verma, I.M. and Cline, M.J. (1984) Expression of cellular oncogenes in human malignancies. Science *224*, 256-262.

Stern, J.B. and Smith, K.A. (1986) Interleukin-2 induction of T-cell G1 progression and c-myb expression. Science *233*, 203-206.

Suhasini, M. and Pilz, R.B. (1999) Transcriptional elongation of c-myb is regulated by NF-kappaB (p50/RelB). Oncogene *18*, 7360-7369.

Suhasini, M., Reddy, C.D., Reddy, E.P., DiDonato, J.A. and Pilz, R.B. (1997) cAMP-induced NF-kappaB (p50/relB) binding to a c-myb intronic enhancer correlates with c-myb up-regulation and inhibition of erythroleukemia cell differentiation. Oncogene *15*, 1859-1870.

Sullivan, J., Feeley, B., Guerra, J. and Boxer, L.M. (1997) Identification of the major positive regulators of c-myb expression in hematopoietic cells of different lineages. J Biol Chem *272*, 1943-1949.

Takahashi, Y., Rayman, J.B. and Dynlacht, B.D. (2000) Analysis of promoter binding by the E2F and pRB families in vivo: distinct E2F proteins mediate activation and repression. Genes Dev *14*, 804-816.

Tanaka, Y., Patestos, N.P., Maekawa, T., and Ishii, S. (1999) B-myb is required for inner cell mass formation at an early stage of development. J Biol Chem *274*, 28067-28070.

Thompson, C.B., Challoner, P.B., Neiman, P.E. and Groudine, M. (1986) Expression of the c-*myb* proto-oncogene during cellular proliferation. Nature *319*, 374-380.

Thompson, M.A., Flegg, R., Westin, E.H. and Ramsay, R.G. (1997) Microsatellite deletions in the c-myb transcriptional attenuator region associated with overexpression in colon tumour cell lines. Oncogene *14*, 1715-1723.

Thompson, M.A. and Ramsay, R.G. (1995) Myb: an old oncoprotein with new roles. Bioessays *17*, 341-350.

Todokoro, K., Watson, R.J., Higo, H., Amanuma, H., Kuramochi, S., Yanagisawa, H. and Ikawa, Y. (1988) Down-regulation of c-myb gene expression is a prerequisite for erythropoietin-induced erythroid differentiation. Proc Natl Acad Sci U S A *85*, 8900-8904.

Torelli, G., Selleri, L., Donelli, A., Ferrari, S., Emilia, G., Venturelli, D., Moretti, L. and Torelli, U. (1985) Activation of c-*myb* expression by phytohemagglutinin stimulation in normal human T lymphocytes. Mol Cell Biol *5*, 2874-2877.

Toscani, A., Mettus, R.V., Coupland, R., Simpkins, H., Litvin, J., Orth, J., Hatton, K.S. and Reddy, E.P. (1997). Arrest of spermatogenesis and defective breast development in mice lacking A-*myb*. Nature *386*, 713-717.

Toth, C.R., Hostutler, R.F., Baldwin, A.S., Jr. and Bender, T.P. (1995) Members of the nuclear factor kappa B family transactivate the murine c-myb gene. J Biol Chem *270*, 7661-7671.

Trauth, K., Mutschler, B., Jenkins, N.A., Gilbert, D. J., Copeland, N.G. and Klempnauer, K.-H. (1994) Mouse A-myb encodes a trans-activator and is expressed in mitotically active cells of the developing central nervous system, adult testis and B lymphocytes. EMBO J *13*, 5994-6005.

Trimarchi, J.M. and Lees, J.A. (2001) Sibling rivalry in the E2F family. Nature Rev Mol Cell Biol *3*, 11-20.

Watson, R.J. (1988a) A transcriptional arrest mechanism involved in controlling constitutive levels of mouse c-myb mRNA. Oncogene *2*, 267-272.

Watson, R.J. (1988b) Expression of the c-myb and c-myc genes is regulated independently in differentiating mouse erythroleukemia cells by common processes of premature transcription arrest and increased mRNA turnover. Mol Cell Biol *8*, 3938-3942.

Watson, R.J., Dyson, P.J. and McMahon, J. (1987). Multiple c-myb transcript cap sites are variously utilized in cells of mouse haemopoietic origin. EMBO J *6*, 1643-1651.

Wells, J., Boyd, K.E., Fry, C.J., Bartley, S.M. and Farnham, P.J. (2000) Target gene specificity of E2F and pocket protein family members in living cells. Mol Cell Biol *20*, 5797-5807.

Wolff, L. (1996) Myb-induced transformation. Crit Rev Oncog *7*, 245-260.

Yanagisawa, H., Nagasawa, T., Kuramochi, S., Abe, T., Ikawa, Y. and Todokoro, K. (1991) Constitutive expression of exogenous c-myb gene causes maturation block in monocyte-macrophage differentiation. Biochim Biophys Acta *1088*, 380-384.

Yuan, W. (2000) Intron 1 rather than 5' flanking sequence mediates cell type-specific expression of c-myb at level of transcription elongation. Biochim Biophys Acta *1490*, 74-86.

Ziebold, U. and Klempnauer, K.H. (1997) Linking Myb to the cell cycle: cyclin-dependent phosphorylation and regulation of A-Myb activity. Oncogene *15*, 1011-1019.

Zwicker, J., Liu, N., Engeland, K., Lucibello, F.C. and Müller, R. (1996) Cell cycle regulation of E2F site occupation in vivo. Science *271*, 1595-1597.

Zwicker, J., Lucibello, F.C., Wolfraim, L.A., Gross, C., Truss, M., Engeland, K. and Müller, R. (1995) Cell cycle regulation of the cyclin A, cdc25c and cdc2 genes is based on a common mechanism of transcriptional repression. EMBO J *14*, 4514-4522.

Zwicker, J. and Muller, R. (1997) Cell-cycle regulation of gene expression by transcriptional repression. Trends Genet *13*, 3-6.

Chapter 11

THE C-MYB DNA BINDING DOMAIN

From Molecular Structure to Function

Kazuhiro Ogata[1], Tahir H. Tahirov[1,3] and Shunsuke Ishii[2]
[1]Department of Biochemistry, Yokohama City University Graduate School of Medicine, 3-9 Fukuura, Kanazawa-ku, Yokohama 236-0004, Japan. [2]Laboratory of Molecular Genetics, RIKEN Tsukuba Institite, 3-1-1 Koyadai, Tsukuba, Ibaraki 305-0074, Japan. [3]RIKEN Harima Institute, 1-1-1 Kouto, Mikazuki, Sayo, Hyogo 679-5148, Japan.

Abstract: c-Myb specifically recognises the consensus DNA sequence AACNG in the promoter regions of haemopoietic target genes. The DNA-binding domain of c-Myb, which is highly conserved from *Drosophila* to man and also with the other mammalian Myb family members A-Myb and B-Myb, consists of three imperfect tandem repeats (R1, R2 and R3), each of which forms a globular architecture containing a helix-turn-helix-related motif. The recognition helices of R2 and R3 cooperatively interact with specific DNA bases, while R1 non-specifically stabilises the R2R3-DNA interactions. c-Myb exhibits synergy with members of the C/EBP family in the transactivation of certain target genes. The R2 region of c-Myb directly interacts with a C-terminal part of the leucine-zipper of C/EBPβ, suggesting that distantly bound c-Myb and C/EBPβ on the promoter DNA form a stereo-specific protein assembly via DNA looping. The mutated points in the R2 region of the oncogenic AMV v-Myb protein are located near the interface between c-Myb and C/EBPβ.

1. INTRODUCTION

Members of the Myb protein family, including c-Myb, A-Myb and B-Myb in higher vertebrates, function as transcriptional regulators. Each family member contains a DNA binding domain (DBD), or 'Myb domain', which is highly conserved within the Myb family and between species from *Drosophila* to man. The Myb DBD is the only domain from Myb proteins that has been extensively analysed from a structural perspective. The Myb DBD plays a critical role both in DNA binding and cooperative interactions with partner proteins such members of the C/EBP family (see Chapter 12). The avian myeloblastosis virus (AMV) derived v-Myb, an oncogenic mutant

J. Frampton (ed.), Myb Transcription Factors: Their Role in Growth, Differentiation and Disease, 223-238.

form of c-Myb, contains point mutations in the Myb DBD that disrupt the functional cooperation with the C/EBP family and influence the phenotype of transformed myeloid cells (Ness et al., 1989; Introna et al., 1990; Kowenz-Leutz et al., 1997).

c-Myb has three functional domains, the DBD, a transactivation domain and a negative regulatory domain (Figure 1a). The c-Myb DBD consists of three imperfectly repeated amino acid sequences of 51 to 52 residues (R1, R2 and R3) and recognises the specific consensus sequence 5'-(T/C)AAC(G/T)G-3' (Figure 1b) (Biedenkapp et al., 1988; Tanikawa et al., 1993; Ogata et al., 1996; Oda et al., 1997). The solution and crystal structures of the c-Myb DBD in the free and DNA-complexed states have been determined (Ogata et al., 1992; Ogata et al., 1994; Ogata et al., 1995; Tahirov et al., 2002). Each of the c-Myb DBD repeats forms a structurally independent globular subdomain in the free state. R2 and R3 co-operatively bind to the specific consensus sequence, while the R1 subdomain non-specifically stabilises the c-Myb R2R3-DNA interactions.

2. STRUCTURE OF THE C-MYB DBD AND ITS SPECIFIC RECOGNITION OF DNA

The solution structures of the free c-Myb R1, R2 and R3 subdomains and their superimposition are shown in Figures 2a and 2b, respectively (Ogata et al., 1995). The overall architectures of these subdomains are very similar to each other. Each repeat has three helical regions with the second and third helices being involved in the helix-turn-helix-related motif (Ogata et al., 1992). The three-helical structure is stabilised by a hydrophobic core containing the three periodically positioned tryptophans, which are characteristic of the Myb domain (marked with asterisks in Figure 1a) (Frampton et al., 1989; Kanei-Ishii et al., 1990). In spite of the time-averaged structural similarity between these subdomains, their dynamic nature, such as the degree of conformational flexibility, is remarkably different. Hence, only the R2 subdomain is conformationally fluctuating (Ogata et al., 1996). This dynamic character of the R2 subdomain could be attributed to the presence of a cavity in the hydrophobic core (Ogata et al., 1995; Ogata et al., 1996) and its functional implications will be discussed later. The subdomains are connected by linkers, resulting in an unfixed orientation between them in the free state. (Ogata et al., 1995; Ogata et al., 1996).

When c-Myb is bound to a specific DNA molecule, the R2 and R3 subdomains fit into the DNA major groove *en bloc*, recognising the consensus DNA sequence mainly through the third 'recognition' helices

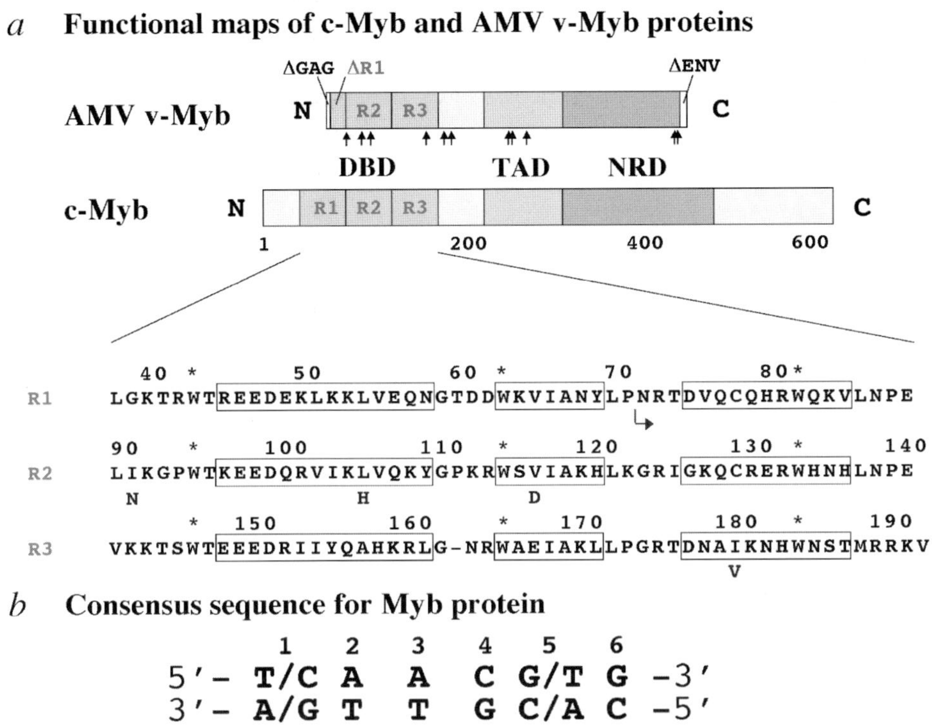

Figure 1

Functional domain maps of c-Myb and AMV v-Myb (a) and consensus binding sequence for Myb proteins (b). (a) In c-Myb the amino acid sequence of the DBD is presented. Three helical regions of each repeat are boxed, and the periodically positioned tryptophans are marked with asterisks. The position of the N-terminal truncation and the four mutated residues in the DBD of AMV v-Myb are shown with an blue arrow and letters below the sequence, respectively. In AMV v-Myb, the truncated R1 and viral Gag and Env protein regions are shown as ΔR1, ΔGAG and ΔENV, respectively. The mutations are indicated by arrows. DBD; DNA-binding domain, TA; *trans*-activation domain, NRD; negative regulatory domain. (see colour section p. xxi)

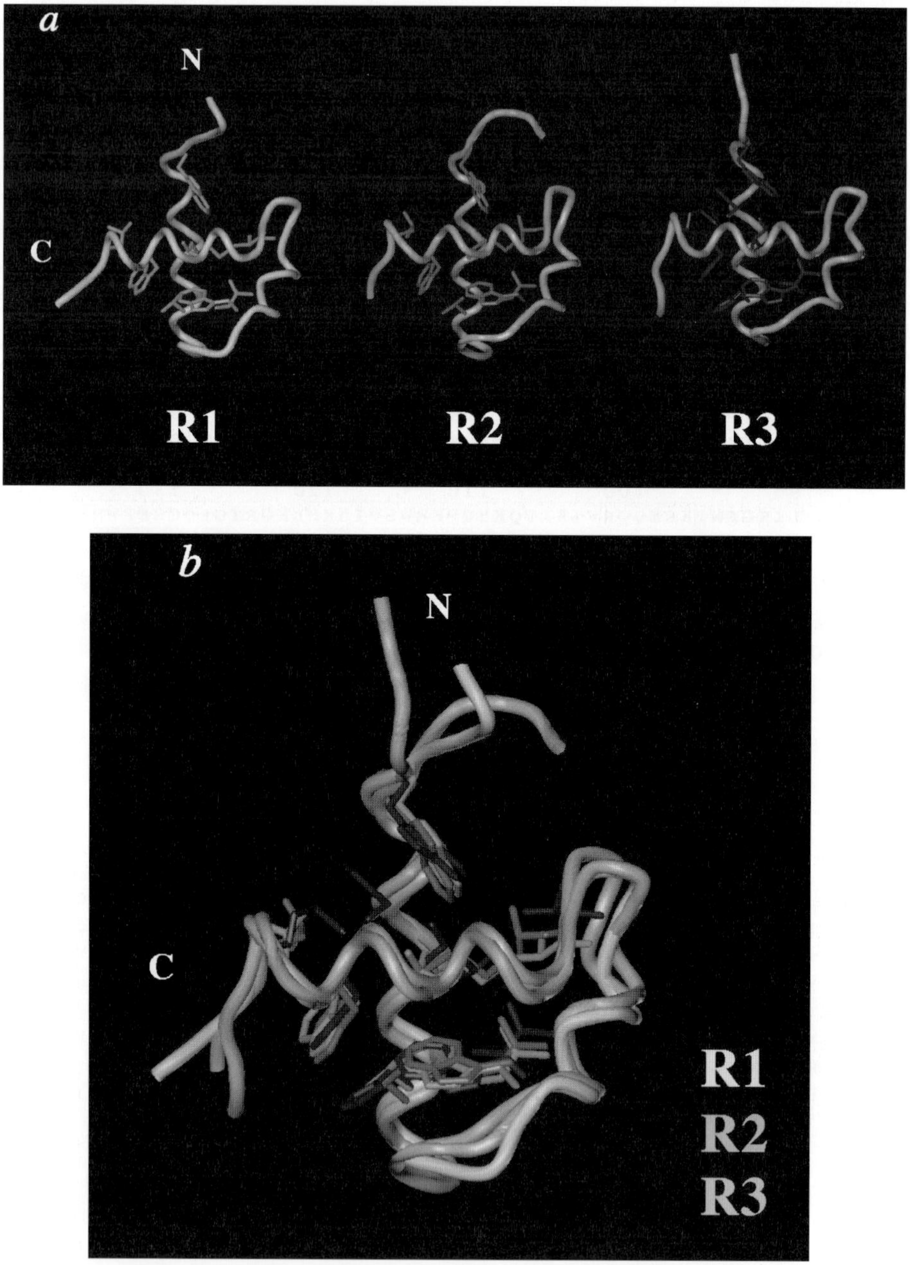

Figure 2

The NMR average structure of the c-Myb DBD consisting of the three subdomains, R1, R2 and R3 (a), and superimposition of them (b). The backbone of each subdomain is shown (R1 - yellow, R2 - magenta and R3 - cyan tubes) and residues in the hydrophobic core (R1 - green, R2 - red and R3 - blue). (see colour section p. xxii) from R2 and R3 (Figures 3a-c) (Ogata et al., 1994; Tahirov et al., 2002). N183, N179 and K182 of R3 form bipartite hydrogen bonds with the adenine bases at positions 2 and 3 and the guanine base at position 4, respectively, while E132 and K128 of R2 form hydrogen bonds with the cytosine base at position 4 and the guanine base at position 6, respectively (The numberings of the DNA base pairs were done according to Figure 1b). The thymine base at position 1 is recognised by the methylene parts of N183 and S187 via van der Waals contacts. These interactions between the protein side chains and the DNA bases are further supported by a network of water-mediated hydrogen bonds. In addition to the DNA base-specific interactions, many non-specific interactions between the protein side chains or backbone and the DNA sugar-phosphate backbones are formed to stabilise the specific R2R3-DNA binding. Amongst these non-specific interactions, a hydrogen bond between the protein backbone of R2 and a DNA phosphate is found in the DNA minor groove and seems to contribute to stabilisation of the multiple protein-DNA assembly with partner proteins such as C/EBP (Ogata et al., 2003). In contrast to the R2R3-DNA interactions, R1 binding to DNA is non-specific involving no DNA sequence-specific hydrogen bonds (Tanikawa et al., 1993; Dini and Lipsick, 1993; Ording et al., 1994; Ogata et al., 1994; Ogata et al., 1995; Tahirov et al., 2002).

3. DYNAMIC ASPECTS OF THE C-MYB DBD STRUCTURE

Although the three subdomains of the c-Myb DBD form very similar three-dimensional structures, their dynamic features were found to be quite different. This conclusion was based on magnetic relaxation measurements from nuclear magnetic resonance (NMR) experiments and on melting temperature (T_m) measurements derived from circular dichroism (CD) spectra and differential scanning calorimetry (DSC) experiments (Sarai et al., 1993; Ogata et al., 1996; Morii et al., 1999). While the R1 and R3 subdomains adopt relatively stable conformations with T_ms of 61°C for R1 and 57°C for R3, the R2 subdomain exhibits slow conformational fluctuations on a time scale of 10^{-4}-10^{-3}s and a T_m of 43°C. Such conformational flexibility of the R2 subdomain could be explained by the presence of a cavity in the hydrophobic core, which is not present in R1 and R3 (Ogata et al., 1995; Ogata et al., 1996). The importance of the cavity for the conformational flexibility of R2 is indicated by the fact that an R2 mutant in which the cavity is filled (V103L) exhibits no significant slow conformational fluctuations and has a T_m of 66°C (Ogata et al., 1996).

The R2 cavity is positioned in the centre of the hydrophobic core and is surrounded by V103, C130 and the methylene part of R133 (Figure 4). These residues are conserved between human c-Myb, A-Myb, B-Myb and *Drosophila* Myb, suggesting that the cavity plays an important role in the functioning of the Myb domain. Consistent with this assumption, conformationally stabilised c-Myb containing the V103L mutation shows a reduced DNA binding activity and a reduced transactivational capacity (Ogata et al., 1996). A comparative structural study of the free and the DNA-complexed states of the c-Myb DBD indicated that the R2 subdomain of c-Myb DBD acquires conformational rearrangements around the cavity upon specific DNA binding (Ogata et al., 1996). In the free state, steric hindrance between the W95 indole ring and one of the V103 methyl groups also appears to contribute to the conformational destabilisation of R2, and hence also facilitates the structural change upon complex formation with DNA (Morii et al., 1999).

4. COMPLEX FORMATION BETWEEN C-MYB AND C/EBPβ ON PROMOTER DNA

Transcriptional regulatory proteins generally cooperate with partner proteins in the regulation of their target genes. Myb family members have been shown to cooperate with various transcriptional regulatory proteins. Partner proteins for c-Myb that have been reported to cooperate in transactivating target genes include members of the C/EBP family, Runx1-CBFβ (Hernandez-Munain, and Krangel, 1994; Bristos-Bray and Friedman, 1997), Ets family proteins (Dudek et al., 1992) and GATA proteins (Zhang et al., (1997)). The cooperation between c-Myb and C/EBP family proteins (C/EBPα, C/EBPβ, C/EBPδ and C/EBPε) is the best characterised and is involved in the regulation of myelomonocytic genes such as *mim-1* (Burk et al., 1993; Ness et al., 1993; Kowenz-Leutz et al., 1997; Tahirov et al., 2002), lysozyme (Ness et al., 1993), *tom-1A* (Burk et al., 1997), myeloperoxidase (Bristos-Bray and Friedman, 1997), neutrophil elastase (Oelgeschläger et al., 1996; Verbeek et al., 1999) and myeloblastin (Lutz et al. 2001).

The crystal structure has been determined of a complex composed of the c-Myb DBD, a homodimer of the C/EBPβ DBD including the basic leucine-zipper motif, and a DNA fragment containing the c-Myb and C/EBP binding sites from the *tom-1A* promoter (Figure 5a) (Tahirov et al., 2002). Unexpectedly, inter-complex interactions were found between the R2 subdomain of c-Myb and the C-terminal part of the leucine-zipper region of

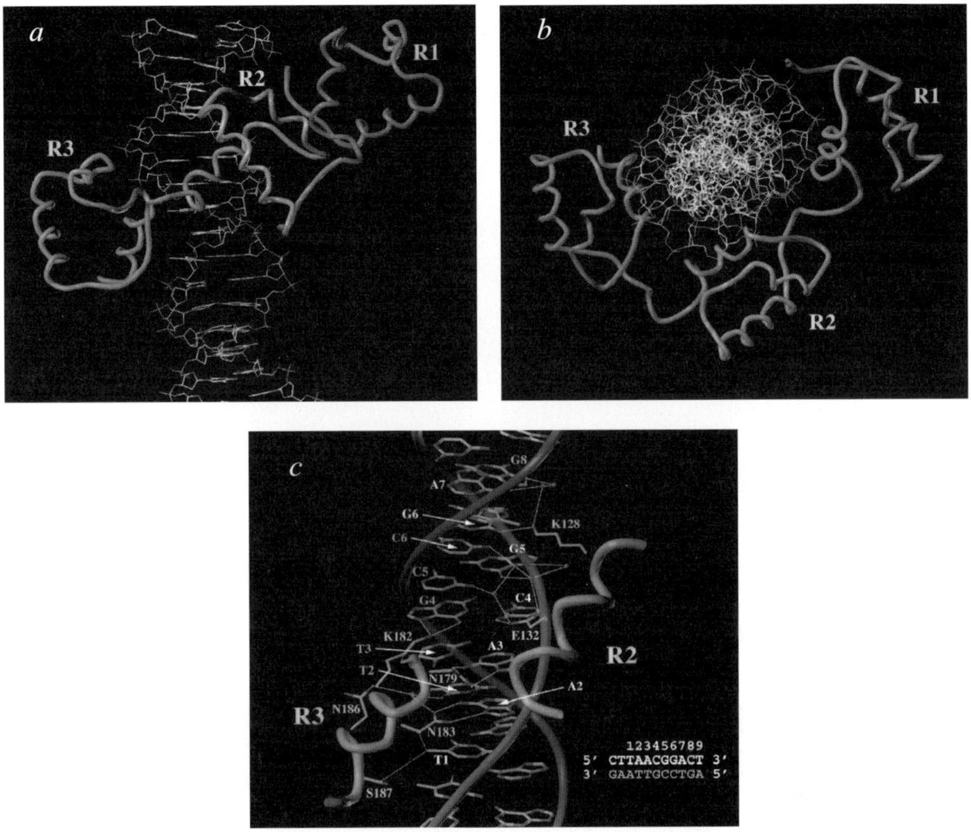

Figure 3

Side and top views of the crystal structure of the c-Myb DBD–DNA complex in the c-Myb–C/EBPβ–DNA ternary complex (a, b), and the specific interactions between c-Myb R2R3 and DNA (c). (a, b) For clarity, the C/EBPβ part has been omitted in these figures. In the c-Myb DBD, only the backbone structure is shown as a tube presentation coloured green, magenta and cyan for R1, R2 and R3, respectively. (c) In c-Myb, two recognition helices from R2 and R3 are shown as tubes coloured magenta and cyan, respectively. In the target DNA, the sugar-phosphate backbones are shown as red and blue tubes. The DNA bases and the side chains of c-Myb R2R3, which are involved in the specific protein–DNA interactions, are shown with capped stick presentations. The water molecules, which mediate the protein–DNA interactions, are shown as red spheres. In the specific protein–DNA interactions, hydrogen bonds and van der Waals interactions are indicated as yellow and orange dotted lines, respectively. The target DNA sequence is shown in the right-bottom corner. (see colour section p. xxiii)

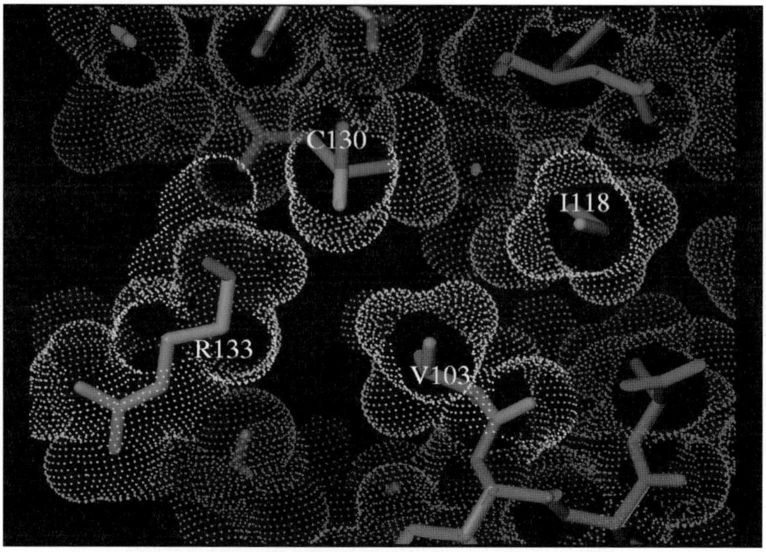

Figure 4

A cavity in the hydrophobic core of c-Myb R2. The side chains of some residues surrounding the cavity are shown as green capped sticks with labellings. The yellow dots represent the van der Waals surfaces of these residues. Other residues are shown as red capped sticks with purple dots of the van der Waals surfaces. (see colour section p. xxiv)

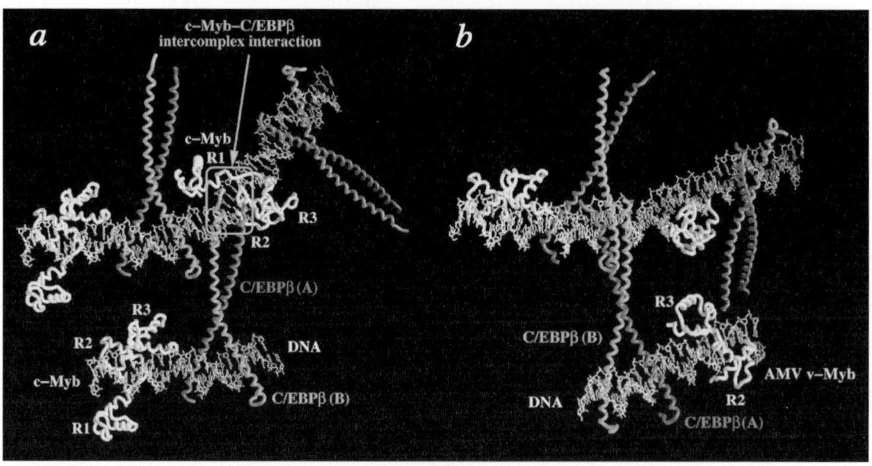

Figure 5

The structures of Myb–C/EBPβ–DNA complexes in crystals. (a) The crystal structure of the c-Myb–C/EBPβ–DNA complex. The backbone structures of c-Myb DBD and two peptide chains of the C/EBPβ homodimer (C/EBPβ(A) and C/EBPβ(B)) are shown as yellow, red and green tubes. The c-Myb–C/EBPβ intercomplex interaction is marked blue. (b) The crystal structure of the AMV v-Myb–C/EBPβ–DNA complex. The backbone structures of the AMV v-Myb DBD and two peptide chains of the C/EBPβ homodimer (C/EBPβ(A) and C/EBPβ(B)) are shown as yellow, red and green tubes. In this structure, intercomplex interaction is not observed. These figures were adopted from Ogata et al. (2003). (see colour section p. xxiv)

C/EBPβ, each protein being bound to separate DNA fragments. These interactions resulted in a four-helix bundle-like structure consisting of the first and second helices of c-Myb R2 and the two helices of the leucine-zipper homodimer of C/EBPβ. The interface between c-Myb and C/EBPβ exhibited a pattern characteristic of functional protein-protein interactions: a hydrophobic interaction core surrounded by a circular hydrogen-bonding network also including a potassium ion and a DNA backbone phosphate. The C-terminal parts of the leucine-zipper regions of C/EBP proteins (C/EBPα, C/EBPβ, C/EBPδ and C/EBPε) contain one additional periodical leucine repeat, which was found to be disordered in the absence of c-Myb. The c-Myb DBD binds to this site in C/EBPβ and induces a C-terminal extension of the coiled coil structure of C/EBPβ (Figure 6) (Tahirov et al., 2002; Ogata et al., 2003). The C-terminal part of the C/EBP leucine-zipper region is well conserved among the family members, and each shows a cooperative interaction with c-Myb. Using a GST-pull down assay it was shown that in solution the c-Myb DBD interacts with the C/EBPβ DBD, and that this interaction can be eliminated by truncation of the C-terminal Myb-binding part of the C/EBPβ DBD (Tahirov et al., 2002).

In contrast, AMV v-Myb DBD, which fails to cooperate with a C/EBPβ, exhibits no interactions with C/EBPβ in the crystal structure including the two proteins and DNA (Figure 5b) (Tahirov et al., 2002; Ogata et al., 2003). AMV v-Myb has three amino acid mutations in its R2 subdomain compared to c-Myb; I91N, L106H and V117D. In the c-Myb-C/EBPβ-DNA complex crystal, I91, L106 and V117 were all mapped near the c-Myb–C/EBPβ interface. Consistent with this it was shown that the I91N or L106H mutations result in loss of C/EBPβ binding capacity in solution (Tahirov et al., 2002). Although the V117D mutation does not significantly affect C/EBPβ binding capacity, it strongly affects recognition of DNA by the Myb DBD when in combination with the I91N and L106H mutations. Together the three point mutations in AMV v-Myb induce conformational changes that impair the hydrogen bonds between the R2 backbone amides and the DNA backbone phosphates including the minor groove position. The hydrogen bonds between the protein backbone amides and DNA backbone phosphates around the DNA minor groove were shown to play an important role in regulation of multi-protein assembly on promoter DNA (Tahirov et al., 2001). In the case of c-Myb, the hydrogen bonds between the R114 and W115 backbone amides of R2 and the DNA backbone phosphate become stabilised by C/EBPβ binding (Tahirov et al., 2002; Ogata et al., 2003). The three point mutations in AMV v-Myb impair this regulatory mechanism by disrupting the critical hydrogen bonds in addition to loss of the C/EBP binding (Figure 7). Furthermore, the water-mediated hydrogen-bonding network between the protein and DNA in the c-Myb-DNA complex (Figure

3c), which seems to stabilise the specific protein-DNA interactions, was also lost in the AMV v-Myb-DNA complex.

Recently, it was reported that S116 of c-Myb R2 is phosphorylated by the protein kinase A (PKA) leading to modulation of the DNA-binding affinity of c-Myb, while the corresponding residue of AMV v-Myb R2 is not phosphorylated because of the presence of the V117D mutation (Andersson et al., 2003). The authors suggested that the V117D mutation in AMV v-Myb is not susceptible to regulation by PKA.

As described above, the interactions between c-Myb and C/EBPβ in the crystal structure were observed to be of an 'inter-complex' type. No 'intra-complex' interactions were seen between c-Myb and C/EBPβ bound to the same DNA fragment. In target gene promoters, the binding sites for c-Myb and C/EBP-binding are usually positioned apart. These two facts suggest that c-Myb and C/EBP might bind separately to the promoter DNA and interact via looping of the DNA between the two binding sites (Figure 8a). In the *mim-1* gene promoter there are three c-Myb-binding sites and two C/EBP-binding sites. Of these, binding sites for c-Myb and C/EBP that are separated by about 80 base pairs function cooperatively in the transactivation of the *mim-1* gene. Atomic force microscopic (AFM) observations showed that a *mim-1* promoter fragment complexed with c-Myb DBD and C/EBPβ DBD exhibits a high frequency of DNA looping (Figure 8b), while a corresponding complex containing AMV v-Myb does not show such looping (Tahirov et al., 2002; Ogata et al., 2003). Transactivation assays showed that while c-Myb and C/EBPβ act synergistically on the *mim-1* gene, AMV v-Myb or a C/EBPβ mutant with a truncation of the c-Myb-binding site do not exhibit this synergism (Tahirov et al., 2002). These observations suggest that transactivation synergy involving interactions between proteins bound at distant DNA sites is achieved by spatial proximity brought about by DNA looping.

5. REGULATION OF C-MYB DBD FUNCTIONS

In eukaryotic cells, the assembly of multiple transcriptional regulatory factors in distinct combinations on target gene promoters is considered to be responsible for the establishment of specific patterns of gene regulation. To date a few crystal structures have been reported for complexes in which multiple transcriptional regulatory factors are bound simultaneously to a specific gene promoter. For example, NFAT-Fos-Jun-DNA (Chen et al., 1998), Ets-1-Pax-5-DNA (Garvie et al., 2001) and PU.1-IRF-4-DNA (Escalante et al., 2002). In addition, a few crystal structures have been reported for complexes composed of a factor bound to DNA together with a

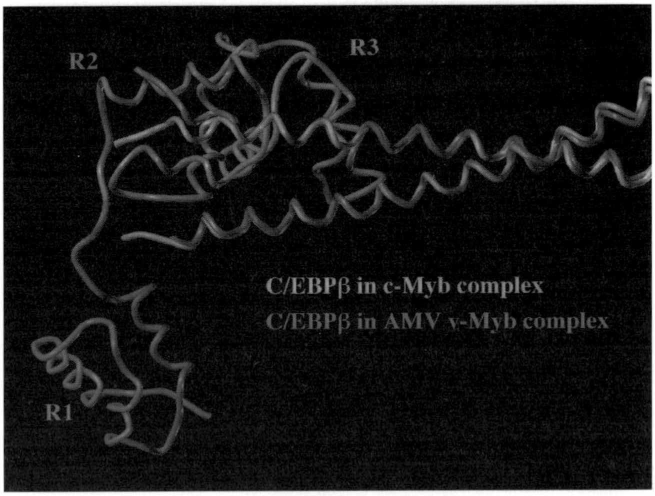

Figure 6

Superimposition of the C-terminal leucine-zipper parts of C/EBPβ in the crystal structures of the c-Myb–C/EBPβ–DNA and AMV v-Myb–C/EBPβ–DNA complexes. The backbones are shown as yellow and orange tubes, respectively. The backbone consisting of the R1, R2 and R3 subdomains of c-Myb DBD in the c-Myb–C/EBPβ–DNA complex is coloured green, magenta and cyan, respectively. The DNA part is excluded for clarity. One of the C-terminal positions of the leucine-zipper parts of C/EBPβ in the AMV v-Myb–C/EBPβ–DNA complex does not take a defined conformation and is not presented. This figure was adopted from Ogata et al. (2003). (see colour section p. xxv)

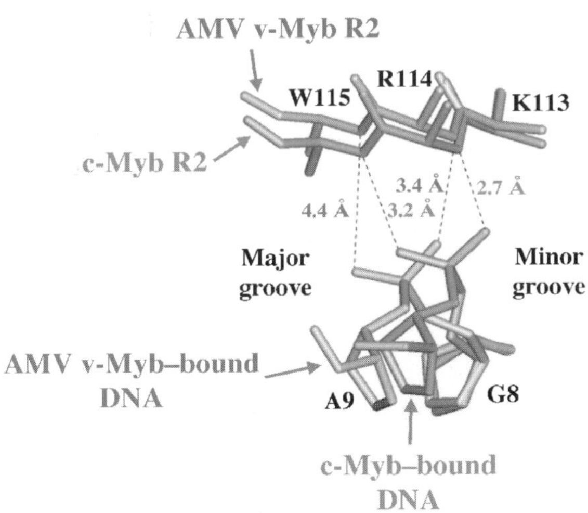

Figure 7

A close-up view of the hydrogen bonds between protein backbones and DNA phosphates in the R2 subdomains of the superimposed c-Myb–C/EBPβ–DNA and AMV v-Myb–C/EBPβ–DNA complex structures. c-Myb and AMV v-Myb are shown as magenta and silver capped sticks, respectively. The DNA base pair positions are labelled according to the numbering in Figure 3 (c). This figure was adopted from Tahirov et al. (2002).
(see colour section p. xxv)

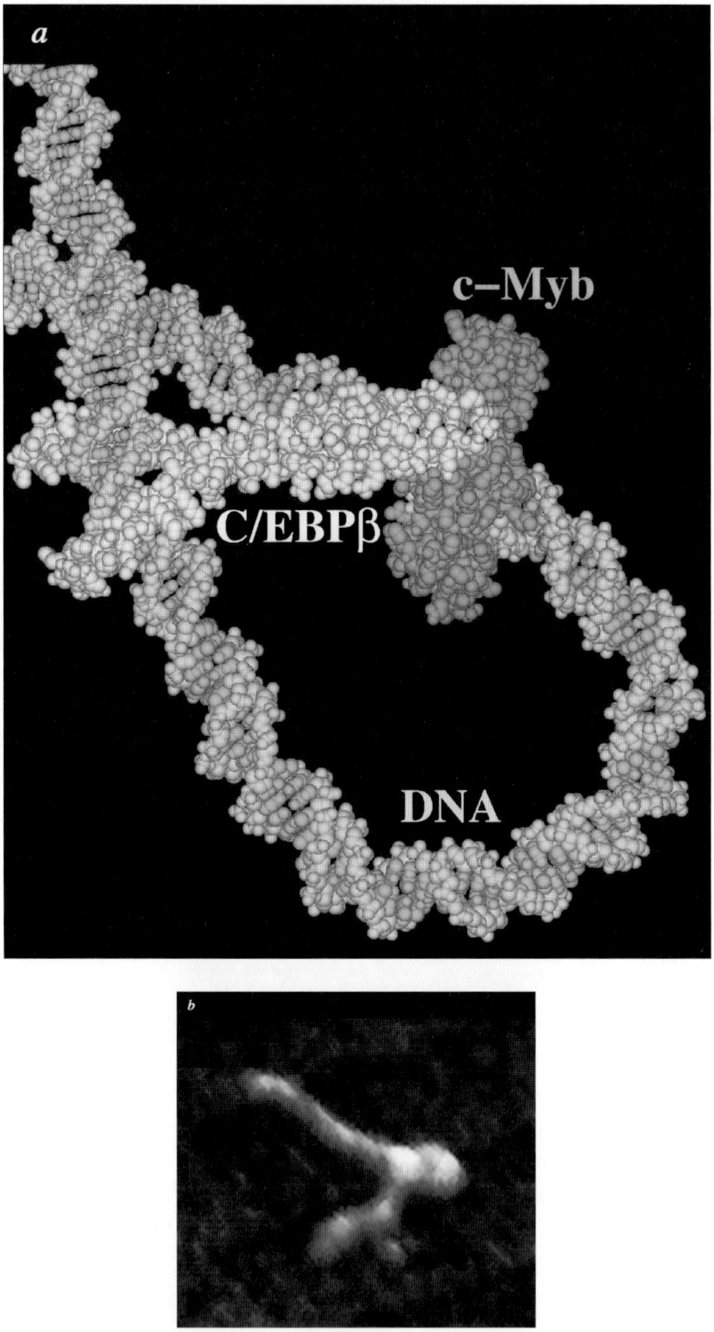

Figure 8

A modelled structure (a) and an AFM image (b) of the complex composed of c-Myb, C/EBPβ and the *mim-1* promoter DNA, showing DNA loop formation. These figures were adopted from Tahirov et al. (2002) (the issue cover) and Ogata et al. (2003). (see colour section p. xxvi) non DNA-binding regulator. For example, GABPα-GABPβ-DNA (Batchelor et al., 1998) and Runx1-CBFβ-DNA (Tahirov et al., 2001; Bravo et al., 2001). In these cases, the non-DNA-binding proteins modulate the DNA-binding factor-DNA interactions in an allosteric fashion. For both types of interaction modes, the underlying regulatory mechanisms are probably mediated through stabilisation of hydrogen bonds between the protein backbone amides and the DNA backbone phosphates (Ogata et al., 2003). Although the c-Myb interaction with C/EBP proteins occurs between proteins bound to sites distant on the DNA, a similar regulatory mechanism of hydrogen bonds stabilisation between the c-Myb R2 backbone and the DNA backbone was observed.

From analyses of the allosteric regulatory mechanism of the interaction between Runx1-DNA and CBFβ, it has been suggested that stabilisation of the conformationally fluctuating protein backbone, which is involved in a hydrogen bond with a DNA backbone phosphate, plays a critical role in the regulation of binding to DNA (Tahirov et al. 2001). In the case of c-Myb, the R2 subdomain is structurally fluctuating due to the presence of a cavity in the hydrophobic core in the free state. When c-Myb binds to a specific DNA site, the R2 structure is partially stabilised by conformational changes. However, even in the DNA-complexed state, the R2 backbone positions of R114 and W115, which are involved in the hydrogen bonds with the DNA backbone phosphates, are still conformationally fluctuating. The DNA phosphate-interacting R2 backbone positions are stabilised by binding of C/EBPβ, which in turn acquires induced fitting to a coiled-coil conformation at the C-terminal part of the leucine-zipper position.

Composite protein binding sites responsible for synergistic transcriptional regulation often consist of a consensus sequence optimised for the binding of one protein, and a non-consensus site for the partner protein, possibly enabling fine control of the cooperative binding to DNA. For example, the IL-2 promoter sequence, which was used for the crystallographic analysis of NFAT-Fos-Jun-DNA complex, has a composite site containing a consensus site for NFAT and a non-consensus site for Fos-Jun. This combination of sites means that a stable DNA-protein complex with the Fos-Jun dimer requires binding of an NFAT molecule (Rao et al., 1997; Chen et al., 1998). Similarly, in the promoter of the *mb*-1 gene (encoding a component of the B cell receptor) that was used for the crystallographic study of Ets-1-Pax5-DNA complex, binding of Ets-1 to a non-consensus Ets site results in cooperation with Pax5 binding to its consensus site (Garvie et al., 2001). In the absence of Pax5, Ets-1 binds to the non-consensus site with a low affinity. Interestingly, in case of the *mim-*

1 promoter, while the c-Myb-binding site is a consensus sequence, the C/EBP-binding sequence is deviated from its respective consensus.

In summary, it is considered that the process through which c-Myb and C/EBP function involves multiple steps of conformational stabilisation of the protein assembly on the promoter DNA. Thus functional regulation of transactivators might be regarded as the regulation of the conformational stability of the proteins. The oncogenic mutations in AMV v-Myb seem to affect various aspects of the c-Myb functional regulatory mechanism including disruption of DNA minor groove recognition and loss of C/EBP binding.

ACKNOWLEDGEMENTS

We thank Drs. H. Morii, T. Okada, T. Kumasaka, M. Yamamoto, H. Nakamura, Y. Nishimura and A. Sarai for collaboration, our laboratory members for their contributions, Dr. M. Shiina for helpful discussions, Dr. S. Akira for C/EBPβ cDNA, Dr. S.A. Ness for the *mim-1* promoter DNA, and Dr. S. Nagakura for his generous support. This work was supported by the Kanagawa Academy of Science and Technology (KAST) and by CREST (Core Research for Evolutional Science and Technology) of the Japan Science and Technology Corporation (JST). We also thank Mr. K. Umehara for his foundation in Yokohama City University Graduate School of Medicine.

REFERENCES

Andersson, K.B., Kowenz-Leutz, E., Brendeford, E.M., Tygsett, A.-H.H., Leuz, A. and Gabrielsen, O.S. (2003) Phosphorylation-dependent down-regulation of c-Myb DNA binding is abrogated by a point mutation in the v-*myb* oncogene. J. Biol. Chem. *278*, 3816-3824.

Batchelor, A.H., Piper, D.E., de la Brousse, F.C., McKnight, S.L. and Wolberger, C. (1998) The structure of GABPα/β: an ETS domain- ankyrin repeat heterodimer bound to DNA. Science *279*, 1037-1041.

Biedenkapp, H., Borgmeyer, U., Sippel, A.E. and Klempnauer, K.-H. (1988) Viral myb oncogene encodes a sequence-specific DNA-binding activity. Nature *335*, 835-837.

Bravo, J., Li, Z., Speck, N.A. and Warren, A.J. (2001) The leukemia-associated AML1 (Runx1)-CBFβ complex functions as a DNA-induced molecular clamp. Nat. Struct. Biol. *8*, 371-378.

Bristos-Bray, M. and Friedman, A.D. (1997) Core binding factor cannot synergistically activate the myeloperoxidase proximal enhancer in immature myeloid cells without c-Myb. Mol. Cell. Biol. *17*, 5127-5135.

Burk, O., Mink, S., Ringwald, M. and Klempnauer, K.-H. (1993) Synergistic activation of the chicken *mim-1* gene by *v-myb* and C/EBP transcription factors. EMBO J. *12*, 2027-2038.

Burk, O., Worpenberg, S., Haenig, B. and Klempnauer, K.-H. (1997) *tom-1*, a novel v-Myb target gene expressed in AMV- and E26-transformed myelomonocytic cells. EMBO J. *16*, 1371-1380.

Chen, L., Glover, J.N., Hogan, P.G., Rao, A. and Harrison, S.C. (1998) Structure of the DNA-binding domains from NFAT, Fos and Jun bound specifically to DNA. Nature *392*, 42-48.

Dini, P.W. and Lipsick, J.S. (1993) Oncogenic truncation of the first repeat of c-Myb decreases DNA binding in vitro and in vivo. Mol. Cell. Biol. *13*, 7334-7348.

Dudek, H., Tantravahi, R.V., Rao, V.N., Reddy, E.S.P. and Reddy, E.P. (1992) Myb and Ets proteins cooperate in transcriptional activation of the *mim-1* promoter. Proc. Nat. Acad. Sci. *89*, 1291-1295.

Escalante, C.R., Brass, A.L., Pongubala, J.M.R., Shatova, E., Shen, L., Singh, H. and Aggarwal, A.K. (2002) Crystal structure of PU.1/IRF-4/DNA ternary complex. Mol. Cell *10*, 1097-1105.

Frampton, J., Leutz, A., Gibson, T. and Graf, T. (1989) DNA-binding domain ancestry. Nature *342*, 134.

Garvie, C.W., Hagman, J. and Wolberger, C. (2001) Structural studies of Ets-1/Pax5 complex formation on DNA. Mol. Cell *8*, 1267-1276.

Hernandez-Munain, C. and Krangel, M.S. (1994) Regulation of the T-cell receptor δ enhancer by functional cooperation between c-Myb and core-binding factors. Mol. Cell. Biol. *14*, 473-483.

Introna, M., Golay, J., Frampton, J., Nakano, T., Ness, S.A. and Graf, T. (1990) Mutations in *v-myb* alter the differentiation of myelomonocytic cells transformed by the oncogene. Cell *63*, 1287-1297.

Kanei-Ishii, C., Sarai, A., Sawazaki, T., Nakagoshi, H., He, D.N., Ogata, K., Nishimura, Y. and Ishii, S. (1990) The tryptophan cluster: A hypothetical structure of the DNA-binding domain of the myb protooncogene product. J. Biol. Chem. *265*, 19990-19995.

Kowenz-Leutz, E., Herr, P., Niss, K. and Leutz, A. (1997) The homeobox gene *GBX2*, a target of the *myb* oncogene, mediates autocrine growth and monocyte differentiation. Cell *91*, 185-195.

Lutz, P.G., Houzel-Charavel, A., Moog-Lutz, C. and Cayre, Y.E. (2001) Myeloblastin is an Myb target gene: mechanisms of regulation in myeloid leukemia cells growth-arrested by retinoic acid. Blood *97*, 2449-2456.

Morii, H., Uedaira, H., Ogata, K., Ishii, S. and Sarai, A. (1999) Shape and energetics of a cavity in c-Myb probed by natural and non-natural amino acid mutations. J. Mol. Biol. *292*, 909-920.

Ness, S.A., Marknell, A. and Graf, T. (1989) The v-*myb* oncogene product binds to and activates the promyelocyte-specific *mim-1* gene. Cell *59*, 1115-1125.

Ness, S.A., Kowenz-Leutz, E., Casini, T., Graf, T. and Leutz, A. (1993) Myb and NF-M: combinatorial activators of myeloid genes in heterologous cell types. Genes Dev. *7*, 749-759.

Oda, M., Furukawa, K., Ogata, K., Sarai, A., Ishii, S., Nishimura, Y. and Nakamura, H. (1997) Investigation of the pyrimidine preference by the c-Myb DNA-binding domain at the initial base of the consensus sequence. J Biol. Chem. *272*, 17966-17971.

Oelgeschläger, M., Janknecht, R., Krieg, J., Schreek, S. and Luscher, B. (1996) Interaction of the coactivator CBP with Myb proteins: effects on Myb-specific transactivation and on the cooperativity with NF-M. EMBO J. *15*, 2771-2780.

Ogata, K., Hojo, H., Aimoto, S., Nakai, T., Nakamura, H., Sarai, A., Ishii, S. and Nishimura, Y. (1992) Solution structure of a DNA-binding unit of Myb: A helix-turn-helix-related motif with conserved tryptophans forming a hydrophobic core. Proc. Natl. Acad. Sci. USA *89*, 6428-6432.

Ogata, K., Morikawa, S., Nakamura, H., Sekikawa, A., Inoue, T., Kanai, H., Sarai, A., Ishii, S. and Nishimura, Y. (1994) Solution structure of a specific DNA complex of the Myb DNA-binding domain with cooperative recognition helices. Cell *79*, 639-648.

Ogata, K., Morikawa, S., Nakamura, H., Hojo, H., Yoshimura, S., Zhang, R., Aimoto, S., Ametani, Y., Hirata, Z., Sarai, A., Ishii, S. and Nishimura, Y. (1995) Comparison of the free and DNA-complexed forms of the DNA-binding domain from c-Myb. Nature Struct. Biol. *2*, 309-320.

Ogata, K., Kanei-Ishii, C., Sasaki, M., Hatanaka, H., Nagadoi, A., Enari, M., Nakamura, H., Nishimura, Y., Ishii, S. and Sarai, A. (1996) The cavity in the hydrophobic core of Myb DNA-binding domain is reserved for DNA recognition and *trans*-activation. Nature Struct. Biol. *3*, 178-187.

Ogata, K., Sato, K. and Tahirov, T.H. (2003) Eukaryotic transcriptional regulatory complexes: cooperativity from near and afar. Curr. Opin. Struct. Biol. *13*, 40-48.

Ording, E., Kravik, W., Bostad, A. and Gabrielsen, O.S. (1994) Two functionally distinct half sites in the DNA-recognition sequence of the c-Myb oncoprotein. Eur. J. Biochem. *222*, 113-120.

Rao, A., Luo, C. and Hogan, P.G. (1997) Transcription factors of the NFAT family: regulation and function. Annu. Rev. Immunol. *15*, 707-747.

Sarai, A., Uedaira, H., Morii, H., Yasukawa, T., Ogata, K., Nishimura, Y. and Ishii, S. (1993) Thermal stability of the DNA-binding domain of the Myb onco-protein. Biochemistry *32*, 7759-7764.

Sasaki, M., Ogata, K., Hatanaka, H. and Nishimura, Y. (2000) Backbone dynamics of the c-Myb DNA-binding domain complexed with a specific DNA. J. Biochem. *127*, 945-953.

Tahirov, T.H., Inoue, T., Sasaki, M., Kimira, K., Morii, H., Fujikawa, A., Shiina, M., Sato, K., Kumasaka, T., Yamamoto, M., Ishii, S. and Ogata, K. (2001) Structural analyses of DNA recognition by the AML1/Runx-1 Runt domain and its allosteric control by CBFβ. Cell *104*, 755-767.

Tahirov, T.H., Sato, K., Ichikawa-Iwata, E., Sasaki, M., Inoue-Bungo, T., Shiina, M., Kimura, K., Takata, S., Fujikawa, A., Morii, H., Kumasaka, T., Yamamoto, M., Ishii, S. and Ogata, K. (2002) Mechanism of c-Myb-C/EBPβ cooperation from separated sites on a promoter. Cell *108*, 57-70.

Tanikawa, J., Yasukawa, T., Enari, M., Ogata, K., Nishimura, Y., Ishii, S. and Sarai, A. (1993) Recognition of specific DNA sequences by the c-*myb* proto-oncogene product: Role of three repeat units in the DNA-binding domain. Proc. Natl. Acad. Sci. USA *90*, 9320-9324.

Verbeek, W., Gombart, A.F., Chumakov, A.M., Muller, C., Friedman, A.D. and Koeffler H.P. (1999) C/EBPε directly interacts with the DNA binding domain of c-myb and cooperatively activates transcription of myeloid promoters. Blood *93*, 3327-3337.

Zhang, X., Xing, G., Fraizer, G.C. and Saunders, G.F. (1997) Transactivation of an intronic hematopoietic-specific enhancer of the human Wilms' tumor 1 gene by GATA-1 and c-Myb. J. Biol. Chem. *272*, 29272-29280.

Chapter 12

MYB PARTNERSHIPS

Xianming Mo, Elisabeth Kowenz-Leutz, and Achim Leutz
Max-Delbrück-Center for Molecular Medicine, Robert-Rössle-Str. 10, 13092 Berlin, Germany.

Abstract: The c-Myb protein is a DNA binding transcription factor that coordinates proliferation and differentiation of haemopoietic cells. Cell-type specific gene activation by c-Myb is achieved via interaction with a number of other transcription factors and multi-protein complexes that determine the function of c-Myb as a regulator of chromatin structure and gene expression downstream of signalling cascades. Mutations that turn c-Myb into a leukaemia gene concomitantly alter the way c-Myb interacts with other proteins. For future consideration of c-Myb or any of its partners as therapeutic targets it will be essential to reveal proteins that it interacts with, their mode of interaction, and the biological consequences of these partnerships.

1. INTRODUCTION

A central issue in understanding how interactions between transcription factors determine their function is to visualise them in 3-dimensions (3-D) on their chromatin template. The structure of the DNA binding domain (DBD) of c-Myb has now been resolved (Ogata et al., 1992; Ogata et al., 1995; Ogata et al., 1994; Tahirov et al., 2002; Tanikawa et al., 1993), however, further 3-D details on a central transactivation domain, a potential coiled-coil domain, or negative regulatory domains at the C-terminus of c-Myb remain to be determined. All of these domains are highly conserved among c-Myb proteins from various species suggesting a high degree of structure/function constraints (Dash et al., 1996).

An increasing number of proteins have been found to interact with c-Myb or its retroviral counterparts (v-Myb) encoded by the avian myeloblastosis virus (AMV) and the avian E26 leukemia virus. Many of these protein interactions have been covered in several excellent recent reviews

J. Frampton (ed.), Myb Transcription Factors: Their Role in Growth, Differentiation and Disease, 239-256.

(Dubendorff and Lipsick, 1999; Introna and Golay, 1999; Ness, 1999; Weston, 1998; Weston, 1999). Here, we concentrate on some of the most recent findings on Myb-protein interactions involving the DNA binding domain (DBD) and activation of the c-Myb protein during gene regulation. The cumulative data suggest that c-Myb is a latent transcription factor that becomes activated by signalling and protein interactions to recognise and to alter the structure and function of chromatin.

2. INTERACTIONS WITH THE DNA BINDING DOMAIN OF C-MYB AND V-MYB

2.1 DNA Binding Domain

The N-terminal c-Myb DBD plays a role not only in the recognition of *cis*-regulatory sites but also in functions that are unrelated to docking to DNA and that supposedly involve protein interactions (Frampton et al., 1991; Introna et al., 1990). It was therefore important to see how the N-terminal Myb repeat motifs fold up when bound to the consensus sequence 5'-YAACNG-3' and what structures remained solvent exposed.

The 3-D structure of the three related 50 amino acid repeats R1, R2, and R3 showed that R2, together with R3, confer specific DNA binding, whereas R1 plays a different role (Ogata et al., 1992; Ogata et al., 1995; Ogata et al., 1994; Tahirov et al., 2002; Tanikawa et al., 1993). All three repeats have very similar folding structures, forming tandemerised sub-domains as previously suggested based on sequence comparison and molecular modelling (Frampton et al., 1991; Frampton et al., 1989). Three conserved tryptophans in each repeat are oriented towards the core of the repeats and turned out to be essential for the structure and the sequence-specific DNA-binding capacity. The first helix is important for the structure of the subdomain and the second and third helix in each repeat form helix-turn-helix motifs (Ogata et al., 1992; Ogata et al., 1995; Ogata et al., 1994). The third helices of R2 and R3 form an extended α-helical structure that fits into the major groove of DNA and specifically binds to *cis*-regulatory Myb sites. R1 does not contribute to sequence recognition, however increases the stability of the DNA-protein complex by interacting loosely with DNA next to the specific Myb-binding site (Tahirov et al., 2002). R2 has a low thermal stability in the absence of DNA which is caused by a solvent exposed hydrophobic region with a cavity inside (Ogata et al., 1996). A cavity-filling mutation that stabilises R2 significantly reduces specific Myb DNA-binding activity and transactivation, implying that inherent conformational flexibility of R2 is associated with the presence of the hydrophobic surface, and could be important for Myb DNA recognition (Ogata et al., 1996).

R2 mutations of AMV v-Myb, I91N and L106H, cause a shift to the C-terminal side along the helical axis. The V117D mutation stabilises a structure that is involved in DNA binding and disrupts a potential phosphorylation site for protein kinase A-type enzymes (Andersson et al., 2003). Thus, structural alteration caused by the AMV-type mutations may explain why v-Myb seems to be less sensitive to SH-specific modifications (Brendeford et al., 1998; Tahirov et al., 2002), and why AMV v-Myb forms protein-DNA complexes with lower stability when compared to c-Myb (Brendeford et al., 1997; Tahirov et al., 2002).

2.2 Myb and CCAAT Enhancer Binding Proteins (C/EBP)

Co-crystallisation of the Myb DBD with the basic leucine zipper and DNA binding domain (bZip) of CCAAT Enhancer Binding Protein beta (C/EBPβ) revealed the structural basis for the synergism observed between both transcription factors. The crystal structure also demonstrated for the first time in 3-D that the Myb DBD may serve as a protein interaction domain when bound to DNA (Tahirov et al., 2002). In addition, the mutations in the AMV v-Myb DBD that alter the structure of the solvent exposed surface were shown to prevent the interaction with C/EBP (Tahirov et al., 2002). Thus, the structural data suggest that some of the biological differences between leukaemogenic and wild type (wt) Myb may result from altered protein interactions with the DBD, and in particular may explain the loss of binding and collaboration with C/EBP (Kowenz-Leutz et al., 1997).

C/EBPs are important partners of Myb that may direct the activity of c-Myb from proliferation support towards differentiation support. C/EBPs were also the first transcription factors found to functionally interact with Myb (Burk et al., 1993; Ness et al., 1993). Some of the C/EBPs have now been shown to be of major importance in haemopoiesis. C/EBPα, β, δ, ε, γ, and ζ comprise a family of transcription factors with highly conserved bZip domains (Landschulz et al., 1989) that bind as homo- and hetero-dimers to specific DNA consensus sequences, 5'-T(T/G)NNGNAA(T/G)-3' (except C/EBPζ). All C/EBPs with the exception of C/EBPγ possess a transactivation domain that maps to their N-terminal end. Internal translation initiation at alternative sites in C/EBPα and β gives rise to full-length and truncated C/EBP isoforms that differ in transactivating and repressive functions (Calkhoven et al., 2000). Both, C/EBPβ and C/EBPε have internal domains that suppress their gene activating potential and may confer repressor functions in the absence of regulatory signals (Kowenz-Leutz et al., 1994; Williams et al., 1995; Williamson et al., 1998). C/EBPγ

(Cooper et al., 1995) and CHOP (Ron and Habener, 1992) are thought to represent dominant inhibitory proteins that neutralise the activity of other C/EBP proteins by dimerisation.

Among other tissues, C/EBPs play important roles in the development and function of granulocytes, monocytes, eosinophils, and B-cells. They cooperate with c-Myb in the activation of a variety of haemopoiteic genes, including mim-1(Burk et al., 1993; Ness et al., 1993), lysozyme (Ness et al., 1993), tom-1A (Burk et al., 1997), myeloperosidase (MPO) (Britos-Bray and Friedman, 1997), neutrophil elastase (Oelgeschlager et al., 1996b; Verbeek et al., 1999), Rag2 (Fong et al., 2000) and myeloblastin (Lutz et al., 2000).

The observation that C/EBP plus c-Myb activate the target genes mim-1 and lysozyme (Burk et al., 1993; Introna et al., 1990; Ness et al., 1993; Ness et al., 1989) showed for the first time that lineage specific gene activation in the haemopoietic system occurs through combinatorial interactions between different transcription factors (Cantor and Orkin, 2001; Sieweke and Graf, 1998). Since c-Myb and C/EBPβ also cooperate in heterologous, non-haemopoietic cell types, such as in fibroblasts, the data also suggested that the combination of both transcription factors is sufficient to set up a haemopoiesis specific genetic switch that could induce commitment to myelopoiesis. However, cooperation between both factors has now also been found in the activation of the human choline acetyltrasferase gene in neuronal cells (Robert et al., 2002).

C/EBPα has been found to be an essential factor in the development of neutrophils and eosinophils (Zhang et al., 1997). During the development of neutrophils, C/EBPα and c-Myb cooperatively activate the defensin genes that are major components of the azurophilic granules, and that contribute to innate and acquired host immunity (Tsutsumi-Ishii et al., 2000). Other myeloid genes, including neutrophil elastase (NE) or the myeloblastin genes (Lutz et al., 2000; Oelgeschlager et al., 1996b) are also activated by a combination of c-Myb and C/EBPα. These data suggest that the combination of c-Myb plus C/EBPα induces differentiation of immature cells.

In contrast to c-Myb, the leukaemogenic version encoded by AMV fails to collaborate with C/EBP to activate mim-1 or lysozyme. AMV encoded Myb, however, constitutively up-regulates the homeobox gene GBX2 that superimposes a myelomonocytic phenoptype in precursor cells (Kowenz-Leutz et al., 1997). Myb induced GBX2 activation depends on either the presence of the three point mutations in the DBD of v-Myb or on c-Myb plus an activated receptor tyrosine/ras/MAP-kinase pathway. Taken together, this suggests that the AMV-v-Myb point mutations are gain-of-function mutations for a signalling event that targets c-Myb, and at the same time, are loss-of-function mutations for the collaboration with C/EBP (Introna et al.,

1990; Kowenz-Leutz et al., 1997; Ness et al., 1989; Ogata et al., 1995). Thus, besides binding to target sites on DNA, the DNA binding domain of Myb is responsible for different synergistic mechanisms in the activation of different genes.

The crystal structures of complexes containing the c-Myb-DBD or the leukaemogenic v-Myb DBD, CEBPβ, and DNA containing binding sites for both factors, has finally revealed similarities and differences between the interaction of these transcription factors bound to their target sites (Tahirov et al., 2002). In the DNA-protein complex the intertwined coiled-coil region of C/EBP dimers interacts with the first two helices of c-Myb R2 subdomain to form a motif resembling a four-helix bundle. The C-terminal part of one of the C/EBPβ chains lies between both c-Myb R2 helices, while the other C/EBP chain extends the interaction with c-Myb bound to the DNA (Tahirov et al., 2002). As C/EBP interacts with both exposed helices of c-Myb R2, the I91N and L106H mutations in AMV v-Myb alter the surface of v-Myb such that it impairs the interaction with C/EBP (Tahirov et al., 2002). This suggests that the AMV v-Myb oncoprotein escapes binding to and activation of differentiation genes that are characterised by Myb and C/EBP sites.

2.3 SANT – Myb

Chromatin remodelling entails structural alterations of nucleosomes induced by ATP-dependent complexes and covalent histone modifications induced by transcriptional cofactors, predominantly on N-terminal histone tails that protrude from nucleosomal core sturctures. Histone tail modifications include dynamic changes by amino acid acetylation, phosphorylation, and methylation, that serve as docking sites for additional co-factors (Strahl and Allis, 2000; Turner, 2002). Distinct domains of such co-factors such as the ubiquitous bromodomains or chromodomains present in many chromatin-associated proteins may interact only with distinct modification patterns and thus read and interpret what has been termed as the "histone code"(Strahl and Allis, 2000).

The Myb DBD has homology to a chromatin interacting domain, termed the SANT domain (Aasland et al., 1996). The SANT domain is found in proteins that participate in various chromatin regulatory complexes, for example in the Swi3, Ada2, TFIIIB, N-CoR, and ISWI proteins. The SANT domain has been described to be functionally involved in acetylation, deacetylation, and remodelling, without displaying enzymatic activity on its own, suggesting that it plays a role in mediating substrate recognition and protein interactions.

The chromatin function of the SANT domain suggests that c-Myb likewise displays related functions that still have to be discovered. What could these "chromatin" functions be? Several reports on SANT functions indicate that the domain serves as a protein interaction domain that binds to histone modification enzymes while simultaneously stimulating substrate recognition and enzymatic activities of complexes. For example, it has been shown that the yeast histone acetyltransferase (HAT) complexes SAGA and ADA (Grant et al., 1997) possess a number of transcriptional relevant subunits, including a trimeric module consisting of the histone acetyltransferase (HAT) Gcn5, Ada3 and the SANT domain protein Ada2p. Mutational analysis revealed that the Ada2p SANT domain is required for Gcn5 interaction, for mediating the full HAT activity in the SAGA complex and for recognition of nucleosomal substrates (Boyer et al., 2002; Sterner et al., 2002). SANT domain proteins are, however, also found in HDAC complexes. The N-CoR co-repressor of nuclear receptors is a SANT domain protein that interacts with the histone deacetylase HDAC3 and stimulates its activity (Zhang et al., 2002). Another SANT domain protein, MTA2, is involved in the modulation of the enzymatic activity of the HDAC1 complex (Zhang et al., 1999). CoREST and Mta-lp proteins also form complexes with HDACs and stimulate their enzymatic activities (Humphrey et al., 2001; You et al., 2001). The SANT domain of the nuclear receptor co-repressors SMRT is a critical component of a deacetylase activation domain (DAD) that binds and activates HDAC3 and, as part of a histone interaction domain (HID), stimulates repression by increasing the affinity of the DAD-HDAC3 enzyme to histone substrate. The fact that the SANT-containing HID preferentially binds to unacetylated histone tails implies that the SANT also participates in the interpretation of the histone code (Yu et al., 2003).

Mutation of the SANT domain in the SWI/SNF-complex protein SWI3p destroys the activity of the complex *in vivo* (Boyer et al., 2002). Mutation of conserved residues required for DNA binding only, however, does not affect the activity of the SWI/SNF complex nor its composition, suggesting a major function in protein-protein interaction (Boyer et al., 2002). In addition, the function of the chromatin remodelling RSC complex also depends on an intact SANT domain. These results suggest an important function for the SANT domain of remodelling complexes in mediating interaction between proteins and in recognising histone modification patterns once the chromatin remodelling complexes have been recruited to specific genes. One therefore wonders whether and how the c-Myb DNA binding domain is also involved in deciphering distinct histone modifications.

3. CBP/P300 AND MYB

The two highly related proteins, cellular CREB binding protein (CBP) and p300 were initially found in association with the adenoviral E1A oncoprotein (Arias et al., 1994; Chrivia et al., 1993; Kwok et al., 1994). Both proteins, CBP and p300, are histone acetylases (Bannister and Kouzarides, 1996; Ogryzko et al., 1996) that are involved in a number of different cellular functions such as gene transcription, cell growth, transformation, and embryo development (Kawasaki et al., 1998; Kung et al., 2000; Yao et al., 1998). The CBP and p300 proteins stimulate activating and repressive functions of many transcription factors. For example, CBP and p300 stimulate transactivation of p53 protein on certain promoters (Avantaggiati et al., 1997; Gu and Roeder, 1997; Lill et al., 1997), while it confers repression on other p53 promoters (Ravi et al., 1998). Acetylation of histones is thought to destabilise the nucleosomal structure and facilitates binding of other transcription factors as well as the basic transcription machinery to DNA. CBP/p300 also acetylate non-histone nuclear proteins, including p53 (Gu and Roeder, 1997; Sakaguchi et al., 1998) or GATA-1(Boyes et al., 1998) and c-Myb (Sano and Ishii, 2001; Tomita et al., 2000). Acetylation of both p53 and GATA-1 increase their DNA binding capacity *in vitro* (Gu and Roeder, 1997) and, in the case of GATA-1, directly stimulates GATA-1-dependent transcription (Boyes et al., 1998). However, GATA-1-dependent transcription is blocked by c-Myb expression presumably by competing for CPB (Takahashi 2000). Thus, cross-talk between haemopoietic transcription factors includes competition of essential co-factors. A dose-dependent role for both co-factors is further supported through studies of CBP/p300 deficient mice. A full complement of CBP, but not p300, is required for normal haemopoietic differentiation (Kung et al., 2000; Yao et al., 1998). Monoallelic inactivation of the CBP gene leads to multilineage defects in haemopoietic differentiation and to an increased incidence of haematological malignancies later on in life (Kung et al., 2000).

c-Myb-dependent transcriptional activation is stimulated by CBP/p300 (Dai et al., 1996; Kiewitz and Wolfes, 1997; Oelgeschlager et al., 1996a), and the presence of C/EBP as another CBP/p300 interacting transcription factor further enhances the effects (Mink et al., 1997; Oelgeschlager et al., 1996a; Robert et al., 2002). Interestingly, the N-terminal region of C/EBPβ also recruits the chromatin remodelling complex SWI/SNF that plays an important role in chromatin remodelling and in the activation of c-Myb-target genes. Transplantation of the SWI/SNF recruiting domain of C/EBPβ onto c-Myb abrogates the C/EBP requirement for the activation of at least some of the c-Myb/C/EBP target genes (Kowenz-Leutz and Leutz, 1999). This suggests that c-Myb and C/EBP orchestrate the recruitment of distinct chromatin remodelling complexes, including CBP/p300 and SWI/SNF, during differentiation and it will be important to see whether sequential effects play a role.

The transcriptional activation domain of c-Myb binds via an amphiphatic helix to the KIX domain of CBP that forms a three-helix structure with a shallow hydrophobic grove (Radhakrishnan et al., 1997). Three hydrophobic residues (I295, L298, and L302) on the surface of the c-Myb helix are important for the interaction with KIX. Mutation of any of the three hydrophobic residues abrogates the interaction with the KIX domain in CBP/p300 and blocks c-Myb target gene activation (Parker et al., 1999). Similarly, mutation of critical protein-interaction residues in the KIX domain of p300 disrupts binding to the surface of c-Myb and affects the development and function of megakaryocytes. In contrast, identical mutations in the KIX domain of CBP do not affect haemopoiesis. Thus, conserved domains in two highly related co-activators p300 and CBP have different roles in c-Myb regulated haemopoiesis (Kasper et al., 2002; Kung et al., 2000).

The negative regulatory domain of c-Myb was also found to bind to the CBP C/H2 domain, which is critical for the acetyltransferase activity. CBP/p300 acetylates c-Myb at five C-terminal lysine residues (438, 441, 467, 476, and 481) and acetylation at any of these sites enhances the association with CBP and its activity (Sano and Ishii, 2001; Tomita et al., 2000). The data indicate that CBP mediates the activation of c-Myb by modifying the molecular structure of c-Myb by acetylation that might enhance the Myb-CBP/p300 functional interaction by a "feed-forward" mechanism.

4. MODULATION OF THE NEGATIVE REGULATORY DOMAIN OF C-MYB

Both the transactivating and the transforming potential of c-Myb are activated by N-and C-terminal truncation suggesting that c-Myb is a latent factor that is inactivated by its own terminal sequences. It was therefore important to see that both ends of c-Myb interact with each other (Dash et al., 1996; Kiewitz and Wolfes, 1997; Twamley-Stein et al., 1996). This immediately suggested that c-Myb folds back on itself and that it is regulated by protein interactions and signalling pathways.

Several reports support the notion that the disruption of the intramolecular interaction of c-Myb plays a critical role. Mutants of c-Myb lacking the negative regulatory domain super-transactivate c-Myb driven reporter expression and unleash its transforming capacity in tissue culture

(Wang et al., 1999). Treatment of purified c-Myb with proteases removes the C-terminal negative regulatory domain and enhances DNA binding (Ness, 1999). Specific DNA-binding of carboxyl-truncated c-Myb is strongly enhanced when compared to the full-length protein (Ramsay et al., 1991; Ramsay et al., 1992). Cyclophilin Cyp-40, a modifier of protein conformational changes, was found to interact with the DBD and with the C-terminal regulatory domain to inhibit DNA binding (Leverson and Ness, 1998). Importantly, Cyp-40 repression is not observed by the AMV derived v-Myb that contains point mutations that abrogate Cyp-40 interaction. Taken together, these findings indicate that a closed structure of c-Myb is enzymatically maintained to block the DNA binding surface of c-Myb. The "closed" c-Myb structure probably masks other docking sites in the DBD. In contrast to c-Myb, the leukaemic avian forms of Myb encoded by the AMV and E26 retroviruses lack the C-terminal negative regulatory domain, suggesting their unhampered binding to DNA and interaction with proteins that bind to the DBD of Myb (Dash et al., 1996; Kowenz-Leutz et al., 1997).

An important regulatory element within the C-terminal negative regulatory region is the PEST/EVES motif (Dash et al., 1996; Kiewitz and Wolfes, 1997; Twamley-Stein et al., 1996). The PEST/EVES motif is a target for phosphorylation and a serine to alanine exchange of the phosphorylation site within the PEST/EVES motif increases the transactivation and DNA binding capacity of Myb concomitantly with the disruption of the interaction between the N-terminal and C-terminal part of the protein (Aziz et al., 1995; Ness, 1999).

Serine 528 of the PEST/EVES motif in the negative regulatory domain of c-Myb serves as a substrate for the p42/44 MAPK signalling pathways (Aziz et al., 1995; Miglarese et al., 1996; Twamley-Stein et al., 1996). Substitution of the serine by alanine (S528A) in c-Myb mediates activation of the CD34 promoter but not the c-myc or mim-1 promoters, suggesting different functions of distinct post-translational modifications on different genes. This amino acid exchange does not affect the DNA binding ability of c-Myb (Miglarese et al., 1996), implicating that phosphorylation of S528 in the negative regulatory domain modulates protein interactions that are important for some but not all target genes.

These findings indicate that the modification at the PEST/EVES motif by a signalling pathway changes the structure of the negative regulatory domain and modifies the architecture of c-Myb. The change in conformation may release interacting surfaces for other molecules and thus permit the activation of c-Myb. In contrast to c-Myb, the retroviral Myb forms of AMV v-Myb and E26 v-Myb lack the C-terminal negative regulatory domain suggesting a Myb conformation that unrestrictedly binds to DNA and lacks functional control through signalling events targetted to the C-terminus (Dash et al., 1996; Kowenz-Leutz et al., 1997). Escape from regulated DNA binding and regulated protein interaction is an important step in activating the transforming potential of Myb.

A similar scenario has been described for the sumoylation of the negative regulatory domain. The targetting sequence in proteins that can be sumoylated is ΦKXE (where Φ is a large hydrophobic amino acid, and X is any residue) (Verger et al., 2003). The 16 kDa SUMO protein, which is related to ubiquitin, is activated in an ATP-dependent manner, transferred to Ubc9 and subsequently attached to the ε amino group of lysine in the consensus sequence (Verger et al., 2003). The SUMO-1-conjugating enzyme Ubc9 was found to interact specifically with PEST/EVES motif within the negative regulatory domain of c-Myb causing sumoylation at two sites, K523 and K499. The single mutation K523R completely abolished modification of c-Myb by SUMO-1 at all target sites and resulted in enhanced transactivation of reporter genes and resident c-Myb target genes. Covalently attached SUMO-1 increased the stability of c-Myb and decreased its transcativation capacity (Bies et al., 2002; Dahle et al., 2003) implying that sumoylation regulates c-Myb function. However, mutations of the SUMO-1 modification sites did not alter its stability, suggesting a mechanism other than competition of ubiquitinylation and sumoylation for the same lysine in the stabilisation of c-Myb. Thus, sumoylation of the PEST/EVES motif is thought to stabilises the intramolecular interaction that leads to suppression of the transactivation function by locking the transactivation domain in an inaccessible conformation.

The PEST/EVES motif was also described as a target of the JAK/STAT pathway linking the protein kinase Pim-1 and the transcriptional co-activator p100 to c-Myb. Pim-1 belongs to a family of serine/threonine protein kinases that enhance haemopoietic survival through inhibition of apoptosis. (White, 2003). The expression of the Pim-1 protein is mediated via a number of cytokines that act through the JAK/STAT-pathway (Wang et al., 2001). The p100 protein was initially identified as a cellular transcriptional co-activator for the Epstein-Barr virus nuclear antigen 2 (EBNA2) (Tong et al., 1995). Subsequently, it was found to function as a co-activator that bridges between STAT6 and RNA polymerase II to enhance transcription. p100 binds to Pim-1 and interacts directly with the DNA binding domains of both c-Myb and v-Myb, suggesting that p100 acts as an important mediator between Myb and upstream regulators like Pim-1. Pim-1 and p100 form a stable complex that stimulates c-Myb transcriptional activity in a cooperative fashion (Leverson et al., 1998). Thus, the transmission of JAK/STAT signalling through Pim-1/p100 might result in a conformational change of the Myb protein exposing its DNA binding domain and making it accessible for cross-talk to the transcriptional machinery.

A few other factors including Pax5, RARα, c-Maf and the proteins p160 and p67 have been assigned to bind to the NRD region of c-Myb and to regulate its activity (Fong et al., 2000; Wang et al., 2000). In B lymphocyte development, c-Myb and Pax5 cooperate in the activation of the Rag-2 promoter via their synergistic DNA-binding. The RAG1/RAG2 recombinase complex plays an important role in the V(D)J recombination during lymphocyte development. The C-terminus of c-Myb was mapped to be responsible for the synergistic interaction with Pax-5, suggesting that Pax-5 may change the structure of c-Myb, thereby exposing a motif in c-Myb that permits the interaction with other molecules. In T cells, c-Myb and GATA3 are essential factors in the activation of Rag2, so both proteins may synergistically interact with each other (Anderson et al., 2002; Wang et al., 2000).

Other factors, that include the retinoic acid receptor alpha (RARα), have been found to induce differentiation and growth arrest by binding and suppressing the function of v-Myb (Vodicka et al., 2000; Zemanova and Smarda, 1998). This interaction requires the DBD of RAR and the C-terminus of c-Myb (Pfitzner et al., 1998). Another inhibitory factor of c-Myb function is c-Maf although it is not known whether c-Maf targets the C-terminus. Expression of c-Maf in human immature myeloblastic cells inhibits CD13/APN-driven reporter gene activity and correlates with its ability to physically associate with c-Myb. Formation of inhibitory Myb-Maf complexes is developmentally regulated with high levels in immature myeloid cell that decrease during differentiation (Hedge et al., 1998; Hegde et al., 1999). However, the main function of c-Maf appears to be in T cell development (Kim et al., 1999).

Finally, the "leucine zipper" motif of v-Myb has been reported to direct development of myeloid progenitors into the macrophage lineage. Mutations in the presumptive coiled-coil region compromise commitment toward myeloid cells and support the development of erythroid cells, thrombocytes, and granulocytes (Karafiat et al., 2001), suggesting a role of the Myb leucine zipper in myelopoiesis. The p160 and p67 proteins are proteins with unknown function in human and mouse (Keough et al., 1999; Tavner et al., 1998) that also bind to the leucine zipper motif in c-Myb. P67, an N-terminal proteolytic fragment of the p160 protein, binds to c-Myb and inhibits its transactivation capacity. However, the functional implications of the p160 interaction with c-Myb remain to be resolved.

5. CONCLUSION

Taken together, it appears that c-Myb is a latent protein, intrinsically inhibited by both of its ends that becomes activated by signalling events such as phosporylation, dephosphorylation, acetylation, or sumoylation. The closed structure of c-Myb initially blocks its DNA binding surfaces and also masks docking sites for other proteins. The modification at the PEST/EVES motif and other residues within the C-terminus, or the interaction with other proteins within this region may change the structure of the C-terminus and presumably modify the whole architecture of c-Myb. Such conformational changes may provide additional and novel interacting surfaces for other proteins and co-factors. This could further promote the activation of c-Myb or modify the way it interacts with the cellular machinery that changes chromatin and that regulates gene expression.

REFERENCES

Aasland, R., Stewart, A. F. and Gibson, T. (1996) The SANT domain: a putative DNA-binding domain in the SWI-SNF and ADA complexes, the transcriptional co-repressor N-CoR and TFIIIB. Trends Biochem Sci *21*, 87-88.

Anderson, M. K., Hernandez-Hoyos, G., Dionne, C. J., Arias, A. M., Chen, D. and Rothenberg, E. V. (2002) Definition of regulatory network elements for T cell development by perturbation analysis with PU.1 and GATA-3. Dev Biol *246*, 103-121.

Andersson, K. B., Kowenz-Leutz, E., Brendeford, E. M., Tygsett, A. H., Leutz, A. and Gabrielsen, O. S. (2003) Phosphorylation-dependent down-regulation of c-Myb DNA binding is abrogated by a point mutation in the v-myb oncogene. J Biol Chem *278*, 3816-3824.

Arias, J., Alberts, A. S., Brindle, P., Claret, F. X., Smeal, T., Karin, M., Feramisco, J. and Montminy, M. (1994) Activation of cAMP and mitogen responsive genes relies on a common nuclear factor. Nature *370*, 226-229.

Avantaggiati, M. L., Ogryzko, V., Gardner, K., Giordano, A., Levine, A. S. and Kelly, K. (1997) Recruitment of p300/CBP in p53-dependent signal pathways. Cell *89*, 1175-1184.

Aziz, N., Miglarese, M. R., Hendrickson, R. C., Shabanowitz, J., Sturgill, T. W., Hunt, D. F., and Bender, T. P. (1995). Modulation of c-Myb-induced transcription activation by a phosphorylation site near the negative regulatory domain. Proc Natl Acad Sci USA *92*, 6429-6433.

Bannister, A. J. and Kouzarides, T. (1996) The CBP co-activator is a histone acetyltransferase. Nature *384*, 641-643.

Bies, J., Markus, J. and Wolff, L. (2002) Covalent attachment of the SUMO-1 protein to the negative regulatory domain of the c-Myb transcription factor modifies its stability and transactivation capacity. J Biol Chem *277*, 8999-9009.

Boyer, L. A., Langer, M. R., Crowley, K. A., Tan, S., Denu, J. M. and Peterson, C. L. (2002) Essential role for the SANT domain in the functioning of multiple chromatin remodeling enzymes. Mol Cell *10*, 935-942.

Boyes, J., Byfield, P., Nakatani, Y. and Ogryzko, V. (1998) Regulation of activity of the transcription factor GATA-1 by acetylation. Nature *396*, 594-598.

Brendeford, E. M., Andersson, K. B. and Gabrielsen, O. S. (1998) Nitric oxide (NO) disrupts specific DNA binding of the transcription factor c-Myb in vitro. FEBS Lett *425*, 52-56.

Brendeford, E. M., Myrset, A. H., Hegvold, A. B., Lundin, M. and Gabrielsen, O. S. (1997) Oncogenic point mutations induce altered conformation, redox sensitivity, and DNA binding in the minimal DNA binding domain of avian myeloblastosis virus v-Myb. J Biol Chem *272*, 4436-4443.

Britos-Bray, M. and Friedman, A. D. (1997) Core binding factor cannot synergistically activate the myeloperoxidase proximal enhancer in immature myeloid cells without c-Myb. Mol Cell Biol *17*, 5127-5135.

Burk, O., Mink, S., Ringwald, M. and Klempnauer, K. H. (1993) Synergistic activation of the chicken mim-1 gene by v-myb and C/EBP transcription factors. EMBO J *12*, 2027-2038.

Burk, O., Worpenberg, S., Haenig, B. and Klempnauer, K. H. (1997) tom-1, a novel v-Myb target gene expressed in AMV- and E26-transformed myelomonocytic cells. EMBO J *16*, 1371-1380.

Calkhoven, C. F., Muller, C. and Leutz, A. (2000) Translational control of C/EBPalpha and C/EBPbeta isoform expression. Genes Dev *14*, 1920-1932.

Cantor, A. B. and Orkin, S. H. (2001) Haemopoietic development: a balancing act. Curr Opin Genet Dev *11*, 513-519.

Chrivia, J. C., Kwok, R. P., Lamb, N., Hagiwara, M., Montminy, M. R. and Goodman, R. H. (1993) Phosphorylated CREB binds specifically to the nuclear protein CBP. Nature *365*, 855-859.

Cooper, C., Henderson, A., Artandi, S., Avitahl, N.,and Calame, K. (1995) Ig/EBP (C/EBP gamma) is a transdominant negative inhibitor of C/EBP family transcriptional activators. Nucleic Acids Res *23*, 4371-4377.

Dahle, O., Andersen, T., Nordgard, O., Matre, V., Del Sal, G. and Gabrielsen, O. S. (2003) Transactivation properties of c-Myb are critically dependent on two SUMO-1 acceptor sites that are conjugated in a PIASy enhanced manner. Eur J Biochem *270*, 1338-1348.

Dai, P., Akimaru, H., Tanaka, Y., Hou, D. X., Yasukawa, T., Kanei-Ishii, C., Takahashi, T. and Ishii, S. (1996) CBP as a transcriptional coactivator of c-Myb. Genes Dev *10*, 528-540.

Dash, A. B., Orrico, F. C. and Ness, S. A. (1996) The EVES motif mediates both intermolecular and intramolecular regulation of c-Myb. Genes Dev *10*, 1858-1869.

Dubendorff, J. W. and Lipsick, J. S. (1999) Transcriptional regulation by the carboxyl terminus of c-Myb depends upon both the Myb DNA-binding domain and the DNA recognition site. Oncogene *18*, 3452-3460.

Fong, I. C., Zarrin, A. A., Wu, G. E. and Berinstein, N. L. (2000) Functional analysis of the human RAG 2 promoter. Mol Immunol *37*, 391-402.

Frampton, J., Gibson, T. J., Ness, S. A., Doderlein, G. and Graf, T. (1991) Proposed structure for the DNA-binding domain of the Myb oncoprotein based on model building and mutational analysis. Protein Eng *4*, 891-901.

Frampton, J., Leutz, A., Gibson, T. and Graf, T. (1989) DNA-binding domain ancestry. Nature *342*, 134.

Grant, P. A., Duggan, L., Cote, J., Roberts, S. M., Brownell, J. E., Candau, R., Ohba, R., Owen-Hughes, T., Allis, C. D., Winston, F., *et al.* (1997) Yeast Gcn5 functions in two multisubunit complexes to acetylate nucleosomal histones: characterization of an Ada complex and the SAGA (Spt/Ada) complex. Genes Dev *11*, 1640-1650.

Gu, W. and Roeder, R. G. (1997) Activation of p53 sequence-specific DNA binding by acetylation of the p53 C-terminal domain. Cell *90*, 595-606.

Hedge, S. P., Kumar, A., Kurschner, C. and Shapiro, L. H. (1998) c-Maf interacts with c-Myb to regulate transcription of an early myeloid gene during differentiation. Mol Cell Biol *18*, 2729-2737.

Hegde, S. P., Zhao, J., Ashmun, R. A. and Shapiro, L. H. (1999) c-Maf induces monocytic differentiation and apoptosis in bipotent myeloid progenitors. Blood *94*, 1578-1589.

Humphrey, G. W., Wang, Y., Russanova, V. R., Hirai, T., Qin, J., Nakatani, Y. and Howard, B. H. (2001) Stable histone deacetylase complexes distinguished by the presence of SANT domain proteins CoREST/kiaa0071 and Mta-L1. J Biol Chem *276*, 6817-6824.

Introna, M. and Golay, J. (1999) How can oncogenic transcription factors cause cancer: a critical review of the myb story. Leukemia *13*, 1301-1306.

Introna, M., Golay, J., Frampton, J., Nakano, T., Ness, S. A. and Graf, T. (1990) Mutations in v-myb alter the differentiation of myelomonocytic cells transformed by the oncogene. Cell *63*, 1289-1297.

Karafiat, V., Dvorakova, M., Pajer, P., Kralova, J., Horejsi, Z., Cermak, V., Bartunek, P., Zenke, M. and Dvorak, M. (2001) The leucine zipper region of Myb oncoprotein regulates the commitment of haemopoietic progenitors. Blood *98*, 3668-3676.

Kasper, L. H., Boussouar, F., Ney, P. A., Jackson, C. W., Rehg, J., Van Deursen, J. M. and Brindle, P. K. (2002) A transcription-factor-binding surface of coactivator p300 is required for haematopoiesis. Nature *419*, 738-743.

Kawasaki, H., Eckner, R., Yao, T. P., Taira, K., Chiu, R., Livingston, D. M. and Yokoyama, K. K. (1998) Distinct roles of the co-activators p300 and CBP in retinoic-acid-induced F9-cell differentiation. Nature *393*, 284-289.

Keough, R., Woollatt, E., Crawford, J., Sutherland, G. R., Plummer, S., Casey, G. and Gonda, T. J. (1999) Molecular cloning and chromosomal mapping of the human homologue of MYB binding protein (P160) 1A (MYBBP1A) to 17p13.3. Genomics *62*, 483-489.

Kiewitz, A. and Wolfes, H. (1997) Mapping of protein-protein interactions between c-myb and its coactivator CBP by a new phage display technique. FEBS Lett *415*, 258-262.

Kim, J. I., Ho, I. C., Grusby, M. J. and Glimcher, L. H. (1999) The transcription factor c-Maf controls the production of interleukin-4 but not other Th2 cytokines. Immunity *10*, 745-751.

Kowenz-Leutz, E., Herr, P., Niss, K. and Leutz, A. (1997) The homeobox gene GBX2, a target of the myb oncogene, mediates autocrine growth and monocyte differentiation. Cell *91*, 185-195.

Kowenz-Leutz, E. and Leutz, A. (1999) A C/EBP beta isoform recruits the SWI/SNF complex to activate myeloid genes. Mol Cell *4*, 735-743.

Kowenz-Leutz, E., Twamley, G., Ansieau, S. and Leutz, A. (1994) Novel mechanism of C/EBP beta (NF-M) transcriptional control: activation through derepression. Genes Dev *8*, 2781-2791.

Kung, A. L., Rebel, V. I., Bronson, R. T., Ch'ng, L. E., Sieff, C. A., Livingston, D. M. and Yao, T. P. (2000) Gene dose-dependent control of haemopoiesis and hematologic tumor suppression by CBP. Genes Dev *14*, 272-277.

Kwok, R. P., Lundblad, J. R., Chrivia, J. C., Richards, J. P., Bachinger, H. P., Brennan, R. G., Roberts, S. G., Green, M. R. and Goodman, R. H. (1994) Nuclear protein CBP is a coactivator for the transcription factor CREB. Nature *370*, 223-226.

Landschulz, W. H., Johnson, P. F. and McKnight, S. L. (1989) The DNA binding domain of the rat liver nuclear protein C/EBP is bipartite. Science *243*, 1681-1688.

Leverson, J. D., Koskinen, P. J., Orrico, F. C., Rainio, E. M., Jalkanen, K. J., Dash, A. B., Eisenman, R. N. and Ness, S. A. (1998) Pim-1 kinase and p100 cooperate to enhance c-Myb activity. Mol Cell *2*, 417-425.

Leverson, J. D. and Ness, S. A. (1998) Point mutations in v-Myb disrupt a cyclophilin-catalyzed negative regulatory mechanism. Mol Cell *1*, 203-211.

Lill, N. L., Grossman, S. R., Ginsberg, D., DeCaprio, J. and Livingston, D. M. (1997) Binding and modulation of p53 by p300/CBP coactivators. Nature *387*, 823-827.

Lutz, P. G., Moog-Lutz, C., Coumau-Gatbois, E., Kobari, L., Di Gioia, Y. and Cayre, Y. E. (2000) Myeloblastin is a granulocyte colony-stimulating factor-responsive gene conferring factor-independent growth to haemopoietic cells. Proc Natl Acad Sci USA *97*, 1601-1606.

Miglarese, M. R., Richardson, A. F., Aziz, N. and Bender, T. P. (1996) Differential regulation of c-Myb-induced transcription activation by a phosphorylation site in the negative regulatory domain. J Biol Chem *271*, 22697-22705.

Mink, S., Haenig, B. and Klempnauer, K. H. (1997) Interaction and functional collaboration of p300 and C/EBPbeta. Mol Cell Biol *17*, 6609-6617.

Ness, S. A. (1999) Myb binding proteins: regulators and cohorts in transformation. Oncogene *18*, 3039-3046.

Ness, S. A., Kowenz-Leutz, E., Casini, T., Graf, T. and Leutz, A. (1993) Myb and NF-M: combinatorial activators of myeloid genes in heterologous cell types. Genes Dev *7*, 749-759.

Ness, S. A., Marknell, A. and Graf, T. (1989) The v-myb oncogene product binds to and activates the promyelocyte-specific mim-1 gene. Cell *59*, 1115-1125.

Oelgeschlager, M., Janknecht, R., Krieg, J., Schreek, S. and Luscher, B. (1996a) Interaction of the co-activator CBP with Myb proteins: effects on Myb-specific transactivation and on the cooperativity with NF-M. Embo J *15*, 2771-2780.

Oelgeschlager, M., Nuchprayoon, I., Luscher, B. and Friedman, A. D. (1996b) C/EBP, c-Myb, and PU.1 cooperate to regulate the neutrophil elastase promoter. Mol Cell Biol *16*, 4717-4725.

Ogata, K., Hojo, H., Aimoto, S., Nakai, T., Nakamura, H., Sarai, A., Ishii, S. and Nishimura, Y. (1992) Solution structure of a DNA-binding unit of Myb: a helix-turn-helix-related motif with conserved tryptophans forming a hydrophobic core. Proc Natl Acad Sci USA *89*, 6428-6432.

Ogata, K., Kanei-Ishii, C., Sasaki, M., Hatanaka, H., Nagadoi, A., Enari, M., Nakamura, H., Nishimura, Y., Ishii, S. and Sarai, A. (1996) The cavity in the hydrophobic core of Myb DNA-binding domain is reserved for DNA recognition and trans-activation. Nat Struct Biol *3*, 178-187.

Ogata, K., Morikawa, S., Nakamura, H., Hojo, H., Yoshimura, S., Zhang, R., Aimoto, S., Ametani, Y., Hirata, Z., Sarai, A. and et al. (1995) Comparison of the free and DNA-complexed forms of the DNA-binding domain from c-Myb. Nat Struct Biol *2*, 309-320.

Ogata, K., Morikawa, S., Nakamura, H., Sekikawa, A., Inoue, T., Kanai, H., Sarai, A., Ishii, S. and Nishimura, Y. (1994) Solution structure of a specific DNA complex of the Myb DNA-binding domain with cooperative recognition helices. Cell *79*, 639-648.

Ogryzko, V. V., Schiltz, R. L., Russanova, V., Howard, B. H. and Nakatani, Y. (1996) The transcriptional coactivators p300 and CBP are histone acetyltransferases. Cell *87*, 953-959.

Parker, D., Rivera, M., Zor, T., Henrion-Caude, A., Radhakrishnan, I., Kumar, A., Shapiro, L. H., Wright, P. E., Montminy, M. and Brindle, P. K. (1999) Role of secondary structure in discrimination between constitutive and inducible activators. Mol Cell Biol *19*, 5601-5607.

Pfitzner, E., Kirfel, J., Becker, P., Rolke, A. and Schule, R. (1998) Physical interaction between retinoic acid receptor and the oncoprotein myb inhibits retinoic acid-dependent transactivation. Proc Natl Acad Sci U S A *95*, 5539-5544.

Radhakrishnan, I., Perez-Alvarado, G. C., Parker, D., Dyson, H. J., Montminy, M. R. and Wright, P. E. (1997) Solution structure of the KIX domain of CBP bound to the transactivation domain of CREB: a model for activator:coactivator interactions. Cell *91*, 741-752.

Ramsay, R. G., Ishii, S. and Gonda, T. J. (1991) Increase in specific DNA binding by carboxyl truncation suggests a mechanism for activation of Myb. Oncogene *6*, 1875-1879.

Ramsay, R. G., Ishii, S. and Gonda, T. J. (1992) Interaction of the Myb protein with specific DNA binding sites. J Biol Chem *267*, 5656-5662.

Ravi, R., Mookerjee, B., van Hensbergen, Y., Bedi, G. C., Giordano, A., El-Deiry, W. S., Fuchs, E. J. and Bedi, A. (1998) p53-mediated repression of nuclear factor-kappaB RelA via the transcriptional integrator p300. Cancer Res *58*, 4531-4536.

Robert, I., Sutter, A. and Quirin-Stricker, C. (2002) Synergistic activation of the human choline acetyltransferase gene by c-Myb and C/EBPbeta. Brain Res Mol Brain Res *106*, 124.

Ron, D. and Habener, J. F. (1992) CHOP, a novel developmentally regulated nuclear protein that dimerizes with transcription factors C/EBP and LAP and functions as a dominant-negative inhibitor of gene transcription. Genes Dev *6*, 439-453.

Sakaguchi, K., Herrera, J. E., Saito, S., Miki, T., Bustin, M., Vassilev, A., Anderson, C. W. and Appella, E. (1998) DNA damage activates p53 through a phosphorylation-acetylation cascade. Genes Dev *12*, 2831-2841.

Sano, Y. and Ishii, S. (2001) Increased affinity of c-Myb for CREB-binding protein (CBP) after CBP-induced acetylation. J Biol Chem *276*, 3674-3682.

Sieweke, M. H. and Graf, T. (1998) A transcription factor party during blood cell differentiation. Curr Opin Genet Dev *8*, 545-551.

Sterner, D. E., Wang, X., Bloom, M. H., Simon, G. M. and Berger, S. L. (2002) The SANT domain of Ada2 is required for normal acetylation of histones by the yeast SAGA complex. J Biol Chem *277*, 8178-8186.

Strahl, B. D. and Allis, C. D. (2000) The language of covalent histone modifications. Nature *403*, 41-45.

Tahirov, T. H., Sato, K., Ichikawa-Iwata, E., Sasaki, M., Inoue-Bungo, T., Shiina, M., Kimura, K., Takata, S., Fujikawa, A., Morii, H., *et al.* (2002) Mechanism of c-Myb-C/EBP beta cooperation from separated sites on a promoter. Cell *108*, 57-70.

Tanikawa, J., Yasukawa, T., Enari, M., Ogata, K., Nishimura, Y., Ishii, S. and Sarai, A. (1993) Recognition of specific DNA sequences by the c-myb protooncogene product: role of three repeat units in the DNA-binding domain. Proc Natl Acad Sci U S A *90*, 9320-9324.

Tavner, F. J., Simpson, R., Tashiro, S., Favier, D., Jenkins, N. A., Gilbert, D. J., Copeland, N. G., Macmillan, E. M., Lutwyche, J., Keough, R. A., *et al.* (1998) Molecular cloning reveals that the p160 Myb-binding protein is a novel, predominantly nucleolar protein which may play a role in transactivation by Myb. Mol Cell Biol *18*, 989-1002.

Tomita, A., Towatari, M., Tsuzuki, S., Hayakawa, F., Kosugi, H., Tamai, K., Miyazaki, T., Kinoshita, T. and Saito, H. (2000) c-Myb acetylation at the carboxyl-terminal conserved domain by transcriptional co-activator p300. Oncogene *19*, 444-451.

Tong, X., Drapkin, R., Yalamanchili, R., Mosialos, G. and Kieff, E. (1995) The Epstein-Barr virus nuclear protein 2 acidic domain forms a complex with a novel cellular coactivator that can interact with TFIIE. Mol Cell Biol *15*, 4735-4744.

Tsutsumi-Ishii, Y., Hasebe, T. and Nagaoka, I. (2000) Role of CCAAT/enhancer-binding protein site in transcription of human neutrophil peptide-1 and -3 defensin genes. J Immunol *164*, 3264-3273.

Turner, B. M. (2002) Cellular memory and the histone code. Cell *111*, 285-291.

Twamley-Stein, G., Kowenz-Leutz, E., Ansieau, S. and Leutz, A. (1996) Regulation of C/EBP beta/NF-M activity by kinase oncogenes. Curr Top Microbiol Immunol *211*, 129-136.

Verbeek, W., Gombart, A. F., Chumakov, A. M., Muller, C., Friedman, A. D. and Koeffler, H. P. (1999) C/EBPepsilon directly interacts with the DNA binding domain of c-myb and cooperatively activates transcription of myeloid promoters. Blood *93*, 3327-3337.

Verger, A., Perdomo, J. and Crossley, M. (2003) Modification with SUMO. EMBO Rep *4*, 137-142.

Vodicka, P., Sevcikova, S., Smardova, J., Soucek, K. and Smarda, J. (2000) The effects of RARalpha and RXRalpha proteins on growth, viability, and differentiation of v-myb-transformed monoblasts. Blood Cells Mol Dis *26*, 395-406.

Wang, D. M., Dubendorff, J. W., Woo, C. H. and Lipsick, J. S. (1999) Functional analysis of carboxy-terminal deletion mutants of c-Myb. J Virol *73*, 5875-5886.

Wang, Q. F., Lauring, J. and Schlissel, M. S. (2000) c-Myb binds to a sequence in the proximal region of the RAG-2 promoter and is essential for promoter activity in T-lineage cells. Mol Cell Biol *20*, 9203-9211.

Wang, Z., Bhattacharya, N., Weaver, M., Petersen, K., Meyer, M., Gapter, L. and Magnuson, N. S. (2001) Pim-1: a serine/threonine kinase with a role in cell survival, proliferation, differentiation and tumorigenesis. J Vet Sci *2*, 167-179.

Weston, K. (1998) Myb proteins in life, death and differentiation. Curr Opin Genet Dev *8*, 76-81.

Weston, K. (1999) Reassessing the role of C-MYB in tumorigenesis. Oncogene *18*, 3034-3038.

White, E. (2003) The pims and outs of survival signaling: role for the Pim-2 protein kinase in the suppression of apoptosis by cytokines. Genes Dev *17*, 1813-1816.

Williams, S. C., Baer, M., Dillner, A. J. and Johnson, P. F. (1995) CRP2 (C/EBP beta) contains a bipartite regulatory domain that controls transcriptional activation, DNA binding and cell specificity. EMBO J *14*, 3170-3183.

Williamson, E. A., Xu, H. N., Gombart, A. F., Verbeek, W., Chumakov, A. M., Friedman, A. D. and Koeffler, H. P. (1998) Identification of transcriptional activation and repression domains in human CCAAT/enhancer-binding protein epsilon. J Biol Chem *273*, 14796-14804.

Yao, T. P., Oh, S. P., Fuchs, M., Zhou, N. D., Ch'ng, L. E., Newsome, D., Bronson, R. T., Li, E., Livingston, D. M. and Eckner, R. (1998) Gene dosage-dependent embryonic development and proliferation defects in mice lacking the transcriptional integrator p300. Cell *93*, 361-372.

You, A., Tong, J. K., Grozinger, C. M. and Schreiber, S. L. (2001) CoREST is an integral component of the CoREST- human histone deacetylase complex. Proc Natl Acad Sci U S A *98*, 1454-1458.

Yu, J., Li, Y., Ishizuka, T., Guenther, M. G. and Lazar, M. A. (2003) A SANT motif in the SMRT corepressor interprets the histone code and promotes histone deacetylation. EMBO J *22*, 3403-3410.

Zemanova, K. and Smarda, J. (1998) Oncoprotein v-Myb and retinoic acid receptor alpha are mutual antagonists. Blood Cells Mol Dis *24*, 239-250.

Zhang, D. E., Zhang, P., Wang, N. D., Hetherington, C. J., Darlington, G. J. and Tenen, D. G. (1997) Absence of granulocyte colony-stimulating factor signaling and neutrophil development in CCAAT enhancer binding protein alpha-deficient mice. Proc Natl Acad Sci USA *94*, 569-574.

Zhang, J., Kalkum, M., Chait, B. T. and Roeder, R. G. (2002) The N-CoR-HDAC3 nuclear receptor corepressor complex inhibits the JNK pathway through the integral subunit GPS2. Mol Cell *9*, 611-623.

Zhang, Y., Ng, H. H., Erdjument-Bromage, H., Tempst, P., Bird, A. and Reinberg, D. (1999) Analysis of the NuRD subunits reveals a histone deacetylase core complex and a connection with DNA methylation. Genes Dev *13*, 1924-1935.

Chapter 13

TARGET GENES OF V-MYB AND C-MYB

Karl-Heinz Klempnauer
Institut für Biochemie, Universität Münster, Wilhelm Klemm Str. 2, D48149 Münster, Germany.

Abstract: Following the observation that Myb act as a *bona fide* transcription factor a number of experimental strategies have been used to search for the direct target genes of v-Myb and c-Myb in haematopoietic cells. To date a substantial number of such genes have been identified. The picture that has emerged from this work implies that Myb performs a complex dual role in the haematopoietic system. On one hand Myb seems to support cellular differentiation by activating the expression of genes that are part of specific haematopoietic differentiation programmes, while on the other hand Myb appears to control proliferation and survival of cells by affecting the expression of genes with known roles in these processes. Analysis of the molecular mechanisms by which Myb affects gene expression has shown that it binds directly to promoter or enhancer regions of most of its targets. In addition, dissection of the Myb-responsive *cis*-acting sequences has led to the identification of several transcription factors cooperating with Myb.

1. INTRODUCTION

This chapter reviews our current knowledge of the genes whose expression is regulated by v-Myb and c-Myb in cells of the haemopoietic system. Historically, the idea that Myb proteins are transcription factors was developed on the basis of three key observations. Biedenkapp et al. (1988) showed that the Myb proteins *in vitro* are able to recognise specific DNA sequences. Around the same time Weston and Bishop (1989) identified a transactivation domain in the Myb protein by assessing the transactivation potential of Gal4-Myb fusion proteins. Finally, by cloning the first Myb target gene, *mim*-1, the work of Ness et al. (1989) provided solid evidence for the function of Myb as a transcriptional activator. Following these discoveries work from many different laboratories has led to the identification of a substantial number of genes whose expression is

J. Frampton (ed.), Myb Transcription Factors: Their Role in Growth, Differentiation and Disease, 257-270.

controlled by Myb proteins. It is now generally believed that the biological functions of Myb are mediated by its effects on the expression of specific cellular target genes, nevertheless, we are only at the very beginning of understanding the roles of individual target genes and the mechanisms by which Myb, in concert with other transcriptional regulators, affects gene expression. I will first give a brief overview of the experimental systems that have been used to identify Myb regulated genes before going on to discuss individual target genes and the mechanisms of their regulation.

2. EXPERIMENTAL SYSTEMS

A number of different approaches have been used to identify genes regulated by v-Myb or c-Myb. Ness et al. (1989) used myelomonocytic cells transformed by a temperature-sensitive mutant of the avian leukaemia virus E26 to screen cDNA libraries for genes differentially expressed at the permissive and non-permissive temperatures. The virus mutant used in these studies (E26*ts21*) contained a point mutation in the Myb DNA-binding domain of the Gag-Myb-Ets fusion protein (Frykberg et al., 1988; Mölling et al., 1985) and it was expected that target genes of v-Myb would show temperature-dependent expression. This work resulted in the identification of the chicken *mim*-1 (myb-inducible myeloid-specific gene 1) as the first *bona fide* Myb target gene. Myelomonocytic cells transformed by the E26*ts21* virus were also instrumental in demonstrating Myb-dependent expression of other genes, including lysozyme (Introna et al., 1990) and *bcl*-2 (Frampton et al., 1996).

As an alternative to the use of E26*ts21* several laboratories have constructed fusion proteins of Myb and the hormone-binding domain of the human oestrogen receptor to generate conditional forms of Myb whose activity can be regulated by administration of oestrogen or tamoxifen. Burk and Klempnauer (1991) used stable expression of a conditional v-Myb/ER fusion protein in the chicken macrophage cell line HD11. Upon addition of oestrogen HD11 cells expressing v-Myb/ER are converted to an immature phenotype resembling v-Myb transformed myeloblasts. This phenotypic shift is fully reversible following withdrawal of the hormone. Although this system has been used quite extensively to identify genes whose expression is up-regulated by Myb the molecular mechanism by which the hormone ligand regulates the activity of the fusion protein is not clear. Subcellular fractionation experiments have shown that a substantial portion of the v-Myb/ER protein localises to the nucleus even in the absence of ligand (Burk and Klempnauer, 1991). It is not known whether binding of the fusion protein to its cognate binding site is regulated by the hormone or whether the protein binds to DNA in the absence of ligand but is unable to activate transcription under these conditions. One useful feature of the v-Myb/ER system is that activation of the fusion protein by oestrogen occurs in the absence of *de novo* protein synthesis (i.e. presynthesised inactive v-Myb/ER protein is converted to an active form after addition of hormone) and is therefore quite fast. Furthermore, activation of target gene expression by oestrogen in the presence of a protein synthesis inhibitor can give an indication whether the effect of Myb is direct or indirect.

In addition to the work mentioned above, Engelke et al. (1997) have generated a retroviral construct expressing a v-Myb/ER fusion protein and used it to infect primary chicken haemopoietic precursor cells. Similarly, a mouse retroviral vector encoding a c-Myb/ER fusion was used to infect primary haemopoietic cells (Bartley et al., 2001; Hogg et al., 1997). Finally, Lyon and Watson (1995) and Schmidt et al. (2000) generated c-Myb/ER fusion constructs that were used to derive stable transfectants in murine erythroid or myeloid cell lines. A somewhat different strategy of obtaining a conditional c-Myb protein has been employed by Kathy Weston and colleagues. They constructed a dominant interfering variant of c-Myb by fusing its DNA-binding domain to the repressor domain of the *Drosophila* Engrailed protein and the hormone-binding domain of the human oestrogen receptor. The resulting protein functions as an oestrogen-dependent repressor of Myb-inducible genes and was used to investigate gene regulation by Myb in the T-cell lineage (Badiani et al., 1994; Taylor et al., 1996). The system has also been used to investigate Myb-dependent gene regulation in myelomonocytic cells (Schmidt et al., 2000).

Several groups have used stable cell transfectants constitutively expressing Myb as a means to demonstrate that Myb is capable of activating candidate target genes. However, in general, constitutive expression systems are not as useful since there can be considerable variation in the level of expression of genes between stable transfectants, necessitating the analysis of many independent clones.

Lastly, searches of known promoter sequences for Myb binding sites in conjunction with transient reporter gene assays have been used extensively to identify Myb-regulated genes. The fact that the Myb recognition motif is quite short and somewhat variable combined with abnormally high concentrations of transactivator proteins associated with the transient transfection procedure means that this approach is problematic if it stands alone.

At present, Myb has been implicated in the regulation of a relatively large number of genes (Table 1). Since the identification of these genes has been based on a variety of approaches, some more stringent and reliable than others, it is likely that some of those listed in Table 1 will turn out not to be regulated by Myb under physiological conditions. It might therefore be useful to consider the criteria that should be met by a candidate gene in order for it to be classified as a genuine Myb target. A reasonable suggestion is that Myb-dependent expression of the candidate target gene should be demonstrated in its native, chromatin-embedded form, preferably by using a conditional Myb expression system. The application of siRNA technology should also facilitate a definitive demonstration of Myb-dependent regulation. Supporting evidence should be provided, such as the analysis of the relevant Myb binding sites by *in vitro* binding studies, *in vivo* chromatin-immunoprecipitation experiments or the analysis of reporter gene constructs.

3. MYB TARGET GENES

A substantial number of Myb targets have been identified using many of the experimental systems and strategies described above and these are summarised in Table 1. To date, much of the work on the identification of Myb target genes has concentrated on the cells of the myelomonocytic lineage since these are natural targets for transformation by oncogenic versions of the protein.

In many cases Myb-dependent expression has been demonstrated at the level of the endogenous gene. Furthermore, for most of these genes binding sites for Myb have been identified and shown to mediate Myb-dependent expression. Taken together, this work allows several interesting conclusions to be drawn. Although Myb is expressed in the more immature cells of most haemopoietic lineages and therefore might have been expected to regulate primarily genes that are expressed in all of these cells, it is clear that Myb activates many genes that are only expressed in a specific haemopoietic lineage. Examples include genes such as *mim*-1, *tom*-1, *gbx-2*, lysozyme and others which are only expressed in the myelomonocytic lineage as well as several lymphoid-specific genes such as *CD4*, *rag*-2 and T cell receptor δ (see Table 1 for references to specific genes). In addition to regulating genes with an expression pattern specific to one or a restricted number of haemopoietic lineages, Myb also activates several genes that are expressed in many different haemopoietic cell types or in non-haemopoietic cells, including c-*kit*, *CD34*, c-*myc*, and *bcl*-2.

An interesting conclusion from this survey of Myb target genes is that they can be loosely grouped into two classes. One class of genes appears to be differentiation-specific and are very likely not involved in controlling cell proliferation or survival, rather they are expressed during differentiation and are important for certain specific functions of the differentiated cell. Typical examples of this class of genes include *mim*-1, which is specifically activated in granulocytic cells and encodes a chemotactic protein secreted from the cells (Bischoff et al., 2001), and the lysozyme gene, whose product is involved in the degradation of bacterial cell walls by granulocytes or macrophages. The second class of Myb target genes encode proteins that are known to play a role in the control of cell proliferation or survival, such as c-Myc, Bcl-2, c-Kit, cyclin A1 and DNA topoisomerase IIα. Interestingly, several of the genes in this second group are expressed in all haemopoietic lineages and in many cases even outside of the haemopoietic system, suggesting that their activation represents a function of Myb that is widely utilised.

Table 1: Known or suspected Myb target genes.

Gene	Target gene criteria I[a]	II[b]	III[c]	IV[d]	Cofactors	Reference
Myelomonocytic						
mim-1	yes	yes	yes	no	C/EBP	Ness et al., 1989
lysozyme	yes	no	no	no	C/EBP	Burk et al., 1993 Introna et al., 1990
tom-1	yes	yes	yes	no	C/EBP, Ets	Burk et al., 1997 Burk and Klempnauer, 1999
MD-1	yes	no	no	no	unknown	Burk and Klempnauer, 1991
gbx-2	yes	no	no	no	unknown	Kowenz-Leutz et al., 1997
C/EBPβ	yes	no	yes	no	C/EBPβ	Mink et al., 1999
CD13/APN[e]	no	yes	yes	no	Ets	Hedge et al., 1998 Shapiro, 1995
neutrophil elastase	no	yes	yes	no	C/EBP, PU.1	Oelgeschlager et al., 1996 Verbeek et al., 1999
myeloblastin	yes	yes	yes	no	C/EBPδ	Lutz et al., 2001
myeloperoxidase	no	yes[f]	yes	no	CBF	Britos-Bray and Friedman, 1997
adenosine receptor 2B[g]	yes	yes	yes	no	unknown	Kattmann and Klempnauer, 2002 Worpenburg et al., 1997
cyclin A1[h]	yes	yes	yes	no	unknown	Müller et al., 1999
T lymphoid						
rag-2[i]	no	yes	yes	yes	unknown	Wang et al., 2000
CD4	no	yes	yes	no	unknown	Siu et al., 1992
TCR δ	no	yes[f]	yes	no	CBF	Hernandez-Munain and Krangel, 1994
pre-TCR α	no	yes[f]	yes	no	unknown	Reizis and Leder, 2001
lck	no	yes	yes	no	Ets	McCraken et al., 1994
adenosine deaminase	no	yes[j]	yes	no	unknown	Ess et al., 1995
Haemopoietic						
Pdcd4	yes	yes	yes	no	unknown	Schlichter et al., 2001a Schlichter et al., 2001b
CD34[k]	yes	yes	yes	no	unknown	Melotti et al., 1994
c-kit[k]	yes	yes	yes	no	Ets	Hogg et al., 1997 Ratajczak et al., 1998
Many						
bcl-2	yes	yes	yes	no	unknown	Frampton et al., 1996 Taylor et al., 1996
c-myc[l]	yes	yes	yes	no	unknown	Cogswell et al., 1993 Hogg et al., 1997 Schmidt et al., 2000
HSP70	no	no[m]	yes	no	unknown	Foos et al., 1993 Kanai-Ishii et al., 1994
DNA topoisomerase IIα	no	yes	yes	no	unknown	Brandt et al., 1997
thrombospondin-2	yes	no[n]	no[n]	no[n]	unknown	Bein et al., 1998

Table footnote: [a]Endogenous gene activation demonstrated; [b]Myb binding sites identified in gene promoter; [c]Activation of a reporter gene in transfection assays; [d]Chromatin immunoprecipitation demonstrated; [e]Myeloid progenitor cells; [f]Myb binding sites in enhancer; [g]Also expressed in erythroid cells; [h]Associated with AML; [i]Also expressed in B-cells; [j]Myb binding sites in thymic locus control region; [k]Expressed in progenitor cells; [l]Not regulated by Myb in all proliferating cells; [m]Binding site-independent mechanism; [n]posttranscriptional regulation.

It therefore seems that Myb functions in two ways in haemopoietic cells: (i) It supports cellular differentiation by activating genes whose expression is turned on as part of a differentiation process, and (ii) It controls proliferation and survival by regulating the expression of genes with known or suspected roles in these processes. In this way, Myb may perform the role of decision-maker, depending on the regulatory input it receives, between a proliferative programme and differentiation. This scenario raises several interesting questions. First, how widely applicable is this proposed dual role for Myb? Unfortunately, our knowledge of the Myb target genes in most haemopoietic lineages apart from myelomonocytic cells is too sparse at present to clearly answer this question. The activation by Myb of genes in T cells, such as *rag*-2, *CD4* and TCR δ, suggests that it can also control lineage-specific genes with specialised functions in differentiated lymphoid cells. Second, what is the critical difference between the oncogenic and non-oncogenic versions of Myb in terms of their influence on proliferation/survival- or differentiation-related target genes? Is the transformation by Myb more related to its effect on proliferation-relevant genes or to the regulation of differentation-associated genes? Intuition argues for the first possibility, however, this raises an additional question, namely why does Myb not transform cells of all haemopoietic lineages if it controls the expression of a set of genes relevant for proliferation in multiple haemopoietic lineages?

At present, we do not have clear answers to these questions. Because of the nature of the Myb target genes known so far it appears reasonable to conclude that Myb is intimately involved in the control of differentiation, proliferation and survival of cells. The biological effects of Myb, such as the transformation of myelomonocytic cells by oncogenic forms of Myb, are therefore probably due to changes in the expression of a large number of genes, affecting all of these processes, rather than to altered expression of one or a few master transforming genes. One would predict that certain aspects of the transformed phenotype might be traced back to particular target genes activated by Myb. This has indeed been shown for the cytokine-independent growth properties of avian myeloblastosis virus (AMV) transformed monoblasts. Activation of the gene encoding the homeobox protein Gbx-2 by AMV v-Myb results in growth factor independence of the transformed cells, presumably through a Gbx-2-mediated increase in the expression of the myelomonocytic growth factor cMGF (Kowenz-Leutz et al., 1997). Another Myb-regulated gene which is interesting in this respect encodes myeloblastin, which has also been shown to confer factor-independent growth to haemopoietic cells (Lutz et al., 2000; Lutz et al., 2001). The *gbx-2* gene also represents an example of a gene whose activation by c-Myb differs compared to that elicited by an oncogenic derivative. Hence, c-Myb activates *gbx-2* expression only in conjunction with an additional signalling event whereas AMV v-Myb activates the gene constitutively (Kowenz-Leutz et al., 1997).

4. MECHANISMS OF GENE ACTIVATION BY MYB PROTEINS

Insight into how Myb actually activates target genes has been obtained especially in the case of those genes that it regulates in myelomonocytic cells. Cooperating transcription factors have been identified in several instances, including C/EBP, c-Ets-1 and CBF, and in some cases clear evidence for combinatorial control with Myb has been obtained. For example, Myb activates the *mim*-1 gene by cooperating with a member of the C/EBP transcription factor family (Burk et al., 1993; Ness et al., 1993). Within the haemopoietic system C/EBP family members are highly expressed only in the myelomonocytic lineage, thus explaining why *mim*-1 is activated by Myb only in cells of this lineage. A similar requirement for a cooperating C/EBP factor has been de*mo*nstrated for several myelomonocyte-specific target genes, including lysozyme, tom-1, C/EBPβ itself, neutrophil elastase and myeloblastin. This suggests that cooperation between Myb and C/EBP family members might be responsible for the activation of a battery of myelomonocyte-specific genes. In most cases, the promoters of these target genes have been shown to contain juxtaposed Myb and C/EBP binding sites (Burk et al., 1993; Ness et al., 1993; Mink et al., 1996). It has been proposed that Myb and C/EBP, when bound to these promoters, communicate via direct protein-protein-interactions (Mink et al., 1996; Tahirov et al., 2002; Verbeek et al., 1999) or through additional proteins such as the coactivator p300/CBP (Mink et al., 1997).

Aside from C/EBP factors, other transactivators that have been implicated in the activation of Myb target genes include Ets family members in the case of *tom*-1 (Burk and Klempnauer, 1999), *CD13/APN* (Shapiro, 1995), c-*kit* (Ratajczak et al., 1998) and *lck* (McCraken et al., 1994) and CBF in the case of the myeloperoxidase gene (Britos-Bray and Friedman, 1997) and the T cell receptor δ gene (Hernandez-Munain and Krangel, 1994). A particularly interesting example is the *CD13/APN* gene, which is cooperatively regulated by Myb and c-Ets-1. Expression of *CD13/APN* decreases during differentiation, apparently because c-Maf, a bZip factor that is up-regulated during differentiation, binds to Myb and thereby inhibits the cooperation between Myb and c-Ets-1 (Hedge et al., 1998).

Cooperation between Myb and additional transcription factors has been demonstrated predominantly in the regulation of genes whose expression is restricted to a certain haemopoietic lineage, the clearest example being between Myb and C/EBP in the control of genes expressed in myelomonocytic cells. The effect of Myb on these genes generally appears to be mediated by one or a few Myb binding sites juxtaposed to binding sites for the cooperating factor. Typical examples are the promoters of the *mim*-1 and *tom*-1 genes whose structure is illustrated schematically in Figure 1. A significantly different promoter organisation is exemplified by the adenosine receptor 2B and Pdcd4 genes that were shown to contain a large number of potential Myb binding sites (Kattmann and Klempnauer, 2002; Schlichter et al., 2001b). A similar situation has been reported for the c-*myc* gene (Nakagoshi et al., 1992; Cogswell et al., 1993). In these cases cooperating transcription factors have not been identified, suggesting that such genes are not activated combinatorially by Myb and lineage-specific transcription factors but rather by multiple Myb molecules binding to the same promoter region. Such an activation mechanism could easily explain how Myb activates a gene in several different or all haemopoietic lineages.

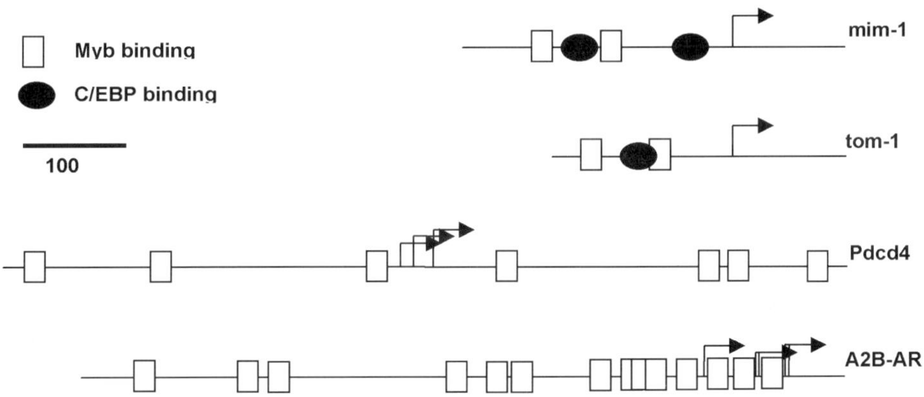

Figure 1

Schematic structure of the promoters of the mim-1, tom-1, Pdcd4 and adenosine receptor 2B (A2B-AR) genes. The transcriptional start sites are indicated by arrows.

There are also several examples of genes that are activated by Myb via upstream enhancing elements, including the TCR δ gene (Hernandez-Munain and Krangel, 1994), the pre-TCR α gene (Reizis and Leder, 2001), the adenosine deaminase (ADA) gene (Ess et al., 1995) and the myeloperoxidase gene (Britos-Bray and Friedman, 1997). An interesting example is presented by the adenosine deaminase gene whose expression in cortical thymocytes is dependent on a locus control region (LCR) located in an intron of the gene. The ADA LCR contains a Myb binding site whose integrity is required for the function of this element (Ess et al., 1995).

Although Myb binding sites have been implicated in the activation of most of the known Myb target genes it appears that Myb is also able to activate certain promoters by a binding site independent mechanism (Foos et al., 1993; Kanai-Ishii et al., 1994; Klempnauer et al., 1989). Such a mechanism has been demonstrated for the promoter of the HSP70 gene. Different promoter elements, such as the TATA-box (Foos et al., 1993) or the heat-shock element (Kanai-Ishii et al., 1994), have been implicated in the effect of Myb on the HSP70 promoter. The physiological meaning of this activation mechanism is unclear, as it has been demonstrated so far only in transfection studies using artificial promoter constructs.

Finally, it should be noted that some genes that have been identified as Myb targets apparently are regulated indirectly by Myb. Activation of the MD-1 gene by Myb requires ongoing protein synthesis, suggesting that its activation is not direct but depends on the prior stimulation of another gene whose product then activates MD-1 (Burk and Klempnauer, 1991). Expression of the thrombospondin-2 gene appears to be regulated by Myb not on the transcriptional level but post-transcriptionally by stabilisation of its RNA (Bein et al., 1998).

5. OPEN QUESTIONS

At present we are just beginning to understand the role of c-Myb in the regulation of cellular gene expression. Many questions remain open and need to be addressed by future work. Although a substantial number of Myb target genes have been identified and characterised we still do not understand how oncogenic forms of Myb transform myelomonocytic cells. As pointed out above, transformation by v-Myb is probably brought about by the deregulation of a large set of target genes affecting cell proliferation, differentiation and survival. Although several studies point to a strong correlation between transcriptional activation of target genes and oncogenic transformation the evidence for a causal relationship is only circumstantial and has been questioned (Lipsick and Wang, 1999).

We are also just beginning to understand how c-Myb proteins are integrated into the transcriptional machinery of the cell and the network of nuclear factors. For example, it would be very interesting to know if Myb cooperates with lineage-specific partners in all haemopoietic lineages or only in myelomonocytic cells. Better knowledge of Myb target genes in different haemopoietic lineages will be necessary to address these questions. Furthermore, effects of c-Myb on the chromatin structure of its targets and its interplay with chromatin remodelling factors have not been studied in any detail.

Finally, one pressing problem waiting to be solved concerns the role of the other members of the Myb family, namely A-Myb and B-Myb, in the regulation of the Myb target genes identified so far. Since the DNA binding specificity of the known Myb family members appear to be rather similar, it is *a priori* not clear whether the genes that have been identified as targets of v-Myb and c-Myb are also regulated by A-Myb or B-Myb. Most of the approaches used to identify Myb-regulated genes probably do not distinguish between targets for different family members and for some genes, such as *mim*-1, *MD*-1 and lysozyme, it has actually been shown that they are also activated by A-Myb (Foos et al., 1994). Therefore the identification and characterisation of targets for A-Myb and B-Myb is a very important goal for future work.

REFERENCES

Badiani, P., Corbella, P., Kioussis, D., Marvel, J. and Weston, K. (1994) Dominant interfering alleles define a role for c-Myb in T-cell development. Genes Dev. *8*, 770-782.

Bartley, P.A., Lutwyche, J.K. and Gonda, T.J. (2001) Identification and validation of candidate Myb target genes. Blood Cells Mol Disease *27*, 409-415

Bein, K., Ware, J.A. and Simons, M. (1998) Myb-dependent regulation of thrombospondin 2 expression. Role of mRNA stability. J. Biol. Chem. *273*, 21423-21429.

Biedenkapp, H., Borgmeyer, U., Sippel, A.E. and Klempnauer, K.-H. (1988) Viral *myb* oncogene encodes a sequence-specific DNA-binding activity. Nature *355*, 835-837.

Bischoff, K.M., Pishko, E.J., Genovese, K.J., Crippen, T.L., Holtzapple, C.K., Stanker, L.H., Nisbet, D.J. and Kogut, M.H. (2001) Chicken mim-1 protein, P33, is a heterophil chemotactic factor present in Salmonella enteritidis immune lymphokine. J. Food Prot. *64*, 1503-1509.

Brandt, T.L., Fraser, D.J., Leal, S., Halandras P.M., Kroll, A.R. and Kroll, D.J. (1997) c-Myb trans-activates the human DNA topoisomerase IIα gene promoter. J. Biol. Chem. *272*, 6278-6284.

Britos-Bray, M. and Friedman, A,D. (1997) Core binding factor cannot synergistically activate the myeloperoxidase proximal enhancer in immature myeloid cells without c-Myb. Mol. Cell. Biol. *17*, 5127-5135.

Burk, O. and Klempnauer, K.-H. (1991) Estrogen-dependent alterations in differentiation state of myeloid cells caused by a v-*myb*/estrogen receptor fusion protein. EMBO J. *10*, 3713-3719.

Burk, O., Mink, S., Ringwald, M. and Klempnauer, K.-H. (1993) Synergistic activation of the chicken *mim*-1 gene by v-*myb* and C/EBP transcription factors. EMBO J. *12*, 2027-2038.

Burk, O., Worpenberg, S., Haenig, B. and Klempnauer, K.-H. (1997) tom-1, a novel v-Myb target gene expressed in AMV- and E26-transformed myelomonocytic cells. EMBO J. *16*, 1371-1380.

Burk, O. and Klempnauer, K.-H. (1999) Myb and Ets transcription factors cooperate at the *myb*-inducible promoter of the *tom*-1 gene. Biochim. Biophys. Acta. *1446*, 243-252.

Chen, J and Bender, T.P. (2001) A novel system to identify Myb target promoters in Friend murine erythroleukemia cells. Blood Cells Mol. Disease *27*, 429-436.

Cogswell, J.P., Cogswell, P.C., Kuehl, W.M., Cuddihy, A.M., Bender, T.M., Engelke, U., Marcu, K.B. and Ting, J.P. (1993) Mechanism of c-*myc* regulation by c-Myb in different cell lineages. Mol. Cell. Biol. *13*, 2858-2869.

Ess, K.C., Whitaker, T.L., Cost, G.J., Witte, D.P., Hutton, J.J. and Aronow, B.J. (1995) A central role for a single c-Myb binding site in a thymic locus control region. Mol. Cell. Biol. *15*, 5707-5715,

Engelke, U., Wang, D.-M. and Lipsick, J.S. (1997) Cells transformed by a v-Myb-estrogen receptor fusion differentiate into multinucleated giant cells. J. Virol. *71*, 3760-3766.

Foos, G., Natour, S. and Klempnauer, K.-H. (1993) TATA-box dependent trans-activation of the human HSP70 promoter by Myb proteins. Oncogene *8*, 1775-1782.

Foos, G., Grimm, S. and Klempnauer, K.-H. (1994) The chicken A-*myb* protein is a transcriptional activator. Oncogene *9*, 2481-2488.

Frampton, J., Ramqvist, T. and Graf, T. (1996) v-Myb of E26 leukemia virus up-regulates *bcl*-2 and suppresses apoptosis in myeloid cells. Genes Dev. *10*, 2720-2731.

Frykberg, L., Metz, T., Brady, G., Introna, M., Beug, H., Vennstrom, B. and Graf, T. (1988) A point mutation in the DNA binding domain of the v-*myb* oncogene of E26 virus confers temperature sensitivity for transformation of myelomonocytic cells. Oncogene Res. *3*, 313-332.

Hedge, S.P., Kumar, A., Kurschner, C. and Shapiro, L.H. (1998) c-Maf interacts with c-Myb to regulate transcription of an early myeloid gene during differentiation. Mol. Cell. Biol. *18*, 2729-2737.

Hernandez-Munain, C. and Krangel, M.S. (1994) Regulation of the T-cell receptor delta enhancer by functional cooperation between c-Myb and core-binding factors. Mol. Cell. Biol. *14*, 473-483.

Hogg, A., Schirm, S., Nakagoshi, H., Bartley, P., Ishii, S., Bishop, J.M. and Gonda, T.J. (1997) Inactivation of a c-Myb/estrogen receptor fusion protein in transformed primary cells leads to granulocyte/macrophage differentiation and down regulation of c-*kit* but not c-*myc* or *cdc2*. Oncogene *15*, 2885-2898.

Introna, M., Golay, J., Frampton, J., Nakano, T., Ness, S.A. and Graf, T. (1990) Mutations in v-*myb* alter the differentiation of myelomonocytic cells transformed by the oncogene. Cell *63*, 1287-1297.

Kanai-Ishii, C., Yasukawa, T., Morimoto, R. and Ishii, S. (1994) c-Myb induced trans-activation mediated by heat shock elements without sequence-specific DNA-binding of c-Myb. J. Biol. Chem. *269*, 15768-15775.

Kattmann, D. and Klempnauer, K.-H. (2002) Identification and characterization of the Myb-inducible promoter of the chicken adenosine receptor 2B gene. Oncogene *21*, 3076-3081.

Klempnauer, K.-H., Arnold, H. and Biedenkapp, H. (1989) Activation of transcription by v-*myb*: evidence for two different mechanisms. Genes Dev. *3*, 1582-1589.

Kowenz-Leutz, E., Herr, P., Niss, K. and Leutz, A. (1997) The homeobox gene GBX2, a target of the *myb* oncogene, mediates autocrine growth and monocyte differentiation. Cell *91*, 185-195.

Lipsick, J.S. and Wang, D.-M. (1999) Transformation by v-Myb. Oncogene *18*, 3047-3055.

Lutz, P.G., Moog-Lutz, C., Coumau-Gatbois, E., Kobari, L., Di Gioia, Y and Cayre, Y.E. (2000) Myeloblastin is a granulocyte colony-stimulating factor-responsive gene conferring factor-independent growth to haemopoietic cells. Proc. Natl Acad. Sci. USA *97*, 1601-1606.

Lutz, P.G., Houzel-Charavel, A., Moog-Lutz, C. and Cayre, Y.E. (2001) Myeloblastin is a Myb target gene: mechanisms of regulation in myeloid leukaemia cells growth arrested by retinoic acid. Blood *97*, 2449-2456.

Lyon, J.J. and Watson, R.J. (1995) Conditional inhibition of erythroid differentiation by c-Myb/estrogen receptor fusion proteins. Differentiation *59*, 171-178.

McCracken, S., Leung, S., Bosselut, R., Ghysdael, J. and Miyamoto, N.G. (1994) Myb and Ets related transcription factors are required for activity of the human *lck* type I promoter. Oncogene *9*, 3609-3615.

Melotti, P., Ku, D.-H. and Calabretta, B. (1994) Regualtion of the expression of the haemopoietic stem cell antigen CD34: role of c-*myb*. J. Exp. Med. *179*, 1023-1028.

Mink, S., Kerber, U. and Klempnauer, K.-H. (1996) Interaction of C/EBPbeta and v-Myb is required for synergistic activation of the *mim*-1 gene. Mol. Cell. Biol. *16*, 1316-1325.

Mink, S., Haenig, B. and Klempnauer, K.-H. (1997) Interaction and functional collaboration of p300 and C/EBPbeta. Mol. Cell. Biol. *17*, 6609-6617.

Mink, S., Jaswal., S., Burk, O. and Klempnauer, K.-H. (1999) The c-Myb oncoprotein activates C/EBPbeta expression by stimulating an autoregulatory loop at the C/EBPbeta promoter. Biochim. Biophys. Acta. *1447*, 175-184.

Moelling, K, Pfaff, E., Beug, H., Beimling, P., Bunte, T., Schaller, H. and Graf, T. (1985) DNA-binding activity is associated with purified Myb proteins from AMV and E26 viruses and is temperature-sensitive for E26 ts mutants. Cell *40*, 983-990.

Müller, C., Yang, R., Idos, G., Tidow, N., Diederichs, S., Koch, O.M., Verbeek, W., Bender, T.P. and Koeffler, P.H. (1999) c-*myb* transactivates the human cyclin A1 promoter and induces cyclin A1 gene expression. Blood *94*, 4255-4262.

Nakagoshi, H., Kanei-Ishii, C., Sawazaki, T., Mizuguchi, G. and Ishii, S. (1992) Transcriptional activation of the c-*myc* gene by the c-*myb* and B-*myb* gene products. Oncogene *7*, 1233-1240.

Ness, S.A., Marknell., A. and Graf, T. (1989) The v-*myb* oncogene product binds to and activates the promyelocyte-specific *mim*-1 gene. Cell *59*, 1115-1125.

Ness, S.A., Kowenz-Leutz, E., Casini, T., Graf, T. and Leutz, A. (1993) Myb and NF-M: combinatorial activators of myeloid genes in heterologous cell types. Genes Dev. *7*, 749-759.

Oelgeschlager, M., Nuchprayoon, I., Luscher, B. and Friedman, A.D. (1996) C/EBP, c-Myb, and PU.1 cooperate to regulate the neutrophil elastase promoter. Mol. Cell. Biol. *16*, 4717-4725.

Ratajczak, M.Z., Pernotti, D., Melotti, P., Powzaniuk, M., Calabretta, B., Onodera, K., Kregenow, D.A., Machalinski, B. and Gewirtz, A.M. (1998) Myb and *ets* proteins are candidate regulators of c-*kit* expression in human haemopoietic cells. Blood *91*, 1934-1946.

Reizis, B. and Leder, P. (2001) The upstream enhancer is necessary and sufficient for the expression of the pre-T cell receptor α gene in immature T lymphocytes. J. Exp. Med. *194*, 979-990.

Schmidt, M., Nazarov, V., Stevens, L., Watson, R. and Wolff, L. (2000) Regulation of the resident chromosomal copy of c-*myc* by c-Myb is involverd in myeloid leukemogenesis. Mol. Cell. Biol. *20*, 1970-1981.

Schlichter, U., Burk., O., Worpenberg, S. and Klempnauer, K.-H. (2001a) The chicken Pdcd4 gene is regulated by v-Myb. Oncogene *20*, 231-239.

Schlichter, U., Kattmann, D., Appl., H., Miethe, J., Brehmer-Fastnacht, A. and Klempnauer, K.-H. (2001b) Identification of the *myb*-inducible promoter of the chicken Pdcd4 gene. Biochim. Biophys. Acta. *1520*, 99-104.

Siu, G., Wurster, A.L., Lipsick, J.S. and Hedrick, S.M. (1992) Expression of the CD4 gene requires a Myb transcription factor. Mol. Cell. Biol. *12*, 1592-1604.

Shapiro, L.H. (1995) Myb and Ets proteins cooperate to transactivate an early myeloid gene. J. Biol. Chem. *270*, 8763-8771.

Tahirov, T.H., Sato, K., Ichikawa-Iwata, E., Sasaki, M., Inoue-Bungo, T., Shiina, M., Kimura, K., Takata, S., Fujikawa, A., Morii, H., Kumasaka, T., Yamamoto, M., Ishii, S. and Ogata,

K. (2002) Mechanism of c-Myb-C/EBP beta cooperation from separated sites on a promoter. Cell *108*, 57-70.

Taylor, D., Badiani, P. and Weston, K. (1996) A dominant negative interfering Myb mutant causes apoptosis in T cells. Genes Dev. *10*, 2732-2744.

Verbeek, W., Gombart, A.F., Chumakov, A.M., Muller, C., Friedman, A.D. and Koeffler, H.P. (1999) C/EBPepsilon directly interacts with the DNA binding domain of c-*myb* and cooperatively activates transcription of myeloid promoters. Blood *93*, 3327-3337.

Wang, Q.F., Lauring, J. and Schlissel, M.S. (2000) c-Myb binds to a sequence in the proximal region of the RAG-2 promoter and is essential for promoter activity in T-lineage cells. Mol. Cell. Biol. *20*, 9203-9211.

Weston, K. and Bishop, J.M. (1989) Transcriptional activation by the v-*myb* oncogene and its cellular progenitor, c-*myb*. Cell *58*, 85-93.

Worpenberg, S., Burk, O. and Klempnauer, K.-H. (1997) The chicken adenosine receptor 2B gene is regulated by v-*myb*. Oncogene *15*, 213-221.

Chapter 14

THE MICROARRAY BIG BANG

Genome-Scale Identification of Myb Regulated Genes

Scott A. Ness

Department of Molecular Genetics and Microbiology, 915 Camino de Salud NE, MSC08 4660, University of New Mexico Health Sciences Center, Albuquerque, New Mexico, 87131-0001, United States of America.

Abstract: Microarrays provide a high-throughput means of identifying Myb-regulated target genes. They also provide a rich assay for measuring changes in Myb protein activity that goes far beyond the qualitative and less relevant results obtained using conventional plasmid-based reporter gene assays. Although there are several significant drawbacks to using these assays, they offer important advantages that can help dissect the complex transcriptional activities displayed by Myb proteins in various cell types.

1. INTRODUCTION

The Myb proteins are DNA-binding transcription factors that regulate the expression of other genes. Identifying those genes, and their roles in proliferation and differentiation, is the central problem in understanding how Myb proteins function in normal and transformed cells. Characterisation of Myb-regulated genes, their promoters and their regulation has led to numerous breakthroughs regarding Myb proteins. These include the identification of high and low affinity binding sites for c-Myb and v-Myb (Ness et al., 1989), the identification of other transcription factors that cooperate with Myb proteins to regulate specific genes (Burk et al., 1993; Mink et al., 1996; Ness et al., 1993), the realisation that minor point mutations in v-Myb alter its ability to regulate specific genes (Introna et al., 1990; Kowenz-Leutz et al., 1997) and the identification of intramolecular auto-inhibitory mechanisms and conformational changes that control the activity of c-Myb (Dash et al., 1996; Leverson and Ness, 1998). These early results set the stage for current studies in numerous laboratories that focus on the biological activities of Myb proteins and the post-translational

J. Frampton (ed.), Myb Transcription Factors: Their Role in Growth, Differentiation and Disease, 271-278.

modifications and interactions with other cofactors that regulate Myb proteins in a variety of cell types. Despite these successes, identification of the biologically relevant target genes that are regulated by Myb proteins has remained the key bottleneck in the Myb field (Ness, 1996; Ness, 1999). It has been particularly difficult to identify the most important Myb-regulated genes in human cells, and to link their regulation to disease processes involving Myb proteins. The advent of microarrays and other high-throughput technologies have changed this equation, suddenly empowering researchers with the ability to identify dozens, if not hundreds of Myb-regulated genes in a single experiment. Here, I will discuss the impact of microarray assays on the study of Myb proteins, and the implications of the first microarray-based results on our understanding of the functions of Myb transcription factors. Although many of the topics discussed here apply equally to Myb proteins in animals and plants, this review will focus only on the Myb proteins and target genes in vertebrates.

2. MYB TARGET GENES: A TRICKLE BECOMES A FLOOD

The first Myb-regulated genes were identified in immature chicken haemopoietic cells transformed by the v-Myb oncoprotein (Burk et al., 1997; Nakano and Graf, 1992; Ness et al., 1989). Since then, a large number of genes from several vertebrate species have been identified that are regulated by A-Myb, B-Myb, c-Myb or v-Myb (Ness, 1996). Nearly all identified target genes have high affinity Myb binding sites in their promoters, although some genes have enhancers containing Myb sites (Hernandez-Munain and Krangel, 1995; Lauzurica et al., 1997). Some promoters that lack Myb binding sites are regulated through indirect mechanisms (Kanei-Ishii et al., 1997; Kanei-Ishii et al., 1994; Klempnauer et al., 1989). Amongst the known Myb-regulated genes, perhaps the most significant are *mim-1*, the first identified and best characterised Myb target gene, which is activated by c-Myb but not v-Myb (Ness et al., 1989), the gene encoding the homeobox transcription factor Gbx2, which is activated by v-Myb but not c-Myb (Kowenz-Leutz et al., 1997), the cell cycle-regulated oncogene c-*myc* (Cogswell et al., 1993; Kumar et al., 2003; Nakagoshi et al., 1992), the gene encoding the cell cycle regulator Cdc2 (Ku et al., 1993), the genes encoding the important haemopoietic cell surface receptors c-Kit and CD34 (Chu and Besmer, 1995; He et al., 1992; Hogg et al., 1997; Melotti and Calabretta, 1994; Melotti et al., 1994) and the anti-apoptotic gene *bcl-2* (Frampton et al., 1996; Taylor et al., 1996; Heckman et al., 2000; Thompson et al., 1998).

The analysis of these and other Myb target genes has led to several important conclusions. First, Myb proteins participate in the regulation of many genes that are expressed in tissue specific patterns. For example, the *mim-1* gene is expressed only in immature myeloid cells, not in other cells that express c-Myb. Thus, Myb proteins must cooperate with other transcription factors, expressed in overlapping tissue-specific patterns, to regulate specific genes. Second, slight changes in Myb proteins have dramatic effects on their ability to regulate specific genes. For example, point mutations in the DNA binding domain of v-Myb permit it to activate the *gbx-2* gene, which c-Myb cannot do. The opposite is true for *mim-1*, which c-Myb, but not v-Myb, can activate. The mutations probably alter protein-protein interactions that are crucial for determining the specificity of the Myb transcription factors. The implication is that other slight differences, such as post-translational modifications, could cause significant differences in the ability of Myb proteins to regulate specific genes in different cellular environments (Ness, 2003). Finally, the analysis of *bona fide* Myb target genes has led to a distinction between the rather promiscuous ability of Myb proteins to activate the transcription of plasmid-based artificial promoters containing Myb binding sites in nearly any cell type, and the much more limited ability of Myb proteins to activate the transcription of chromatin-embedded natural genes only in the correct context. For example, although c-Myb and v-Myb have different activities on the endogenous *mim-1* gene, both are able to activate transcription of the isolated *mim-1* promoter when it is in the context of a plasmid-based reporter construct. Thus, although transfected reporter genes are convenient, they yield results that may fail to represent the complexity and specificity of Myb transcription factors observed with chromosomal genes. As a consequence, confirming that a gene can be regulated by Myb proteins *in vivo* is often a much more stringent and more relevant test that merely demonstrating that an isolated promoter containing Myb binding sites fused to a reporter gene in a plasmid can be activated by co-expressed Myb proteins.

3. USING MICROARRAYS TO ASSESS MYB PROTEIN ACTIVITIES

Microarray assays of gene expression offer researchers an expanded tool set for following the activities of Myb proteins or other transcription factors. Essentially, screening microarrays provides a means of following the expression of up to 30,000 endogenous genes at a time. Thus, if two different Myb proteins, for example v-Myb and c-Myb, were expressed in cells, a subsequent microarray assay should be able to detect all the changes in gene expression that occur as a result. This approach offers several advantages, but also has several drawbacks, when compared to more conventional plasmid-based reporter gene assays.

3.1 Reporter genes versus endogenous genes

Unlike plasmid-based reporter gene assays, microarray experiments measure changes in the expression of endogenous genes, expressed in their normal chromatin context. This is extremely important, as recent studies have implicated a variety of enzymes that induce changes in chromatin structure, such as histone deacetylases, methylases and ubiquitin ligases, in the regulation of transcription. By studying the genes in their normal chromatin context, it is more likely the genes will be regulated normally, so the results will have greater biological relevance. An additional factor concerns gene copy number. In transfection experiments using plasmid-based reporter genes, the transfected cells often contain hundreds of copies of the promoter being studied. In such a situation, any limiting factors will be titrated away, making the results much less relevant to the normal situation. Since Myb proteins often act in a combinatorial manner with other transcription factors (Ness et al., 1993), a vast overabundance of promoters is sure to upset the normal balance of DNA binding proteins that should be interacting.

Studying the regulation of endogenous genes also has significant drawbacks. Perhaps the biggest problem concerns the populations of cells being studied. In typical transfection assays, reporter plasmids and expression vectors are introduced into a relatively small fraction of the cells in a culture dish. However, since the assay only follows the reporter plasmids, the fact that 95% or more of the cells remain untransfected is irrelevant since they are simply invisible in the assay. If such a culture were used for microarray assays, only about 5% of the cells would have been transfected by plasmids expressing, for example, c-Myb. A microarray assay could be used to follow changes in gene expression, but the vast majority of untransfected cells would interfere. On average, a gene induced 10-fold by c-Myb in the 5% of cells that were transfected would appear to be up-regulated only about 0.5-fold in the entire culture. There are two ways to overcome this problem. One approach is to use an expression vector that also expresses a fluorescent-detectable marker, such as Green Fluorescent Protein (GFP). This allows the GFP-expressing transfected cells to be enriched, for example by flow sorting, before preparing RNA for the microarray studies. Two drawbacks are that the flow sorting could induce changes in gene expression and GFP is toxic in some cell types. Another approach is to use recombinant adenoviruses or some other more efficient means of expressing the Myb proteins in the cells of interest. The drawbacks are that constructing multiple adenoviruses can be quite labour-intensive, the viruses do not infect all cell types equally, they are poorly-suited for long-term assays and they can induce changes in gene expression on their own. Nevertheless, we recently used recombinant adenoviruses to successfully study the effects of Myb proteins in MCF7 cells (Rushton et al., 2003).

3.2 The first lessons from microarray assays using Myb proteins

To test the usefulness of microarray assays for the identification of Myb-regulated target genes, our laboratory used this technology to characterise the changes in gene expression induced by infecting MCF7 mammary epithelial cells with recombinant adenoviruses expressing A-Myb, B-Myb or c-Myb proteins. After 16 hours of infection, RNA was isolated and used for preparing fluorescently-tagged cRNA probes that were hybridised to Affymetrix U95 genome arrays, containing probe sets for approximately 65,000 human genes (Rushton et al., 2003). The major result from these experiments was the demonstration that overexpression of each Myb protein led to the activation of a distinct set of human genes. Thus, each protein interacted with the cellular milieu in a unique way to activate a specific set of endogenous genes. Although some genes were activated by more than one Myb protein, most of the genes activated by A-Myb were not affected by expression of B-myb or c-Myb, or vice-versa. However, all three Myb proteins can bind the same DNA sequence and activated the same plasmid-based reporter genes in transfection assays (Rushton and Ness, 2001). Thus, although all three Myb proteins have nearly identical DNA binding domains, and bind the same plasmid-based reporters, they had completely different effects on the expression of endogenous genes.

The results described above confirm that plasmid-based reporter genes fail to measure the differences in Myb protein activity that are evident when endogenous target gene activation is assayed. The results also confirm that the A-Myb, B-Myb and c-Myb proteins have unique effects on gene expression, consistent with their different biological functions. Thus, the results with microarray assays are much more complex, and more closely reflect the biological differences characteristic of the different Myb proteins.

4. WHERE WILL MICROARRAY ASSAYS LEAD?

Microarray assays provide two different advantages to researchers studying Myb proteins, or other transcription factors. First, they provide a high-throughput means of identifying potential target genes that are regulated, either directly or indirectly, by changes in Myb protein expression or activity. In the example cited above, ectopic expression of A-Myb, B-Myb or c-Myb led to significant changes in expression of over 400 endogenous genes, or roughly 0.6% of the genes represented on the microarrays (Rushton et al., 2003). Some of these are certain to be regulated directly by Myb proteins that bind to their promoters, but many others are likely to be regulated indirectly. Thus, the gene expression patterns represent fingerprints characteristic of different Myb protein activities. The fact that these fingerprints are so extensive belies the importance of Myb transcription factors, and their ability to induce numerous gene expression changes in cells.

The second advantage that microarray-based assays of gene expression offer is as a replacement for standard plasmid-based reporter gene assays. Although there exist significant drawbacks to this approach, the ability of microarray assays to measure activities that more closely mimic the biological activities of Myb proteins is significant. The microarray studies cited above uncovered a depth and complexity in Myb protein function that was not detected using standard assays. Dissecting the molecular mechanisms at work to create that complexity remains a major challenge in the Myb protein field.

REFERENCES

Burk, O., Mink, S., Ringwald, M. and Klempnauer, K.-H. (1993) Synergistic activation of the chicken mim-1 gene by v-myb and C/EBP transcription factors. EMBO J *12*, 2027-2038.

Burk, O., Worpenberg, S., Haenig, B. and Klempnauer, K. H. (1997) tom-1, a novel v-Myb target gene expressed in AMV- and E26-transformed myelomonocytic cells. EMBO Journal *16*, 1371-1380.

Chu, T. Y. and Besmer, P. (1995) Characterization of the promoter of the proto-oncogene c-kit. Proceedings of the National Science Council, Republic of China - Part B, Life Sciences *19*, 8-18.

Cogswell, J. P., Cogswell, P. C., Kuehl, W. M., Cuddihy, A. M., Bender, T. M., Engelke, U., Marcu, K. B. and Ting, J. P. (1993) Mechanism of c-myc regulation by c-Myb in different cell lineages. Mol Cell Biol *13*, 2858-2869.

Dash, A.B., Orrico, F.C. and Ness, S.A. (1996) The EVES motif mediates both intermolecular and intramolecular regulation of c-Myb. Genes Dev *10*, 1858-1869.

Frampton, J., Ramqvist, T. and Graf, F. (1996) v-Myb of E26 leukemia virus up-regulates bcl-2 and suppresses apoptosis in myeloid cells. Genes Dev *10*, 2720-2731.

He, X. Y., Antao, V. P., Basila, D., Marx, J. C. and Davis, B. R. (1992) Isolation and molecular characterization of the human CD34 gene. Blood *79*, 2296-2302.

Heckman, C. A., Mehew, J. W., Ying, G. G., Introna, M., Golay, J. and Boxer, L. M. (2000) A-Myb up-regulates Bcl-2 through a Cdx binding site in t(14;18) lymphoma cells. J Biol Chem *275*, 6499-6508.

Hernandez-Munain, C. and Krangel, M. S. (1995) c-Myb and core-binding factor/PEBP2 display functional synergy but bind independently to adjacent sites in the T-cell receptor delta enhancer. Mol Cell Biol *15*, 3090-3099.

Hogg, A., Schirm, S., Nakagoshi, H., Bartley, P., Ishii, S., Bishop, J. M. and Gonda, T. J. (1997) Inactivation of a c-Myb/estrogen receptor fusion protein in transformed primary cells leads to granulocyte/macrophage differentiation and down regulation of c-kit but not c-myc or cdc2. Oncogene *15*, 2885-2898.

Introna, M., Golay, J., Frampton, J., Nakano, T., Ness, S. A. and Graf, T. (1990) Mutations in v-myb alter the differentiation of myelomonocytic cells transformed by the oncogene. Cell *63*, 1287-1297.

Kanei-Ishii, C., Tanikawa, J., Nakai, A., Morimoto, R. I. and Ishii, S. (1997) Activation of heat shock transcription factor 3 by c-Myb in the absence of cellular stress. Science *277*, 246-248.

Kanei-Ishii, C., Yasukawa, T., Morimoto, R. I. and Ishii, S. (1994) c-Myb-induced trans-activation mediated by heat shock elements without sequence-specific DNA binding of c-Myb. J Biol Chem *269*, 15768-15775.

Klempnauer, K. H., Arnold, H. and Biedenkapp, H. (1989) Activation of transcription by v-myb: evidence for two different mechanisms. Genes Dev *3*, 1582-1589.

Kowenz-Leutz, E., Herr, P., Niss, K. and Leutz, A. (1997) The homeobox gene GBX2, a target of the myb oncogene, mediates autocrine growth and monocyte differentiation. Cell *91*, 185-195.

Ku, D. H., Wen, S. C., Engelhard, A., Nicolaides, N. C., Lipson, K. E., Marino, T. A. and Calabretta, B. (1993) c-myb transactivates cdc2 expression via Myb binding sites in the 5'-flanking region of the human cdc2 gene. J Biol Chem *268*, 2255-2259.

Kumar, A., Lee, C. M., and Reddy, E. P. (2003). C-Myc is essential but not sufficient for c-Myb-mediated block of granulocytic differentiation. J Biol Chem. *278*, 11480-11488.

Lauzurica, P., Zhong, X. P., Krangel, M. S. and Roberts, J. L. (1997) Regulation of T cell receptor delta gene rearrangement by CBF/PEBP2. J Exp Med *185*, 1193-1201.

Leverson, J.D. and Ness, S.A. (1998) Point Mutations in v-Myb Disrupt a Cyclophilin-Catalyzed Negative Regulatory Mechanism. Mol Cell *1*, 203-211.

Melotti, P. and Calabretta, B. (1994) Ets-2 and c-Myb act independently in regulating expression of the hematopoietic stem cell antigen CD34. J Biol Chem *269*, 25303-25309.

Melotti, P., Ku, D. H. and Calabretta, B. (1994) Regulation of the expression of the hematopoietic stem cell antigen CD34: role of c-myb. J Exp Med *179*, 1023-1028.

Mink, S., Kerber, U. and Klempnauer, K.-H. (1996) Interaction of C/EBPb and v-Myb is required for synergistic activation of the mim-1 gene. Mol Cell Biol *16*, 1316-1325.

Nakagoshi, H., Kanei-Ishii, C., Sawazaki, T., Mizuguchi, G. and Ishii, S. (1992) Transcriptional activation of the c-myc gene by the c-myb and B-myb gene products. Oncogene *7*, 1233-1240.

Nakano, T. and Graf, T. (1992) Identification of genes differentially expressed in two types of v-myb-transformed avian myelomonocytic cells. Oncogene *7*, 527-534.

Ness, S. A. (2003) Myb Protein Specificity: Evidence of a Context-Specific Transcription Factor Code. Blood Cells Mol Dis in press.

Ness, S. A., Kowenz-Leutz, E., Casini, T., Graf, T. and Leutz, A. (1993) Myb and NF-M: Combinatorial activators of myeloid genes in heterologous cell types. Genes Dev *7*, 749-759.

Ness, S. A., Marknell, Å. and Graf, T. (1989) The v-myb oncogene product binds to and activates the promyelocyte-specific mim-1 gene. Cell *59*, 1115-1125.

Ness, S.A. (1996) The myb oncoprotein: regulating a regulator. BBA Reviews on Cancer *1288*, F123-F139.

Ness, S.A. (1999) Myb Binding Proteins: Regulators and Cohorts in Transformation. Oncogene *18*, 3039-3046.

Rushton, J. J., Davis, L. M., Lei, W., Mo, X., Leutz, A. and Ness, S. A. (2003) Distinct changes in gene expression induced by A-Myb, B-Myb and c-Myb proteins. Oncogene *22*, 308-313.

Rushton, J. J. and Ness, S. A. (2001) The Conserved DNA Binding Domain Mediates Similar Regulatory Interactions for A-Myb, B-Myb, and c-Myb Transcription Factors. Blood Cells Mol Dis *27*, 459-463.

Taylor, D., Badiani, P. and Weston, K. (1996) A dominant interfering Myb mutant causes apoptosis in T cells. Genes Dev *10*, 2732-2744.

Thompson, M. A., Rosenthal, M. A., Ellis, S. L., Friend, A. J., Zorbas, M. I., Whitehead, R. H. and Ramsay, R. G. (1998) c-Myb down-regulation is associated with human colon cell differentiation, apoptosis, and decreased Bcl-2 expression. Cancer Res *58*, 5168-5175.

Chapter 15

THE V-MYB ONCOGENE

Two Models for Activation

Fan Liu and Scott A. Ness
Department of Molecular Genetics and Microbiology, 915 Camino de Salud NE, MSC08 4660, University of New Mexico Health Sciences Center, Albuquerque, New Mexico 87131-0001, Unites States of America.

Abstract: The *v-myb* oncogenes, the oncogenic components of two different avian leukaemia viruses, encode proteins that are mutated and truncated versions of c-Myb. Although derived from chicken c-*myb* gene, the biological effects of v-Myb are strikingly different from that of c-Myb. While c-Myb is essential for haemopoietic development, overexpression of v-Myb transforms cytokine-dependent immature haemopoietic cells in culture and induces acute leukaemias in animals. This has led to the speculation that v-Myb specific mutations and truncations unmask the normally latent transforming activity of c-Myb. In this chapter, we critically review some important aspects of v-Myb including its transforming activities, haemopoietic specificity and the structure and function of the oncoprotein. Our analysis will emphasise the molecular mechanisms of how v-Myb specific mutations and truncations lead to its oncogenic activation.

1. INTRODUCTION

The v-*myb* oncogene has been identified twice, as the oncogenic components of two different avian leukaemia viruses: Avian Myeloblastosis Virus (AMV) and E26, both of which induce acute leukaemias in chickens and transform immature haemopoietic cells in tissue culture. The v-*myb* oncogenes are truncated, mutated and constitutively expressed versions of c-*myb*, the founding member of a large family of transcription factors characterized by a highly conserved DNA binding domain (Introna et al., 1990; Lipsick, 1996; Lipsick and Wang, 1999; Ness, 1996; Oh and Reddy, 1999). All Myb proteins are DNA-binding transcription factors that are localised predominantly in the nucleus and that regulate the expression of genes involved in growth control and differentiation. The c-Myb

J. Frampton (ed.), Myb Transcription Factors: Their Role in Growth, Differentiation and Disease, 279-306.

transcription factor usually plays an important role in regulating differentiation and proliferation in a variety of cell types. However, the v-Myb proteins are mutated and although the nature of the mutations required for converting c-Myb into a transforming protein have been known for some time, recent evidence has shed new light on the consequences of such mutations. Here, we will review the functional evidence in order to better understand the transforming activity of v-Myb and other transcription factor oncogenes, and we will investigate two models of how v-Myb is activated by mutations: quantitative changes that disrupt negative regulatory mechanisms to create an activated version of c-Myb versus qualitative changes that generate a v-Myb transcription factor with unique activities.

2. THE ORIGINS OF V-MYB: AMV AND E26

2.1 AMV v-Myb

Avian Myeloblastosis Virus (AMV) was isolated in 1939 from two 11-week old chickens with Marek's disease, a T cell lymphoma caused by an avian herpes virus (Lipsick and Wang, 1999). AMV is able to induce acute myeloblastoid leukaemias in chickens and of transforming immature myeloid cells derived from bone marrow or yolk sac in tissue culture. Subsequent molecular cloning and sequence analysis of the viral genomes revealed that AMV is likely to be derived from Myeloblastosis Associated Virus type 1 (MAV-1) (Kan et al., 1985) and has a typical retroviral genomic structure including a long terminal repeat region (LTR) present at both ends of the proviral DNA and intact genes encoding group specific antigens (Gag) and reverse transcriptase (Pol). In AMV, most of the envelope gene (*env*) except the last 11 codons are replaced by the v-*myb* oncogene. The loss of the *env* gene renders the AMV virus replication defective, so it can only replicate in cells that are co-infected with an intact helper virus that complements its deficiencies. The AMV virus encodes two major transcripts. A full-length genomic mRNA that encodes Gag and Pol and that is packaged in the infectious virions and a sub-genomic spliced mRNA that encodes a protein containing the first six amino acids from the N-terminus of Gag fused to v-Myb.

AMV v-*myb* is derived from the chicken c-*myb* gene, which has at least 17 exons and encodes a 75 kDa nuclear phosphoprotein, while AMV v-*myb* gene includes 1199 base pairs from c-*myb*: 86 bp from the intron between exons 3 and 4, all of the region represented by exons 4 to 9 plus 120 bp of c-*myb* exon 10 (Baluda and Reddy, 1994). Therefore, the truncated v-Myb protein encoded by AMV is only 48 kDa and lacks 71 amino acids at its amino terminus, which are replaced by six amino acids from Gag, as well as 199 amino acids at the carboxyl terminus. Furthermore, in the region homologous to c-Myb, the v-Myb of AMV contains ten (or eleven in some clones) substitutions, which are thought to strongly affect the phenotype of the transformed cells and the transforming ability of the virus (Lipsick and Wang, 1999).

2.2 E26: Gag-Myb-Ets

The E26 virus, which also encodes a version of the v-*myb* oncogene, was isolated several years after the discovery of AMV (Oh and Reddy, 1999). The E26 virus encodes a single tripartite mRNA transcript in which the retroviral gag gene at the N-terminus is fused to the central v-*myb* gene, which is in turn fused to a second oncogene, v-*ets*. The fusion transcript encodes a 135,000 kDa Gag-Myb-Ets fusion protein. As will be discussed below, both the Myb and Ets portions of this unique fusion oncoprotein contribute to the transforming properties of E26. Since part of the *gag* gene and all of the *pol* and *env* genes are missing, E26 is also replication defective and needs a helper virus for formation of infectious virions.

2.3 Comparisons to c-Myb

The major translational product of the c-*myb* proto-oncogene is a 75 kDa nuclear protein with 636 amino acids (or 89 kDa with an additional 121 amino acids from alternative splicing site in some cases) (Oh and Reddy, 1999). Similar to other transcription factors, all the Myb proteins have a modular structure, so that domains responsible for various functions like DNA binding or transcriptional transactivation can be exchanged between proteins. Full-length c-Myb consists of an N-terminal DNA binding domain, a centrally-located transactivation domain, and a C-terminal negative regulatory domain (Sakura et al., 1989). Whereas, in the v-Myb from either AMV or E26, the very N-terminus of the protein, part of the DNA binding domain, and the C-terminal domain is conspicuously absent. In addition, v-Myb of AMV contains a number of acquired point mutations that contribute to its oncogenic activity (Weston, 1990). These amino acid substitutions can be placed into three groups on the basis of the functional domains in which they occur: four substitutions occur within the highly conserved DNA binding domain, three of them have proved to be important for the interactions between Myb and other regulatory proteins, two substitutions are located within a proline-rich putative hinge region, and the last three substitutions reside in the transactivation domain (Dini et al., 1995).

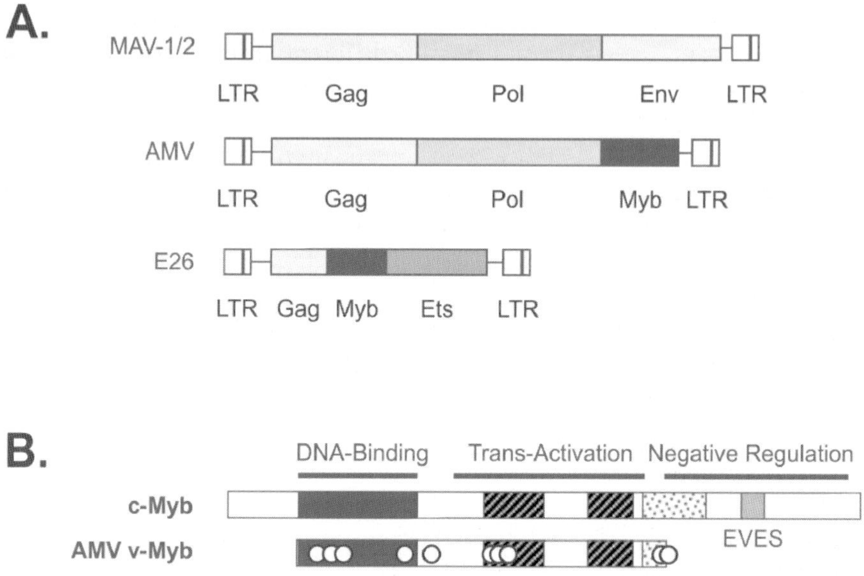

Figure 1

Structures of the Myb retroviruses and oncoproteins. (A) Genomic structure of parent retrovirus MAV-1/2, Avian Myeloblastosis Virus (AMV) and E26. The various viral gene products are labeled and indicated by shading. (B) The structure of vertebrate c-Myb protein is compared to v-Myb protein encoded by AMV leukaemia retrovirus. The v-Myb protein is truncated at both ends relative to c-Myb, and has 10 amino acid substitutions in the region shared with c-Myb. The functional domains of c-Myb described in text have been labelled above the diagram. The AMV specific residues are indicated in the regions where they occur.

3. TRANSFORMATION BY V-MYB

The *c-myb* gene is normally highly expressed in immature, proliferating haemopoietic cells and its expression declines as cells stop proliferating and undergo terminal differentiation. This pattern implies an important role for *c-myb* in haemopoietic cell differentiation and proliferation. In support of this conclusion, homozygous c-*myb* mutant mice produced by gene knock-out techniques died *in utero* by day 15 with a severe anaemia resulting from failure of foetal liver erythropoiesis and myelopoiesis (Mucenski et al., 1991), and anti-sense oligonucleotides that block c-Myb protein expression disrupted haemopoietic cell differentiation *in vitro* and blocked the proliferation of cultured T-cells (Gewirtz et al., 1989; Gewirtz and Calabretta, 1988). Although c-*myb* genes were first discovered as the cellular counterpart of retroviral oncogenes, a role for c-Myb in transformation or oncogenesis remains obscure. For example, the constitutive expression of full-length c-Myb can block the induced differentiation of haemopoietic cells (Clarke et al., 1988), suggesting that c-Myb has the potential to block differentiation and contribute to transformation. However, the normal c-Myb protein has little or no transforming activity (Gonda et al., 1989; Todokoro et al., 1988) and ectopic overexpression of full-length c-Myb in transgenic mice leads to numerous abnormalities but does not induce tumours or leukaemias (Furuta et al., 1993), suggesting that the c-Myb protein is not oncogenic. Analysis of both *in vivo* and *in vitro* transformation by mutants of c-Myb implied that truncation of either the N- and/or C-terminus is required for tumourigenesis, and that the additional amino acid substitutions in v-Myb increased its oncogenic activity (Dini et al., 1995; Grässer et al., 1991). These results have led to a model in which the deletions and mutations that distinguish v-Myb are required to unmask the latent transforming activity of c-Myb, which is otherwise under tight negative regulation. Thus, c-Myb can transform immature haemopoietic cells that are cultured in optimized conditions (Fu and Lipsick, 1997), suggesting that activated cytokine signalling pathways may be able to activate or enhance the otherwise repressed transforming activity of wild type c-Myb. The combined results could be interpreted to mean that haemopoietic cell transformation results from subtle shifts in the balance of pathways leading to differentiation or proliferation, and that c-Myb likely plays a role in regulating the ratio of these two competing cell fates.

3.1 The v-*myb* Oncogenes Transform Primary, Cytokine-Dependent Cells

Unlike many other oncogenes, the v-Myb proteins are unable to transform fibroblasts by inducing anchorage-independent growth or reduced serum requirements. Instead, they are only able to transform immature haemopoietic cells that remain dependent on specific cytokines for their survival and proliferation. For example, immature haemopoietic cells isolated from chick bone marrow or yolk sac are able to form small colonies in semi-solid medium and, in the appropriate conditions, to differentiate into mature erythroid or myeloid cells. In contrast, cells transformed by AMV or E26 remain immature and proliferate, forming large colonies that can be isolated and expanded for several weeks in liquid culture. The v-Myb transformed cells do not become immortalised, but undergo senescence and stop proliferating after 4 to 8 weeks. During their brief lifespan, the transformed cells divide rapidly and remain dependent on the chicken cytokine cMGF, a distant relative of mammalian IL-6 or G-CSF (Leutz et al., 1984; Leutz et al., 1989). In colony assays using defined medium, immature cells transformed by AMV appear to be at least partially cMGF independent. However, careful analysis showed that the AMV version of v-Myb induces the expression of the cMGF gene (Kowenz-Leutz et al., 1997), allowing the cells to produce small amounts of the growth factor and to stimulate their own proliferation via an autocrine mechanism. The v-Myb oncogenes have been shown to cooperate with a variety of oncogenes that activate signal transduction pathways, all of which lead to activation of the cMGF gene (Sterneck et al., 1992a; Sterneck et al., 1992b). Furthermore, recombinant retroviruses expressing both v-Myb and cMGF are highly oncogenic in birds (Sterneck et al., 1992a; Sterneck et al., 1992b), suggesting that transformation by v-Myb requires a cMGF-activated growth or survival signal. It is clear that transformation by v-Myb is completely cMGF dependent, although the unique nature of the signal provided by cMGF has not been investigated. This remains one of the important unanswered questions regarding the biology and transforming activities of the v-Myb oncogenes.

3.2 Differences Between Transformation by E26 and AMV

The Gag-Myb-Ets fusion protein encoded by E26 includes 272 amino acids from Gag, 283 from Myb, and 491 from Ets. The v-Myb region of E26 is smaller than the one in AMV, and has only one amino acid substitution relative to c-Myb. The Ets domain is derived from c-ets, another gene that encodes a transcription factor involved in the regulation of haemopoietic cell differentiation (Blair and Athanasiou, 2000). The v-Ets domain encoded by E26 contains three amino acid mutations and differs from chick c-ets at both the 5' and 3' ends. Interestingly, since the c-*myb* and c-*ets* proto-oncogenes are present on different chromosomes, creation of the E26 virus must have required two successive recombination steps with cellular DNA, or a transcript from a pre-existing chromosomal translocation that was captured by the virus (Symonds et al., 1984).

Unlike AMV, which transforms immature myeloid cells in culture and causes an acute myeloblastosis leukaemia in animals, the E26 virus is able to transform multipotent haemopoietic progenitors and to induce leukaemia in animals involving both erythroid and myeloid cells. In fact, the E26 transformed cells can be induced to differentiate into erythrocytes, thrombocytes, eosinophils and myeloblasts (McNagny and Graf, 2002; McNagny et al., 1992). The Myb-Ets oncoprotein includes two DNA binding domains: one from c-Myb and one from c-Ets. Studies with mutants in which one of the two domains was rendered temperature-sensitive for

DNA binding suggested that a functional Myb domain is required to block both thrombocytic and macrophage differentiation, while an active Ets domain is necessary to block erythroid differentiation (Beug et al., 1984; Frampton et al., 1995; Kraut et al., 1994).

Although both Myb and Ets are oncogenic transcription factors, a fusion between these two proteins is required for the leukaemogenesis by E26 (Metz and Graf, 1991). Viral constructs expressing Myb and Ets as separate proteins failed to induce leukaemia in animals, while cells infected by this virus exhibited a phenotype different from that induced by E26. A few animals that were injected with the same virus did eventually get leukaemia, but in each case an internal deletion had occurred in the viral genome, resulting in the re-construction of novel Myb-Ets fusion proteins. The required fusion between Myb and Ets proteins for transformation by E26 suggests that the fusion viral protein has acquired unique, qualitatively different biological properties relative to its cellular counterparts.

3.3 The Transforming Activity of v-Myb is Cell Type Dependent

As described above, the AMV and E26 retroviruses transform only a limited number of specialised, primary cells. However, v-Myb has been shown to transform several other cell types under certain conditions. For example, B- and T-cell lymphomas induced by v-Myb or other v-Myb like proteins have been reported. Transgenic mice in which v-Myb (AMV) oncoprotein is expressed in a T-cell-specific fashion developed high grade T-cell lymphomas with a long period of latency in a significant portion of animals (Badiani et al., 1996). Ectopic expression of AMV v-Myb affected the ratio of helper to cytotoxic T-cells and inhibited thymic involution, suggesting that Myb protein may play an important role in the regulation of T-cell development. Insertional mutagenesis of the c-*myb* locus led to rapid-onset B-cell lymphomas in chick embryos infected with the RAV-1 leukosis virus. (Kanter et al., 1988). Moreover, transformations of non-haemopoietic cells by v-Myb have also been implicated. Oncogenic activation of c-Myb by insertional mutagenesis, in which the c-Myb protein is truncated by only 20 amino acids, induced a high incidence of sarcomas and adenocarcinomas, as well as B-cell lymphomas (Jiang et al., 1997). The v-Myb of E26, in cooperation with the erbB oncogene or high levels of EGF signaling through its receptor, transformed chick embryonic cells that most closely resembled melanocyte precursors or melanoblasts (Bell and Frampton, 1999). Attempts have been made to transform other cells, but in general the transformation capacity of v-Myb is restricted to only a few cell types.

The mechanism responsible for the cell type-specificity of v-Myb transformation is unclear. It is possible that the v-Myb protein cooperates with or requires tissue specific co-factors that effectively limit its effectiveness in other cell types. As noted above, AMV and E26 only transform cells whose growth is supported by the chicken cytokine cMGF, so the necessary co-factors may lie downstream or be activated by a cMGF-activated signaling pathway. Identification of such co-factors would undoubtedly yield important information about the oncogenicity of v-Myb and the specific mechanisms through which it transforms cells.

3.4 Cooperation Between Myb and Other Oncogenes

The tissue specificity of transformation by v-Myb suggests that its oncogenic activity depends on cooperation with other oncogenes. As noted above, activated kinase-type oncogenes can cooperate with v-Myb by activating signal transduction pathways leading to the expression of the necessary cytokine, cMGF. Another intriguing issue about v-Myb transformation is the possible cooperation between Myb and Myc. The *myc* oncogene was identified originally as the transforming component in the avian myelocytomatosis virus MC29, but has been identified in several other retroviruses as well (Prendergast, 1999a; Prendergast, 1999b). The *myc* gene encodes an oncogenic transcription factor, and when overexpressed is capable of transforming a variety of haemopoietic and other cell types. Interestingly, results with v-Myb induced T-cell lymphomas suggests the existence of cooperative links between Myb and Myc (Davies et al., 1999). When neonatal transgenic mice expressing the v-Myb protein in a T-cell-specific fashion were infected with the slow-transforming retrovirus Moloney murine leukaemia virus (M-MuLV), which causes tumours by inserting in the vicinity of cellular oncogenes, T-cell lymphomas developed with a latency of 13 weeks, compared to 60 weeks in uninfected transgenic animals. Analysis of DNA recovered from the transgenic tumours showed that the M-MuLV provirus had integrated into either the c-*myc* or the N-*myc* genes with high frequency, suggesting that v-Myb and activated c-Myc can cooperate in the genesis of T-cell lymphomas. In addition, the other v-Myb virus, E26, has also been shown to cooperate with v-Myc in transformation of chicken myelomonocytic and neuroretina cells (Amouyel et al., 1989).

The cooperative link between Myb and Myc is further supported by some indirect evidence. First: the c-*myc* gene promoter has been shown to be a target for activation by c-Myb (Cogswell et al., 1993; Evans et al., 1990). Ectopic overexpression of c-Myb activates both the endogenous, chromosomal c-*myc* gene and a plasmid-based reporter gene driven by the *myc* promoter. Second, in many haemopoietic neoplasms, both c-*myb* and c-

myc are co-overexpressed. Knocking out the activity of either gene by antisense technology reduces proliferation and viability of the cells (Gewirtz et al., 1998). Finally, if Myc and Myb do cooperate with each other, Myb seems to be an ideal candidate to compensate for the deficiencies of oncogenic c-Myc in transformation. Overexpression of c-Myc drives cells into the cell cycle, but induces apoptosis at the same time. The activated Myb proteins may act as survival factors, perhaps by activating specific genes that block apoptosis (Weston, 1999).

4. THE V-MYB PROTEIN

The v-*myb* oncogene is unique among known oncogenes in that it causes only acute leukaemia in animals and transforms only haemopoietic cells in culture, suggesting that transformation by the *myb* gene family is dependent on cell type or context, and further indicating the important role of interactions between these oncogenic transcription factors and other cell type specific co-factors in tumourigenesis. Although a variety of proteins have been identified which can interact with c-Myb or v-Myb, and some of those have been shown to alter the activity or specificity of the Myb protein, the co-factors which decide the cell type specific transformation of v-Myb remain elusive (Ness, 1999). Another extremely intriguing issue about v-Myb is the role of its mutations and truncations in oncogenesis. Compared to c-Myb, v-Myb is truncated at both ends, and in the case of AMV, it contains ten amino acid substitutions in the region shared with c-Myb. Ectopically overexpressed c-Myb is only a weak oncogene in tissue culture and fails to induce tumours in transgenic animals, suggesting that the v-Myb type mutations and truncations are required to reveal its oncogenic activity. However, the mechanisms by which these mutations and truncations convert c-Myb to a stronger transforming protein are still controversial.

4.1 The Myb DNA Binding Domain

The highly conserved DNA binding domain of c-Myb is composed of three imperfect repeats of approximately 50 amino acids each, referred to as R1, R2 and R3, each of which is a variant of the homeo domain or helix-turn-helix motif (Frampton et al., 1989; Saikumar et al., 1990). R2 and R3 together comprise the minimal DNA binding region, while R1 loosely covers the DNA position next to the R2 binding site. The solution structure of the R2R3 bound to DNA showed that the third helix of each repeat lies in the major groove of DNA, making site-specific contacts important for sequence recognition (Ogata et al., 1994; Ogata et al., 1995). In v-Myb from either

AMV or E26, two thirds of the R1 repeat has been deleted. Although the function of R1 is still obscure, deletion of R1 as well as the very N-terminus of the protein is definitely associated with the oncogenic activation of c-Myb. In addition to its DNA binding activity, the R2R3 domain of Myb is believed to form a crucial surface for interaction with other cellular regulatory proteins, and may contribute to the regulation of target gene specificity of Myb protein (Ness, 1996). For example, the DNA binding domain of AMV v-Myb contains three amino acid substitutions that are responsible for the failure of v-Myb to activate the endogenous, chromosomal *mim*-1 gene. The AMV specific residues, which lie in a hydrophobic patch within R2 region, face away from the DNA. Thus, instead of disrupting the DNA binding activity, the mutations in the AMV Myb protein may disrupt a crucial interaction surface, suggesting that interactions with other cellular proteins in its DNA binding domain must affect the transcription of some unique sets of genes by Myb proteins.

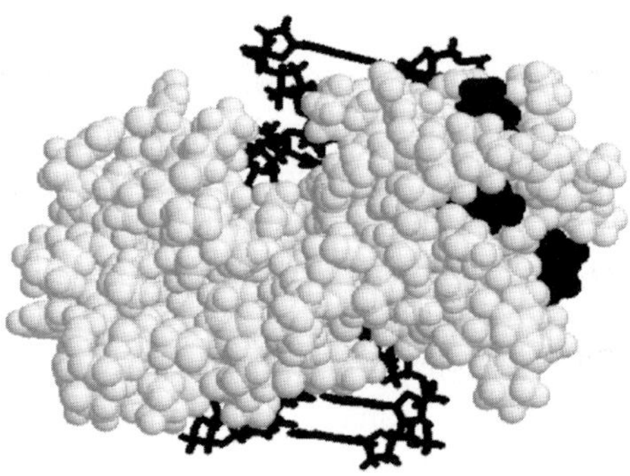

Figure 2

The solution structure of c-Myb R2R3 DNA binding domain. The c-Myb R2R3 portion of the DNA binding domain is shown as a space-filling model in white. The DNA strands, shown as sticks and mostly obscured, are in black. The AMV specific residues (I91N, L106H and V117D), which are located on the surface of R2 and face away from the DNA, are shaded black. This image was produced using the program RasMol and the published structure coordinates (Ogata et al., 1994; Ogata et al., 1995).

4.2 Other Conserved Functional Domains

The extreme N-terminus of c-Myb contains a 15 residue acidic region, 10 of which are either aspartic or glutamic acid. Similar acidic regions found in a number of other transcription factors are thought to play a role in protein-protein interactions that regulate their transcription activity (Ptashne, 1988). A site for phosphorylation by casein kinase II (CKII) has been mapped immediately upstream of this negatively charged region. Phosphorylation of c-Myb at this site has been shown to affect DNA binding and transcriptional cooperativity with NF-M, the chick version of the transcription factor C/EBP or NF-IL6 (Oelgeschläger et al., 1995). The acidic region and the CKII phosphorylation site are deleted in v-Myb.

The primary nuclear localisation signal of 29 amino acids for v-Myb is located within the R2 region of v-Myb. This highly conserved region is absolutely required for nuclear transport of v-Myb. A second motif, which lies approximately 130 amino acids downstream of the first one, may function to stabilise v-Myb in the nucleus through protein-protein interactions (Ibanez and Lipsick, 1988).

The transcriptional activation domains of both v-Myb and c-Myb are located in the middle of the proteins, downstream of the DNA binding domain (Klempnauer et al., 1989; Lane et al., 1990). A series of deletion experiments established that approximately 50 amino acids (from amino acids 204 to 254) constitutes the minimal region sufficient for v-Myb to activate promoters containing its DNA binding sites linked to report genes (Bortner and Ostrowski, 1991; Weston and Bishop, 1989). This transactivation domain contains a number of charged residues, and because of the modular structure of Myb proteins, it can function with heterologous DNA binding domains in a reporter assay, such as the DNA binding domain from budding yeast GAL4 proteins and the E.coli LexA proteins (Dubendorff et al., 1992; Kalkbrenner et al., 1990). Although this minimal transactivation region is capable of activating artificial substrates, a much larger region is required for v-Myb to transform cells, suggesting that activation of endogenous, chromosomal v-Myb specific genes is much more complicated. For example, a heptad leucine repeat, also referred as the "leucine zipper repeat", and a conserved FAETL motif at its carboxyl-terminus are also required for the transformation ability of v-Myb (Chen et al., 1995). Furthermore, the transactivation activity of v-Myb is not always correlated with its transforming ability. For example, fusing the Myb DNA binding domain to a strong transactivation domain from VP16 leads to an increased capacity of this hybrid protein to activate a reporter gene linked to a promoter containing several Myb binding sites in both animals and yeast (Chen et al., 1995; Ibanez and Lipsick, 1990). However, the same protein only moderately activates the *mim*-1 gene (SAN, unpublished) and fails to transform cells (Engelke et al., 1995). In addition, some mutations in the transactivation domain of v-Myb increase its ability to activate transcription of a reporter gene, but on the other hand decrease its ability to transform (Wang and Lipsick, 2002). All these observations suggest that the role of the transactivation domain may be complex, most likely mediating protein-protein interactions that lead to activation of reporter genes but also affecting the specificity of the Myb proteins when activating target genes in the chromatin. A transcriptional coactivator protein CBP (CREB binding protein) has been shown to interact with the transactivation domain of v-Myb as well as c-Myb, and is involved in regulation of protein activity in transformation and tumourigenesis (Dai et al., 1996; Oelgeschläger et al., 1996). CBP and the related protein, p300, serve as a transcriptional adapter for several transcription factors, such as CREB (cAMP Response Element Binding Protein), Ets gene family Spi-B and PU.1, by direct bridging between basic transcription factors and several sequence-specific transcriptional coactivators (Yamamoto et al., 2002). Therefore, CBP as well as other transcriptional adapters may function as key factors mediating positive or negative cross-talk between Myb and other transcription factors during growth and differentiation of haemopoietic cells.

Another highly studied domain in v-Myb is a heptad leucine repeat consisting of a stretch of 44 amino acids with an isoleucine and 7 leucine residues at the appropriate heptad spacing (Baluda and Reddy, 1994). This region resembles the leucine zipper structures of b-ZIP proteins, and is a part of the C-terminal negative regulatory domain of c-Myb, leading to the suggestion that the repeat could mediate hetero- or homo-dimerisation involved in negative regulation. Mutation of the third and forth leucines to prolines or alanines, which should affect either zipper secondary structure or hydrophobicity, resulted in activation of c-Myb protein activity measured both in transcriptional regulation and oncogenic transformation assays (Kanei-Ishii et al., 1992). The results are consistent with a model in which this region serves a negative regulatory function in c-Myb, but the mechanisms underlying the negative regulation are still under investigation. It has been suggested that this region might mediate the formation of Myb homodimers which are unable to bind DNA (Nomura et al., 1993). A second model proposes a role for the leucine zipper repeat in mediating interactions with other cellular proteins, such as c-Jun, leading to the negative regulation of c-Myb (Favier and Gonda, 1994). However, in the context of v-Myb, the leucine repeat, particularly the FAETL region, which is in the region of leucine repeat, appears to have a positive role for both transcriptional activity and oncogenic transformation (Fu and Lipsick, 1996). The deletion of the FAETL motif from AMV v-Myb renders the protein incapable of transactivation and transformation in culture. A more recent study indicated that the leucine zipper region may also play a role in the regulation of commitment of haemopoietic progenitors (Karafiat et al., 2001). Wild-type AMV v-Myb with the intact leucine zipper repeat directs development of progenitors into macrophage lineage. Mutations in this region compromise commitment toward myeloid cells and cause v-Myb also to support erythroid cells, thrombocytes, and granulocytes, similar to the c-Myb protein, in which the accessibility of leucine zipper repeat for binding of other regulatory proteins is under the control of intramolecular interaction between C-terminus and N-terminus of the protein. These observations suggest that Myb leucine zipper modifies the activity of the transcription factor, probably by interacting with context-specific cofactors, thereby regulating specific genes involved in regulating differentiation.

The rest of the c-Myb C-terminus is absent in both the AMV and E26 Myb proteins, leading to the suggestion that truncation of this region is required for oncogenic activation of c-Myb (Kalkbrenner et al., 1990; Sakura et al., 1989). Loss of the C-terminus correlates best with the oncogenic activation of c-Myb (Ness, 1996). In addition, the EVES domain near the C-terminus has been shown to interact with the DNA binding domain of c-Myb, leading to a model in which the activity of c-Myb is regulated by intramolecular interactions (Dash et al., 1996; Ness, 1996). However, the C-terminus of c-Myb also increases transcriptional activation in budding yeast (Chen and Lipsick, 1993), suggesting that the C-terminal domain, like the rest of the Myb protein, may have multiple and complex activities.

5. V-MYB AND C-MYB: QUALITATIVE DIFFERENCES?

In normal cells, c-Myb regulates the proliferation and differentiation of haemopoietic progenitors and the transcriptional activity of c-Myb is likely to be regulated by upstream signaling pathways. Several distinct regulatory mechanisms have been shown to affect c-Myb protein activity. For example, the N-terminal DNA binding domain can interact with the C-terminal EVES domain, suggesting that c-Myb activity is regulated by an intramolecular mechanism intrinsic to the protein itself (Dash et al., 1996). The c-Myb protein is subject to several types of post-translational modifications including acetylation (Sano and Ishii, 2001; Tomita et al., 2000) and sumoylation (Bies et al., 2002). It can also be phosphorylated at multiple sites in response to certain upstream signals, and phosphorylation of some sites has been shown to reduce either the DNA binding activity or transcriptional activity, or both (Andersson et al., 2002; Lüscher et al.,

1990). Some of these phosphorylation sites are disrupted by the amino acid substitutions acquired by v-Myb, suggesting that phosphorylation could contribute to negative regulation (Andersson et al., 2002). The same mutations also disrupt interactions with other cellular proteins, such as the peptidyl-prolyl isomerase Cyp40, which can disable the DNA binding activity of c-Myb (Leverson and Ness, 1998), and C/EBP proteins, transcription factors that cooperate with Myb to activate specific target genes like mim-1 (Ness et al., 1993; Tahirov et al., 2002). These results have generally led to a model in which c-Myb activity is auto-inhibited or negatively-regulated while the mutated and oncogenic v-Myb is constitutively activated. In this view, the activities of c-Myb and v-Myb are similar and the differences between them are quantitative in nature, since the de-regulated v-Myb is equivalent to greater activity of c-Myb. However, there is also ample evidence that the differences between c-Myb and v-Myb are qualitative in nature, and that v-Myb has a very different transcriptional specificity than c-Myb.

5.1 Disruption of Negative Regulatory Mechanisms

The c-*myb* gene is a frequent target for retrovirus insertional activation in myeloid leukaemias and T- and B- cell lymphomas in chickens and mice. Analysis of tumours with activated c-*myb* genes revealed that retroviral insertions always result in expression of N- or C-terminus truncated forms of c-Myb protein, suggesting that truncations lead to deletions of negative regulatory domains and oncogenic activation of Myb protein. These observations led to the hypothesis that both ends of the Myb protein are involved in negative regulation (Dash et al., 1996). This hypothesis was supported by data showing that the two ends of c-Myb protein can interact with one another in a yeast two-hybrid assay. The interacting domains were mapped to the conserved DNA binding domain at the N-terminus and a small C-terminal motif containing the highly conserved string of amino acids EVES (in the single-letter amino acid code). Interestingly, a p42MAPK (mitogen-activated protein kinase) phosphorylation site, serine 532 (in the chick c-Myb numbering scheme), has been mapped to the same EVES domain (Aziz et al., 1995; Aziz et al., 1993), suggesting that the intramolecular interactions mediated by the EVES domain are under control of a specific signal transduction pathway leading to p42MAPK. Although mutation of serine 532 to alanine does not convert c-Myb into a transforming protein, it does increase the transcription activity in transfection assays, suggesting that phosphorylation of the EVES domain is involved in negative regulation.

The identification of the EVES domain, which interacts with the Myb DNA binding domain, led to the finding that p100, a transcriptional coactivator with a related EVES domain, is also a Myb-binding protein (Dash et al., 1996). In addition to Myb proteins, p100 has been shown to interact with other proteins such as the EBNA2 transcriptional activator protein encoded by Epstein-Barr virus and Transcription Factor IIE (TFIIE), a component of the general transcription machinery (Tong et al., 1995). The p100 protein also interacts with the oncogenic serine/threonine kinase Pim-1 (Leverson et al., 1998). Pim-1 has been shown to phosphorylate the DNA binding domains of c-Myb and v-Myb (Winn et al., 2003), and Pim-1 and p100 cooperate to stimulate the transcriptional activity of Myb proteins in haemopoietic cells (Leverson et al., 1998; Winn et al., 2003), suggesting that p100 acts as a coactivator for Myb transcription factors. A model for autoregulation of c-Myb activity has been proposed: phosphorylation at the EVES domain stabilises interactions between the C-terminal EVES domain and the N-terminal DNA binding domain in c-Myb, rendering the protein inactive. The p100 protein mediates c-Myb function by competing with the C-terminus for interaction sites in the DNA binding domain (Dash et al., 1996; Ness, 1996). This model explains several questions related to the regulation of Myb activity by C-terminal negative regulatory domain. For example, how does phosphorylation at the C-terminus affect the activity of the DNA binding domain, located at the N-terminus?

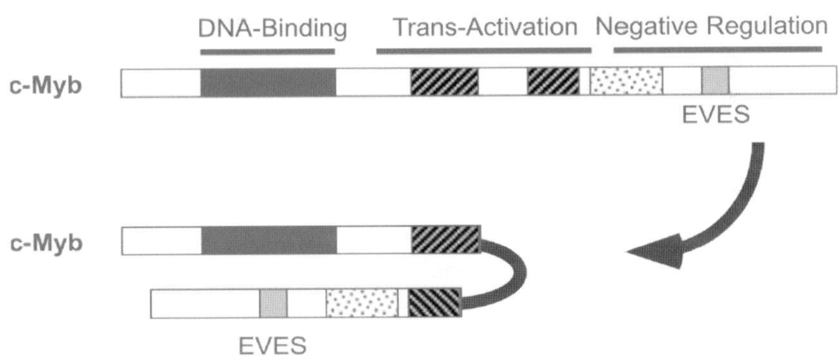

Figure 3

Model of c-Myb autoregulation. The C-terminal EVES domain in c-Myb has been shown to interact with the N-terminal DNA binding domain, suggesting that Myb could be negatively-regulated through intramolecular interactions. Such interactions could be regulated by modifications, such as phosphorylation or sumoylation, or through the binding of other regulatory proteins.

Conversely, intramolecular interactions with the N-terminal DNA binding domain would also regulate interactions between the C-terminus and other cofactors that could regulate the activity or specificity of c-Myb.

Although the EVES domain mediating the interactions between two termini provides an elegant model for Myb regulation, deletion of this region from the C-terminus is not sufficient to fully unmask the protein's transactivation and transformation potential, suggesting that it may not be the only element in the C-terminal domain that functions in the negative regulation of c-Myb. An earlier study using a series of deletions and reporter assays showed that the negative regulatory activity of the C-terminal domain resides in at least two regions spanning amino acids 428 to 462 and 499 to 558, respectively (Dubendorff et al., 1992). Negative regulation by these two sequences within the C-terminus of chicken c-Myb also functions in the context of a heterologous DNA binding domain, indicating this negative regulation is Myb DNA binding domain independent. In contrast, they further pointed out that a sequence containing part of the minimal Myb transcriptional activation domain rather than DNA binding domain is the target for negative regulation by C-terminus in trans. Thus, instead of interacting with the sequence within the DNA binding domain, the negative regulatory regions in C-terminal domain may interact directly or indirectly with the central transactivation domain of Myb protein.

5.2 Qualitative Changes in Transcriptional Activity

As transcription factors, Myb proteins exert their regulatory and transforming activity by regulating other genes. Identification of these target genes is crucial to fully understand the activities of Myb proteins. Unfortunately, although many Myb–regulated genes have been characterised so far, regulation of the identified ones cannot account for the biological importance of Myb proteins, especially the transforming activity of v-Myb (Ness, 1999).

The hallmark of all Myb related proteins is their unique DNA binding domain, composed of at least two tandem repeats which constitute the minimal DNA binding region (Frampton et al., 1989). Subtle changes in the DNA binding domain may result in dramatic differences in target gene selection. For example, c-Myb and the related proteins A- and B-Myb have nearly identical DNA binding domains. Indeed, they have been shown to bind the same or very similar DNA sequences and to activate the same reporter gene constructs in animal cells. However, when tested in northern or microarray assays, each protein induces a unique and specific set of genes (Rushton and Ness, 2001). The DNA binding domain of AMV v-Myb differs from that of c-Myb by deletion of the N-terminal R1 and four acquired point mutations in R2R3. Several types of evidence suggest that AMV specific mutations contribute to an altered target gene selection by v-Myb.

5.3 Regulation of the mim-1 Gene by Myb and C/EBP□

Perhaps the best-characterised target of Myb transcriptional regulation is the chicken *mim*-1 gene (Ness et al., 1989), which encodes a secretable component of chick promyelocyte granules that has been implicated in bone morphogenesis (Falany et al., 2001) and has been shown to have acetyltransferase activity (Allen and Hebbes, 2003). Interestingly, the bovine paralogue of *mim*-1 is *lect2*, a component of the fetal bovine serum used to grow most tissue culture cells (Yamagoe et al., 1996), raising questions about the role of *mim*-1/*lect2* in cell proliferation. The *mim*-1 gene promoter contains three Myb binding sites and, when fused to a reporter gene and tested in transfection assays, can be activated by c-Myb and v-Myb from either E26 or AMV. In contrast, only c-Myb and E26 v-Myb can activate the endogenous, chromosomal *mim*-1 gene. The contrast between the activation of the real *mim*-1 gene and the plasmid-based reporter genes has several important implications: First, the ability of Myb proteins to regulate reporter genes is not necessarily an accurate reflection of their activity on the natural target genes embedded in the chromatin; Second, regulation of the chromosomal genes may involve a more complex mechanism, perhaps requiring specific protein-protein interactions.

The DNA binding domain of AMV v-Myb contains four amino acid substitutions, which have been shown to be important for the transformed phenotype (Introna et al., 1990). The three mutations located within the R2 repeat (I91N, L106H and V117D) are responsible for the inability of AMV v-Myb to activate the natural *mim*-1 gene (Introna et al., 1990; Ness et al., 1989). Analysis of the solution structure of the Myb DNA binding domain showed that these three point mutations lie in a hydrophobic patch of amino acids which face away from the DNA and toward the solvent (Ogata et al., 1995; Ogata et al., 1994). Thus, instead of disrupting the DNA binding activity, mutations in the AMV DNA binding region may disrupt a crucial protein-protein interaction surface, suggesting that activation of the endogenous *mim*-1 gene requires cooperation between Myb and other cellular proteins. At least one cellular protein, C/EBPβ (or NF-M, the chicken version of C/EBPβ) has been identified as a co-regulator that synergises with c-Myb/E26 v-Myb in activation of the *mim*-1 gene (Burk et al., 1993; Ness et al., 1993).

The C/EBPβ (CCAAT-enhancer binding protein β) protein belongs to the b-Zip family of transcription factors, characterised by a basic DNA–binding region linked to a leucine zipper dimerization motif. The family contains three closely related members: C/EBPα, C/EBPβ and C/EBPδ (Birkenmeier et al., 1989; Kageyama et al., 1991; Landschulz et al., 1988). Members of the C/EBP family are frequent partners for Myb proteins, cooperating to regulate the transcription of a number of haemopoietic cell-specific genes such as *mim*-1 (Burk et al., 1993; Ness et al., 1993), lysozyme (Burk et al., 1997; Ness et al., 1993), neutrophil elastase (Oelgeschlager et al., 1996), myeloperoxidase (MPO) (Britos-Bray and Friedman, 1997) and *rag*-2 (Fong et al., 2000). Detailed structural analyses have shown that the C/EBPβ DNA binding domain interacts directly with the R2 region of the c-Myb DNA binding domain when the two proteins are both bound to the *mim*-1 gene promoter (Tahirov et al., 2002), and these interactions may induce looping of the intervening DNA (Tahirov et al., 2002). To assess the effect of AMV type point mutations on C/EBPβ binding activity in solution, the GST pull-down assays using c-Myb mutants with various combinations of the AMV type mutations demonstrated that the I91N and L106H each drastically impair c-Myb-C/EBPβ binding, while V117D does not. Comparison of DNA-bound c-Myb R2R3 with AMV v-Myb R2R3 revealed a significant structural difference in R2 (a positional shift of α2 helix) caused by I91N and L106H, and this structural difference would be expected to alter the C/EBPβ binding surface, impairing the interaction between C/EBPβ and the mutant c-Myb (Tahirov et al., 2002). However, some earlier evidence suggests that the AMV type mutations may not be able to completely block the interactions between Myb and C/EBPβ. For example, although AMV does not induce the chromosomal *mim*-1 gene, it synergises quite efficiently with NF-M to activate *mim*-1 promoter-driven reporter genes in transfection assays (Ness et al., 1993) (Burk et al., 1993). This cooperation would not be expected if the proteins were unable to interact. In support of this concept, AMV protein has been isolated from transformed cells in a complex with NF-M (Mink et al., 1996). Moreover, a myeloblast cell line transformed by AMV v-Myb (AMV BM2) expresses high levels of *mim*-1 (Dini et al., 1995; Mink et al., 1996), suggesting that the AMV protein can activate the transcription of *mim*-1 gene in a correct cell context. It is possible that the AMV mutations alter Myb's interactions with NF-M through an allosteric mechanism which prevents *mim*-1 gene activation without disrupting the physical contacts between the two proteins (Dash et al., 1996).

Synergistic activation of chicken *mim*-1 gene by Myb and C/EBP transcription factors provides a model that explains how subtle changes in the DNA binding domain of a transcription factor affect its ability to regulate specific genes. In addition to C/EBPβ, the Myb DNA binding domain has been reported to interact with several other factors, including p100 (Ness, 1999), HSF3 (Kanei-Ishii et al., 1997), and D-type cyclins (Ganter et al., 1998). Furthermore, some promoters of Myb target genes contain the binding sites for Ets (Hernandez-Munain and Krangel, 1994), AML1/Runx-1 (Britos-Bray and Friedman, 1997), or GATA-1(Altschul et al., 1997). Therefore, it is possible that the Myb DNA binding domain may function as a DNA/protein docking surface capable of mediating a wide variety of protein-protein interactions.

5.4 The gbx-2 Gene is Specifically Activated by v-Myb

A homeobox gene essential for normal development of the central nervous system (CNS) (Tour et al., 2001; Tour et al., 2002) encodes the chicken transcription factor Gbx-2. This protein is also an essential regulator of the gene that encodes the cytokine cMGF (Kowenz-Leutz et al., 1997), required for growth by v-Myb transformed myeloblasts. As noted above, AMV-transformed cells produce small quantities of cMGF that stimulates their growth in an autocrine fashion, suggesting that Gbx-2 is involved in transformation by AMV v-Myb. Indeed, ectopic expression of AMV v-Myb stimulates the expression of the endogenous *gbx*-2 gene (Kowenz-Leutz et al., 1997). However, neither c-Myb nor E26 v-Myb are able to activate *gbx*-2, and the important differences map to the AMV-specific substitutions in the DNA binding domain (Kowenz-Leutz et al., 1997). Interestingly, c-Myb was able to activate the *gbx*-2 gene when co-expressed with additional oncogenes, such as EJ-ras, or when the cells were treated with phorbol ester to activate signal transduction pathways. These results suggest that c-Myb requires the cooperation of additional signalling pathway-activated cofactors to stimulate *gbx*-2 expression. On the other hand, AMV v-Myb has acquired point mutations that render it able to activate *gbx*-2 in the absence of additional signals (Kowenz-Leutz et al., 1997).

The differential regulation of *gbx*-2 by v-Myb and c-Myb illustrates how the acquired mutations in v-Myb can contribute to its oncogenic potential, by relieving the requirement for additional signal activation to stimulate the expression of cMGF. However, it is not clear whether the mutations alter the transcriptional properties of v-Myb, for example by allowing it to activate genes that c-Myb is unable to, or whether the mutations merely remove the regulatory controls that prevent c-Myb from activating the *gbx*-2 gene in the absence of the additional signals.

5.5 Deletion of R1 in Alteration of Target Gene Selection

In addition to the three point mutations in DNA binding domain, other mutations or truncations, such as deletion of R1, have also been implicated in the altered expression pattern of Myb-regulated genes (Dini and Lipsick, 1993). R1, the first repeat in the c-Myb DNA binding domain, has been highly conserved throughout vertebrate evolution, both in c-Myb and in the closely related proteins A-Myb and B-Myb. Similar to the other two repeats, structure studies confirmed that R1 folds to form three helices (Ogata et al., 1995). Although the physiological function of R1 remains elusive, its deletion is associated with the oncogenic activation of c-Myb (Grässer et al., 1991). This observation raises at least one question: how is R1 able to affect the transforming ability of Myb proteins?

First, R1 might serve as a supplementary DNA binding domain, capable of stabilising the binding of c-Myb to low affinity binding sites. Truncation of R1 could render the v-Myb protein able to regulate only a subset of the genes normally regulated by the full-length c-Myb (Dini and Lipsick, 1993; Lipsick and Wang, 1999). Evidence for this hypothesis comes from the observation that oncogenic truncation of the first repeat non-specifically decreases the DNA binding activity of Myb protein (Dini and Lipsick, 1993). The authors speculated that c-Myb might have two functions, one to drive proliferation of progenitor cells and the second to permit the activation of genes required for terminal differentiation, whereas, v-Myb and other oncogenic forms of c-Myb would cause transformation by carrying out the first but not the second function of normal c-Myb (with the exception of *gbx*-2 perhaps).

Second, R1 might play little or no role in DNA binding, instead serving as a supplemental protein-binding domain (Ness, 1996). As noted above, the Myb DNA binding domains have been highly conserved and have been shown to interact with a number of cellular proteins. The extra R1 domain could allow c-Myb to interact with additional proteins involved in negative regulation, or additional cofactors that influence its target gene specificity.

These two models are not exclusive. Indeed, the R1 region could be both DNA- and protein-binding, like the rest of the DNA binding domain. Both models explain how mutations, such as truncation of R1, could affect target gene specificity and increase the transforming potential of Myb proteins. So far, since only a few genes, such as *mim*-1 (Ness et al., 1989), *tom*-1 (Burk et al., 1997), *bcl*-2 (Frampton et al., 1996), c-*kit* (Hogg et al., 1997) and *gbx*-2 (Kowenz-Leutz et al., 1997) have been identified as v-Myb (E26 and AMV) target genes, it is not possible to distinguish whether R1 participates as a protein interaction domain or an extension that affects the specificity of c-Myb.

6. QUESTIONS STILL UNANSWERED

The biggest unanswered question about v-Myb is: Why is it an oncogene? Although a number of genes that can be regulated by v-Myb have been identified, none can explain why v-Myb is able to induce leukaemias in animals and transform immature haemopoietic cells in tissue culture. It certainly seems likely that the transforming activity of v-Myb is linked to its ability to activate specific genes, so identification of additional Myb-regulated genes, especially in the immature myeloid cells that are the targets of transformation by the AMV and E26 viruses, will provide important clues about how v-Myb works. Recent microarray experiments have shown that ectopic expression of Myb transcription factors can lead to altered expression of hundreds of cellular genes (Rushton et al., 2003). Thus, it is likely that there is no single transformation-relevant target of v-Myb. Rather, v-Myb probably transforms cells by causing simultaneous changes in the expression of multiple, perhaps dozens of genes. Thus, it will be important to understand how networks of v-Myb regulated genes interact with one another and contribute in various ways to the transformed phenotype.

A second major question concerns the relationship between v-Myb and c-Myb, and how v-Myb specific mutations lead to oncogenic activation of c-Myb. A large body of evidence has shown that the mutations in v-Myb cause both qualitative and quantitative changes in activity, relative to c-Myb. As noted above, c-Myb protein is subject to several types of post-translational modifications, including phosphorylation, acetylation and sumoylation. Thus it is likely that mutations of v-Myb affect its transcriptional activity and transformation ability by altering such post-translational medications. Understanding the nature of the differences in v-Myb and c-Myb activity, and the role of the R1 region and other conserved domains, will be crucial to understanding how the normal c-Myb protein was converted to an oncogenic v-Myb.

Finally, a major question remains regarding the importance of activated Myb proteins as human oncogenes. Amplified or rearranged alleles of c-*myb* have been detected in numerous human tumours (Alitalo et al., 1984; Pelicci et al., 1984; Wallrapp et al., 1997; Welter et al., 1990; Winqvist et al., 1985). However, despite the fact that c-*myb* is expressed in nearly all leukaemias and lymphomas, and that rearranged or mutated alleles of c-*myb* induce tumours and leukaemias in chickens and mice, there is little evidence that mutations in c-*myb* play a causal role in the development of human haemopoietic lesions. Recently, a relatively large fraction of human leukaemias were found to have point mutations in specific transcription factors such as GATA-1, AML1/Runx1 and PU.1 (Mueller et al., 2002;

Rainis et al., 2003; Roumier et al., 2003), all of which have been shown to interact with or cooperate with c-Myb or v-Myb to regulate haemopoietic cell-specific genes. To date, no systematic study of point mutations in the c-*myb* alleles expressed by human leukaemias have been reported, but it is certainly possible that the Myb proteins expressed by leukaemias or other tumours are actually mutated versions which, like AMV v-Myb, have altered target gene specificities or inactivated negative regulatory circuits. In light of the mutations identified in other transcription factors, a study of Myb proteins expressed in human tumours is highly warranted.

REFERENCES

Alitalo, K., Winqvist, R., Lin, C.C., De la Chapelle, A., Schwab, M. and J.M., Bishop. (1984) Aberrant expression of an amplified c-myb oncogene in two cell lines from a colon carcinoma. Proc Natl Acad Sci USA *81*, 4535-4538.

Allen, S.C. and Hebbes, T.R. (2003) Myb induced myeloid protein 1 (Mim-1) is an acetyltransferase. FEBS Lett *534*, 119-124.

Altschul, S.F., Madden, T.L., Schaffer, A.A., Zhang, J., Zhang, Z., Miller, W. and Lipman, D.J. (1997) Gapped BLAST and PSI-BLAST: a new generation of protein database search programs. Nucl Acids Res *25*, 3389-3402.

Amouyel, P., Laudet, V., Martin, P., Li, R.P., Quatannens, B., Stehelin, D. and Saule, S. (1989) Two nuclear oncogenic proteins, P135gag-myb-ets and p61/63myc, cooperate to induce transformation of chicken neuroretina cells. J Virol *63*, 3382-3388.

Andersson, K.B., Kowenz-Leutz, E., Brendeford, E.M., Tygsett, A.H., Leutz, A. and Gabrielsen, O.S. (2002). Phosphorylation dependent down-regulation of c-Myb DNA-binding is abrogated by a point mutation in the v-myb oncogene. J Biol Chem. *278*, 3816-3824

Aziz, N., Miglarese, M.R., Hendrickson, R.C., Shabanowitz, J., Sturgill, T.W., Hunt, D.F. and Bender, T. P. (1995) Modulation of c-Myb-induced transcription activation by a phosphorylation site near the negative regulatory domain. Proc Natl Acad Sci USA *92*, 6429-6433.

Aziz, N., Wu, J., Dubendorff, J.W., Lipsick, J.S., Sturgill, T.W. and Bender, T. P. (1993). c-Myb and v-Myb are differentially phosphorylated by p42mapk in vitro. Oncogene *8*, 2259-2265.

Badiani, P.A., Kioussis, D., Swirsky, D.M., Lampert, I.A. and Weston, K. (1996) T-cell lymphomas in v-Myb transgenic mice. Oncogene *13*, 2205-2212.

Baluda, M.A. and Reddy, E.P. (1994) Anatomy of an integrated avian myeloblastosis provirus: structure and function. Oncogene *9*, 2761-2774.

Bell, M.V. and Frampton, J. (1999) v-Myb can transform and regulate the differentiation of melanocyte precursors. Oncogene *18*, 7226-7233.

Beug, H., Leutz, A., Kahn, P. and Graf, T. (1984) Ts mutants of E26 leukaemia virus allow transformed myeloblasts, but not erythroblasts or fibroblasts, to differentiate at the nonpermissive temperature. Cell *39*, 579-588.

Bies, J., Markus, J. and Wolff, L. (2002) Covalent Attachment of the SUMO-1 Protein to the Negative Regulatory Domain of the c-Myb Transcription Factor Modifies Its Stability and Transactivation Capacity. J Biol Chem *277*, 8999-9009.

Birkenmeier, E.H., Gwynn, B., Howard, S., Jerry, J., Gordon, J.I., Landchulz, W.H. and McKnight, S.L. (1989) Tissue-specific expression, developmental regulation, and genetic mapping of the gene encoding CCAAT/enhancer binding protein. Genes Dev *3*, 1146-1156.

Blair, D. G. and Athanasiou, M. (2000) Ets and retroviruses - transduction and activation of members of the Ets oncogene family in viral oncogenesis. Oncogene *19*, 6472-6481.

Bortner, D.M. and Ostrowski, M.C. (1991) Analysis of the v-myb structural components important for transactivation of gene expression. Nucl Acids Res *19*, 1533-1539.

Britos-Bray, M. and Friedman, A.D. (1997) Core binding factor cannot synergistically activate the myeloperoxidase proximal enhancer in immature myeloid cells without c-Myb. Mol Cell Biol *17*, 5127-5135.

Burk, O., Mink, S., Ringwald, M. and Klempnauer, K.-H. (1993) Synergistic activation of the chicken mim-1 gene by v-myb and C/EBP transcription factors. EMBO J *12*, 2027-2038.

Burk, O., Worpenberg, S., Haenig, B. and Klempnauer, K.-H. (1997) tom-1, a novel v-Myb target gene expressed in AMV- and E26-transformed myelomonocytic cells. EMBO J *16*, 1371-1380.

Chen, R.-H., Fields, S. and Lipsick, J.S. (1995) Dissociation of transcriptional activation and oncogenic transformation by v-Myb. Oncogene *11*, 1771-1779.

Chen, R.-H. and Lipsick, J.S. (1993) Differential transcriptional activation by v-Myb and c-Myb in animal cells and Saccharomyces cervisiae. Mol Cell Biol *13*, 4423-4431.

Clarke, M.F., Kukowska-Latallo, J.F., Westin, E., Smith, M. and Prochownik, E.V. (1988) Constitutive expression of a c-myb cDNA blocks friend murine erythroleukaemia cell differentiation. Mol Cell Biol *8*, 884-892.

Cogswell, J.P., Cogswell, P.C., Kuehl, W.M., Cuddihy, A.M., Bender, T.M., Engelke, U., Marcu, K.B. and Ting, J.P. (1993) Mechanism of c-myc regulation by c-Myb in different cell lineages. Mol Cell Biol *13*, 2858-2869.

Dai, P., Akimaru, H., Tanaka, Y., Hou, D.-X., Yasukawa, T., Kanei-Ishii, C., Takahashi, T. and Ishii, S. (1996) CBP as a transcriptional coactivator of c-Myb. Genes Dev *10*, 528-540.

Dash, A.B., Orrico, F.C. and Ness, S.A. (1996) The EVES motif mediates both intermolecular and intramolecular regulation of c-Myb. Genes Dev *10*, 1858-1869.

Davies, J., Badiani, P. and Weston, K. (1999) Cooperation of Myb and Myc proteins in T cell lymphomagenesis. Oncogene *18*, 3643-3647.

Dini, P.W., Eltman, J.T. and Lipsick, J.S. (1995) Mutations in the DNA-binding and transcriptional activation domains of v-Myb cooperate in transformation. J Virol *69*, 2515-2524.

Dini, P.W. and Lipsick, J.S. (1993) Oncogenic truncation of the first repeat of c-Myb decreases DNA binding in vitro and in vivo. Mol Cell Biol *13*, 7334-7348.

Dubendorff, J.W., Whittaker, L.J., Eltman, J.T. and Lipsick, J.S. (1992) Carboxy-terminal elements of c-Myb negatively regulate transcriptional activation in cis and in trans. Genes Dev *6*, 2524-2535.

Engelke, U., Whittaker, L. and Lipsick, J.S. (1995) Weak transcriptional activation is sufficient for transformation by v-Myb. Virol *208*, 467-477.

Evans, J.L., Moore, T.L., Kuehl, W.M., Bender, T. and Ting, J.P. (1990) Functional analysis of c-Myb protein in T-lymphocytic cell lines shows that it trans-activates the c-myb promoter. Mol Cell Biol *10*, 5747-5752.

Falany, M.L., Thames, A.M., 3rd, McDonald, J.M., Blair, H.C., McKenna, M.A., Moore, R.E., Young, M.K. and Williams, J.P. (2001) Osteoclasts secrete the chemotactic cytokine mim-1. Biochem Biophys Res Commun *281*, 180-185.

Favier, D. and Gonda, T.J. (1994) Detection of proteins that bind to the leucine zipper motif of c-Myb. Oncogene *9*, 305-311.

Fong, I.C., Zarrin, A.A., Wu, G.E. and Berinstein, N.L. (2000) Functional analysis of the human RAG 2 promoter. Mol Immunol *37*, 391-402.

Frampton, J., Leutz, A., Gibson, T. and Graf, T. (1989) DNA binding domain ancestry. Nature *342*, 134.

Frampton, J., McNagny, K., Sieweke, M., Philip, A., Smith, G., and Graf, T. (1995) v-Myb DNA binding is required to block thrombocytic differentiation of Myb-Ets-transformed multipotent haematopoietic progenitors. EMBO J *14*, 2866-2875.

Frampton, J., Ramqvist, T. and Graf, F. (1996) v-Myb of E26 leukaemia virus up-regulates bcl-2 and suppresses apoptosis in myeloid cells. Gene Dev *10*, 2720-2731.

Fu, S-L. and Lipsick, J.S. (1996) FAETL Motif Required for Leukemic Transformation by v-Myb. J Virol *70*, 5600-5610.

Fu, S.L. and Lipsick, J.S. (1997) Constitutive expression of full-length c-Myb transforms avian cells characteristic of both the monocytic and granulocytic lineages. Cell Growth Diff *8*, 35-45.

Furuta, Y., Aizawa, S., Suda, Y., Ikawa, Y., Nakasgoshi, H., Nishina, Y. and Ishii, S. (1993) Degeneration of skeletal and cardiac muscles in c-myb transgenic mice. Transgen Res *2*, 199-207.

Ganter, B., Fu, Sl and Lipsick, J.S. (1998) D-type cyclins repress transcriptional activation by the v-Myb but not the c-Myb DNA-binding domain. EMBO J *17*, 255-268.

Gewirtz, A.M., Anfossi, G., Venturelli, D., Valpreda, S., Sims, R. and Calabretta, B. (1989) G1/S transition in normal human T-lymphocytes requires the nuclear protein encoded by c-myb. Science *245*, 180-183.

Gewirtz, A.M., Sokol, D.L. and Ratajczak, M.Z. (1998) Nucleic acid therapeutics: state of the art and future prospects. Blood *92*, 712-736.

Gewirtz, A.M. and Calabretta, B. (1988) A c-myb antisense oligodeoxynucleotide inhibits normal human haemopoiesis in vitro. Science *242*, 1303-1306.

Gonda, T.J., Buckmaster, C. and Ramsay, R.G. (1989) Activation of c-myb by carboxy-terminal truncation: relationship to transformation of murine haemopoietic cells in vitro. EMBO J *8*, 1777-1783.

Grässer, F.A., Graf, T. and Lipsick, J. (1991) Protein Truncation is Required for the Activation of the c-myb Proto-oncogene. Mol Cell Biol *11*, 3987-3996.

Hernandez-Munain, C. and Krangel, M.S. (1994) Regulation of the T-cell receptor delta enhancer by functional cooperation between c-Myb and core-binding factors. Mol Cell Biol *14*, 473-483.

Hogg, A., Schirm, S., Nakagoshi, H., Bartley, P., Ishii, S., Bishop, J.M. and Gonda, T.J. (1997) Inactivation of a c-Myb/estrogen receptor fusion protein in transformed primary cells leads to granulocyte/macrophage differentiation and down regulation of c-kit but not c-myc or cdc2. Oncogene *15*, 2885-2898.

Ibanez, C.E. and Lipsick, J.S. (1990) trans activation of gene expression by v-myb. Mol Cell Biol *10*, 2285-2293.

Ibanez, Carlos E. and Lipsick, J.S. (1988) Structural and functional domains of the myb oncogene: Requirements for nuclear transport, myeloid transformation, and colony formation. J Virol *62*, 1981-1988.

Introna, M., Golay, J., Frampton, J., Nakano, T., Ness, S.A. and Graf, T. (1990) Mutations in v-myb alter the differentiation of myelomonocytic cells transformed by the oncogene. Cell *63*, 1287-1297.

Jiang, W., Kanter, M.R., Dunkel, I., Ramsay, R.G., Beemon, K.L. and Hayward, W.S. (1997) Minimal truncation of the c-myb gene product in rapid-onset B-cell lymphoma. J Virol *71*, 6526-6533.

Kageyama, R., Sasai, Y. and Nakanishi, S. (1991) Molecular characterization of transcription factors that bind to the cAMP responsive region of the substance P precursor gene. J Biol Chem *226*, 15525-15531.

Kalkbrenner, F., Guehmann, S. and Moelling, K. (1990) Transcriptional activation by human c-myb and v-myb genes. Oncogene *5*, 657-661.

Kan, N.C., Baluda, M.A. and Papas, T.S. (1985) Sites of recombination between the transforming gene of avian myeloblastosis virus and its helper virus. Virol *145*, 323-329.

Kanei-Ishii, C., MacMillan, E.M., Nomura, T., Sarai, A., Ramsay, R.G., Aimoto, S., Ishii, S. and Gonda, T.J. (1992) Transactivation and transformation by Myb are negatively regulated by a leucine-zipper structure. Proc Natl Acad Sci USA *89*, 3088-3092.

Kanei-Ishii, C., Tanikawa, J., Nakai, A., Morimoto, R. I. and Ishii, S. (1997) Activation of heat shock transcription factor 3 by c-Myb in the absence of cellular stress. Science *277*, 246-248.

Kanter, M.R., Smith, R.E. and Hayward, W.S. (1988) Rapid Induction of B-Cell Lymphomas: Insertional Activation of c-myb by Avian Leukosis Virus. J Virol *62*, 1423-1432.

Karafiat, V., Dvorakova, M., Pajer, P., Kralova, J., Horejsi, Z., Cermak, V., Bartunek, P., Zenke, M. and Dvorak, M. (2001) The leucine zipper region of Myb oncoprotein regulates the commitment of haemopoietic progenitors. Blood *98*, 3668-3676.

Klempnauer, K.-H., Arnold, H. and Biedenkapp, K. (1989) Activation of transcription by v-myb: evidence for two different mechanisms. Genes Dev *3*, 1582-1589.

Kowenz-Leutz, E., Herr, P., Niss, K. and Leutz, A. (1997) The homeobox gene GBX2, a target of the myb oncogene, mediates autocrine growth and monocyte differentiation. Cell *91*, 185-195.

Kraut, N., Frampton, J., McNagny, K.M. and Graf, T. (1994) A functional Ets DNA-binding domain is required to maintain multipotency of hematopoietic progenitors transformed by Myb-Ets. Genes Dev. *8*, 33-44.

Landschulz, W.H., Johnson, P.F. and McKnight, S.L. (1988) The Leucine Zipper: a Hypothetical Structure Common to a New Class of DNA Binding Proteins. Science *240*, 1759-1764.

Lane, T.N., Ibanez, C.E., Garcia, A., Graf, T. and Lipsick, J.S. (1990) Transformation by v-myb correlates with trans-activation of gene expression. Mol Cell Biol *10*, 2591-2598.

Leutz, A., Beug, H. and Graf, T. (1984) Purification and characterization of cMGF, a novel chicken myelomonocytic growth factor. EMBO J *3*, 3191-3197.

Leutz, A., Damm., K., Sterneck, E., Kowenz, E., Ness, S., Frank, R., Gausepohl, H., Pan, A.-C.E., Smart, J., Hayman, M.J. and Graf, T. (1989) Molecular cloning of the chicken myelomonocytic growth factor (cMGF) reveals relationship to interleukin 6 and granulocyte colony stimulating factor. EMBO J *8*, 175-181.

Leverson, J.D., Koskinen, P.J., Orrico, F.C., Rainio, E.-M., Jalkanen, K.J., Dash, A.B., Eisenman, R.N. and Ness, S.A. (1998) Pim-1 Kinase and p100 Cooperate to Enhance c-Myb Activity. Mol Cell *2*, 417-425.

Leverson, J.D. and Ness, S.A. (1998) Point Mutations in v-Myb Disrupt a Cyclophilin-Catalyzed Negative Regulatory Mechanism. Mol Cell *1*, 203-211.

Lipsick, J.S. (1996) One Billion Years of Myb. Oncogene *13*, 223-235.

Lipsick, J. S. and Wang, D.M. (1999) Transformation by v-Myb. Oncogene *18*, 3047-3055.

Lüscher, B., Christenson, E., Litchfield, D.W., Krebs, E.G. and Eisenman, R.N. (1990) Myb DNA binding inhibited by phosphorylation at a site deleted during oncogenic activation. Nature *344*, 517-522.

McNagny, K.M. and Graf, T. (2003). E26 leukaemia virus converts primitive erythroid cells into cycling multilineage progenitors. Blood *101*, 1103-1110.

McNagny, K.M., Lim, F., Grieser, S. and Graf, T. (1992) Cell surface proteins of chicken haemopoietic progenitors, thrombocytes and eosinophils detected by novel monoclonal antibodies. Leukaemia *6*, 975-984.

Metz, T. and Graf, T. (1991) v-myb and v-ets transform chicken erythroid cells and cooperate both in trans and in cis to induce distinct differentiation phenotypes. Genes Dev *5*, 369-380.

Mink, S., Kerber, U. and Klempnauer, K.-H. (1996) Interaction of C/EBPb and v-Myb is required for synergistic activation of the mim-1 gene. Mol Cell Biol *16*, 1316-1325.

Mucenski, M.L., McLain, K., Kier, A.B., Swerdlow, S.H., Schreiner, C.M., Miller, T.A., Pietryga, D.W., Scott, W.J. Jr. and Potter, S.S. (1991) A functional c-myb gene is required for normal murine fetal hepatic haemopoiesis. Cell *65*, 677-689.

Mueller, B.U., Pabst, T., Osato, M., Asou, N., Johansen, L.M., Minden, M.D., Behre, G., Hiddemann, W., Ito, Y. and Tenen, D.G. (2002) Heterozygous PU.1 mutations are associated with acute myeloid leukaemia. Blood *100*, 998-1007.

Ness, S.A., Kowenz-Leutz, E., Casini, T., Graf, T. and Leutz, A. (1993) Myb and NF-M: Combinatorial activators of myeloid genes in heterologous cell types. Genes Dev *7*, 749-759.

Ness, S.A., Marknell, Å. and Graf, T. (1989) The v-myb oncogene product binds to and activates the promyelocyte-specific mim-1 gene. Cell *59*, 1115-1125.

Ness, S.A. (1996) The myb oncoprotein: regulating a regulator. BBA Reviews on Cancer 1288, F123-F139.

Ness, S.A. (1999) Myb Binding Proteins: Regulators and Cohorts in Transformation. Oncogene *18*, 3039-3046.

Nomura, T., Sakai, N., Sarai, A., Sudo, T., Kanei, Ishii C., Ramsay, R.G., Favier, D., Gonda, T.J. and Ishii, S. (1993) Negative autoregulation of c-Myb activity by homodimer formation through the leucine zipper. J Biol Chem *268*, 21914-21923.

Oelgeschläger, M., Janknecht, R., Krieg, J., Schreek, S. and Lüscher, B. (1996) Interaction of the co-activator CBP with Myb proteins: effects on Myb-specific transcription and on the cooperativity with NF-M. EMBO J *15*, 2771-2780.

Oelgeschläger, M., Krieg, J., Lüscher-Firzlaff, J.M. and Lüscher, B. (1995) Casein kinase II phosphorylation site mutations in c-Myb affect DNA binding and transcriptional cooperativity with NF-M. Mol Cell Biol 15, 5966-5974.

Oelgeschlager, M., Nuchprayoon, I., Luscher, B. and Friedman, A.D. (1996). C/EBP, c-Myb, and PU.1 cooperate to regulate the neutrophil elastase promoter. Mol Cell Biol *16*, 4717-4725.

Ogata, K., Morikawa, S., Nakamura, H., Hojo, H., Yoshimura, S., Zhang, R., Aimoto, S., Ametani, Y., Hirata, Z., Sarai, A., Ishii, S. and Nishimura, Y. (1995) Comparison of the free and DNA-complexed forms of the DNA-binding domain from c-Myb. Nat Struct Biol *2*, 309-320.

Ogata, K., Morikawa, S., Nakamura, H., Sekikawa, A., Inoue, T., Kanai, H., Sarai, A., Ishii, S. and Nishimura, Y. (1994) Solution structure of a specific DNA complex of the Myb DNA-binding domain with cooperative recognition helices. Cell *79*, 639-648.

Oh, I.H. and Reddy, E.P. (1999) The myb gene family in cell growth, differentiation and apoptosis. Oncogene *18*, 3017-3033.

Pelicci, P.G., Lanfrancone, L., Brathwaite, M.D., Wolman, S.R. and Dalla-Favera, R. (1984) Amplification of the c-myb oncogene in a case of human acute myelogenous leukaemia. Science *224*, 1117-1121.

Prendergast, G. C. (1999a) Mechanisms of apoptosis by c-Myc. Oncogene *18*, 2967-2987.

Prendergast, G.C. (1999b) Myc and Myb: are the veils beginning to lift? Oncogene *18*, 2914-2915.

Ptashne, M. (1988) How eukaryotic transcriptional activators work. Nature *335*, 683-689.

Rainis, L., Bercovich, D., Strehl, S., Teigler-Schlegel, A., Stark, B., Trka, J., Amariglio, N., Biondi, A., Muler, I., Rechavi, G., Kempski, H., Haas, O.A. and Izraeli, S. (2003) Mutations in exon 2 of GATA1 are early events in megakaryocytic malignancies associated with trisomy 21. Blood *102*, 981-986.

Roumier, C., Fenaux, P., Lafage, M., Imbert, M., Eclache, V. and Preudhomme, C. (2003) New mechanisms of AML1 gene alteration in hematological malignancies. Leukaemia *17*, 9-16.

Rushton, J.J., Davis, L.M., Lei, W., Mo, X., Leutz, A. and Ness, S.A. (2003) Distinct changes in gene expression induced by A-Myb, B-Myb and c-Myb proteins. Oncogene *22*, 308-313.

Rushton, J.J. and Ness, S.A. (2001) The conserved DNA binding domain mediates similar regulatory interactions for A-Myb, B-Myb, and c-Myb transcription factors. Blood Cells Mol Dis *27*, 459-463.

Saikumar, P., Murali, R. and Reddy, E.P. (1990) Role of tryptophan repeats and flanking amino acids in Myb-DNA interactions. Proc Natl Acad Sci USA *87*, 8452-8456.

Sakura, H., Kanai-Ishii, C., Nagase, T., Nakagoshi, H., Gonda, T.J. and Ishii, S. (1989) Delineation of three functional domains of the transcriptional activator encoded by the c-myb protooncogene. Proc Natl Acad Sci USA *86*, 5758-5762.

Sano, Y. and Ishii, S. (2001) Increased affinity of c-Myb for CREB-binding protein (CBP) after CBP- induced acetylation. J Biol Chem *276*, 3674-3682.

Sterneck, E., Blattner, C., Graf, T. and Leutz, A. (1992a) Structure of the chicken myelomonocytic growth factor gene and specific activation of its promoter in avian myelomonocytic cells by protein kinases. Mol Cell Biol *12*, 1728-1735.

Sterneck, E., Müller, C., Katz, S. and Leutz, A. (1992b) Autocrine growth induced by kinase-type oncogenes in myeloid cells requires AP-1 and NF-M, a myeloid-specific, C/EBP-like factor. EMBO J *11*, 115-126.

Symonds, G., Stubblefield, E., Guyaux, M. and Bishop, J.M. (1984) Cellular oncogenes (c-erb-A and c-erb-B) located on different chicken chromosomes can be transduced into the same retroviral genome. Mol Cell Biol *4*, 1627-1630.

Tahirov, T. H., Sato, K., Ichikawa-Iwata, E., Sasaki, M., Inoue-Bungo, T., Shiina, M., Kimura, K., Takata, S., Fujikawa, A., Morii, H., Kumasaka, T., Yamamoto, M., Ishii, S. and Ogata, K. (2002) Mechanism of c-Myb-C/EBP beta cooperation from separated sites on a promoter. Cell *108*, 57-70.

Todokoro, K., Watson, R.J., Higo, H., Amanuma, H., Kuramochi, S., Yanagisawa, H. and Ikawa, Y. (1988) Down-regulation of c-myb gene expression is a prerequisite for erythropoietin-induced erythroid differentiation. Proc Natl Acad Sci USA *85*, 8900-8904.

Tomita, A., Towatari, M., Tsuzuki, S., Hayakawa, F., Kosugi, H., Tamai, K., Miyazaki, T., Kinoshita, T. and Saito, H. (2000) c-Myb acetylation at the carboxyl-terminal conserved domain by transcriptional co-activator p300. Oncogene *19*, 444-451.

Tong, X., Drapkin, R., Yalamanchili, R., Mosialos, G. and Kieff, E. (1995) The Epstein-Barr Virus nuclear protein 2 ccidic domain forms a complex with a novel cellular coactivator that can interact with TFIIE. Mol Cell Biol *15*, 4735-4744.

Tour, E., Pillemer, G., Gruenbaum, Y. and Fainsod, A. (2001) The two Xenopus Gbx2 genes exhibit similar, but not identical expression patterns and can affect head formation. FEBS Lett *507*, 205-209.

Tour, E., Pillemer, G., Gruenbaum, Y. and Fainsod, A. (2002) Gbx2 interacts with Otx2 and patterns the anterior-posterior axis during gastrulation in Xenopus. Mech Dev *112*, 141-151.

Wallrapp, C., Muller-Pillasch, F., Solinas-Toldo, S., Lichter, P., Friess, H., Buchler, M., Fink, T., Adler, G. and Gress, T.M. (1997) Characterization of a high copy number amplification at 6q24 in pancreatic cancer identifies c-myb as a candidate oncogene. Cancer Res *57*, 3135-3139.

Wang, D.M. and Lipsick, J.S. (2002) Mutational analysis of the transcriptional activation domains of v-Myb. Oncogene *21*, 1611-1615.

Welter, C., Henn, W., Theisinger, B., Fischer, H., Zang, K.D. and Blin, N. (1990) The cellular myb oncogene is amplified, rearranged and activated in human glioblastoma cell lines. Cancer Lett *52*, 57-62.

Weston, K. (1999) Reassessing the role of C-MYB in tumourigenesis. Oncogene *18*, 3034-3038.

Weston, K. and Bishop, J.M. (1989) Transcriptional activation by the v-myb oncogene and its cellular progenitor, c-myb. Cell *58*, 85-93.

Weston, K. (1990) The myb genes. Seminars in Cancer Biology *1*, 371-382.

Winn, L.M., Lei, W. and Ness, S.A. (2003) Pim-1 Phosphorylates the DNA Binding Domain of c-Myb. Cell Cycle *2*, 258-262.

Winqvist, R., Knuutila, S., Leprince, D., Stehelin, D. and Alitalo, K. (1985) Mapping of amplified c-myb oncogene, sister chromatid exchanges, and karyotypic analysis of the COLO 205 colon carcinoma cell line. Cancer Genet Cytogenet *18*, 251-264.

Yamagoe, S., Yamakawa, Y., Matsuo, Y., Minowada, J., Mizuno, S. and Suzuki, K. (1996) Purification and primary amino acid sequence of a novel neutrophil chemotactic factor LECT2. Immunol Lett *52*, 9-13.

Yamamoto, H., Kihara-Negishi, F., Yamada, T., Suzuki, M., Nakano, T. and Oikawa, T. (2002) Interaction between the haemopoietic Ets transcription factor Spi-B and the coactivator CREB-binding protein associated with negative cross-talk with c-Myb. Cell Growth Diff *13*, 69-75.

Chapter 16

C-MYB AND LEUKAEMOGENESIS

Juraj Bies[1] and Linda Wolff[2]
[1]*Laboratory of Molecular Virology, Cancer Research Institute, Slovak Academy of Sciences, 833 91 Bratislava, Slovakia,* [2]*Laboratory of Cellular Oncology, National Cancer Institute, NIH, Bethesda, MD 20892-4255, United States of America..*

Abstract: The c-*myb* proto-oncogene has repeatedly been a target of retroviral insertional mutagenesis in murine and avian haemopoietic neoplasms. The most common mechanism by which avian and murine retroviruses activate c-*myb*'s oncogenic potential is promoter insertional mutagenesis where the viral LTR function replaces the endogenous transcriptional control resulting in constitutive expression of the *myb* mRNA. Another mechanism of activation of c-Myb is achieved through the integration of retroviruses into the 3' region of the c-*myb* locus. The 3' untranslated region is replaced with viral polyA causing an increased stability of the c-*myb* mRNA. In addition, this type of activation results in the carboxyl-terminal truncation of the c-Myb protein providing increased proteolytic stability and transactivation capacity. Several virus integration sites were also mapped within the genomic region surrounding the c-*myb* locus suggesting that retrovirus integrations outside of the coding region can also impose activation via the long-range effect of retroviral regulatory elements.

1. INTRODUCTION

The proto-oncogene c-*myb* encodes a transcription factor that is expressed in cells of the haemopoietic lineage and regulates the transcription of genes involved in proliferation, differentiation and apoptosis. Interest in this gene has been stimulated by the numerous examples of its involvement in haemopoietic neoplasias of the myeloid and lymphoid lineages in animals.

In this review, we will summarise the different modes of oncogenic activation of c-*myb* by retroviral insertional mutagenesis in chickens and mice and their potential role in Myb-induced leukaemogenesis with respect to known functions and modes of regulation of the c-Myb transcription factor. Our emphasis will be primarily on murine model systems. Because

J. Frampton (ed.), Myb Transcription Factors: Their Role in Growth, Differentiation and Disease, 307-329.

several other review articles about v-Myb proteins have been published (Introna and Golay, 1999; Lipsick and Wang, 1999), aspects of AMV and E-26 v-Myb induced transformation will be mentioned only briefly, to underscore some unifying mechanistic concepts. A more comprehensive description of its protein structure and molecular properties can be found in several excellent reviews published recently (Ness, 1999; Oh and Reddy, 1999). At the end we also discuss the possible role of the c-Myb oncoprotein in human leukaemia.

2. TRANSFORMATION OF HAEMOPOIETIC CELLS

c-Myb was identified more than 30 years ago through the discovery of the transforming v-Myb protein encoded by two avian retroviruses that induce leukaemia in chickens. The viruses, AMV and E-26, which encode different versions of the c-Myb oncoprotein, transform cells of the myeloid lineage. *In vitro* assays, using cells from tissues rich in haemopoietic progenitor cells, confirmed the role of v-Myb in transformation of myeloid lineage cells (Lipsick and Wang, 1999). Sequence analysis showed that both v-Myb proteins suffered severe truncations at the amino and carboxy-termini. Although there are 10 amino acid substitutions in AMV v-Myb compared to c-Myb, none of them are required for transformation. Some of these substitutions affect the phenotype of the transformed cells (Lipsick and Wang, 1999).

c-*myb* can be activated by retroviral insertional mutagenesis following inoculation of animals with replication competent retroviruses. For a review of insertional mutagenesis see Jonkers and Berns, 1996. In the animals, viral DNA integrates randomly into the genomic DNA during its life cycle and the mutagenic effects of the integrated provirus on c-*myb* are selected due to a growth advantage conferred by the virus. These experimentally induced leukaemias in chickens and mice have facilitated our identification of the alterations in c-*myb* gene and its protein product that lead to transformation of haemopoietic cells.

In chickens, insertional mutagenesis at the c-*myb* locus has been associated with lymphoid neoplasms. Inoculation of embryos with either RAV-1 or EU-8 results in rapid induction of B-cell lymphomas that have activated c-*myb* (Kanter et al., 1988; Piser and Humphries, 1989; Piser et al., 1992; Press et al., 1995). In addition, activation of c-*myb* by a RAV-1 provirus was discovered in an avian T-cell lymphoma cell line derived from chickens inoculated with Marek's disease virus (Le Rouzic and Perbal, 1996).

In mice, c-*myb* is activated with high frequency in a model for promonocytic leukaemia due to integration of Moloney murine leukaemia virus (M-MuLV), Friend MuLV or amphotropic virus 4070A (Wolff, 1996). These leukaemias are often referred to as MML for murine myeloid leukaemias. In this model, leukaemia develops in virus-inoculated susceptible strains of mice undergoing a pristane-induced chronic inflammation in the peritoneal cavity (Wolff et al., 1988; Wolff et al., 1991). There is a rather long latency for disease development in this animal model that suggests there is a multi-step process in the transformation of myeloid cells. Activated c-*myb,* therefore, must collaborate with other oncogenic events. Intriguingly, leukaemias with a mutagenised c-*myb* locus develop only in mice that have the inflammatory granuloma induced by pristane. In its absence, virus-inoculated mice do not develop leukaemia, despite the fact that, in haemopoietic tissues, 100% of animals have detectable viral *gag-myb* transcripts, shown to be a consequence of proviral integration in the c-*myb* locus (Nason-Burchenal and Wolff, 1992; Nason-Burchenal and Wolff, 1993). These results also suggest that, in this model, c-*myb* activation is probably one of the first oncogenic events in disease progression, because cells expressing aberrant *myb* messages are detected within the first 2-3 weeks following virus inoculation and can be found in the bone marrow and spleen (Nason-Burchenal and Wolff, 1993). Interestingly, expression of the aberrant fusion message is also detected in the thymus, but lymphomas never develop (Belli et al., 1995); it is believed this is due to a lack of a sufficient number of cooperating events. Thus, putative mutagenic events involved in progression of the promonocytic leukaemia seem to be strictly tissue-specific. Promonocytic leukaemias, induced by retroviruses in mice with activated c-*myb* expression can be adapted to growth *in vitro* in the absence of growth factors (Wolff et al., 1988) although initial growth is facilitated with the addition of granulocyte-macrophage colony stimulating factor (GM-CSF).

In other experimental murine models involving retroviruses, the c-*myb* region has been identified as a target of insertional mutagenesis. These include myeloid leukaemias induced by Cas-Br-M (Shen-Ong et al., 1986; Joosten et al., 2002) and myeloid leukaemias in BXH2 mice (Blaydes et al., 2001).

Similar to the avian system, a transformation assay for murine Myb has been developed (Gonda et al., 1993; MacMillan and Gonda, 1994; Ferrao et al., 1995). Murine retroviruses expressing wild-type c-Myb or carboxy-terminally truncated c-Myb can transform foetal liver cells, but only in the presence of required growth factors. Transformed cells have morphological characteristics of myeloid progenitors and respond to growth factors such as GM-CSF and interleukin-3 (IL-3). Interestingly, these experiments also showed that under specific experimental conditions including growth at high density, enforced expression of full-length c-Myb has transforming potential (Ferrao et al.1995).

Although most of the *myb* related leukaemias studied in mice have been myeloid, a T-cell lymphoma model involving Myb was reported by Kathy Weston and colleagues. In v-Myb transgenic mice, with T-cell-specific expression, high-grade T-cell lymphomas develop in older animals (Badiani et al., 1996).

3. MECHANISMS THAT ACTIVATE MYB'S ONCOGENIC POTENTIAL

Experiments from many laboratories have shown that truncation of either end of the c-Myb protein contributes to increased transforming potential (Lipsick and Wang, 1999). During activation of c-*myb* by retroviral insertional mutagenesis, for example, the protein is truncated on or other end depending on the location of the provirus (Figure 1). In addition, regulatory LTR sequences of the provirus enhance or promote transcription or cause termination of transcription. In the following sections, we will discuss in more detail the alterations caused by integrated proviruses in the c-*myb* gene as well as their impact on regulation of the transcription factor.

3.1 Activation of c-Myb by Retroviral Integration into the 5' End of the Gene

Proviruses are found at the 5' end of c-*myb* in at least 95% of murine promonocytic leukaemias, which are induced by the combination of replication competent virus and pristane, and in an equally high percentage of avian retrovirus-induced B-cell lymphomas. The integration sites are found in the first, second, or third introns. In these leukaemias transcription of aberrant c-*myb* mRNA is initiated in the retroviral 5' LTR. Read-through c-*myb* transcripts, in both the murine and avian neoplasms, are spliced. In the mouse promonocytic leukaemias, the splice sites utilised a cryptic *gag* donor splice site and one of the normal splice acceptors of c-*myb* at the next available exon (Shen-Ong and Wolff, 1987). In the chicken lymphomas, splicing is similar except a normal donor splice site in *gag* is utilised instead of a cryptic donor site (Kanter et al., 1988).

Interestingly, integrated proviruses at the 5' end of c-*myb* are positioned in the genome in a manner that allows a transcriptional pause site to be bypassed (Mukhopadhyaya and Wolff, 1992; Piser et al., 1992; Jiang et al., 1997). It should be emphasised that the normal down-regulation of c-*myb* expression observed during maturation of haemopoietic cells does not occur at the level of transcriptional initiation, but rather at the level of transcriptional elongation. The endogenous c-*myb* promoter does not have a TATA motif, and transcription appears to initiate at multiple sites within a CpG island located in the promoter region (Bender and Kuehl, 1986; Dvorak et al., 1989). This type of TATA-less promoter is usually characterised by low-level, constitutive activity and is not subject to rapid transcription factor-induced regulation arising from extracellular signals. Therefore, the transcriptional pause site within the first intron is an important and major mechanism of regulation of c-*myb* expression (Watson, 1988; Reddy and Reddy, 1989; Wang et al., 1994). The bypassing of the c-*myb* elongation block is likely to be the most critical event in the transformation process, because it permits escape from maturation-associated down-regulation of c-*myb* expression. In addition, the viral LTR provides strong promoter/enhancer elements that contribute to the constitutive transcription. As shown in *in vitro* studies, this constitutive expression of c-Myb provides continued growth of cells even during the G-CSF induced 32D cells differentiation to the granulocyte lineage and IL-6-induced macrophage maturation (Bies et al., 1995; Bies et al., 1996). Interestingly, there was no difference observed between truncated and full-length protein in these experiments, emphasising the role of inappropriate expression in transformation (Selvakumaran et al., 1992; Bies et al., 1995). Similarly, others found that under specific conditions, enforced expression of full-length c-Myb transforms foetal liver cells (Ferrao et al., 1995).

The observations that proviruses cause constitutive expression of c-Myb in leukaemic cells that have undergone insertional mutagenesis and that full length c-Myb can block differentiation and induce transformation *in vitro*, might suggest that mutation of c-*myb* alleles is not required for Myb-specific transformation. However, leukaemia-associated c-*myb* alleles with proviral integrations at the 5' end, always undergone some alteration in sequences affecting the amino-terminal coding region of c-Myb (Mukhopadhyaya and Wolff, 1992, Piser et al., 1992; Jiang et al., 1997), and it is, therefore, possible that these alterations do facilitate the transformation of haemopoietic cells. In murine promonocytic leukaemias, the most common sites of integration are located in introns 2 and 3 with removal of 47 or 71 amino acids from amino-terminus of c-Myb, respectively (Figure 1). More rarely integrations have been found in intron 1 and result in truncation of 20 amino acids (Shen-Ong and Wolff, 1987; Mukhopadhyaya and Wolff, 1992). This is in contrast to chicken B-cell tumours, where the most common integration site is intron 1 (Piser et al., 1992). All of these deletions remove the conserved acidic region at the amino-terminus including phosphorylation sites Ser 11 and 12 (Figure 1).

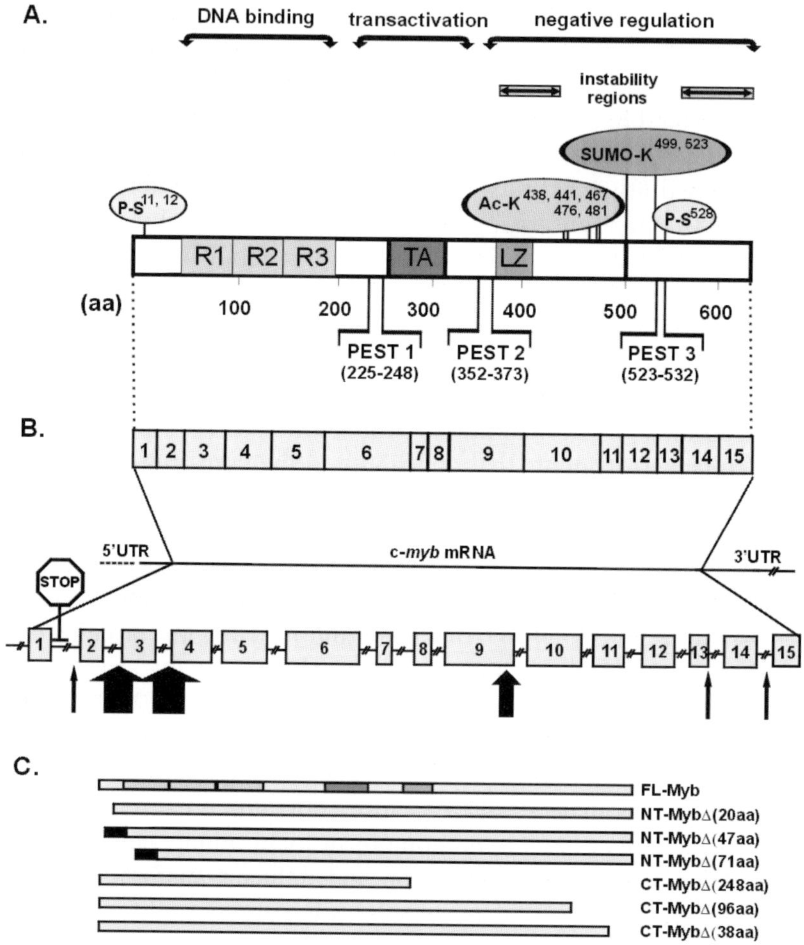

Figure 1

(A) Structure of the c-Myb protein: R1, R2, R3, imperfect tandem repeats composing the DNA-binding domain; TA, acidic and highly hydrophilic domain that is part of the transactivation region; LZ, putative leucine zipper structure, PEST1, 2, and 3 regions identified by PEST-FIND program. Posttranslational modification of c-Myb: SUMO-K, SUMO-1-conjugated lysines; Ac-K, acetylated lysines; and P-S, phosphorylated serines. (B) Activation of the c-Myb by retroviral integration into c-myb locus. Schematic diagram of the genomic structure of c-myb with transcription pause site (STOP sign), mRNA, 5' and 3' untranslated regions (UTRs), and exon structure. Black arrows represent locations of proviruses identified in murine myeloid leukaemias (MML); thickness of arrows reflects frequency of integration sites in MML. (C) Wild type and truncated forms of the protein detected in MML. Black regions in NT-MybΔ(47aa) and NT-MybΔ(71aa) represent viral Gag protein sequences.

Phosphorylation of both serines by casein kinaseII (CKII) was implicated in the negative regulation of DNA-binding affinity of c-Myb (Lüscher et al.,

1990; Oelgeschlager et al., 1995). Truncation of c-Myb by 47 or 71 amino acids removes part of the R1 repeat of the DNA binding domain. While the R1 repeat is not involved in direct contact of c-Myb with DNA, it was suggested that this repeat can either affect DNA-binding affinity of c-Myb (Dini and Lipsick, 1993) or it can facilitate intramolecular interaction with the carboxy-terminus (Dash et al., 1996). More recent studies with murine proteins with similar size amino-terminal truncations revealed that the most extreme truncation of the amino-terminus disrupted the activation of chromatin embedded target genes in collaboration with the transcription factor C/EBPβ (Oelgeschlager et al., 2001). This result suggests a new role for the R1 repeat domain in cooperation with other transcription factors in the transcriptional regulation of resident genes in intact chromatin. In addition, Oelgeschlager and coauthors also suggested an inverse correlation between activation of chromatin embedded genes and leukaemogenic potential of the amino-terminally truncated Myb proteins (Oelgeschlager et al., 2001). Since the products of studied genes are associated with the differentiation process rather than with proliferation, the inability of some oncogenic forms of c-Myb to activate these types of genes may to some extent enhance the transforming capability of the c-Myb proteins.

At present, the only evidence that an amino-terminal truncation is in itself oncogenic was provided by experiments in chickens where embryos were infected with retroviral vectors expressing wild type c-Myb or c-Myb truncated by 20 amino acids. This truncated protein produced a high incidence of rapid onset tumours while the wild-type c-Myb was only weakly oncogenic (Jiang et al., 1997).

3.2 Activation of c-Myb by Retroviral Integration into the Middle or 3' End of the Gene

Integration of retroviruses into the middle or at the 3' end of the c-*myb* gene has been found less frequently than integration at the 5' end (Wolff, 1996) and only in murine myeloid leukaemias. The most frequent site of virus integration in this category is within exon 9. The virus LTR at this site causes premature termination of transcription and translation of truncated c-*myb* RNA is terminated at stop codon within the LTR sequence. This produces a c-Myb protein that is severely truncated at the carboxy-terminus by 240 to 248 amino acids. Examples of myeloid leukaemia cell lines with these truncations are a myelomonocytic leukaemia cell line NFS60 (Shen-Ong et al., 1986) and a promonocytic leukaemic cell line R1-4-11 (Mukhopadhyaya and Wolff, 1992). A very similar truncation was also detected in an IL-3-dependent cell line, VFLJ2, generated *in vitro* by retroviral infection (Weinstein et al., 1987). Additional integration sites in c-

myb were also reported and include sites in exon 13 and the intron 14. These integrations result in carboxy-terminal truncations by 96 and 38 amino acids, respectively (Nazarov and Wolff, 1995; Bies et al., 1999).

3.2.1 mRNA stabilisation

Integration of viral promoter/enhancer sequences into the 3' end of the c-*myb* gene could potentially affect the endogenous enhancer elements and transcriptional pause site, thus mimicking the effects of proviruses integrated at the 5'end of the gene (see above). However, we were unable to show that integration of retroviruses in the down-stream region could prevent down-regulation of c-myb expression in cells treated with differentiation-inducing agents (Haviernik et al., 2002). Therefore, viruses must employ a different strategy to keep sufficient levels of c-Myb to prevent growth arrest in these cells when they are exposed to differentiation-inducing cytokines *in vivo*. Changes in the stability of mRNA or protein might provide sufficient steady-state levels of c-Myb to keep cells in a proliferative state until additional mutagenic events occur that promote progression of leukaemia.

c-*myb* mRNA in myeloid cells has a short half-life of around 45 minutes (Figure 2) and its turnover can be rapidly accelerated during the initial stages of differentiation (Watson, 1988). Decay of many unstable mRNAs is controlled via *cis*-acting structural elements, located in the 3' untranslated region (3'-UTR). These elements can bind *trans*-acting factors and accelerate mRNA decay. The best characterised *cis*-acting destabilising elements in 3'-UTR are AU-rich elements (ARE) found in many short-lived mRNAs (Guhaniyogi and Brewer, 2001). The most common AREs found in unstable mRNAs consist of multiple pentamers of AUUUA, or AU- or U-rich elements (Mitchell and Tollervey, 2000). c-*myb* mRNA contains five copies of the AUUUA sequence in its 3'UTR (Figure 2).

Interestingly, integration of a retrovirus into the 3' end of the c-*myb* locus causes aberrant termination of c-*myb* transcription in 5'LTR and replaces the endogenous c-*myb* 3'-UTR containing AREs with viral LTR sequences. This leads to the dramatic increase in the stability of c-*myb* RNAs. The half-life of c-*myb* mRNA in M1 cells is around 45 minutes, while the half-lives of aberrant c-myb mRNAs in cell lines RI-4-11 (provirus integration in exon 9) and 45-16 (integration in exon 13) is substantially longer. As shown in Figure 2, we detected only 20% of full-length c-*myb* mRNA in M1 cells treated with actinomycin D for 120 minutes, while we measured 90% of the two oncogenic mRNAs forms in RI-4-11 and 45-16 cells. Importantly, the half-life of c-*myc* mRNA in all three cell lines was very similar, suggesting that the dramatic difference of c-*myb* mRNA stability is indeed an intrinsic feature and is not cell line-specific.

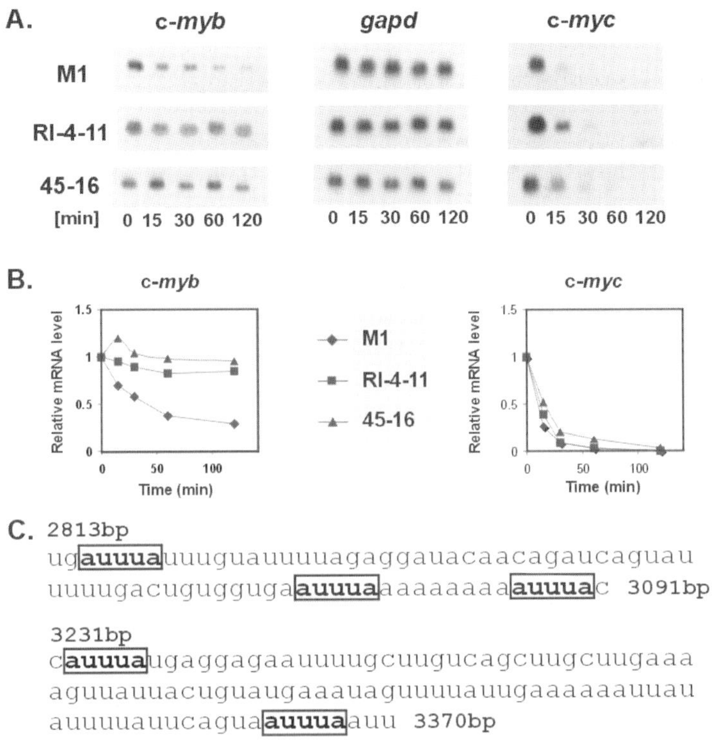

Figure 2

Stabilisation of c-*myb* mRNA as a consequence of proviral integration into 3' end region. (A) Cell lines expressing wild type c-*myb* (M1) as well as *myb* mRNAs truncated at the 3' end due to retrovirus integration (RI-4-11 - provirus integrated into exon 9; 45-16 - provirus integrated into exon 13) were cultivated in the presence of the RNA synthesis inhibitor actinomycin D (5µg/ml) for the indicated times. Isolated RNA samples were analyzed by northern blotting using radioactively labelled cDNAs for murine c-*myb*, c-*myc* and visualised by radiography. Hybridisation with a GAPDH probe was used as a control for loading and the integrity of RNA samples. (B) Quantitative analysis was performed on a PhosphoImager 425 using IMAGEQUANT software. Levels of c-*myb* and c-*myc* at each time point were normalised to GAPDH level and plotted on the graph. Each point represents the relative amount of either c-*myb* or c-*myc* mRNA at different times. RNA levels were assigned to 100% at the beginning of actinomycin D treatment. (C) c-*myb* 3' UTR removed from oncogenic *myb* RNAs due to integration of proviruses into the 3' end of c-*myb* gene. Destabilising heptamers sequences "auuua" are boxed.

3.2.2 Protein stabilisation

Removal of the 3'end of the c-*myb* gene causes not only loss of sequences important for targetting mRNA for decay, but it also removes protein sequences important for targetting the protein for degradation. We have shown that c-Myb is a very unstable protein rapidly degraded by the ubiquitin/26S proteasome pathway (Bies and Wolff, 1997). Attachment of polyubiquitin chains serves as a recognition signal for 26S proteasome machinery ultimately leading to proteolysis. Recently we found that the targetting of the protein to the proteasome depends on Ser/Thr phosphorylation (Bies et al., 2000; Bies et al., 2001). The previously identified phosphorylation sites, Ser 11, 12 and 528, however, are not involved. We located two independent instability determinants, one in the extreme carboxy-terminus and one overlapping the putative leucine zipper (Bies et al., 1999). The ubiquitin/26S proteasome pathway is the only proteolytic system described for tightly controlling the amount of c-Myb in cells. The half-life does not change during proliferation and differentiation of myeloid cells (Feikova et al., 2000) and we have not detected a situation where degradation of c-Myb is induced. This suggests that its turnover is a constitutive process. However, the fact that inhibition of Ser/Thr protein phosphatases rapidly causes hyperphosphorylation-induced conformational changes in the carboxy-terminus and accelerated proteolytic breakdown of c-Myb, suggests that there may be a signal transduction pathway that regulates proteolysis of this protein.

As mentioned above, deletion of the carboxy-terminal negative regulatory domain of c-Myb results in protein stabilisation. Evidence suggests that carboxy-terminally truncated proteins are less efficiently ubiquitinated and, therefore, have increased resistance to degradation (Bies and Wolff, 1997). c-Myb is not the only protein found to be activated to become oncogencic by stabilisation. Examples of other transcription factors where stablisation occurs in conjunction with oncogenic activation are c-Myc, c-Fos and c-Jun (Hershko and Ciechanover, 1998). Therefore, stabilisation of proteins with oncogenic potential may be a common mechanism for increasing transforming potential.

3.2.3 Increased transactivation

Carboxy-terminal truncation of c-Myb increases transforming activity (Gonda et al., 1989; Grasser et al., 1991; Ferrao et al., 1995) and this is related, in part, to the fact that the carboxy-terminal portion of the protein, beyond the transactivation domain, negatively regulates c-Myb activity as a transcription factor. Early studies with carboxy-terminal deletion mutants suggested that carboxy-terminal negative regulatory domain (NRD) decreases the transactivation capacity (Sakura et al., 1989; Kalkbrenner et al.1990; Hu et al., 1991), as well as the DNA-binding affinity of c-Myb (Ramsay et al., 1991; Tanaka et al., 1997). Although removal of protein instability determinants in the NRD alone may account in part for the observed increases in transactivation upon removal of the NRD, removal of other regulatory elements are likely to account for these increases in activity as well.

The NRD of c-Myb is subjected to post-translational modifications. Phosphorylation of serine 528 by MAPK kinase or other cellular proline-directed kinase has been implicated in the negative regulation of c-Myb transactivation on some promoters without affecting DNA-binding affinity of Myb (Aziz et al., 1995; Miglarese et al., 1996).

Recently, we described a post-translational modification of c-Myb by a ubiquitin-like protein SUMO-1. We showed that sumolation negatively regulates c-Myb transactivation and increases its stability (Bies et al., 2002). SUMO-1 is covalently attached to the NRD of c-Myb on lysine residues 499 and 523 and another lysine residue that has not been specifically identified. Sumolation of Lys523 is necessary for subsequent covalent attachment of a second molecule of SUMO-1 to Lys499, and modification of both Lys523 and Lys499 is required for conjugation of the third molecule of SUMO-1 to the unidentified lysine (Bies et al., 2002). It is assumed that sumolation of the first Lys523 in c-Myb can affect conformation of the carboxy-terminus so that it allows covalent attachment of SUMO-1 to other lysines. We provided evidence for the involvement of SUMO-1 in negative regulation of transactivation, using the K523R mutant that is completely deficient in sumolation. This mutant was shown to have an increased transactivation capacity on a Myb-responsive promoter. One proposed mechanism by which SUMO-1 could inhibit transactivation is by steric hindrance of acetylation at the carboxy-terminus of c-Myb (Figure 1), as it was reported that acetylation of several lysine residues by p300/CBP in carboxy-terminus can positively regulate its transactivation capacity (Tomita et al., 2000; Sano and Ishii, 2001).

Several other components of the NRD with a specific role in negative regulation of c-Myb have been identified. First, a leucine zipper-like (LZ) sequence, located in a region spanning amino acids 375-403, has negative regulatory activity. Deletion or point mutation of the LZ increased transactivation and transforming activity of c-Myb (Kanei-Ishii et al., 1992). It was proposed that the negative activity of the leucine zipper is due to the binding of specific inhibitor proteins. The first inhibitor that was suggested was c-Myb itself. It was shown that the leucine zipper-like sequence of c-Myb is capable of forming homodimers *in vitro* (Nomura et al., 1993). However, there is no evidence so far for the presence of full-length c-Myb homodimers in cells. Use of a GST-LZ-Myb pull-down assay led to the identification and molecular characterisation of a c-Myb binding protein p67/p160. It was shown that p67 (the amino-terminal portion of p160 protein) represses transactivation of c-Myb (Tavner et al., 1998). The LZ can be interrupted by alternative splicing and the alternatively spliced form identified in many normal and tumour cell lines encodes a protein with increased transactivation activity (Shen-Ong, 1987; Woo et al., 1998). Interestingly, the LZ structure also overlaps with one of the two instability regions identified in the negative regulatory domain suggesting its role in proteolytic processing (Bies at al., 1999).

Two regions within the c-Myb NRD have been shown to be important in the negative regulation of DNA-binding affinity and this effect would be predicted to influence transactivation (Tanaka et al., 1997). A region containing the PEST3/EVES motif within the NRD was shown to be capable of interacting with the DNA-binding domain (Dash et al., 1996). The p100 protein contains an identical EVES motif through which it can also interact with the DNA binding domain of c-Myb (Dash et al., 1996), and compete with the EVES region binding (Ness, 1999). Therefore, transactivation may be inhibited by the intramolecular interaction and activated by binding of the p100. More recently, an adenovirus E1A-associated protein BS69, was shown to interact with carboxy-terminus of c-Myb and inhibited its transactivation capacity (Ladendorff et al., 2001).

Thus, deletion of the negative regulatory domain, which is observed frequently in activated forms of c-Myb, increases stability of truncated proteins as well as facilitates an escape from multiple negative regulations imposed by post-translational modifications and protein-protein interactions.

4. RETROVIRUSES INTEGRATED 30-100 KB UPSTREAM AND DOWNSTREAM OF THE C-MYB TRANSCRIPTIONAL UNIT

In murine and feline leukaemias, sites of integrated proviruses have been mapped as far as 25-100 kbp upstream or downstream of the c-*myb* gene transcriptional unit on chromosome 10. *Mml1*, *Mml2, and Mml3* are sites that have been identified in our laboratory in myeloid promonocytic leukaemias and map 25-70 kbp upstream of the c-*myb* locus (Koller et al., 1996, Haviernik et al., 2002) (Figure 3). Mapping even further upstream, by approximately 30 kbp, is another feline leukaemia virus-common integration site in thymic lymphomas *Fit-1* (*Fti1*) (Tsujimoto et al., 1993; Hanlon et al., 2003). Analysis of the region encompassing these proviruses did not reveal the presence of any gene, but led to the identification of two potential scaffold (matrix) attachment regions (SARs/MARs) (Haviernik et al., 2002). SARs/MARs are sites where chromatin attaches to the nuclear matrix. These sequences play an important role in organising and regulating nuclear processes including transcription (Deppert, 2000). The important regulatory function of the nuclear matrix makes it a likely target for structural alteration during neoplastic transformation.

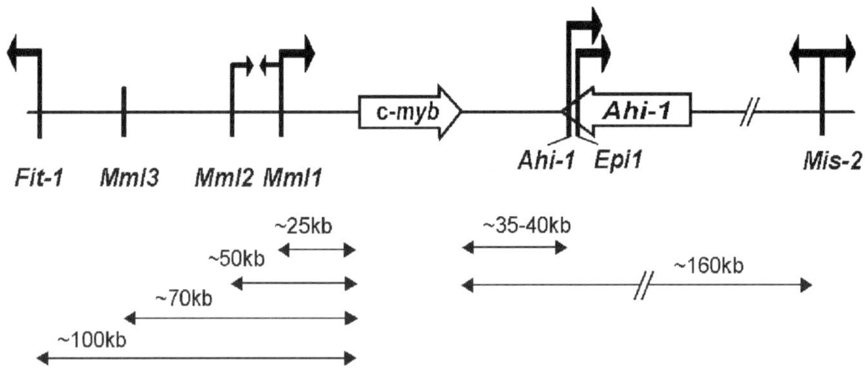

Figure 3
Common integration sites upstream and downstream of c-myb transcription unit. Positions of proviruses in genomic DNA on mouse chromosome 10 surrounding c-myb and Ahi-1 genes are marked by vertical black lines. Open arrows show the orientation of the two genes. The orientations of the proviruses at different sites are showed by black arrows above the integration sites, and the approximate distance of common integration sites from c-myb gene are marked under the diagram.

There is evidence to suggest that alterations in chromatin may be important in cancer. For example, the transforming potential of simian virus 40 (SV40) large T antigen is closely associated with perturbation of chromatin structure and changes in nuclear architecture (Malyapa et al., 1996). Also, oncoproteins with an A/T hook domain can bind directly to A/T rich sequences in SARs. The mixed lineage leukaemia (MLL) protein is an example of a protein with oncogenic potential and it contains an A/T hook domain. Competitive binding of proteins like MLL and nuclear matrix for binding to MARs could modulate chromatin structure and ultimately lead to altered regulation of gene expression (Caslini et al., 2000). Interestingly, the gene that encodes MLL is frequently involved in translocations in human acute myeloid and lymphoid leukaemias and the translocations encode fusion proteins between the SAR binding portion of MLL and other protein domains.

It is conceivable that integration of the proviral genome into regions that bind nuclear matrix could also perturb normal regulation of gene expression and ultimately lead to cancer. Because no new transcription units were found in the upstream genomic region, the most probable candidate affected by these proviruses remains to be c-*myb* itself (Haviernik et al., 2002). Although, promonocytic leukaemias with provirus in *Mml1* do not always express c-*myb*, it is possible that c-Myb expression was crucial for leukaemia progression, but only during earlier stages of development. Development of promonocytic leukaemia may involve an early immature stage in which regulation of c-*myb* transcription requires the upstream region. A recent paper by Hanlon and colleagues strongly indicates that c-*myb* is a key target gene affected by long-range transcriptional activation (Hanlon et al., 2003). They suggest that c-*myb* expression may become dispensable during cultivation of cells *in vitro* or *in vivo* during the progression of virus-induced leukaemias in mice.

Two other common murine leukaemia virus insertion loci, *Ahi-1* (Jiang et al., 1994) and *Mis-2* (Villeneuve et al., 1993), were mapped approximately 35 and 160 kbp downstream of c-*myb* in murine lymphomas. Recently, a somatically acquired common retroviral integration site, *Epi1,* located just 30-40 kbp downstream of c-*myb* was described in murine myeloid leukaemias in BXH2 mice (Blaydes et al., 2001) and may overlap *Ahi-1* sites. Although these proviruses are integrated at the end of a recently described gene *Ahi-1* (Jiang et al., 2002), this gene does not seem to be over-expressed in association with retrovirus integration. Therefore, it remains possible that downstream proviruses affect c-*myb* expression as well. Interestingly, proviruses found in *Epi1* and *Ahi-1* were integrated in the same transcriptional orientation as the c-*myb* gene, and this observation is in complete agreement with the theory of viral enhancement, where the enhancer in 5' LTR activates a gene located upstream of provirus integration site (Jonkers and Berns, 1996). Northern analyses did not reveal increased c-*myb* expression in tumours with viruses integrated in *Epi1*. However, it is possible c-*myb* plays a role in leukaemogenesis in these tumours during an earlier stage of development as suggested by Hanlon and colleagues (Hanlon et al., 2003).

That c-*myb* is a target of retrovirus activation in leukaemia has been further emphasised in high throughput retroviral tagging of genes (Joosten et al., 2002; Lund et al., 2002; Mikkers et al., 2002; Suzuki et al., 2002). The c-*myb* locus was among the most frequently targetted genes identified in genome-wide screenings involving murine myeloid and lymphoid leukaemias. Whether these sites of integrated proviruses are located within or outside the immediate c-*myb* transcriptional unit has not been reported. However, this further implicates for c-Myb as having a crucial role in the induction of these diseases.

5. C-MYB TARGET GENES IMPLICATED IN MYELOID CELL TRANSFORMATION

The ability of c-Myb to both promote transcription and transform cells (Lane et al., 1990; Kanei-Ishii et al., 1992) indicates that it induces myeloid and lymphoid leukaemia through activation of specific target genes. An increasing amount of evidence from the past two decades has confirmed a direct role of c-Myb in regulation of cellular processes such as proliferation, differentiation and programmed cell death (Wolff, 1996; Oh and Reddy, 1999). Several Myb-target genes with "leukaemogenic" potential have been identified and will be briefly discussed.

A gene regulated by c-Myb that is critical to its transforming capacity and involved in the regulation of the cell cycle is c-*myc*. Initially, it was shown that the c-*myc* promoter is responsive to c-Myb (Cogswell et al., 1993; Evans et al., 1990; Zobel et al., 1992) with its highest level of activity in myeloid cells. However, until recently there was a lack of evidence that expression of the endogenous c-*myc* promoter is controlled by this transcription factor. Studies employing conditional expression of Myb and dominant negative forms of Myb demonstrated that the resident chromosomal c-*myc* gene is regulated directly by c-Myb in myeloid leukaemic cells (Schmidt et al., 2000, Wolff et al., 2001, Chen et al., 2002).

Other genes, proposed to be targets of c-Myb and essential for proliferation of haemopoietic cells, include p34*cdc2* (Ku et al., 1993), DNA topoisomerase IIα (Brandt et al., 1997), c-*kit* (Hogg et al., 1997; Ratajczak et al., 1998 Vandenbark et al., 1996) and *cyclinA1* (Muller et al., 1999).

More recently, it was reported that c-Myb inhibits the expression of the proposed tumour suppressor gene $p15^{INK4b}$, which encodes an inhibitor of cyclin dependent kinases cdk4/6 (Schmidt et al., 2001). Ectopic expression either full-length c-Myb or its truncated forms prevented the induction of $p15^{INK4b}$ mRNA in M1 cells during interleukin-6-induced monocytic differentiation. The effect of c-Myb on $p15^{INK4b}$ expression appears to be indirect and not due to the action of c-Myb on the Ink4b promoter (our own unpublished data). However, this function of c-Myb is probably an important mechanism in the transformation process, because the majority of c-Myb-induced tumours (in contrast to Myc-induced monocytic tumours) do not express this gene (Schmidt et al., 2001). Therefore, the inhibition of a growth arrest pathway involved in monocyte differentiation is another mechanism by which c-Myb promotes proliferation.

Modulation of programmed cell death by c-Myb represents another important mechanism involved in transformation of haemopoietic cells by Myb. Several groups identified *bcl-2* gene, which encodes an anti-apoptotic protein as a direct target for the c-Myb transcription factor in lymphoid and myeloid cells (Taylor et al., 1996; Frampton et al., 1996; Schmidt et al., 2001), and placed c-Myb oncoprotein into a family of transcription factors with "survival" function. It is important to note, however, that under some conditions c-Myb can actually promote apoptosis rather than prevent it (Selvakumaran et al., 1994, Sala et al., 1996).

Based on the evidence mentioned above, transformation of haemopoietic cells by c-Myb seems to be achieved through modulation of at least two distinct pathways, one involving proliferation, and one related to programmed cell death. Therefore, deregulation of c-Myb via oncogenic activation, which increases its stability or its ability to transactivate genes, plays an important role maintaining cell cycle progression and preventing programmed cell death. Since c-Myb induces c-*myc* expression and this oncogene is known to activate the p53 pathway leading to apoptosis, it is to c-Myb's advantage as an oncogene to counteract c-Myc's anti-tumour effects by preventing c-Myc induction of apoptosis.

6. C-MYB AND HUMAN LEUKAEMIA

The human c-*myb* oncogene is located on chromosome 6q22-24 (Harper et al., 1983). Abnormalities at this locus, such as amplification or deletion, have been observed in leukaemic cells with over-expression of unaltered c-*myb* (Pelicci et al., 1984; Barletta et al., 1987; Okada et al., 1990). The only activated form of c-Myb in human leukaemia detected to date was a carboxy-terminally truncated c-Myb (Tomita et al., 1998). It was observed in the TK-6 cell line, which was established from a patient with chronic myelogenous leukaemia and resulted from a large deletion in the chromosome. This abnormality was associated with late progression, because it was acquired after T-cell blast crisis. A recent screening for activating mutations in the negative regulatory domain of c-Myb in patients with myeloid leukaemia did not reveal any abnormalities (Lutwyche et al., 2001). Therefore, so far it is not clear whether full length c-Myb over-expression or timing of expression, due to altered regulation, plays a role in human leukaemia.

Although little evidence of c-Myb's involvement in leukaemia has been reported so far there are ways one may envision that the transcription factor could positively affect transformation of human haemopoietic cells. The ideas are based on what we have learned from animal model systems. It is clear that c-Myb is required for proliferation of haemopoietic cells and the only examples of proliferating haemopoietic cells that lack c-Myb expression, as far as we know, are those with deregulated c-Myc. The first mechanism to consider would be a disturbance in the abundance of the protein. Alterations in the abundance of critical proteins are frequently observed defects in cancer cells. Among the primary mechanisms used by cells to adjust protein concentrations are gene dosage, mRNA abundance, and protein stability. We have shown that increased RNA and protein stability significantly contribute to increased levels of c-Myb in murine myeloid leukaemia. Interestingly, increased stability of c-*myb* RNA was observed in patients with acute myeloid leukaemia (Baer et al., 1992). In addition, microsatellite deletions in the c-*myb* transcriptional attenuator located in the first intron was associated with overexpression of normal c-Myb in colon carcinomas providing more evidence for an oncogenic potential of deregulated full-length c-Myb (Thompson et al., 1997). In regard to mRNA regulation in animal models, several retroviral integrations at a distance from the c-*myb* locus have been hypothesised to positively affect transcription (Koller et al., 1996, Haviernik et al., 2002, Hanlon et al., 2003). This leaves open the possibility that alterations that disturb these distal chromosomal regulatory mechanisms, could affect transformation of human cells as well. Alterations that cause changes in abundance may be subtle and difficult to detect, however, at present, they cannot be ruled out. The second mechanism to consider, based on evidence in animal models, is an alteration that would increase directly or indirectly c-Myb's transactivation potential. Several post-translational modifications such as phosphorylation, acetylation and sumolation are involved in the control of c-Myb activity and/or proteolytic processing. Altered regulation of these pathways in leukaemic cells could theoretically result in potentiating c-Myb's oncogenic ability to drive proliferation and prevent apoptosis.

ACKNOWLEDGEMENTS

This work was supported in part by the grant No. 2/1108/23 from the VEGA Slovak Academy of Sciences.

REFERENCES

Aziz, N., Miglarese, M.R., Hendrickson, R.C., Shabanowitz, J., Sturgill, T.W., Hunt, D.F. and Bender, T.P. (1995) Modulation of c-Myb-induced transcription activation by a phosphorylation site near the negative regulatory domain. Proc. Natl. Acad. Sci. USA *92*, 6429-6433.

Badiani, P.A., Kioussis, D., Swirsky, D.M., Lampert, I.A. and Weston, K. (1996) T-cell lymphomas in v-Myb transgenic mice. Oncogene *13*, 2205-2212.

Baer, M.R., Augustinos, P. and Kinniburgh, A.J. (1992) Defective c-myc and c-myb RNA turnover in acute myeloid leukaemia cells. Blood *79*, 1319-1326.

Barletta, C., Pelicci, P.G., Kenyon, L.C., Smith, S.D. and Dalla-Favera, R. (1987) Relationship between the c-myb locus and the 6q-chromosomal aberration in leukaemias and lymphomas. Science *235*, 1064-1067.

Belli, B., Wolff, L., Nazarov, V. and Fan, H. (1995) Proviral activation of the c-myb proto-oncogene is detectable in preleukaemic mice infected neonatally with Moloney murine leukaemia virus but not in resulting end stage T lymphomas. J. Virol. *69*, 5138-5141.

Bender, T.P. and Kuehl, W.M. (1986) Murine myb protooncogene mRNA: cDNA sequence and evidence for 5' heterogeneity. Proc. Natl. Acad. Sci. USA *83*, 3204-3208.

Bies, J., Feikova, S., Bottaro, D.P. and Wolff, L. (2000) Hyperphosphorylation and increased proteolytic breakdown of c-Myb induced by the inhibition of Ser/Thr protein phosphatases. Oncogene *19*, 2846-2854.

Bies, J., Feikova, S., Markus, J. and Wolff, L. (2001) Phosphorylation-dependent conformation and proteolytic stability of c-Myb. Blood Cells Mol. Dis. *27*, 422-428.

Bies, J., Hoffman, B., Amanullah, A., Giese, T. and Wolff, L. (1996) B-Myb prevents growth arrest associated with terminal differentiation of monocytic cells. Oncogene *12*, 355-363.

Bies, J., Markus, J. and Wolff, L. (2002) Covalent attachment of the SUMO-1 protein to the negative regulatory domain of the c-Myb transcription factor modifies its stability and transactivation capacity. J. Biol. Chem. *277*, 8999-9009.

Bies, J., Mukhopadhyaya, R., Pierce, J. and Wolff, L. (1995) Only late, nonmitotic stages of granulocyte differentiation in 32Dcl3 cells are blocked by ectopic expression of murine c-myb and its truncated forms. Cell. Growth Differ. *6*, 59-68

Bies, J., Nazarov, V. and Wolff, L. (1999) Identification of protein instability determinants in the carboxy-terminal region of c-Myb removed as a result of retroviral integration in murine monocytic leukaemias. J. Virol. *73*, 2038-2044.

Bies, J. and Wolff, L. (1997) Oncogenic activation of c-Myb by carboxyl-terminal truncation leads to decreased proteolysis by the ubiquitin-26S proteasome pathway. Oncogene *14*, 203-212.

Blaydes, S.M., Kogan, S.C., Truong, B.T., Gilbert, D.J., Jenkins, N.A., Copeland, N.G., Largaespada, D.A. and Brannan, C.I. (2001) Retroviral integration at the Epi1 locus cooperates with Nf1 gene loss in the progression to acute myeloid leukaemia. J. Virol. *75*, 9427-9434.

Brandt, T.L., Fraser, D.J., Leal, S., Halandras, P.M., Kroll, A.R. and Kroll, D.J. (1997) c-Myb trans-activates the human DNA topoisomerase IIalpha gene promoter. *J. Biol. Chem. 272*, 6278-6284.

Brown, K.E., Kindy, M.S. and Sonenshein, G.E. (1992) Expression of the c-myb proto-oncogene in bovine vascular smooth muscle cells. J. Biol. Chem. *267*, 4625-4630.

Caslini, C., Alarcon, A.S., Hess, J.L., Tanaka, R., Murti, K.G. and Biondi, A. (2000) The amino terminus targets the mixed lineage leukaemia (MLL) protein to the nucleolus, nuclear matrix and mitotic chromosomal scaffolds. Leukaemia *14*, 1898-1908.

Chenm J,, Kremerm C,S. and Bender, T,P. (2002) A Myb dependent pathway maintains Friend murine erythroleukaemia cells in an immature and proliferating state. Oncogene *21*, 1859-1869.

Chen, R.H., Fields, S. and Lipsick, J.S. (1995) Dissociation of transcriptional activation and oncogenic transformation by v-Myb. Oncogene *11*, 1771-1779.

Cogswell, J.P., Cogswell, P.C., Kueh,l W.M., Cuddihy, A.M., Bender, T.M., Engelke, U., Marcu, K.B. and Ting, J.P. (1993) Mechanism of c-myc regulation by c-Myb in different cell lineages. Mol. Cell. Biol. *13*, 2858-2869.

Dash, A.B., Orrico, F.C. and Ness, S.A. (1996) The EVES motif mediates both intermolecular and intramolecular regulation of c-Myb. Genes Dev. *10*, 1858-1869.

Deppert, W. (2000) The nuclear matrix as a target for viral and cellular oncogenes. Crit. Rev. Eukaryot. Gene Expr. *10*, 45-61.

Dini, P.W. and Lipsick, J.S. (1993) Oncogenic truncation of the first repeat of c-Myb decreases DNA binding *in vitro* and *in vivo*. Mol. Cell. Biol. *13*, 7334-7348.

Dvorak, M., Urbanek, P., Bartunek, P., Paces, V., Vlach, J., Pecenka, V., Arnold, L., Travnicek, M. and Riman, J. (1989) Transcription of the chicken myb proto-oncogene starts within a CpG island. Nucleic Acids Res. *17*, 5651-5664.

Evans, J.L., Moore, T.L., Kuehl, W.M., Bender, T.P. and Ting, J.P. (1990) Functional analysis of c-Myb protein in T-lymphocytic cell lines shows that it trans-activates the c-myc promoter. Mol. Cell. Biol. *10*, 5747-5752.

Feikova, S., Wolff, L. and Bies, J. (2000) Constitutive ubiquitination and degradation of c-myb by the 26S proteasome during proliferation and differentiation of myeloid cells. Neoplasma *47*, 212-218.

Ferrao, P., Macmillan, E.M., Ashman, L.K. and Gonda, T.J. (1995) Enforced expression of full length c-Myb leads to density-dependent transformation of murine haemopoietic cells. Oncogene *11*, 1631-1638.

Frampton, J., Ramqvist, T. and Graf, T. (1996) v-Myb of E26 leukaemia virus up-regulates bcl-2 and suppresses apoptosis in myeloid cells. Genes Dev. *10*, 2720-2731.

Gonda, T.J., Buckmaster, C. and Ramsay, R.G. (1989) Activation of c-myb by carboxy-terminal truncation: relationship to transformation of murine haemopoietic cells *in vitro*. EMBO J. *8*, 1777-1783.

Gonda, T.J., Macmillan, E.M., Townsend, P. and Hapel, A.J. (1993) Differentiation state and responses to haemopoietic growth factors of murine myeloid cells transformed by myb. Blood *82*, 2813-2822.

Grasser, F.A., Graf, T. and Lipsick, J.S. (1991) Protein truncation is required for the activation of the c-myb proto-oncogene. Mol. Cell. Biol. *11*, 3987-3996.

Guhaniyogi, J. and Brewer, G. (2001) Regulation of mRNA stability in mammalian cells. Gene *265*, 11-23.

Hanlon, L., Barr, N.I., Blyth, K, Stewart, M., Haviernik, P., Wolff, L., Weston, K., Cameron, E.R. and Neil, J.C. (2003) Long-range effects of retroviral insertion on c-myb: overexpression may be obscured by silencing during tumour growth *in vitro*. J. Virol. *77*, 1059-1068.

Harper, M.E., Franchini, G., Love, J., Simon, M.I., Gallo, R.C. and Wong-Staal, F. (1983) Chromosomal sublocalization of human c-myb and c-fes cellular onc genes. Nature *304*, 169-171.

Haviernik, P., Festin, S.M., Opavsky, R., Koller, R.P., Barr, N.I., Neil, J.C. and Wolff, L. (2002) Linkage on chromosome 10 of several murine retroviral integration loci associated with leukaemia. J. Gen. Virol. *83*, 819-827.

Hershko, A, and Ciechanover, A. (1998) The ubiquitin system. Annu. Rev. Biochem. *67*, 425-479.

Hogg, A., Schirm, S., Nakagoshi, H., Bartley, P., Ishii, S., Bishop, J.M. and Gonda, T.J. (1997) Inactivation of a c-Myb/estrogen receptor fusion protein in transformed primary cells leads to granulocyte/macrophage differentiation and down regulation of c-kit but not c-myc or cdc2. Oncogene *15*, 2885-2898.

Hu, Y.L, Ramsay, R.G., Kanei-Ishii, C., Ishii, S. and Gonda, T.J. (1991) Transformation by carboxyl-deleted Myb reflects increased transactivating capacity and disruption of a negative regulatory domain. Oncogene *6*, 1549-1553.

Introna, M. and Golay, J. (1999) How can oncogenic transcription factors cause cancer: a critical review of the myb story. Leukaemia *13*, 1301-1306.

Jiang, X., Hanna, Z., Kaouass, M., Girard, L. and Jolicoeur, P. Ahi-1, a novel gene encoding a modular protein with WD40-repeat and SH3 domains, is targeted by the Ahi-1 and Mis-2 provirus integrations. J. Virol. *76*, 9046-9059.

Jiang, W., Kanter, M.R., Dunkel, I., Ramsay, R.G., Beemon, K.L. and Hayward, W.S. (1997) Minimal truncation of the c-myb gene product in rapid-onset B-cell lymphoma. J. Virol. *71*, 6526-6533.

Jiang, X., Villeneuve, L., Turmel, C., Kozak, C.A. and Jolicoeur, P. (1994) The Myb and Ahi-1 genes are physically very closely linked on mouse chromosome 10. Mamm. Genome *5*, 142-148.

Joosten, M., Vankan-Berkhoudt, Y., Tas, M., Lunghi, M., Jenniskens, Y., Parganas, E., Valk, P.J., Lowenberg, B., van den Akker, E. and Delwel, R. (2002) Large-scale identification of novel potential disease loci in mouse leukaemia applying an improved strategy for cloning common virus integration sites. Oncogene *21*, 7247-7255.

Jonkers, J. and Berns, A. (1996) Retroviral insertional mutagenesis as a strategy to identify cancer genes. Biochim. Biophys. Acta *1287*, 29-57.

Kalkbrenner, F., Guehmann, S. and Moelling, K. (1990) Transcriptional activation by human c-myb and v-myb genes. Oncogene *5*, 657-661.

Kanei-Ishii, C., MacMillan, E.M., Nomura, T., Sarai, A., Ramsay, R.G., Aimoto, S., Ishii, S. and Gonda, T.J. (1992) Transactivation and transformation by Myb are negatively regulated by a leucine-zipper structure. Proc. Natl. Acad. Sci. USA *89*, 3088-3092.

Kanter, M.R., Smith, R.E. and Hayward, W.S. (1988) Rapid induction of B-cell lymphomas: insertional activation of c-myb by avian leukosis virus. J. Virol. *62*, 1423-1432.

Koller, R., Krall, M., Mock, B., Bies, J., Nazarov, V. and Wolff, L. (1996) Mml1, a new common integration site in murine leukaemia virus-induced promonocytic leukaemias maps to mouse chromosome 10. Virology *224*, 224-234.

Ku, D.H., Wen, S.C., Engelhard, A., Nicolaides, N.C., Lipson, K.E., Marino, T.A. and Calabretta, B. c-myb transactivates cdc2 expression via Myb binding sites in the 5'-flanking region of the human cdc2 gene. J. Biol. Chem. *268*, 2255-2259.

Ladendorff, N.E., Wu, S. and Lipsick, J.S. (2001) BS69, an adenovirus E1A-associated protein, inhibits the transcriptional activity of c-Myb. Oncogene *20*, 125-132.

Lane, T., Ibanez, C., Garcia, A., Graf, T. and Lipsick, J. (1990) Transformation by v-myb correlates with trans-activation of gene expression. Mol. Cell. Biol. *10*, 2591-2598.

Le Rouzic, E. and Perbal, B. (1996) Retroviral insertional activation of the c-myb proto-oncogene in a Marek's disease T-lymphoma cell line. J. Virol. *70*, 7414-7423.

Lipsick, J.S. and Wang, D.M. (1999) Transformation by v-Myb. Oncogene *18*, 3047-3055.

Lund, A.H., Turner, G., Trubetskoy, A., Verhoeven, E., Wientjens, E., Hulsman, D,. Russell, R., DePinho, RA., Lenz, J. and van Lohuisen, M. (2002) Genome-wide retroviral insertional tagging of genes involved in cancer in Cdkn2a-deficient mice. Nat. Genet. *32*, 160-165.

Lüscher, B., Christenson, E., Litchfield, D.W., Krebs, E.G. and Eisenman, R.N. (1990) Myb DNA binding inhibited by phosphorylation at a site deleted during oncogenic activation. Nature *344*, 517-522.

Lutwyche, J.K., Keough, R.A., Hughes, T.P. and Gonda, T.J. (2001) Mutation screening of the c-MYB negative regulatory domain in acute and chronic myeloid leukaemia. Br. J. Haematol. *114*, 632-634.

Macmillan, E.M. and Gonda, T.J. (1994) Murine myeloid cells transformed by myb require fibroblast-derived or autocrine growth factors in addition to granulocyte-macrophage colony-stimulating factor for proliferation. Blood *83*, 209-216.

Malyapa, R.S., Wright, W.D., Taylor, Y.C. and Roti JL. (1996) DNA supercoiling changes and nuclear matrix-associated proteins: possible role in oncogene-mediated radioresistance. Int. J. Radiat. Oncol. Biol. Phys. *35*, 963-973.

Mikkers, H., Allen, J., Knipscheer, P., Romeijn, L., Hart, A., Vink, E., Berns, A. and Romeyn L. (2002) High-throughput retroviral tagging to identify components of specific signalling pathways in cancer. Nat. Genet. *32*, 153-159.

Miglarese, M.R., Richardson, A.F., Aziz, N. and Bender, T.P. (1996) Differential regulation of c-Myb-induced transcription activation by a phosphorylation site in the negative regulatory domain. J. Biol. Chem. *271*, 22697-22705.

Mitchell, P. and Tollervey, D. (2000) mRNA stability in eukaryotes. Curr. Opin. Genet. Dev. *10*, 193-198.

Mukhopadhyaya, R. and Wolff, L. (1992) New sites of proviral integration associated with murine promonocytic leukaemias and evidence for alternate modes of c-myb activation. J. Virol. *66*, 6035-6044.

Muller, C., Yang, R., Idos, G., Tidow, N., Diederichs ,S., Koch, O.M., Verbeek, W., Bender, T.P. and Koeffler, H.P. (1999) c-myb transactivates the human cyclin A1 promoter and induces cyclin A1 gene expression. Blood *94*, 4255-4262.

Nason-Burchenal, K. and Wolff, L. (1992) Involvement of the spleen in preleukaemic development of a murine retrovirus-induced promonocytic leukaemia. Cancer Res. *52*, 5317-5322.

Nason-Burchenal, K. and Wolff, L. (1993) Activation of c-myb is an early bone-marrow event in a murine model for acute promonocytic leukaemia. Proc. Natl. Acad. Sci. USA *90*, 1619-1623.

Nazarov, V. and Wolff, L. (1995) Novel integration sites at the distal 3' end of the c-myb locus in retrovirus-induced promonocytic leukaemias. J. Virol. *69*, 3885-3888.

Ness, S.A. (1999) Myb binding proteins: regulators and cohorts in transformation. Oncogene *18*, 3039-3046.

Nomura, T., Sakai, N., Sarai, A., Sudo, T., Kanei-Ishii, C., Ramsay, R.G., Favier, D., Gonda, T.J. and Ishii, S. (1993) Negative autoregulation of c-Myb activity by homodimer formation through the leucine zipper. J. Biol. Chem. *268*, 21914-21923.

Oelgeschlager, M., Kowenz-Leutz, E., Schreek, S., Leutz, A. and Lüscher, B. (2001) Tumourigenic N-terminal deletions of c-Myb modulate DNA binding, transactivation, and cooperativity with C/EBP. Oncogene *20*, 7420-7424.

Oelgeschlager, M., Krieg, J., Lüscher-Firzlaff, J.M. and Lüscher, B. (1995) Casein kinase II phosphorylation site mutations in c-Myb affect DNA binding and transcriptional cooperativity with NF-M. Mol. Cell. Biol. *15*, 5966-5974.

Oh, I.H. and Reddy, E.P. (1999) The myb gene family in cell growth, differentiation and apoptosis. Oncogene *18*, 3017-3033.

Okada, M., Tada, M., Kanda, N., Masuda, M., Mizoguchi, H., Kazuma, M., Wada, E., Kubota, K. and Nomura, Y. (1990) c-myb gene analysis in T-cell malignancies with del(6q). Cancer Genet. Cytogenet. *48*, 229-236.

Pelicci, P.G., Lanfrancone, L., Brathwaite, M.D., Wolman, S.R. and Dalla-Favera, R. (1984) Amplification of the c-myb oncogene in a case of human acute myelogenous leukaemia. Science *224*, 1117-1121.

Piser, E. and Humphries, E.H. (1989) RAV-1 insertional mutagenesis: disruption of the c-myb locus and development of avian B-cell lymphomas. J. Virol. *63*, 1630-1640.

Piser, E.S., Baba, T.W. and Humphries, E.H. (1992) Activation of the c-myb locus is insufficient for the rapid induction of disseminated avian B-cell lymphoma. J. Virol. *66*, 512-523.

Press, R.D., Wisner, T.W. and Ewert, D.L. (1995) Induction of B cell lymphomas by overexpression of a Myb oncogene truncated at either terminus. Oncogene *11*, 525-535.

Ramsay, R.G, Ishii, S. and Gonda, T.J. (1991) Increase in specific DNA binding by carboxyl truncation suggests a mechanism for activation of Myb. Oncogene *6*, 1875-1879.

Ratajczak, M.Z., Perrotti, D., Melotti, P., Powzaniuk, M., Calabretta, B., Onodera, K., Kregenow, D.A., Machalinski, B. and Gewirtz, A.M. (1998) Myb and ets proteins are candidate regulators of c-kit expression in human haemopoietic cells. Blood *91*, 1934-1946.

Reddy, C.D. and Reddy, E.P. (1989) Differential binding of nuclear factors to the intron 1 sequences containing the transcriptional pause site correlates with c-myb expression. Proc. Natl. Acad. Sci. USA *86*, 7326-7330.

Sala, A., Casella, I., Grasso, L., Bellon, T., Reed, J.C., Miyashita, T. and Peschle, C. (1996) Apoptotic response to oncogenic stimuli: cooperative and antagonistic interactions between c-myb and the growth suppressor p53. Cancer Res. *56*, 1991-1996.

Sakura, H., Kanei-Ishii, C., Nagase, T., Nakagoshi, H., Gonda ,T.J. and Ishii, S. (1989) Delineation of three functional domains of the transcriptional activator encoded by the c-myb protooncogene. Proc. Natl. Acad. Sci. USA *86*, 5758-5762.

Sano, Y. and Ishii, S. (2001) Increased affinity of c-Myb for CREB-binding protein (CBP) after CBP-induced acetylation. J. Biol. Chem. *276*, 3674-3682.

Schmidt, M., Nazarov, V., Stevens, L., Watson, R.J. and Wolff, L. (2000) Regulation of the resident chromosomal copy of c-myc by c-Myb is involved in myeloid leukaemogenesis. Mol. Cell. Biol. *20*, 1970-1981.

Schmidt, M., Koller, R., Haviernik, P., Bies, J., Maciag, K. and Wolff, L. (2001) Deregulated c-Myb expression in murine myeloid leukaemias prevents the up-regulation of p15(INK4b) normally associated with differentiation. Oncogene *20*, 6205-6214.

Selvakumaran, M., Liebermann, D.A., Hoffman-Liebermann, B. (1992) Deregulated c-myb disrupts interleukin-6- or leukaemia inhibitory factor-induced myeloid differentiation prior to c-myc: role in leukaemogenesis. Mol. Cell. Biol. *12*, 2493-2500.

Selvakumaran, M., Lin, H.K., Sjin, R.T., Reed, J.C., Liebermann, D.A. and Hoffman, B. (1994) The novel primary response gene MyD118 and the proto-oncogenes myb, myc, and bcl-2 modulate transforming growth factor beta 1-induced apoptosis of myeloid leukaemia cells. Mol. Cell. Biol. *14*, 2352-2360.

Shen-Ong, G.L. (1987) Alternative internal splicing in c-myb RNAs occurs commonly in normal and tumour cells. EMBO J. *6*, 4035-4039.

Shen-Ong, G.L. Morse, H.C. 3rd, Potter, M. and Mushinski, J.F. (1986) Two modes of c-myb activation in virus-induced mouse myeloid tumours. Mol. Cell. Bio*l*. *6*, 380-392.

Shen-Ong, G.L. and Wolff, L. (1987) Moloney murine leukaemia virus-induced myeloid tumours in adult BALB/c mice: requirement of c-myb activation but lack of v-abl involvement. J. Virol. *61*, 3721-3725.

Suzuki, T., Shen, H., Akagi, K., Morse, H.C., Malley, J.D., Naiman, D.Q., Jenkins, N.A. and Copeland, N.G. (2002) New genes involved in cancer identified by retroviral tagging. Nat. Genet. *32*, 166-174.

Tanaka, Y., Nomura, T. and Ishii, S. (1997) Two regions in c-myb proto-oncogene product negatively regulating its DNA-binding activity. FEBS Lett. *413*, 162-168.

Tavner, F.J., Simpson, R., Tashiro, S., Favier, D., Jenkins, N.A., Gilbert, D.J., Copeland, N.G., Macmillan, E.M., Lutwyche, J., Keough, R.A, Ishii, S. and Gonda, T.J. (1998) Molecular cloning reveals that the p160 Myb-binding protein is a novel, predominantly nucleolar protein which may play a role in transactivation by Myb. Mol. Cell. Bio*l*. *18*, 989-1002.

Taylor, D., Badiani, P. and Weston, K. (1996) A dominant interfering Myb mutant causes apoptosis in T cells. Genes Dev. *10*, 2732-2744.

Thompson, M.A., Flegg, R., Westin, E.H. and Ramsay, R.G. (1997) Microsatellite deletions in the c-myb transcriptional attenuator region associated with over-expression in colon tumour cell lines. Oncogene *14*, 1715-1723.

Tomita, A., Towatari, M., Tsuzuki, S., Hayakawa, F., Kosugi, H., Tamai, K., Miyazaki, T., Kinoshita, T. and Saito, H. (2000) c-Myb acetylation at the carboxyl-terminal conserved domain by transcriptional co-activator p300. Oncogene *19*, 444-451.

Tomita, A., Watanabe, T., Kosugi, H., Ohashi, H., Uchida, T., Kinoshita, T., Mizutani, S., Hotta, T., Murate, T., Seto, M. and Saito, H. (1998) Truncated c-Myb expression in the human leukaemia cell line TK-6. Leukaemia *12*, 1422-1429.

Tsujimoto, H., Fulton, R., Nishigaki, K., Matsumoto, Y., Hasegawa, A., Tsujimoto, A., Cevario, S., O'Brien, S.J., Terry, A. and Onions, D. (1993) A common proviral integration region, fit-1, in T-cell tumours induced by myc-containing feline leukaemia viruses. Virology *196*, 845-848.

Vandenbark, G.R., Chen, Y., Friday, E., Pavlik, K., Anthony, B., deCastro, C. and Kaufman, R.E. (1996) Complex regulation of human c-kit transcription by promoter repressors, activators, and specific myb elements. Cell. Growth Differ. *7*, 1383-1392.

Villeneuve, L., Jiang, X., Turmel, C., Kozak, C.A. and Jolicoeur, P. (1993) Long-range mapping of Mis-2, a common provirus integration site identified in murine leukaemia virus-induced thymomas and located 160 kilobase pairs downstream of Myb. J. Virol. *67*, 5733-5739.

Wang, L.G., Liu, X.M., Li, Z.R., Denstman, S. and Bloch, A. (1994) Differential binding of nuclear c-ets-1 protein to an intron I fragment of the c-myb gene in growth versus differentiation. Cell. Growth Differ. *5*, 1243-1251.

Watson, R.J. (1988) Expression of the c-myb and c-myc genes is regulated independently in differentiating mouse erythroleukaemia cells by common processes of premature transcription arrest and increased mRNA turnover. Mol. Cell. Biol. *8*, 3938-3942.

Weinstein, Y., Cleveland, J.L., Askew, D.S., Rapp, U.R. and Ihle, J.N. (1987) Insertion and truncation of c-myb by murine leukaemia virus in a myeloid cell line derived from cultures of normal haemopoietic cells. J. Virol. *61*, 2339-2343.

Wolff, L., Koller, R. and Davidson, W. (1991) Acute myeloid leukaemia induction by amphotropic murine retrovirus (4070A): clonal integrations involve c-myb in some but not all leukaemias. J. Virol. *65*, 3607-316.

Wolff, L. (1996) Myb-induced transformation. Crit. Rev. Oncog. *7*, 245-260.

Wolff, L, Schmidt, M., Koller, R., Haviernik, P., Watson, R., Bies, J. and Maciag, K. (2001) Three genes with different functions in transformation are regulated by c-Myb in myeloid cells. Blood Cells Mol. Dis. *27*, 483-488.

Wolff, L., Mushinski, J.F., Shen-Ong, G.L. and Morse, H.C. 3rd. (1988) A chronic inflammatory response. Its role in supporting the development of c-myb and c-myc related promonocytic and monocytic tumours in BALB/c mice. J. Immunol. *141*, 681-689.

Woo, C.H., Sopchak, L. and Lipsick, J.S. (1998) Overexpression of an alternatively spliced form of c-Myb results in increases in transactivation and transforms avian myelomonoblasts. J. Virol. *72*, 6813-6821.

Zobel, A., Kalkbrenner, F., Vorbrueggen, G., Moelling, K. (1992) Transactivation of the human c-myc gene by c-Myb. Biochem. Biophys. Res. Commun. *186*, 715-722.

Chapter 17

THE INVOLVEMENT OF C-MYB IN VASCULAR INJURY

Cathy M. Holt and Nadim Malik
Cardiovascular Research Group, School of Medicine, University of Manchester, United Kingdom

Abstract: Coronary heart disease is a multifactorial disease that results in progressive narrowing of the arteries that supply blood to the heart. One of the major treatments of coronary heart disease is to open the occluded arteries using percutaneous procedures such as balloon dilatation (angioplasty) and implantation of metal scaffolds (stents) into the artery across the narrowing. A consequence of these procedures is the occurrence of restenosis or re-occlusion of the treated artery in approximately 20% of cases. Several of the key events involved in the pathogenesis of restenosis include cell differentiation, proliferation, apoptosis and matrix deposition. C-myb is a transcription factor with diverse roles in various cellular events including those leading to restenosis and is therefore a likely key player in its pathogenesis. In addition, c-myb may be inhibited, for instance, using antisense oligonucleotides, as a potential mechanism for the prevention of restenosis.

1. CORONARY ARTERY DISEASE AND ITS TREATMENT

Atherosclerosis is the deposition of lipid and cells within the wall of the artery, especially the intima. This accumulation known as plaque, results in progressive narrowing of the arterial lumen (Ross, 1993). Coronary artery disease results from the progressive blockage of arteries by plaques and sometimes, more abruptly, by thrombus. Clinical syndromes such as angina and myocardial infarction are the subsequent result of an imbalance between the supply and demand for blood and therefore oxygen. Inadequate perfusion of the myocardium due to a significant narrowing of the lumen of the epicardial artery, in the face of an increased metabolic demand, results in ischemic symptoms (Carboni et al., 1987; Gage et al., 1986). Coronary

J. Frampton (ed.), Myb Transcription Factors: Their Role in Growth, Differentiation and Disease, 331-349.

artery disease is initially treated by pharmacological means. Patients that are refractory to maximal medical therapy, are treated by either non-surgical intervention including percutaneous transluminal coronary angioplasty or PTCA, with or without stent implantation, or coronary artery bypass grafting using minimally invasive or open heart surgery technique (Cowley et al., 1984). PTCA and stenting, together known as percutaneous coronary intervention (PCI) has emerged as a treatment of choice for many patients with flow limiting atherosclerotic coronary artery disease. This is a minimally invasive procedure performed under local anaesthesia and x-ray screening. The procedure involves the insertion of a guide wire followed by an angioplasty balloon across the coronary stenosis. The PTCA balloon is then inflated at the site of lesion, with the aim of breaking or rearranging the plaque that is causing the narrowing of the lumen, with subsequent increase in vessel lumen. In present day clinical practice the majority of PTCA procedures involve the placement of a coronary stent into the artery (Odell et al., 2002). This is a permanent, metal, implant that is designed to hold open the vessel wall. Although patients undergoing PTCA and stenting generally show alleviation of symptoms of coronary artery disease, unfortunately, a common complication of both of these procedures is the occurrence of restenosis or arterial narrowing at the site of PCI (Welt and Rogers, 2002). Despite a better understanding of the underlying pathological processes leading up to restenosis, the rate remains unacceptably high. In selected subsets of patients, the angiographic restenosis rates after PTCA alone are 30 to 50 percent and this is associated with the recurrence of ischemic symptoms requiring further intervention to improve blood flow across the lesion (Popma et al., 1991; Serruys et al., 1988). Typically, restenosis at the site of the initial balloon dilatation occurs within six months of the procedure and is a multifactorial process. Briefly, the mechanisms leading to the process of restenosis include acute elastic recoil, intima hyperplasia and chronic arterial wall remodelling (Bennett and O'Sullivan, 2001). Intimal hyperplasia is due to vascular smooth muscle cell proliferation and migration and extracellular matrix accumulation with some contribution from the adventitia. Restenosis after PTCA, remains a challenging clinical problem (Serruys et al., 1994; Fischman et al., 1994; Versaci et al., 1997). Several pharmacological clinical trials aimed at inhibiting post PTCA restenosis have been carried out without much success (Popma et al.,1991). This lack of success in clinical applications has enhanced interest in, and the use of, intracoronary stents for the prevention and treatment of post PTCA restenosis. Indeed, three landmark trials in recent years have suggested that in selected patients with large or medium sized vessels and localised disease, stenting reduced restenosis rates when compared with PTCA alone (Serruys et al., 1994; Fischman et al., 1994; Versaci et al., 1997). This is due to their ability to provide a larger lumen and to prevent elastic recoil. However, stenting is now known to result in greater neointima formation although the exact mechanisms occurring in in-stent restenosis remain poorly understood.

An alternative treatment for coronary heart disease is coronary artery bypass grafting. During this procedure, autologous vein or artery is grafted onto the patient's aorta and coronary artery, distal to the occlusion, thereby "bypassing" the blockage. Despite the fact that patients undergoing coronary bypass surgery experience an immediate alleviation of symptoms, up to 50% of bypass grafts become occluded ten years following surgery (Campeau et al., 1984). Late occlusion of bypass grafts is due to the formation of a neointima and remodelling of the vessel wall. Thus many of the characteristics of restenosis following angioplasty and stenting also limit the long term success of bypass grafting (Shuhaiber et al., 2002). There remains, therefore, a requirement for further understanding of the cellular and molecular events occurring in restenosis following PCI and failure of saphenous vein bypass grafts.

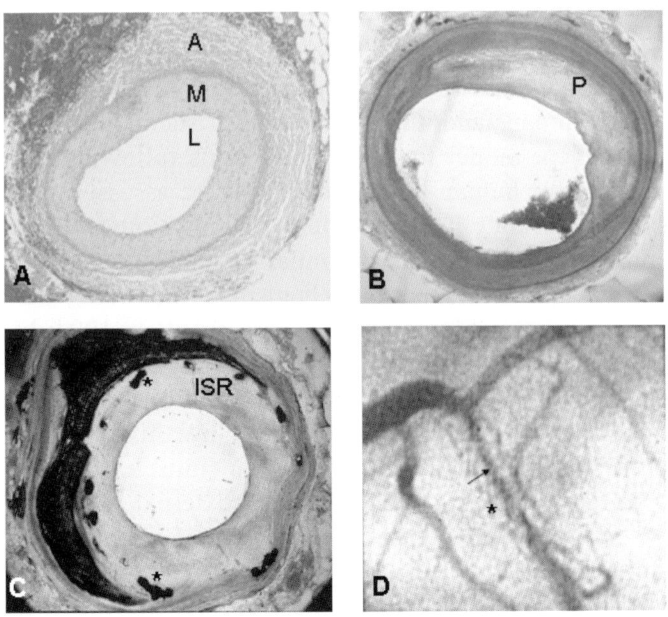

Figure 1

Histology of normal and diseased arteries. A. Transverse histological cross-section of normal artery showing the different layers of the vessel wall; lumen (L), media (M) and adventitia

(A). In a normal artery, the intimal layer consists of a single layer of endothelial cells lining the vessel wall and is not visible in this photomicrograph. B. Transverse histological cross section of an artery showing features of atherosclerosis. The plaque (P) region of the vessel wall is contained within the thickened intima. Thinning of the medial layer is also present. C. Transverse histological cross section of an atherosclerotic artery that has been implanted with a stent. Asterisks represent stent struts and ISR represents in-stent restenosis encroaching on the vessel lumen. D. Coronary angiogram showing in-stent restenosis. For an angiogram image, the patient's artery is injected with a radio-opaque contrast media and visualised under X-ray. The vessel lumen appears in black. The arrow indicates restricted blood flow caused by narrowing of the blood vessel lumen. The asterisk represents a region of in-stent restenosis. The shaded appearance surrounding the vessel lumen represents the stent struts that are radio-opaque and indicates the original vessel lumen prior to onset of in-stent restenosis. (see colour section p. xxvii)

2. RESTENOSIS, THE PROBLEM

Most of what is known about restenosis has been obtained from animal models of vascular injury. A considerable amount of research has been undertaken using the domestic Yorkshire White Pig (*Sus scrofa*) as an experimental species. The advantage of using such a model is that the size of pig coronary artery allows for technically feasible evaluation of devices intended for use in humans. The size and anatomy of pig coronary arteries are comparable to humans and allow for usage of standard clinical devices. The pig is an excellent model for stent evaluations following implantation, since stents suitable for human implantation may be assessed after standard implantation as per current clinical practice. The physiology of the response to injury in porcine arteries is very similar to the human response as it incorporates all the essential mechanisms, including thrombosis, inflammation and neointimal hyperplasia (Schwartz et al., 1994; Malik et al., 1998). Pig coronary arteries consist of a single layer of endothelial cells lining the vessel lumen. This comprises the intima of a normal vessel and lies directly upon the internal elastic lamina. Below this layer is the medial layer composed mainly of vascular smooth muscle cells within an extracellular matrix. This is bordered by the external elastic lamina and outside of this is the adventia that is composed of connective tissue and fibroblasts. Immediately following angioplasty, the endothelium becomes injured and may be removed in parts (Figure 2). Platelets are deposited on the exposed sub endothelial layer and leucocytes become attached. Medial injury occurs, as does acute elastic recoil. By two days following injury, the thrombus becomes organised and leucocytes that have adhered to the exposed sub endothelial layer start to infiltrate within the vessel wall. At this time medial smooth muscle cell apoptosis is also observed (Malik et al., 1998). Between three and seven days following injury the intimal layer becomes thickened. Thrombus has now become incorporated into this intimal layer and smooth muscle cell migration from the media to the intima has occurred. Smooth muscle cell proliferation also occurs at this stage and smooth muscle cells change phenotype from a quiescent to a synthetic phenotype. Endothelial cells are seen to migrate from the wound edge and the proliferation of endothelial cells is also observed. In addition, fibroblasts within the adventitia differentiate into myofibroblasts and migrate from the adventitia into the noeintima (Shi et al., 1997). From ten to fourteen days following injury reendothelialisation is observed, however, endothelial dysfunction at the wound edge may be apparent. Smooth muscle cell proliferation has now subsided and extracellular matrix deposition is occurring. By twenty eight days, chronic downsize remodelling is seen to occur in parallel with continued deposition of extracellular matrix. In summary, the key features leading to restenosis are endothelial damage, apoptosis and proliferation of vascular smooth muscle cells, differentiation of fibroblasts into myofibroblasts and the deposition of extracellular matrix (Bennett and O'Sullivan, 2001; Welt and Rogers, 2002).

3. C-MYB: A MULTIFUNCTIONAL GENE WITH A ROLE IN RESTENOSIS

c-Myb is a transcription factor that shares homology with the transforming gene product of avian myeloblastosis virus (Gonda and Bishop, 1983; Weston, 1998; Lipsick, 1996). It consists of a DNA binding domain, transactivation domain and a negative regulatory domain (Ness, 1996). c-Myb has diverse roles, several of which are described in preceding chapters of this book. These include differentiation, proliferation, apoptosis and affects on extracellular matrix (Badiani et al., 1994; Taylor et al., 1996; Lee et al., 1995; Oh and Reddy, 1999). Expression of c-Myb in components of the blood vessel wall means that it is likely to be involved in the major processes occurring in restenosis.

The expression of *c-myb* in vascular smooth muscle cells has been shown to vary with the cell cycle, being low in quiescence and increased in mid to late G_1 (Brown et al., 1992). Inhibition of its expression prevents entry into S phase. Its role in vascular smooth muscle cell proliferation has been demonstrated by several investigators and will be discussed in more detail later. Signalling events, such as those that eventually lead to proliferation, often involve the movement of calcium across the plasma membrane (Berridge, 1995). Intracellular calcium levels have been shown to decrease in immortalised vascular smooth muscle cells in mid G_1 of the cell cycle

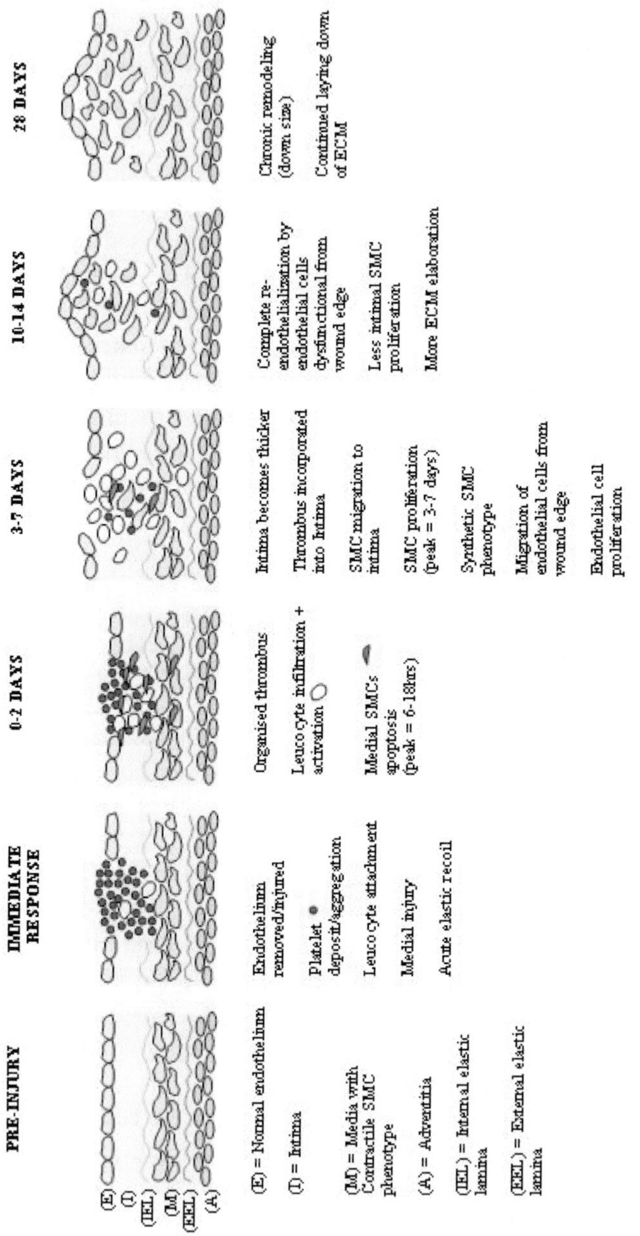

Figure 2

The possible mechanisms and time course of restenosis following percutaneous coronary intervention. (see colour section p. xxviii)

following a two-fold increase at the G_2/S interface (Simons et al., 1993). When *c-myb* is transfected into rat vascular smooth muscle cells, a two-fold increase in intracellular calcium and a corresponding two-fold decrease in calcium efflux is observed (Simons et al., 1995). Using both stable and inducible expression of dominant negative c-Myb in rat vascular smooth muscle cells, Husain et al (1997) investigated Myb-dependent intracellular calcium changes and identified a role for the plasma membrane calcium ATPase. The decrease in intracellular calcium arising from the decrease in c-Myb was shown to be due to a ten-fold increase in calcium extrusion from the cell.

In addition to its effects on cell proliferation, c-Myb is also involved in migration and the deposition of collagen and thus may be important in the remodelling that occurs post angioplasty (Nikkari et al., 1994). Although few studies have been performed in vascular smooth muscle cells, the expression of collagen has been associated with c-Myb in various cell types including embryonic vascular smooth muscle cells (Lee et al., 1995; Gadson et al., 1997; Piccinini et al., 1999).

Furthermore, c-Myb is also thought to be anti-apoptotic. Transfection of a neuroblastoma cell line with an antisense orientated *c-myb* expression construct induced apoptosis (Piacentini et al., 1994) and dominant negative c-Myb induced apoptosis in K562 cells (Yi et al., 2002). In addition, our own work discussed later has shown that dysregulation of c-Myb induces apoptosis in vascular smooth muscle cells. In summary, there is a large body of evidence implicating c-Myb in the vascular response to injury, in particular differentiation, proliferation, apoptosis and extracellular matrix deposition.

4. C-MYB EXPRESSION FOLLOWING ANGIOPLASTY IN PIG CORONARY ARTERIES

Work from our laboratory has investigated the induction of *c-myb* following PTCA in pigs (Gunn et al., 1997; Lambert et al., 2001). *c-myb* mRNA could not be detected by RT-PCR in control or injured pig coronary arteries. A few hours following balloon angioplasty low levels of *c-myb* expression were seen. The levels of RNA were maximal at eighteen hours following injury and declined to basal levels by twenty eight days (Gunn et al., 1997). Segments of balloon injured pig coronary artery were also processed for histological analysis. Arterial blocks showing maximum balloon injury, defined as maximum disruption to the internal elastic lamina were selected for further analysis. c-Myb immunostaining was then performed on transverse paraffin sections of control and angioplasty pig coronary arteries obtained at different time intervals following procedure. Positive immunostaining for c-Myb was determined semiquantitatively using a grading system from 0 to 3 (Table 1). The region of trauma was defined as the cross sectional area of the artery adjacent to the breached internal elastic lamina including remnants of the media, adventitia and loose connective tissue (Lambert et al., 2001). Control and uninjured arteries showed very low levels of c-Myb. Following balloon angioplasty c-Myb was detected by immunohistochemistry (Figure 3). One and 6 hours after PTCA, positive staining for c-Myb was detected within the adventia and this appeared to localise within lesions containing an inflammatory infiltrate. At 18 hours, strong positive staining was detected within the media and adventitial immunostaining was still observed in regions with an inflammatory infiltrate. Three days after injury adventitial staining was increased and c-Myb was also detected in microvascular endothelium and adjacent inflammatory cells. Luminal endothelial and neointimal cells also expressed c-Myb. At 7 days following PTCA, staining was similar to that seen in specimens analysed at three days with c-Myb present within the media and neointima and α-actin positive cells within the adventia. At 14 days the distribution of c-Myb was similar to that at 7 days but was less intense and at 28 days minimal c-Myb staining was observed.

Table 1. c-Myb Expression at Various Time Points After Angioplasty in Different Regions of Pig Cornonary Artery

	Arterial tissue						
Time after injury	Endo-thelium	Intima	Media	Adventitia	Micro-vessel	Inflam-matory cells	Total Score*
0	0	0	1	1	0	0	2
1 h	0	0	1	2	0	0	3
6 h	0	0	1	1	0	1	3
18 h	0	1	2	2	1	2	8
3 d	0	1	2	3	1	0	7
7 d	2	2	2	3	1	0	10
14 d	0	1	2	2	1	0	6
28 d	0	1	2	2	0	0	5

Scores are average values obtained from 3-6 vessels per time point. *Highest possible score is 18.

In order to identify localisation of c-Myb within specific cell types of the blood vessel wall, double immunohistochemistry was performed. Six hours following PTCA c-Myb was localised mainly within inflammatory cells within areas of media and overlying thrombus at sites of trauma and within the adventitia (Figure 3). Cell types that were not stained with the phenotypic markers used were also positive for c-Myb. These may have been fibroblasts. At 18 hours, co-localisation of c-Myb with inflammatory cells was still evident and in addition positive cells were also identified within the media where co-localisation with α-actin was observed. In vascular smooth muscle cells maximal c-Myb expression was identified at 3 to 7 days and at these time points c-Myb was also detected within advential microvascular endothelium. At 7 days following PTCA the advential cells expressing c-Myb were now α-actin positive.

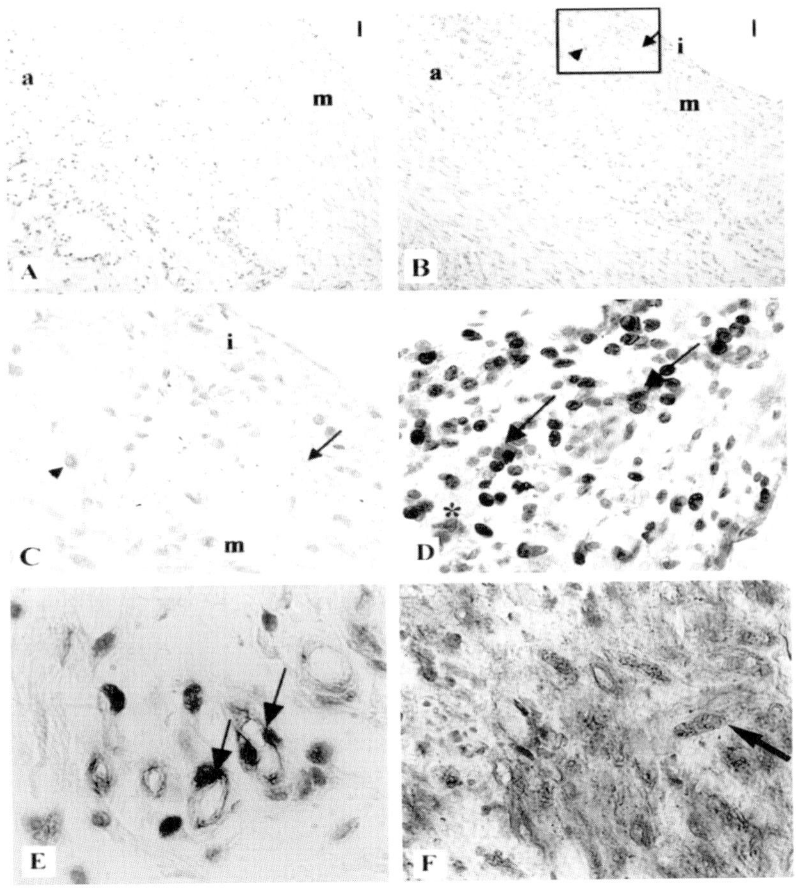

Figure 3

c-Myb expression and control in balloon injured pig coronary artery. A. Transverse histological section of a control pig coronary artery immunostained for c-Myb. Note the minimal positive staining. l indicates lumen; m, media; and a, adventitia (original magnification x20). B. Seven days after angioplasty. Numerous c-Myb-positive cells can be seen within the media (m, arrowhead) and are also present within the intima (i, brown). The arrow indicates internal elastic lamina (original magnification x20). C. High-power view of boxed area, shown in panel B, 7 days after angioplasty (original magnification x100). D. Six hours after angioplasty. A marked inflammatory infiltrate of CD68-positive cells (brown) with some positive for c-Myb staining is shown (red, arrow). Some c-Myb-positive cells are

CD68 negative (*). The area of inflammation is localised to a region of trauma (original magnification x100). E. Three days after angioplasty showing adventitial microvessel stained positive for dolichos biflorus–lectin (brown). Some of the endothelial cells are c-Myb positive (red, arrows) (original magnification x100). (Taken from Lambert et al., 2001). (see colour section p. xxix)

5. INHIBITION OF C-MYB IN VASCULAR SMOOTH MUSCLE CELLS

Various studies have been undertaken in isolated vascular cells to determine the role of c-Myb in vascular smooth muscle cell proliferation. The use of antisense oligonucleotide sequences directed against specific factors may result in their effective elimination from biological processes, either by interfering with translation or by causing destruction of mRNA by RNAse H. The antisense therapeutic approach has been successfully employed *in vitro* for the suppression of vascular smooth muscle cell proliferation, via downregulation of *c-myb* (Gunn et al., 1997; Villa et al., 1995; Edelman et al., 1995; Pitsch et al., 1996). Vascular smooth muscle cells from various species have been employed in these studies and different oligonucleotide sequences investigated. In addition, catalytic hammerhead ribozymes directed against c-*myb* have also been shown to suppress proliferation of rat vascular smooth muscle cells *in vitro* with an associated decrease in intact *c-myb* mRNA (Jarvis et al., 1996). Adenovirus mediated expression of a ribozyme to c-*myb* mRNA was also shown to inhibit vascular smooth muscle cell proliferation *in vitro* (Macejak et al., 1999).

In addition to an effect on suppressing vascular cell proliferation, work from our own group has demonstrated that antisense oligonucleotides to c-*myb* induce apoptosis of vascular smooth muscle cells. This has been demonstrated using several methods of detection including TUNEL (Figure 4), cell death ELISAs (Lambert et al., 2001) and caspase 3 activation (S. Withers, unpublished). Interestingly, a similar observation was not made in vascular endothelial cells, suggesting that if used in the setting of clinical restenosis, c-*myb* antisense would induce apoptosis of vascular smooth muscle cells but allow the endothelial cells to regrow and thus re-establish a continuous monolayer. Despite encouraging results with antisense oligonucleotides, several workers, including ourselves, have reported non-specific effects that may account, at least partially, for their inhibitory effects on restenosis (Gunn et al., 1997; Burgess et al., 1995; Villa et al., 1995; Lee et al., 1999).

An alternative approach to antisense strategies is to block protein function using dominant negative proteins. Dominant negative c-Myb expression constructs incorporating the c-Myb binding domain alone or coupled to the *Drosophila* engrailed repressor domain have been used to examine c-Myb function in T cell development (Badiani et al., 1994). Engrailed is a *Drosophila* homeodomain-containing protein that contributes to segmental patterning. Its repressor activity can be transferred to a heterologous DNA binding domain (Janes and O'Farrell, 1991). These constructs have been transiently transfected into rat, rabbit and human vascular smooth muscle cells resulting in dramatic reductions in proliferation and increased apoptosis (Schmitt et al.,1999).

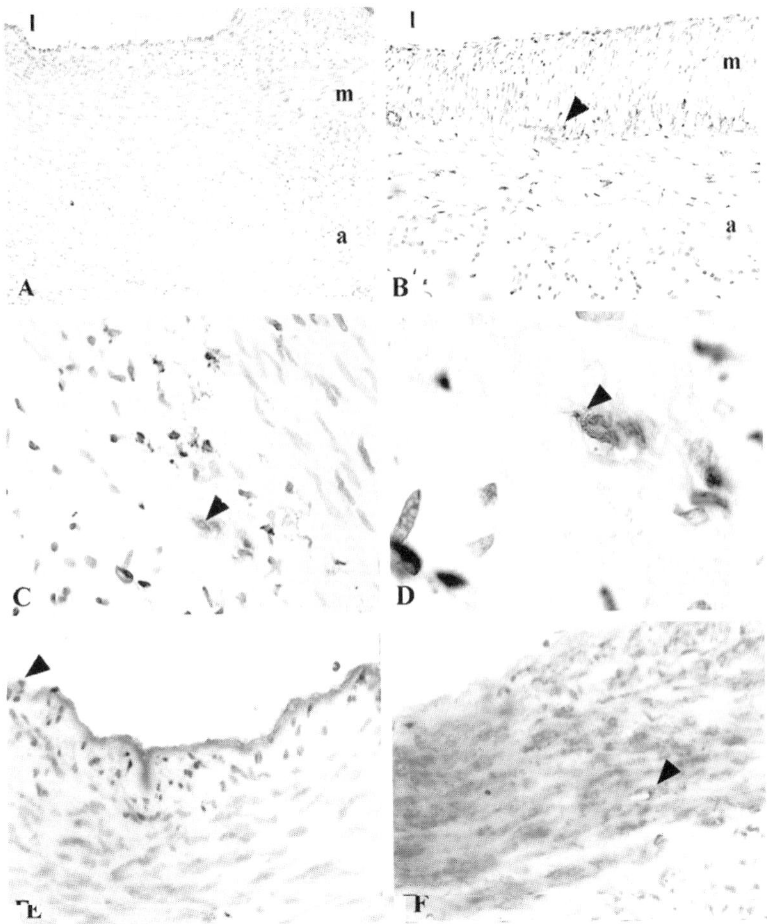

Figure 4

Pig coronary artery obtained 6 hours after balloon injury and delivery of c-myb antisense. A. Control artery that has undergone the TUNEL procedure. Note the lack of TUNEL-positive cells. l indicates lumen; m, media; and a, adventitia (original magnification x20). B. TUNEL-positive cells in the balloon-injured vessel showing brown staining and characteristic nuclear condensation. The majority of TUNEL-positive cells are located within the outer media (arrowhead) (original magnification x20). C. Macrophage stained with Mac387 (red, arrowhead) and TUNEL (brown) (original magnification x40). D. High-power view of area indicated by arrowhead in panel C (original magnification x100). E. von Willebrand factor antigen staining showing the vascular endothelial layer and a TUNEL-positive cell (arrowhead) (original magnification x40). F. α-actin-stained artery showing TUNEL-positive smooth muscle cell (arrowhead) (original magnification x40). (Taken from Lambert et al, 2001). (see colour section p. xxx)

6. INHIBITION OF C-MYB: IN VIVO STUDIES OF VASCULAR INJURY

A key study published in 1992 by Simons and colleagues demonstrated that antisense *c-myb* oligonucleotides inhibit intimal arterial smooth muscle cell accumulation *in vivo* (Simons et al., 1992). These investigators used a rat carotid artery injury model that has been extensively used as a model of vascular injury. In this study, *c-myb* antisense oligonucleotides were delivered via a pluronic gel which was administered to the adventitial surface of the surgically exposed carotid arteries. On contact with the blood vessel at 37°C, the pluronic solution gels and generates a layer enveloping the treated region. Animals were allowed to recover and were subsequently killed two weeks after injury. Northern blot analysis showed that injured arteries treated with antisense *c-myb* oligonucleotides had no detectable levels of *c-myb* mRNA in contrast with sense oligonucleotide treated vessels. These investigators then determined the effect of antisense oligonucleotide *c-myb* on intimal smooth muscle cell accumulation. Morphometric analysis of histological cross sections of the vessel wall identified suppression of intimal smooth muscle cell accumulation following delivery of antisense oligonucleotide *c-myb* without apparent effect on medial smooth muscle cell viability. This study is claimed to be the first reported use of antisense oligonucleotide to inhibit synthesis of a normal gene product under *in vivo* conditions, with a subsequent effect on a cellular process. The authors suggest that the results obtained were probably due to the local delivery used which allowed high concentrations of antisense oligonucleotide to be directed to the specific site of injury. This landmark study promoted further investigations using local delivery of antisense oligonucleotides to the vessel wall. A subsequent study by these investigators demonstrated that antisense *c-myb* oligonucleotides inhibit smooth muscle cell accumulation within the blood vessel wall of injured rat carotid arteries whether they were administered for only the first few hours after injury or for the full duration of the experiment when released from ethylene vinyl acetate copolymer (EVAC) matrixes (Edelman et al., 1995). EVAC matrices were investigated since they are able to provide a more sustained release of antisense compared with pluronic gel. One of the major problems relating to treatment and prevention of restenosis is delivery of the therapeutic agent. Another study performed by our group analysed the effect of local delivery of *c-myb* antisense oligonucleotides in a porcine coronary angioplasty model using a clinical device known as the transport catheter, previously developed for local drug delivery to the coronary artery (Gunn et al., 1997). This is a dual inflation and delivery catheter that has been specifically designed to perform local drug delivery within human coronary arteries (Gunn and Cumberland, 1996). Unmodified *c-myb* antisense or sense oligonucleotides, saline or nothing was delivered immediately following balloon dilation using the transport device. Use of fluorescent labelled oligonucleotides confirmed that delivery throughout the vessel wall had been achieved. Morphometric analysis of histological cross sections of vessel obtained 4 weeks after PTCA demonstrated that maximal intimal medial cross sectional area was reduced with *c-myb* antisense oligonucleotide by 79% compared with saline alone; 82% compared with sense oligonucleotide and 63% compared with nothing. This study confirmed, therefore, that local delivery of *c-myb* antisense oligonucleotide via the transport catheter reduced neointimal formation in a porcine angioplasty model. Another study by Azrin and colleagues also used a porcine model, but they performed angioplasty of peripheral arteries (Azrin et al., 1997). These investigators showed successful delivery and intramural persistence of oligonucleotides for over 24 hours following delivery of oligonucleotide with hydrogel coated balloons. In their investigations they looked at the effect on the vessel at 7 days following injury. At this time point, very little neointima formation was present and no significant difference in the sense and antisense oligonucleotide treated vessels was found. However, following immunohistochemical staining of proliferating cell nuclear antigen smooth muscle cell proliferation was found to be significantly reduced in antisense oligonucleotide treated, compared to control treated, vessels. As well as the antisense oligonucleotide approach, ribozymes to *c-myb* mRNA have also been investigated for their ability to inhibit neointimal formation *in vivo*. Macejak et al (1999) generated a recombinant adenovirus expressing ribozymes against *c-myb* mRNA and tested these both *in vitro* and *in vivo*. The adenovirus ribozyme inhibited smooth muscle cell proliferation in culture and neointimal formation in a rat carotid artery balloon injury model of restenosis.

As well as an inhibitory effect on vascular proliferation *in vivo* we have also identified that inhibition of *c-myb* using antisense oligonucleotides causes the induction of apoptosis of vascular smooth muscle cells *in vivo*. (Lambert et al., 2001). Again using a porcine angioplasty model local delivery of *c-myb* antisense via the transport catheter was performed and apoptosis quantified 6 hours following the procedure (Lambert et al., 2001). The incidence of apoptosis was significantly enhanced following delivery of antisense oligonucleotide compared to control. The majority of apoptotic cells were localised within the media of the vessel wall (Lambert et al., 2001).

Antisense oligonucleotides to c-*myb* have also been investigated in animal models of intimal hyperplasia in experimental vein grafts. Fulton et al (1997) surgically inserted vein into the carotid artery of rabbits. Different groups of vein were inserted including control vein, vein coated with puronic gel and vein coated with pluronic gel containing either *c-myb* antisense or control sense oligonucleotides. A 38% reduction in intimal thickness, no difference in medial thickness and preservation of acetylcholine-induced endothelium-dependent relaxation was observed in the *c-myb* antisense oligonucleotide treated group. Thus, it appears that c-Myb is also important in the neointima that forms following vein grafting.

Table 2. Summary of c-myb antisense and ribozyme data

Mode of inhibition of c-myb	Species	Model	Model of delivery	Inhibition of intimal:medial ratio	Reference
AS ODN[1]	Rat	Carotid balloon injury	Pluronic gel	94%	Simons et al (1992)
AS ODN	Rat	Carotid balloon injury	EVAC	72%	Edelman et al (1995)
AS ODN	Pig	Coronary angioplasty	Transport catheter	82%	Gunn et al (1997)
AS ODN	Pig	Peripheral angioplasty	Hydrogen catheter	NS[2]	Azrin et al (1997)
AS ODN	Rabbit	Interposition vein graft	Pluronic gel	38% (intimal thickness)	Fulton et al (1997)
Adenoviral Ribozyme	Rat	Carotid balloon injury	Catheter instillation	53%	Macejak et al (1999)

[1]AS ODN, antisense oligodeoxynucleotide, [2]Not significant

7. NOVEL MODES OF DELIVERY OF ANTISENSE OLIGONUCLEOTIDES IN VIVO

Unfortunately, the promise of various local drug delivery devices in delivering therapeutic agents to coronary arteries has not been upheld in clinical studies. Many of the devices developed have been shown to cause injury to the vessel wall (Holt et al., 1999). This might be from the delivery device itself; for example, some devices cause a jetting effect, or actually use fine needles to inject agents into the vessel wall. Alternatively, the volume of fluid delivered via these devices could cause additional damage to the vessel wall. Because of these problems, and also because the majority of patients undergoing angioplasty procedures now involve stent implantation, the use of drug eluting stents has recently experienced a surge of interest. Various preclinical and clinical studies have now been performed with encouraging results (Babapulle and Eisenberg 2002a; Babapulle and Eisenberg 2002b). We are currently investigating the safe and efficient delivery of antisense oligonucleotides via stent coatings. By modifying the phosphoryline polymer coating to contain cationic moieties we are able to load the stents with antisense oligonucleotides. These loaded stents can then be deployed into vessels and deposition of antisense oligonucleotide is observed in the area of the vessel wall directly surrounding the stent (J. Armstrong, Chan, J. Gunn and CMH, unpublished). Current studies are aimed at investigating the efficacy of these novel coated stents for the prevention of in-stent restenosis in pigs.

8. FUTURE AREAS OF INVESTIGATION

This chapter has summarised what is currently known about the role of c-Myb in blood vessels, however many areas require further investigation. Mounting evidence suggests that c-Myb is anti-apoptotic. Intricate pathways are involved in determining whether a cell survives or undergoes apoptosis. It is unknown at what point c-Myb influences these pathways and this will be an area for future research. Once there is a better understanding of the mechanisms involved in induction of apoptosis following the dysregulation of c-Myb, use of c-*myb*-based agents for anti-restenosis could be explored in the clinical setting. Further understanding regarding the downstream targets of c-Myb in vascular cells is required. With the advent of microarray screening, investigations into the expression of genes that are activated in the presence or absence of c-Myb would greatly add to our understanding in this area. Mouse models of vascular injury and atherosclerosis are now widely used (Drew, 2001). By creating inducible c-*myb* knockout mice and performing vascular injury or crossing these with mice that develop vascular lesions, we will be able to gain a better understanding regarding the role of this important gene in the blood vessel wall in health and disease.

ACKNOWLEDGEMENTS

We would like to thank our colleagues with an interest in Myb: Julian Gunn, Paul Sheridan, Jo Armstrong, Sarah Withers and the late Darren Lambert, to whom this chapter is dedicated. We are also grateful to Ros

Poulton for secretarial support. Work described in this chapter was funded by the British Heart Foundation and Medical Research Council.

REFERENCES

Azrin, M.A., Mitchel, J.F., Bow, L.M., Pederson, C.A., Cartun, R.W., Aretz T.H., Waters, D.D. and McKay, R.A. (1997) Local delivery of c-myb antisense oligonucleotides during balloon angioplasty. Cathet Cardiovasc Diagn *41*, 232-240.

Babapulle, M.N. and Eisenberg, M.J. (2002a) Coated stents for the prevention of restenosis: Part I. Circulation *106*, 2734-2740.

Babapulle, M.N., and Eisenberg, M.J. (2002b) Coated stents for the prevention of restenosis: Part II. Circulation *106*, 2859-2866.

Badiani, P., Corbella, P., Kiossis, D., Marvel, J. and Weston, K. (1994) Dominant interfering alleles define a role for c-myb in T-cell development. Genes Dev *8*, 770-782.

Bennett, M.R. and O'Sullivan, M. (2001) Mechanisms of angioplasty and stent restenosis: implications for design of rational therapy. Pharmacology & Therapeutics *91*, 149-166.

Berridge, M.J. (1995) Calcium signalling and cell proliferation. BioEssays *17*, 491-500.

Brown, K.E., Kindy, M.S. and Sonenshein, G.E. (1992) Expression of the c-myb proto-oncogene in bovine vascular smooth muscle cells. J Biol Chem *267*, 4625-30.

Burgess, T. L., Fisher, E.F., Ross, S.L., Bready, J.V., Qian, Y.X., Bayewitch, L.A., Cohen, A.M., Herrera, C.J., Hu, S.S.F., Kramer, T.B., Lott, F.D., Martin, F.H., Pierce, G.F., Simonet, L. and Farrell, C.L. (1995) The antiproliferative activity of c-myb and c-myc antisense oligonucleotides in smooth-muscle cells is caused by a nonantisense mechanism. Proc Natl Acad Sci USA *92*, 4051-4055.

Campeau, L., Enjalbert, M., Lesperance, J., Bourassa M.G., Kwiterovich P. Jnr, Wacholder, S. and Sniderman, A. (1984) The relation of risk factors to the development of atherosclerosis in saphenous-vein bypass grafts and the progression of disease in the native circulation. A study 10 years after aortocoronary bypass surgery. New Engl J Med *311*, 1329-1332.

Carboni, G.P., Lahiri, A., Cashman, P.M.M. and Raftery, E.B. (1987) Ambulatory heart rate and ST-segment depression during painful and silent myocardial ischemia in chronic stable angina pectoris. Am J Cardiol *59*, 1029-1034.

Cowley, M.J., Dorros, G., Kelsey, S.F., van Raden, M. and Detre, K.M. (1984) Acute coronary events associted with pericutaneous transluminal coronary angioplasty. Am J Cardiol *53*, 12C-16C.

Drew, A.F. (2001) Atherosclerosis Experimental Methods and Protocols, Methods in Molecular Medicine. Totowa New Jersey: Humana Press.

Edelman, E.R., Simons, M., Sirois, M.G. and Rosenberg, R.D. (1995) c-myc in Vasculoproliferative disease. Circ Res *76*, 176-182.

Fulton, G.J., Davies, M.G., Koch, W.J., Dalen, H., Svendson, E. and Hagen, P.O. (1997) Antisense oligonucleotide to proto-oncogene c-myb inhibits the formation of intimal hyperplasia in experimental vein grafts. J Vasc Surg *25*, 453-463.

Fischman, D.L., Leon, M.B., Baim, D.S., Schatz, R.A., Savage, M.P., Penn, I.M., Detre, K.M., Veltri, L., Ricci, D.R., Nobuyoshi, M., Clemen M.R., Heuser, D., Almond, P.S., Teirstein, R.D., Fish, A., Colombo, A., Brinker, J.A., Moses, J.W.., Shankovitch, A., Hirshfeld, J., Bailey, S., Ellis, S. and Rake, R. (1994) A Randomised comparison of coronary-stent placement and balloon angioplastry in the treatment of coronary artery disease. New Engl J Med *331*, 496-501.

Gadson, P. Jr, Dalton, M.L. and Patterson, E. (1997) Differential response of mesoderm and neural crest-derived smooth muscle to TGF-β1: regulation of c-myb and alpha1 (I) procollagen genes. Exp Cell Res *230*, 168-180.

Gage, J.E., Hess, O.M., Murakami, T., Ritter, M., Grimm J., and Krayenbuehl, H.P. (1986) Vasoconstriction of stenotic coronary arteries during dynamic exercise in patients with chronic angina pectoris: reversibility by nitro-glycerine. Cirulation *73*, 865-876.

Gonda, T.J. and Bishop, M. (1983) Structure and transcription of the cellular homolog (c-myb) of the avian myeloblastosis virus transformign gene (v-myb). J Gen Virol *46*, 212-220.

Gunn, J. and Cumberland, D.C. (1996) Dual balloon catheter. Semin Interv Cardiol *1*, 31-33.

Gunn, J., Holt, C.M., Francis, S.E., Shepherd, L., Grohmann, M., Newman, C.M., Crossman, D.C. and Cumberland, D.C. (1997) The effect of oligonucleotides to c-myb on vascular smooth muscle cell proliferation and neointima formation after porcine coronary angioplasty. Circ Res *80*, 520-531.

Holt, C.M., Gunn, J., Lambert, D.L., Cumberland, D.D. and Crossman, C.M. (1999) Delivery of antisense oligonucleotides sto the vascular wall. In Vascular Disease Molecular Biology and Gene Therapy Protocols, Baker, A.H. (ed) Totowa, New Jersey: Humana Press.

Husain, M.,L. Jiang, V. See, K. Bein, M. Simons, Alper, S.L. and Rosenberg, R.D. (1997) Regulation of vascular smooth muscle cell proliferation by plasma membrane Ca^{2+}-ATPase. Am Physiol Soc C1947-C1959.

Jaynes, J.B., and O'Farrell, P.H. (1991) Active repression of transcription by the engrailed homeodomain protein. EMBO J *10*, 1427-1433.

Jarvis, T.C., Beaudry, A.A., Wincott, F.E., Beigelman, L., McSwiggen, J.A., Usman, N. and Stinchcomb, D.T. (1996) Inhibition of vascular smooth muscle cell proliferation by ribozymes that cleave *c-myc* mRNA. RNA *2*, 419-428.

Lambert, D.L., Malik, N., Shepherd, L., Gunn, J., Francis, S.E., King, A., Crossman, D.C., Cumberland, D.C. and Holt, C.M. (2001) Localisation of c-Myb and induction of apoptosis by antisense oligonucleotide c-Myb after angioplasty of porcine coronary arteries. Arterioscler Thromb Vasc Biol *21*, 1727-1732.

Lee, K.S., Buck, M., Houglum, K. and Chojkier, M. (1995) Activation of hepatic stellate cells by TGF alpha and collagen type 1 is mediated by oxidative stress through c-myb expression. J Clin Invest *96*, 2461-2468.

Lee, M., Simon, A.D., Stein, C.A. and Rabbani, L.E. (1999) Antisense strategies to inhibit restenosis. Antisense Nucl Acid Drug Dev *9*, 487-492.

Lipsick, J.S. (1996) One billion years of Myb. Oncogene *13*, 223-235.

Macejak, D.G., Lin, H., Webb, S., Chase, J., Hensen, K., Jarvis, T.C., Leiden, J.M. and Couture, L. (1999) Adenovirus-mediated expression of a ribozyme to c-*myb* mRNA inhibits smooth muscle cell proliferation and neotintima formation *in vivo*. J Virol *73*, 7745-7751.

Malik, N., Francis, S.E., Holt, C.M., Gunn, J., Thomas, G.L., Shepherd, L., Chamberlain, J., Newman, C.M., Cumberland, D.C. and Crossman, D.C. (1998) Apoptosis and cell proliferation after porcine coronary angioplasty. Circulation *98*, 1657-1665.

Ness, S.A. (1996) The Myb oncoprotein: regulating a regulator. Biochim Biophys Acta *1288*, F123-139.

Nikkari, S.T., Jarvelainen, H.T., Wight, T.N., Ferguson, M.W. and Clowes, A.W. (1994) Smooth muscle cell expression of extracellular matrix genes after arterial injury. Am J Pathol *144*, 1348-1355.

Odell, A.,T. Gudnason, T. Andersson, H. Jidbratt, and Grip, L. (2002). One-year outcome after percutaneous coronary intervention for stable and unstable angina pectoris with or without application of general usage of stents in unselected European pateint groups. Am J Cardiol *90*, 112-118.

Oh, H., and Reddy, E.P. (1999) The myb gene family in cell growth, differentiation and apoptosis. Oncogene *18*, 3017-3033.

Piacentini, M., Rashella, G., Calabretta, B. and Melino, G. (1994) C-myb down regulation is associated with apoptosis in human neuroblastoma cells. Cell Death Diff *1*, 85-92.

Piccinini, G., Golay, J. and Flora, A. (1999) C-myb but not B-myb upregulates type 1 collagen gene expression in human fibroblasts. J Invest Dermatol *112*, 191-196.

Pitsch, R.J., Goodman, G.R., Minion, D.J., Madura, J.A., Fox, P.L. and Graham, L.M. (1996) Inhibition of smooth muscle cell proliferation and migration in vitro by antisense oligonucleotide to *c-myb*. J Vasc Surg *23*, 783-791.

Popma, J.J., Califf, R.M. and Topol, E.J. (1991) Clinical trials of restenosis after coronary angioplasty. Circulation *84*, 1426-1436.

Ross, R. (1993) The pathogenesis of atherosclerosis: a perspective for the 1990s. Nature *362*, 801-809.

Schmitt, J.F., Keogh, M.C. and Dennehy, U. (1999) Tissue-selective expression of dominant-negative proteins for the regulation of vascular smooth muscle cell proliferation. Gene Ther *6*, 1184-1191.

Schwartz, R.S., Edwards, W.D., Bailey, K.R., Camrud, A.R., Jorgenson, M. and Holmes D.R. Jnr. (1994) Differential neotintimal response to coronary artery injury in pigs and dogs. Implications for restensosis models. Arterioscler Thromb Vasc Biol *14*, 395-400.

Serruys, P. W., de Jaegere, P., Kiemeneij, F., Macaya, C., Rutsch, W., Heyndrickx, G., Emanuelsson, H., Macro, J., Legrand, V., Materne, P., Belardi, J., Sigwart, U., Colombo, A., Goy, J.J., van den Heuvel, P., Delcan, J.L. and Morel, M.A. (1994) A comparison of balloon-expandable-stent implantation with balloon angioplasty in patients with coronary artery disease. New Engl J Med *331*, 489-495.

Serruys, P.W., Luijten, H.E., Beatt, K.J., Geuskens, R., De Feyter, P.J., van den Brand, M., Reiber, J.H., ten Katen, H.J., van Es, G.A. and Hugenholtz, P.G. (1988) Incidence of restenosis after suffessful coronary angioplasty: a time-related phenomenon. A quantitative angiographic study in 342 consecutive patients at 1, 2, 3 and 4 months. Circulation *77*, 361-371.

Shi, Y., O'Brien, J.E.Jnr, Mannion, J.D., Morrison, R.C., Chung, W., Fard, A. and Zalewski, A. (1997) Remodeling of autologous saphenous vein grafts. Circulation *95*, 2684-2693.

Shuhaiber, J.H., Evans, A.N., Massad, M.G. and Geha, A.S. (2002) Mechanisms and future directions for prevention of vein graft failure in coronary bypass surgery. Cardiothor Surg *22*, 387-396.

Simons, M., Ariyoshi, H., Salzman, E.W. and Rosenberg, R.D. (1995) C-*myb* affects intracellular calcium handling in vascular smooth muscle cells. Am Physiol Soc C856-C868.

Simons, M., Edelman, E.R., Dekeyser, J.L., Langer, R. and Rosenberg, R.D. (1992) Antisense c-myb oligonucleotides inhibit intimal arterial smooth-muscle cell accumulation in vivo. Nature *359*, 67-70.

Simons, M., Morgan, K.G., Parker, C., Collins, E. and Rosenberg, R.D. (1993) The proto-oncogene c-*myb* mediates an intracellular calcium rise during the late G_1 phase of the cell cycle. J Biol Chem *268*, 627-632.

Taylor, D., Badiani, P. and Weston, K. (1996) A dominant interfering Myb mutant causes apoptosis in T cells. Genes Dev *10*, 2732-2744.

Versaci, F., Gaspardone, A., Tomai, F., Crea, F., Chiariello, L. and Gioffre, P.A. (1997) A comparison of coronary-artery stenting with angioplasty for isolated stenosis of the proximal left anterior descending coronary arter. New Engl J Med *336*, 817-822.

Villa, A. E., Guzman, L.A., Poptic, E.J., Labhasetwar, V., Dsouza, S., Farrell, C.L., Plow, E.F., Levy, R.J., Dicorleto, P.E. and Topol, E.J. (1995) Effects of antisense c-myb oligonucleotides on vascular smooth-muscle cell-proliferation and response to vessel wall injury. Circ Res *76*, 505-513.

Welt, F.G.P., and Rogers, C. (2002) Inflammation and restenosis in the stent era. Arterioscler Thromb Vasc Biol *22*, 1769-1776.

Weston, K. (1998) Myb proteins in life, death and differentiation. Curr Opin Genet Dev *8*, 76-81.

Yi, H.K., Sang, Y.N., Kim, J.C., Kim, J.S., Lee, D.Y. and Hwang, P.H. (2002) Induction of apoptosis in K562 cells by dominant negative c-myb. Exp Hematol *30*, 1139-1146.

Chapter 18

REPRESSION OF MATRIX GENE EXPRESSION BY B-MYB

Claudia S. Hofmann and Gail E. Sonenshein
Department of Biochemistry, Boston University School of Medicine, Boston Massachusetts 02118 United States of America.

Abstract: Vascular smooth muscle cells (SMCs) synthesise collagens type I and V matrix proteins, which are major constituents of the arterial wall. In culture, matrix gene expression varies inversely with the rate of SMC proliferation. Previously we showed that B-*myb*, a member of the *myb* gene family, is expressed in SMCs in a cell-cycle dependent fashion, and that it is a negative regulator of matrix gene transcription. Phosphorylation by cyclin A/cdk2 relieved B-Myb-mediated repression of α2 (V) collagen gene transcription, and the sites of phosphorylation were distinct from those affecting activation by B-Myb. The domain responsible for repression mapped to residues 491 to 582 of the C-terminal region of B-Myb. Transgenic mice over-expressing B-Myb displayed significantly reduced collagen expression in the aorta. Thus, B-Myb functions *in vivo* as a repressor of collagen gene expression in vascular SMCs.

1. INTRODUCTION

Smooth muscle cells (SMCs), which are the major cellular constituents of the medial layer of the artery, synthesise and deposit matrix proteins responsible for structural framework and for vascular tone. The major extracellular matrix components include the fibrillar collagens types I and V/XI, and elastin as well as other basement membrane proteins and enzymes involved in matrix deposition. The most abundant collagen species produced by the SMC is type I, which is composed of a heterotrimer of two α1(I) and one α2(I) chains (Vuorio and de Crombrugghe, 1990). Type V collagen consists of three members, i.e., α1, α2 and α3 (Birk, 2001), and can also form heterotrimers with chains of the closely related type XI collagen, e.g., α1(XI) (Brown et al., 1991). Type V/XI collagen is the least abundant

J. Frampton (ed.), Myb Transcription Factors: Their Role in Growth, Differentiation and Disease, 351-366.

of the fibrillar collagens in the vessel wall (Kypreos et al., 2000; Murata et al., 1987). Elastin is one of the major structural proteins of large arteries, contributing the physical properties of extensibility and elastic recoil. Elastin is synthesised as a soluble monomeric precursor tropoelastin, that is processed and secreted from the cell where it is assembled into a highly stable, insoluble, branched polymeric structure in the extracellular matrix through covalent cross-links derived from lysine residues (Debelle and Tamburro, 1999; Smith-Mungo and Kagan, 1998).

Once the artery has been fully formed, SMCs differentiate from a synthetic state to a quiescent, contractile phenotype in which they normally remain (Campbell and Campbell, 1990). As a response to injury and in certain disease states, however, SMCs are activated and dedifferentiate and migrate to the intimal layer. In this environment, modest rounds of proliferation are followed by expression of abundant levels of matrix components, mainly collagens, elastin and proteases that modify the surrounding matrix (Ross, 1993; Sanz-Gonzalez et al., 2000). These synthetic responses of SMCs, in association with deposition of lipids and minerals, can result in formation of an atherosclerotic plaque, and lead to the occlusion of the vessel. In an analogous fashion, after injury caused by mechanical intervention, a vascular healing response is elicited which can stimulate SMC proliferation and excessive matrix deposition, leading to postangioplasty restenosis.

2. MATRIX GENE EXPRESSION VARIES INVERSELY WITH RATE OF SMC PROLIFERATION IN CULTURE

In culture, vascular SMCs appear to revert to a synthetic phenotype, and we and others have shown that synthesis of collagen and elastin proteins by cultured SMCs varies inversely with the growth rate. Thus, mRNA levels of the chains of types I and V/XI collagen, and of elastin are low when SMCs are subconfluent and growing exponentially, and increase as cells become more dense and their growth rate slows (Ang et al., 1990; Barone et al., 1988; Beldekas et al., 1981; Brown et al., 1991; Liau and Chan, 1989; Stepp et al., 1986; Tajima, 1996; Toselli et al., 1992; Wachi et al., 1995). Serum deprivation, which renders subconfluent SMCs quiescent (Kindy and Sonenshein, 1986), induces type I collagen mRNA levels 2- to 15-fold by 48 hours (Kindy et al., 1988). These changes in mRNA levels are due, in part, to increases in the rates of $\alpha 1(I)$ and $\alpha 2(I)$ gene transcription (Kindy et al., 1988). Similar effects were seen with type V collagen gene expression

(Brown et al., 1991). Induction of type I collagen gene expression is also observed when SMC cultures are rendered quiescent by amino acid deprivation (Chang and Sonenshein, 1991), or upon treatment with transforming growth factor (TGF)-β1 (Davidson et al., 1993; Lawrence et al., 1994). Conversely, treatment of bovine aortic SMCs with basic fibroblast growth factor (bFGF), a member of the FGF family that promotes SMC proliferation (Speir et al., 1991), decreases α1(I) and α2(V) collagen gene expression at the transcriptional level (Kennedy et al., 1995; Kypreos et al., 1998; Kypreos and Sonenshein, 1998; Majors and Ehrhart, 1993). Serum deprivation also results in a reversible increase in elastin mRNA expression in confluent vascular SMCs (Wachi et al., 1995). Similarly, treatment with heparin or retinoic acid, which are potent inhibitors of vascular SMC proliferation (Hayashi et al., 1995; Tajima, 1996; Wachi et al., 1995), induces elastin expression. Conversely, epidermal growth factor (EGF), angiotensin II, and bFGF, all potent stimulators of vascular SMC proliferation, inhibit elastin synthesis and decrease elastin mRNA levels (Tajima, 1996; Tokimitsu et al., 1994). A similar inverse relationship between collagen and elastin production and proliferation occurs *in vivo*. In human atherosclerotic plaques, SMC proliferation and collagen synthesis are independent events (Rekhter and Gordon, 1994), whereas active TGF-β1 co-localises with type I collagen gene expression (Bahadori et al., 1995). Cell proliferation and elastin mRNA levels are inversely correlated throughout embryogenesis (James et al., 1998). Furthermore, during intimal thickening in a balloon injury model in rabbit carotid arteries, the low level of elastin gene expression increased only as the SMC proliferative rate decreased (Aoyagi et al., 1997). Overall, these studies indicate that collagen and elastin gene expression vary inversely with the proliferative rate of the vascular SMC.

3. B-MYB IS EXPRESSED IN SMCS AND FUNCTIONS AS A NEGATIVE REGULATOR

Several years ago, we demonstrated that B-*myb* mRNA is expressed in SMCs in a cell cycle-dependent fashion (Marhamati and Sonenshein, 1996), as seen in other cell types (reviewed in Sala and Watson, 1999). The B-*myb* mRNA levels were low in quiescent serum-deprived or confluent SMCs and increased in G1 and S phase upon addition of serum growth factors, EGF, phorbol ester plus IGF-1 or bFGF (Kypreos et al., 1998; Marhamati and Sonenshein, 1996). Interestingly, ectopic expression of B-*myb* was unable to push quiescent SMCs into S phase (Marhamati et al., 1997). Specifically, quiescent SMCs failed to enter S phase upon co-microinjection with vectors expressing the competence factor c-Myc and B-Myb, whereas up to 75% of cells entered S phase within 20 hours following the microinjection of vectors expressing c-Myc plus A-Myb or c-Myb. Thus, B-*myb* gene expression relates to the cell cycle in SMCs, but does not appear to control progression of the SMC through the cycle, in contrast to results obtained with fibroblasts and haemopoietic cells (Arsura et al., 1992; Sala and Calabretta, 1992). These findings are consistent with the failure of B-*myb* to induce transcription in vascular SMCs of growth-promoting genes containing Myb binding sites (MBS), such as c-*myc* and c-*myb* (Marhamati et al., 1997).

4. B-MYB NEGATIVELY REGULATES TYPE I COLLAGEN GENE EXPRESSION IN BOVINE AORTIC SMCS

B-Myb has been reported to function as a transcriptional repressor as well as an activator in a cell-type specific fashion (Foos et al., 1992; Mizuguchi et al., 1990; Tashiro et al., 1995; Watson et al., 1993). In bovine aortic SMCs, co-transfection of either a bovine or human B-*myb* expression vector decreased activity of a 9 copy MBS element-driven thymidine kinase (TK) promoter CAT reporter construct (KHK-CAT-dAX), indicating that B-Myb functions as a negative regulator of transcription in SMCs (Marhamati and Sonenshein, 1996).

Computer analysis of the genes encoding the two chains of type I collagen genes detected multiple upstream putative MBS elements. The inverse correlation between matrix gene expression and proliferative rate of bovine SMCs in culture led us to hypothesise several years ago that B-Myb plays a role in controlling expression of matrix genes in SMCs. Co-transfection experiments were conducted with an α1(I) collagen reporter vector containing 3.6 kbp of the promoter as well as all of exon 1 and intron 1 and a bovine B-Myb expression vector. A dose-dependent downregulation of promoter activity was observed, with maximal inhibition of ~8.8-fold (Marhamati and Sonenshein, 1996). Similar results were obtained with the α2(I) collagen promoter containing 3.5 kbp of sequences upstream of the transcription start site and 58 bp of exon 1. Reporter activity was reduced in a dose-dependent manner, with maximum inhibition of 72% and 82% seen with bovine and human B-Myb vectors, respectively (Marhamati and Sonenshein, 1996). Thus, B-Myb down regulates the activity of the promoters of both type I collagen genes, α1(I) and α2(I), consistent with the coordinated regulation commonly observed for these two collagen chains.

Basic FGF potently induces vascular SMC proliferation, both *in vivo* and *in vitro*, and decreases the expression of type I collagen (Kennedy et al.,

1995; Majors and Ehrhart, 1993). We first showed this decrease was due to a drop in the rate of transcription of the two genes encoding type I collagen chains (Kypreos et al., 1998). We then selected the α1(I) chain for further study. Pretreatment of SMC cultures with B-*myb* antisense oligonucleotide inhibited the drop in the α1(I) collagen mRNA levels induced by bFGF, indicating that B-*myb* mediates the decrease in α1(I) collagen expression (Kypreos et al., 1998). We next tested whether B-*myb* expression can reduce endogenous collagen mRNA levels in bovine SMCs, which can be transfected with ~60% efficiency. Ectopic B-Myb expression for 48 hours caused a modest decrease in α1(I) collagen levels (~20%) in exponentially growing cells, consistent with the known long half-life of this mRNA (Kypreos et al., 1998). Importantly, B-*myb* almost completely ablated the induction of α1(I) collagen mRNA levels that would normally occur upon serum deprivation (Kypreos et al., 1998). Taken together these findings indicate that B-*myb* mediates the changes in transcription of type I collagen mRNA in SMCs upon serum withdrawal or bFGF treatment.

Interestingly, the role of B-Myb in matrix regulation appears to be cell type dependent. We have observed that B-Myb alone cannot repress collagen promoter activity in NIH 3T3 fibroblasts. In scleroderma fibroblasts, co-transfection with B-Myb had no effect on α2(I) collagen promoter activity (Luchetti et al., 2003; Piccinini et al., 1999). In these cells, c-Myb was able to transactivate the α2(I) collagen promoter via an MBS-containing region, and B-Myb was able to inhibit this transactivation in a dose-dependent manner (Luchetti et al., 2003), indicating that B-Myb might function as a repressor of c-Myb-mediated transactivation in these cells.

5. B-MYB NEGATIVELY REGULATES TYPE V COLLAGEN GENE EXPRESSION IN BOVINE AORTIC SMCS

Expression of the α2(V) collagen gene also occurs inversely with growth (Brown et al., 1991). Thus, we next tested whether α2(V) collagen gene transcription is similarly downregulated by B-*myb*. Co-transfection of B-*myb* caused a 3.8-fold decrease in activity of an α2(V) collagen promoter construct, which contains 2350 bp of upstream sequence and 150 bp of exon 1. Upon overexpression of ectopic B-*myb* in exponentially growing SMCs, a significant decrease in α2(V) collagen mRNA levels (1.6-fold) was observed (Kypreos et al., 1999). Furthermore, overexpression of B-*myb* completely prevented the 2-fold induction of α2(V) collagen mRNA normally observed in SMCs upon serum deprivation (Kypreos et al., 1999).

To begin to assess the mechanism of B-*myb*-mediated repression of the α2(V) collagen gene, we mapped the responsive region to a fragment including 100 bp of promoter and 150 bp of exon 1 sequences (pST0.25 construct). Two CRE-like elements, implicated in regulation by c-Myb (Dai et al., 1996; Oelgeschlager et al., 1996) were noted within exon 1. Mutation of these elements significantly decreased basal levels of α2(V) collagen promoter activity and ablated inhibition by B-Myb in bovine SMCs (Kypreos et al., 1999). However, competition with excess unlabelled oligonucleotide containing a CRE element was unsuccessful; therefore, factors other than CBP are implicated in regulation of type V collagen gene transcription by B-Myb, and we have termed this factor MRF-V. Binding of factors to these elements was ablated upon addition of B-Myb-glutathionine S-transferase (GST) fusion protein. These results are consistent with a model in which downregulation of type V collagen gene expression by B-Myb involves protein-protein interactions that inhibit positive transactivation signals mediated via binding of MRF-V to two positive elements in exon 1 (Kypreos et al., 1999). Thus, B-Myb is a key intracellular mediator of signals driving the inverse relationship between type I and type V collagen gene expression and growth state of the SMC in culture.

6. CYCLIN A REDUCES B-MYB-MEDIATED REPRESSION OF α2(V) COLLAGEN GENE EXPRESSION

B-Myb is a target for cyclin A/cyclin dependent kinase (cdk)2 kinase activity, and this phosphorylation enhances its transactivation potential (Johnson et al., 2002; Sala et al., 1997; Ziebold et al., 1997). A total of 22 putative cyclin A phosphorylation sites have been identified, with most of these sites located in the C-terminal half of the B-Myb protein (Johnson et al., 1999). Mutation of ten of these sites significantly reduced the ability of cyclin A to augment transactivation by B-Myb (Johnson et al., 1999). More recently, an additional five sites have been implicated (Johnson et al., 2002). Physical interaction between cyclin A and the C-terminal portion of B-Myb was demonstrated, suggesting phosphorylation is direct (Muller-Tidow et al., 2001). Somewhat paradoxically with enhancing transactivation activity, cyclin A has also been shown to decrease the half-life of B-Myb protein, via a ubiquitination- and proteosome-mediated pathway (Charrasse et al., 2000). Given the ability of phosphorylation to enhance transactivation by B-Myb, we investigated the effects of cyclin A on B-Myb-mediated repression of α2(V) collagen gene transcription. Co-expression of cyclin A was found to completely relieve the repression of the α2(V) collagen promoter mediated by B-Myb. B-Myb plus cyclin A yielded an average of 1.1 ± 0.4-fold activity relative to the control value rather than the 3 to 4-fold repression normally seen with B-Myb alone. Conversely, inhibition of cdk2 upon expression of a dominant negative cdk2 (cdk2-DN) enhanced B-Myb-mediated repression of the α2(V) collagen promoter activity to 5.9-fold. Overall, these studies indicate that cyclin A/cdk2 kinase activity ablates repression of the α2(V) collagen promoter by B-Myb (Petrovas et al., 2003).

To begin to map the specific sites of phosphorylation, we used a construct expressing a B-Myb protein with mutation of ten identified cyclin A phosphorylation sites (pCDNA3-10Mut expressing B-Myb-10Mut). Mutation of these sites substantially reduced the positive effects of cyclin A on transactivation by B-Myb (Johnson et al., 1999). Expression of B-Myb-10Mut inhibited α2(V) collagen promoter activity to the same extent as wild type B-Myb. Furthermore, cotransfection of a cyclin A expression vector effectively prevented repression of the α2(V) collagen promoter activity by B-Myb-10Mut (0.9-fold of basal level). Thus, the ten sites of cyclin A/cdk2 phosphorylation in B-Myb-10Mut do not appear to be required for repression. Furthermore, while cyclin A increased the rate of proteasome-mediated turnover of wild type B-Myb in SMCs, it had no detectable effect on the half-life of decay of the B-Myb-10Mut molecule. Taken together, these data indicate that these ten phosphorylation sites are essential for the acceleration of the rate of B-Myb turnover by cyclin A. However, these sites are not sufficient to block the repressive effect of B-Myb on the α2(V) collagen promoter activity, suggesting more rapid degradation is not responsible for the observed loss of repression (Petrovas et al., 2003).

7. C-TERMINUS IS REQUIRED FOR REPRESSION BY B-MYB AND REGULATION BY CYCLIN A

A mutant of B-Myb that is lacking amino acids 374-582 (B-Myb-Mut3) (Figure 1A) activates, rather than represses, the α2(V) collagen promoter (Kypreos et al., 1999). For example, as shown in Figure 1B, B-Myb-Mut3 caused a 2.7-fold induction of the α2(V) collagen promoter activity, instead of the 3- to 4-fold repression typically observed with wild type B-Myb. Using both cellular fractionation and indirect immunofluorescence, B-Myb-Mut3 was found to display a nuclear and cytoplasmic localization (Petrovas et al., 2003). Furthermore, we observed that cyclin A was not able to alter the effects of B-Myb-Mut3 on the α2(V) collagen promoter (Figure 1B). Cyclin A also did not appear to alter the stability of the protein (Petrovas et al., 2003). Taken together these results suggest that nuclear B-Myb-Mut3 transactivates the α2(V) collagen promoter, indicating that the critical domain for B-Myb-mediated suppression activity is within the C-terminal amino acids 373 to 382.

To further localise the region of B-Myb that mediates repression of the α2(V) collagen promoter, we used additional constructs encoding C-terminal truncated B-Myb molecules containing amino acids 1-422 and 1-491 (Figure 1A). We performed dose-response curves to assess their effects on α2(V) collagen promoter activity and examined their nuclear localisation using cellular fractionation. For B-Myb-491, a dose-dependent induction of α2(V) collagen promoter activity was observed, with a maximal activation of 5.9-fold with 3 μg B-Myb-491 expression vector (Figure 2A). An exclusive

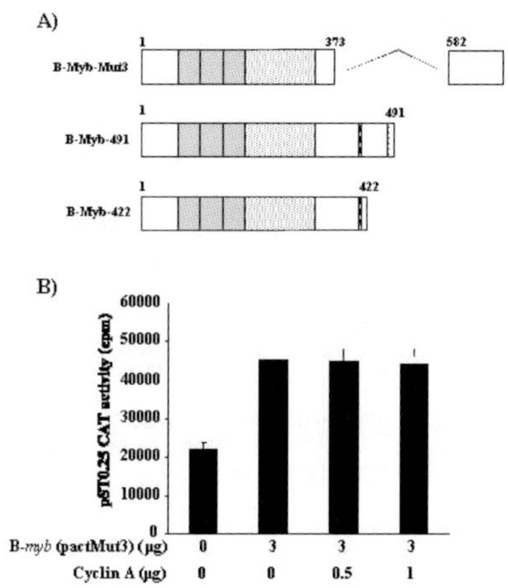

Figure 1

Transactivation activity and stability of a B-Myb mutant lacking amino acids 374-581 is not affected by cyclin A. (A) Schematic representation of the truncated B-Myb derivatives B-Myb-Mut3, B-Myb-491 and B-Myb-422. (B) SMCs were transfected with pST0.25-CAT α2(V) collagen promoter construct together with pactMut3, expressing B-Myb-Mut3, and pCMVcyclinA expression vector, as indicated, and CAT activity measured.

nuclear localisation of B-Myb-491 was detected by immunoblotting (Figure 2A, inset). In contrast, only a minimal induction of α2(V) collagen promoter activity was observed with the highest dose of B-Myb-422 expression vector (1.3-fold), although, it similarly displayed nuclear localisation (Figure 2B). Thus, we focused on the B-Myb-491 protein. Co-expression of cyclin A had only a slight effect on the positive regulation of α2(V) collagen promoter activity by B-Myb-491 (Figure 3A). Cyclin A had no effect on B-Myb-491 protein levels as judged by immunoblot analysis of nuclear extracts from co-transfected cells (Figure 3B). Thus, the failure of cyclin A to alter the ability of B-Myb-491 to repress α2(V) collagen promoter is not due to reduced levels of B-Myb protein.

Overall, our findings indicate that the ability of B-Myb to function as a repressor of matrix promoter activity is abolished by cyclin A. Thus, cyclin A/cdk 2 appears to function as the master switch, alleviating repression and promoting transactivation. Our studies have also mapped the sites mediating negative regulation by B-Myb, as well as the effects of cyclin A/cdk2 phosphorylation, to the C-terminal region including amino acids 491-582. Rather than altering DNA binding activity, cyclin A/cdk2 phosphorylation likely affects B-Myb protein-protein interactions (Bessa et al., 2001; Ziebold et al., 1997). These interactions have been described to either positively or negatively regulate B-Myb activity (Cervellera and Sala, 2000; Horstmann et al., 2000; Li and McDonnell, 2002; Masselink et al., 2001).

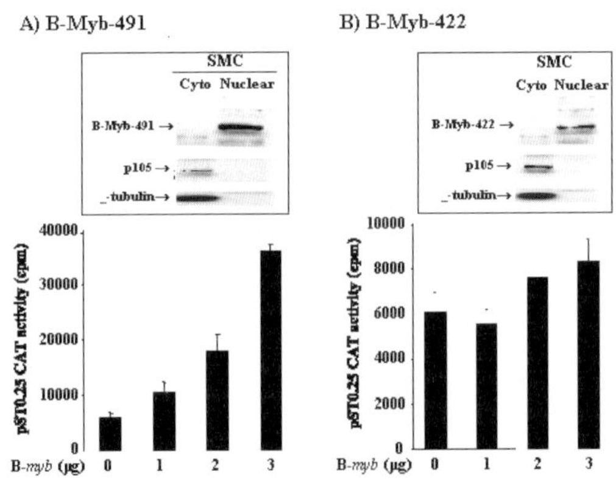

Figure 2

B-Myb-491 C-terminal truncated B-Myb transactivates the α2(V) collagen promoter. SMCs were plated at a density of 2 x 10^5 cells in 6-well dishes and transfected 24 h later with 1 µg pST0.25 CAT α2(V) collagen promoter construct plus the indicated dose of either (A) B-Myb-491 or (B) B-Myb-422 expression vector. Enough pCDNA3 empty vector was added to make a total final concentration of 5 µg DNA. After 72 h, samples were assessed for CAT activity. (A) B-Myb-491; (B) B-Myb-422. Insets, Nuclear and cytoplasmic extracts were prepared from the transfected cells and assessed for levels of B-Myb-491 and B-Myb-422, in panels A and B, respectively, as well as for p105 precursor of p50 and β-tubulin, which show an exclusive cytoplasmic localisation, as control.

As discussed above, B-Myb-mediated repression of the α2(V) collagen promoter also seemed to occur by an indirect mechanism, via interaction and inactivation of MRF-V, a positive regulatory factor (Kypreos et al., 1999). Thus, the effects of cyclin A/cdk2 on relieving repression of the α2(V) collagen promoter by B-Myb might occur by altering the interaction of B-Myb with other proteins, e.g., interaction with MRF-V, or other factors implicated in regulation of α2(V) collagen promoter activity (Penkov et al., 2000). The effects of cyclin A/cdk2 could be mediated by phosphorylation of either B-Myb or the interacting factor. If the observed effects relate to phosphorylation of B-Myb, our results suggest that mutation of the critical phosphorylation site(s) would lead to a super repressor of α2(V) collagen promoter activity. Furthermore, Masselink et al. (2001) and Li and McDonnell (2002) have demonstrated association of B-Myb with the N-CoR and SMRT co-repressor proteins; although, differing results were obtained with respect to the effects of cyclin A/cdk2 on this interaction. Experiments are in progress to elucidate the mechanism of the effects of cyclin A phosphorylation on repression by B-Myb.

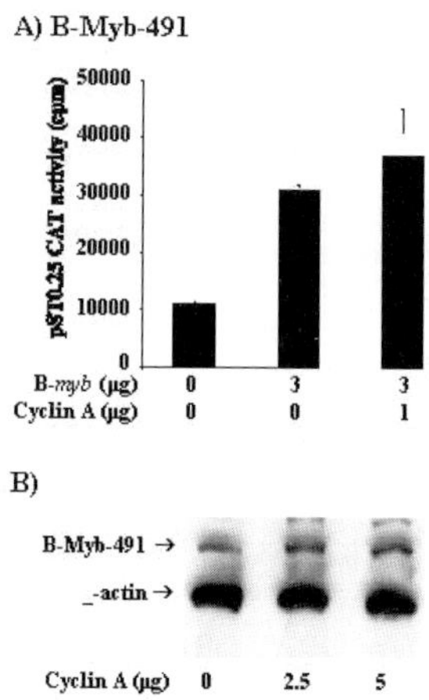

Figure 3

Cyclin A does not affect the activity or stability of B-Myb-491 C-terminal truncated B-Myb. (A) SMCs were transfected with 1 µg pST0.25 CAT α2(V) collagen promoter construct in the absence (0) or presence of the indicated amounts of B-Myb-491 and pCMVcyclinA expression vector, and CAT activity determined. (B) SMCs were transfected with 10 µg B-Myb-491 expression vector in the absence (0) or presence of 2.5 or 5 µg pCMVcyclinA vector. Nuclear extracts were prepared, and samples (30 µg) subjected to immunoblot analysis for B-Myb-491 and β-actin, which confirmed equal loading.

8. B-MYB REDUCES COLLAGEN AND ELASTIN GENE EXPRESSION *IN VIVO*

To test whether B-Myb mediates the repression of matrix gene expression by the vascular SMCs *in vivo*, we have constructed a mouse model in which the human B-*myb* cDNA is driven by the basal Cytomegalovirus (CMV) promoter. Three FVB mouse founders have been identified, termed line 2, 4 and 16. The mice developed normally. The expression of human B-*myb* transgene RNA was detected by RT-PCR in all tissues tested in the transgenic mice, including the heart, liver, brain, and aorta. Importantly, B-Myb protein expression was elevated an average of 3.3-fold in the aorta of the three transgenic lines, with the highest level in line 16. The expression of matrix genes in the CMV-B-*myb* and wild type mice was examined. RNA was isolated from the aortas of male and female mice at 6 weeks of age, and Northern analysis was performed for mRNA expression of α1(I) collagen and GAPDH, to test for RNA integrity and equal loading. Densitometry indicated that both the female and male mice display decreased aortic α1(I) collagen mRNA levels, although, the extent of the decrease seemed somewhat greater in the female mice. The normalised level of α1(I) collagen mRNA in the transgenic lines compared to wild type levels was 41%, 35.1% and 33.6% for female mice of lines 2, 4 and 16. In the males, lines 4 and 16 displayed 77% and 50.4% relative to the wild type, respectively, while in line 2, levels were essentially identical to those observed in the wild type mouse. Overall, the level of B-Myb overexpression correlated inversely with the degree of decrease in α1(I) collagen mRNA expression. These findings are consistent with *in vivo* B-Myb-mediated repression of α1(I) collagen gene transcription.

Although numerous factors have been shown to modulate elastin expression, their mechanisms of action are still poorly understood. For example, okadaic acid, EGF, phorbol ester, TNF-α, IL-β and bFGF have all been shown to repress elastin synthesis (Berk et al., 1996; Carreras et al., 2002; Ichiro et al., 1990; Parks et al., 1992; Rich et al., 1996). Interestingly, many of these agents have been shown to induce B-Myb, including EGF, phorbol ester, IGF-1, and bFGF (Kypreos et al., 1998; Marhamati and Sonenshein, 1996), suggesting that B-Myb may play a role in mediating some of these signals. To begin to assess the effects of B-Myb on expression of the elastin gene, the blots of aortic RNA from 6-week old mice were probed for elastin mRNA levels. Elastin mRNA expression was substantially decreased in the aorta of 6-week old transgenic male and female mice as compared to the wild type animals. Again the most dramatic effect was seen in line 16, which expressed the highest amounts of B-Myb (males and females displayed 2.2% and 7.1% of wild type, respectively).

Overall, the reduction in elastin mRNA levels in the aorta of the transgenic mice correlated inversely with B-Myb overexpression and paralleled the decreases in $\alpha 1(I)$ collagen mRNA. Importantly, aortic SMCs isolated from the transgenic animals showed a similar increase in B-*myb* mRNA levels, and decrease in elastin mRNA expression as compared to SMCs from wild type mice. Furthermore, co-transfection of a human B-*myb* expression vector caused an ~67% decrease in activity of a 2.2 kbp rat elastin promoter-CAT reporter construct in bovine aortic SMCs. These findings identify elastin as another target of repression by B-Myb in vascular SMCs, suggesting that B-Myb negatively regulates expression of multiple matrix genes in the vascular SMC.

Surprisingly, Verhoff van Gieson staining for elastin detected no differences in the elastic layering of the vessel wall using a in the CMV-B-*myb* transgenic animals as compared to wild type controls at 5 weeks and 2 months of age. Quantitative analysis of insoluble elastin protein in the aorta of transgenic animals as compared to wild type controls is currently underway. In the hemizygous elastin (+/-) mouse, a 50% decrease in elastin mRNA and protein was observed (Li et al., 1998). The arterial compliance at physiologic pressure was nearly normal in these mice, and this phenomenon was explained by a 35% increase in the number of elastic lamellae in the aortas of mice hemizygous for elastin as compared to the wild type. The increase in lamellae apparently acts to compensate for the decrease in elastin gene expression in the individual layers. Several explanations are possible for the differences between our model and the hemizygous elastin mouse. It is conceivable that the reduction in elastin protein was not as dramatic at the very earliest stages of development as the reduction observed in the elastin (+/-) mouse, and hence the effects in our model are more subtle. Alternatively, a mechanism may exist in the B-*myb* overexpressing mice to enhance elastin deposition via a post-transcriptional pathway. Also, the decrease in collagen gene expression may play a compensatory role. Since collagen provides rigidity and elastin gives elasticity to the aorta, the inhibition of collagen gene expression may have ameliorated the decrease in elastin in our model. Remodelling of the ECM of blood vessels, and in particular, the degradation of the internal elastic lamina in early stages of atherosclerosis, plays an important role in the formation of the atherosclerotic lesion (Davies, 1996). Interestingly, upon injury within the vessel wall, an inverse relationship is seen between elastin and collagen gene expression and SMC proliferation (Aoyagi et al., 1997; Rekhter and Gordon, 1994; Yamamoto et al., 1995). Thus, we are in the process of testing the role of overexpression of B-Myb following loss of integrity of the vessel wall upon mechanical injury in the CMV-B-*myb* versus wild type mice.

REFERENCES

Ang, A.H., Tachas, G., Campbell, J.H., Bateman, J.F. and Campbell, G.R. (1990) Collagen synthesis by cultured rabbit aortic smooth-muscle cells. Alteration with phenotype. Biochem J *265*, 461-469.

Aoyagi, M., Yamamoto, M., Azuma, H., Niimi, Y., Tajima, S., Hirakawa, K. and Yamamoto, K. (1997) Smooth muscle cell proliferation, elastin formation, and tropoelastin transcripts during the development of intimal thickening in rabbit carotid arteries after endothelial denudation. Histochem Cell Biol *107*, 11-17.

Arsura, M., Introna, M., Passerini, F., Mantovani, A. and Golay, J. (1992) B-myb antisense oligonucleotides inhibit proliferation of human hematopoietic cell lines. Blood *79*, 2708-2716.

Bahadori, L., Milder, J., Gold, L. and Botney, M. (1995) Active macrophage-associated TGF-beta co-localizes with type I procollagen gene expression in atherosclerotic human pulmonary arteries. Am J Pathol *146*, 1140-1149.

Barone, L.M., Wolfe, B.L., Faris, B. and Franzblau, C. (1988) Elastin mRNA levels and insoluble elastin accumulation in neonatal rat smooth muscle cell cultures. Biochemistry *27*, 3175-3182.

Beldekas, J.C., Smith, B., Gerstenfeld, L.C., Sonenshein, G.E. and Franzblau, C. (1981) Effects of 17 beta-estradiol on the biosynthesis of collagen in cultured bovine aortic smooth muscle cells. Biochemistry *20*, 2162-2167.

Berk, J.L., Massoomi, N., Krupsky, M. and Goldstein, R.H. (1996) Effect of okadaic acid on elastin gene expression in interstitial lung fibroblasts. Am J Physiol *271*, L939-948.

Bessa, M., Saville, M.K. and Watson, R.J. (2001) Inhibition of cyclin A/Cdk2 phosphorylation impairs B-Myb transactivation function without affecting interactions with DNA or the CBP coactivator. Oncogene *20*, 3376-3386.

Birk, D.E. (2001) Type V collagen: heterotypic type I/V collagen interactions in the regulation of fibril assembly. Micron *32*, 223-237.

Brown, K.E., Lawrence, R. and Sonenshein, G. E. (1991) Concerted modulation of alpha 1(XI) and alpha 2(V) collagen mRNAs in bovine vascular smooth muscle cells. J Biol Chem *266*, 23268-23273.

Campbell, G.R. and Campbell, J.H. (1990) The phenotypes of smooth muscle expressed in human atheroma. Ann N Y Acad Sci *598*, 143-158.

Carreras, I., Rich, C.B., Panchenko, M.P. and Foster, J.A. (2002) Basic fibroblast growth factor decreases elastin gene transcription in aortic smooth muscle cells. J Cell Biochem *85*, 592-600.

Cervellera, M.N. and Sala, A. (2000) Poly(ADP-ribose) polymerase is a B-MYB coactivator. J Biol Chem *275*, 10692-10696.

Chang, C.J. and Sonenshein, G.E. (1991) Increased collagen gene expression in vascular smooth muscle cells cultured in serum or isoleucine deprived medium. Matrix *11*, 242-251.

Charrasse, S., Carena, I., Brondani, V., Klempnauer, K.H. and Ferrari, S. (2000) Degradation of B-Myb by ubiquitin-mediated proteolysis: involvement of the Cdc34-SCF(p45Skp2) pathway. Oncogene *19*, 2986-2995.

Dai, P., Akimaru, H., Tanaka, Y., Hou, D.X., Yasukawa, T., Kanei-Ishii, C., Takahashi, T. and Ishii, S. (1996) CBP as a transcriptional coactivator of c-Myb. Genes Dev *10*, 528-540.

Davidson, J.M., Zoia, O. and Liu, J.M. (1993) Modulation of transforming growth factor-beta 1 stimulated elastin and collagen production and proliferation in porcine vascular smooth muscle cells and skin fibroblasts by basic fibroblast growth factor, transforming growth factor-alpha, and insulin-like growth factor-I. J Cell Physiol *155*, 149-156.

Davies, M.J. (1996) Stability and Instability: Two Faces of Coronary Atherosclerosis: The Paul Dudley White Lecture 1995. Circulation *94*, 2013-2020.

Debelle, L. and Tamburro, A. M. (1999) Elastin: molecular description and function. Int J Biochem Cell Biol *31*, 261-272.

Foos, G., Grimm, S. and Klempnauer, K.H. (1992) Functional antagonism between members of the myb family: B-myb inhibits v-myb-induced gene activation. Embo J *11*, 4619-4629.

Hayashi, A., Suzuki, T. and Tajima, S. (1995) Modulations of elastin expression and cell proliferation by retinoids in cultured vascular smooth muscle cells. J Biochem (Tokyo) *117*, 132-136.

Horstmann, S., Ferrari, S. and Klempnauer, K.H. (2000) An alternatively spliced isoform of B-Myb is a transcriptional inhibitor. Oncogene *19*, 5428-5434.

Ichiro, T., Tajima, S. and Nishikawa, T. (1990) Preferential inhibition of elastin synthesis by epidermal growth factor in chick aortic smooth muscle cells. Biochem Biophys Res Commun *168*, 850-856.

James, M.F., Rich, C.B., Trinkaus-Randall, V., Rosenbloom, J. and Foster, J.A. (1998) Elastogenesis in the developing chick lung is transcriptionally regulated. Dev Dyn *213*, 170-181.

Johnson, L.R., Johnson, T.K., Desler, M., Luster, T. A., Nowling, T., Lewis, R.E. and Rizzino, A. (2002) Effects of B-Myb on gene transcription: phosphorylation-dependent activity ans acetylation by p300. J Biol Chem *277*, 4088-4097.

Johnson, T.K., Schweppe, R.E., Septer, J. and Lewis, R.E. (1999) Phosphorylation of B-Myb regulates its transactivation potential and DNA binding. J Biol Chem *274*, 36741-36749.

Kennedy, S.H., Qin, H., Lin, L. and Tan, E.M. (1995) Basic fibroblast growth factor regulates type I collagen and collagenase gene expression in human smooth muscle cells. Am J Pathol *146*, 764-771.

Kindy, M.S., Chang, C.J. and Sonenshein, G.E. (1988) Serum deprivation of vascular smooth muscle cells enhances collagen gene expression. J Biol Chem *263*, 11426-11430.

Kindy, M.S. and Sonenshein, G.E. (1986) Regulation of oncogene expression in cultured aortic smooth muscle cells. Post-transcriptional control of c-myc mRNA. J Biol Chem *261*, 12865-12868.

Kypreos, K.E., Birk, D., Trinkaus-Randall, V., Hartmann, D.J. and Sonenshein, G.E. (2000) Type V collagen regulates the assembly of collagen fibrils in cultures of bovine vascular smooth muscle cells. J Cell Biochem *80*, 146-155.

Kypreos, K.E., Marhamati, D.J. and Sonenshein, G.E. (1999) B-Myb represses trans-activation of the Col5A2 collagen promoter indirectly via inhibition of binding of factors interacting with positive elements within the first exon. Matrix Biol *18*, 275-285.

Kypreos, K.E., Nugent, M.A. and Sonenshein, G.E. (1998) Basic fibroblast growth factor-induced decrease in type I collagen gene transcription is mediated by B-myb. Cell Growth Differ *9*, 723-730.

Kypreos, K.E. and Sonenshein, G.E. (1998) Basic fibroblast growth factor decreases type V/XI collagen expression in cultured bovine aortic smooth muscle cells. J Cell Biochem *68*, 247-258.

Lawrence, R., Hartmann, D.J. and Sonenshein, G.E. (1994) Transforming growth factor beta 1 stimulates type V collagen expression in bovine vascular smooth muscle cells. J Biol Chem *269*, 9603-9609.

Li, D.Y., Faury, G., Taylor, D.G., Davis, E.C., Boyle, W.A., Mecham, R.P., Stenzel, P., Boak, B., and Keating, M.T. (1998) Novel arterial pathology in mice and humans hemizygous for elastin. J Clin Invest *102*, 1783-1787.

Li, X. and McDonnell, D.P. (2002) The transcription factor B-Myb is maintained in an inhibited state in target cells through its interaction with the nuclear corepressors N-CoR and SMRT. Mol Cell Biol *22*, 3663-3673.

Liau, G. and Chan, L.M. (1989) Regulation of extracellular matrix RNA levels in cultured smooth muscle cells. Relationship to cellular quiescence. J Biol Chem *264*, 10315-10320.

Luchetti, M.M., Paroncini, P., Majlingova, P., Frampton, J., Mucenski, M., Svegliati Baroni, S., Sambo, P., Golay, J., Introna, M., and Gabrielli, A. (2003) c-Myb regulates the transcription of the human α(2) type I collagen gene. J Biol Chem *278*, 1533-1541.

Majors, A. and Ehrhart, L.A. (1993) Basic fibroblast growth factor in the extracellular matrix suppresses collagen synthesis and type III procollagen mRNA levels in arterial smooth muscle cell cultures. Arterioscler Thromb *13*, 680-686.

Marhamati, D.J., Bellas, R.E., Arsura, M., Kypreos, K.E. and Sonenshein, G.E. (1997) A-myb is expressed in bovine vascular smooth muscle cells during the late G1-to-S phase transition and cooperates with c-myc to mediate progression to S phase. Mol Cell Biol *17*, 2448-2457.

Marhamati, D.J. and Sonenshein, G.E. (1996) B-Myb expression in vascular smooth muscle cells occurs in a cell cycle-dependent fashion and down-regulates promoter activity of type I collagen genes. J Biol Chem *271*, 3359-3365.

Masselink, H., Vastenhouw, N. and Bernards, R. (2001) B-myb rescues ras-induced premature senescence, which requires its transactivation domain. Cancer Lett *171*, 87-101.

Mizuguchi, G., Nakagoshi, H., Nagase, T., Nomura, N., Date, T., Ueno, Y. and Ishii, S. (1990) DNA binding activity and transcriptional activator function of the human B-myb protein compared with c-MYB. J Biol Chem *265*, 9280-9284.

Muller-Tidow, C., Wang, W., Idos, G.E., Diederichs, S., Yang, R., Readhead, C., Berdel, W.E., Serve, H., Saville, M., Watson, R. and Koeffler, H.P. (2001) Cyclin A1 directly interacts with B-myb and cyclin A1/cdk2 phosphorylate B-myb at functionally important serine and threonine residues: tissue-specific regulation of B-myb function. Blood *97*, 2091-2097.

Murata, K., Kotake, C. and Motoyama, T. (1987) Collagen species in human aorta: with special reference to basement membrane-associated collagens in the intima and media and their alteration with atherosclerosis. Artery *14*, 229-247.

Oelgeschlager, M., Janknecht, R., Krieg, J., Schreek, S., and Luscher, B. (1996) Interaction of the co-activator CBP with Myb proteins: effects on Myb- specific transactivation and on the cooperativity with NF-M. EMBO J *15*, 2771-2780.

Parks, W.C., Kolodziej, M.E. and Pierce, R.A. (1992) Phorbol ester-mediated downregulation of tropoelastin expression is controlled by a posttranscriptional mechanism. Biochemistry *31*, 6639-6645.

Penkov, D., Tanaka, S., Di Rocco, G., Berthelsen, J., Blasi, F. and Ramirez, F. (2000) Cooperative interactions between PBX, PREP, and HOX proteins modulate the activity of the alpha 2(V) collagen (COL5A2) promoter. J Biol Chem *275*, 16681-16689.

Petrovas, C., Jeay, S., Lewis, R.E. and Sonenshein G.E. (2003) B-Myb repressor function in vascular smooth muscle cells is regulated by cyclin A phosphorylation and sequences within the C-terminal domain. Oncogene, in press.

Piccinini, G., Golay, J., Flora, A., Songia, S., Luchetti, M., Gabrielli, A. and Introna, M. (1999) C-myb, but not B-myb, upregulates type I collagen gene expression in human fibroblasts. J Invest Dermatol *112*, 191-196.

Rekhter, M.D. and Gordon, D. (1994) Cell proliferation and collagen synthesis are two independent events in human atherosclerotic plaques. J Vasc Res *31*, 280-286.

Rich, C.B., Nugent, M.A., Stone, P. and Foster, J.A. (1996) Elastase release of basic fibroblast growth factor in pulmonary fibroblast cultures results in down-regulation of elastin gene transcription. A role for basic fibroblast growth factor in regulating lung repair. J Biol Chem *271*, 23043-23048.

Ross, R. (1993) The pathogenesis of atherosclerosis: a perspective for the 1990s. Nature *362*, 801-809.

Sala, A. and Calabretta, B. (1992) Regulation of BALB/c 3T3 fibroblast proliferation by B-myb is accompanied by selective activation of cdc2 and cyclin D1 expression. Proc Natl Acad Sci U S A *89*, 10415-10419.

Sala, A., Kundu, M., Casella, I., Engelhard, A., Calabretta, B., Grasso, L., Paggi, M.G., Giordano, A., Watson, R.J., Khalili, K. and Peschle, C. (1997) Activation of human B-MYB by cyclins. Proc Natl Acad Sci U S A *94*, 532-536.

Sala, A. and Watson, R. (1999) B-Myb protein in cellular proliferation, transcription control, and cancer: latest developments. J Cell Physiol *179*, 245-250.

Sanz-Gonzalez, S.M., Poch, E., Perez-Roger, I., Diez-Juan, A., Ivorra, C. and Andres, V. (2000) Control of vascular smooth muscle cell growth by cyclin-dependent kinase inhibitory proteins and its implication in cardiovascular disease. Front Biosci *5*, D619-628.

Smith-Mungo, L.I. and Kagan, H.M. (1998) Lysyl oxidase: properties, regulation and multiple functions in biology. Matrix Biol *16*, 387-398.

Speir, E., Sasse, J., Shrivastav, S. and Casscells, W. (1991) Culture-induced increase in acidic and basic fibroblast growth factor activities and their association with the nuclei of vascular endothelial and smooth muscle cells. J Cell Physiol *147*, 362-373.

Stepp, M.A., Kindy, M.S., Franzblau, C. and Sonenshein, G.E. (1986) Complex regulation of collagen gene expression in cultured bovine aortic smooth muscle cells. J Biol Chem *261*, 6542-6547.

Tajima, S. (1996) Modulation of elastin expression and cell proliferation in vascular smooth muscle cells in vitro. Keio J Med *45*, 58-62.

Tashiro, S., Takemoto, Y., Handa, H. and Ishii, S. (1995) Cell type-specific trans-activation by the B-myb gene product: requirement of the putative cofactor binding to the C-terminal conserved domain. Oncogene *10*, 1699-1707.

Tokimitsu, I., Kato, H., Wachi, H. and Tajima, S. (1994) Elastin synthesis is inhibited by angiotensin II but not by platelet-derived growth factor in arterial smooth muscle cells. Biochim Biophys Acta *1207*, 68-73.

Toselli, P., Faris, B., Sassoon, D., Jackson, B.A. and Franzblau, C. (1992) In-situ hybridization of tropoelastin mRNA during the development of the multilayered neonatal rat aortic smooth muscle cell culture. Matrix *12*, 321-332.

Vuorio, E. and de Crombrugghe, B. (1990) The family of collagen genes. Annu Rev Biochem *59*, 837-872.

Wachi, H., Seyama, Y., Yamashita, S. and Tajima, S. (1995) Cell cycle-dependent regulation of elastin gene in cultured chick vascular smooth-muscle cells. Biochem J *309*, 575-579.

Watson, R.J., Robinson, C. and Lam, E.W. (1993) Transcription regulation by murine B-myb is distinct from that by c-myb. Nucleic Acids Res *21*, 267-272.

Yamamoto, K., Aoyagi, M. and Yamamoto, M. (1995) Changes in elastin-binding proteins during the phenotypic transition of rabbit arterial smooth muscle cells in primary culture. Exp Cell Res *218*, 339-345.

Ziebold, U., Bartsch, O., Marais, R., Ferrari, S. and Klempnauer, K.H. (1997) Phosphorylation and activation of B-Myb by cyclin A-Cdk2. Curr Biol *7*, 253-260.

Chapter 19

THE ROLE OF C-MYB IN GASTROINTESTINAL TRACT DEVELOPMENT AND CARCINOGENESIS

Robert G. Ramsay, Daniel Ciznadji and [1]Gabriella Zupi
Peter MacCallum Cancer Institute, Melbourne, Australia and [1]Experimental Chemotherapy Laboratory, Regina Elena Cancer Institute, Rome, Italy

Abstract: The list of cancers in which c-Myb appears to play a role includes tumours of the upper gastrointestinal (GI) tract, colon, pancreas, melanocytes, brain and breast. This review focuses on the involvement of c-Myb in normal and neoplastic colon cells. Expression of c-*myb* RNA in rat, mouse and human colons is similar to, or exceeds, that observed in haemopoietic cells. In the colon c-Myb is essential for development, Bcl-2 expression and recovery from cyto-toxic damage. As in the blood system, elongation of c-*myb* transcripts is tightly regulated in the GI tract. Over expression of c-Myb in colon cell lines blocks differentiation, thus satisfying an important foundation for malignant transformation. c-*myb* expression increases during colon and oesophageal adenocarcinoma progression and in colon cancer. The level of c-*myb* expression in GI tract tumours has prognostic significance for patient survival.

1. INTRODUCTION

c-Myb has been studied extensively in the haemopoietic system and blood cell malignancies thereof. This is due to its high expression in immature haemopoietic progenitor cells and the finding that over-expression of c-Myb, or expression of activated forms of c-Myb, transforms haemopoietic cells (Fu & Lipsick, 1997; Gonda et al., 1989a; 1989b). The list of malignancies in which c-Myb appears to play an important role is growing and includes upper gastrointestinal (GI), colon and pancreatic tumours, melanoma, neurological cancers and some sub-sets of breast cancer. It was the anecdotal reports of c-Myb expression in GI cells that drove our early interest in exploring the role of c-Myb outside the haemopoietic compartment.

J. Frampton (ed.), Myb Transcription Factors: Their Role in Growth, Differentiation and Disease, 367-388.

We are convinced that c-Myb is essential for normal colon development and biology in general. Basic expression data show that rat, mouse and human colons have relative levels of c-*myb* expression similar to, or exceeding, those observed in thymus and bone marrow (Alexander et al., 1992; Rosenthal et al., 1996; Thompson & Ramsay, 1995; Thompson et al., 1998). As in haemopoietic cells, when c-Myb is over-expressed in colon cell lines it blocks differentiation (see below) thus satisfying an important foundation for malignant transformation. The final key point that has encouraged our studies of c-Myb expression and the GI tract is the finding that the level of c-*myb* increases during colon (Ramsay et al., 1992) and oesophageal adenocarcinoma progression (Brabender et al., 2001). In colon cancer at least, this rise has great importance as the level of *c-myb* expression has prognostic significance for patient survival (Biroccio et al., 2001).

Consistent reports of *c-myb* expression in colon cancer cell lines have appeared in the literature dating back to the 1980s (Torelli et al., 1987; Trainer et al., 1988). Reports on colon dominate perhaps because gastric (Park & Gazdar, 1996; Whitehead et al., 1989) and oesophageal cancer cell lines (de Both et al., 2001) are less well characterised and relatively uncommon. It was through many discussions with Robert Whitehead (then at the Ludwig Melbourne Branch and now at Vanderbilt, Tennessee, USA) that our fascination with the parallels between the proliferative potential of bone marrow cells and the colonic crypt cells intensified. This prompted our *c-myb* studies on the biology of the colon that have now spanned more than a decade of research.

We initially examined transformed human colon cell lines to convince ourselves that *c-myb* was indeed expressed at the protein level. Firstly, we determined if the mRNA was actually translated and secondly we were very mindful of the frequent incidence of c-Myb amino or carboxyl terminal truncations implicated in transformation (Thompson & Ramsay, 1995; Weston, 1990). At that time we did not appreciate how important the mechanism underlying the high level of *c-myb* transcripts in colon would be to our research effort. This mechanism will be discussed more fully later. Regarding the protein, to date we have only observed full-length c-Myb in colon tumours of mice and humans. Subsequently we have investigated the role of c-Myb during development and tumourigenesis, identified its target genes in epithelial cells, studied colon cancer patient prognosis, and our most recent work focuses upon the role of c-Myb in the regulation of differentiation within the colonic crypt. The key findings in each of these areas have been distilled and will be discussed.

2. C-MYB AND GASTROINTESTINAL DEVELOPMENT

Understanding the architecture of the normal crypt is very instructive when drawing parallels to the islands of haemopoiesis found in the bone marrow. Both units generate an astonishing number of cells during the life span of a mammal. These highly proliferative factories are driven by stem cells and their progeny are prime candidates for transformation. Much of what we know about crypt stem cells and progenitors in the small and large intestine comes from exacting studies by Chris Potten and colleagues.

The mammalian colon is lined with a rapidly renewing and highly regulated epithelium consisting of columnar enterocytes, mucin producing goblet cells and enteroendocrine cells compartmentalised into crypts. The crypt serves as the functional unit in the colon. Mouse crypts consist of approximately 500 cells where stem cells located at the base of the crypt give rise to a dividing population of cells that differentiate along the three lineages (Potten, 1998). This process takes 4-7 days during which transient amplifying (progenitor) cells progressively differentiate while migrating towards the top of the crypts where many of the cells undergo apoptosis (Gavrieli et al., 1992). During development, differentiation of the murine intestine occurs in a proximal to distal wave between embryonic days 15 and 19. Crypts continue to lengthen and divide by fission until 28 days postnatal, after which time homeostasis is attained (Gordon & Hermiston, 1994). This rapid expansion of multiple lineages building up to homeostasis in the adult, parallels exactly what happens in the haemopoietic system where blood stem cells also give rise to progenitor cells and finally terminally differentiated cells. A model of a human colonic crypt depicting the zones of proliferation, differentiation and apoptosis is shown in Figure 1.

Several transgenes and gene disruptions in mice have produced alterations in gut formation or biology. These include mice lacking one allele of the homeobox gene *cdx2* that were prone to tumour-like growths in the colon (Chawengsaksophak et al., 1997) and mice homozygous for a null allele ("nullizygous") of the transcription factor *fkh6* displaying epithelial abnormalities of the stomach and small intestine (Kaestner et al., 1997). Targetted deletion of the cytokeratin 8 gene altered gut development leading to colonic hyperplasia (Baribault et al., 1994) while over-expression of the adhesion molecule APC affected small intestinal epithelial cell migration (Wong et al., 1996). Conversely, E-cadherin over-expression suppressed small intestine crypt cell proliferation, induced apoptosis and slowed migration (Hermiston et al., 1996). Later, expression of a stable form of β-catenin was shown to have similar effects (Wong et al., 1998). Targetted disruption of Trefoil proteins leads to gastric or intestinal abnormalities

(Lefebvre et al., 1996; Mashimo et al., 1996) as does the selective ablation of STAT signalling through gp130 (Tebbutt et al., 2002).

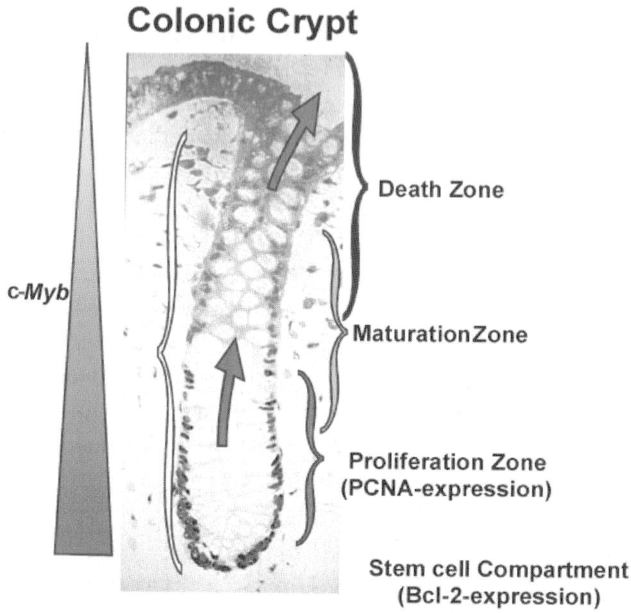

Figure 1

A human colonic crypt with regions of proliferation, differentiation and apoptosis. c-Myb expression is highest at the base of the crypt and declines as cells migrate towards the lumen of the colon (Thompson *et al.*, 1998a). Bcl-2 expression is tightly restricted to the base of the crypt where stem cells reside (Merritt *et al.*, 1995).

In the mouse embryo, *c-myb* is expressed early in the development of neuronal and many epithelial tissues, including liver, kidney and colonic mucosa. Expression persists in the adult colonic epithelia in both mouse and human (Rosenthal et al., 1996) where it is expressed at higher levels than anywhere else in the animal (Thompson and Ramsay, 1995). We used the *c-myb*$^{-/-}$ mouse (Mucenski *et al*, 1991) to directly investigate the role of *c-myb* in normal colon development. The effect of the *c-myb* deletion on intestinal development could not be assessed directly since mice with targetted disruptions of both *c-myb* alleles die at embryonic day 15 due to a severe defect in foetal liver haemopoiesis (Mucenski et al., 1991). However, by dissecting colon and small intestine from *c-myb*$^{-/-}$ embryos at embryonic day 14 and transplanting the tissue under the kidney capsule of recipient adult mice development was allowed to proceed.

We found that colon tissue from the *c-myb*$^{-/-}$ embryos developed with profound epithelial disorganisation (Zorbas et al, 1999). In contrast, we were very surprised that the small intestine transplants were indistinguishable from the wild type even though *c-myb* is also expressed in the small intestinal crypt. One of the most striking observations was the elevated apoptosis in the nullizygous tissue, which we attributed at that time to the almost total absence of Bcl-2 protein expression in the foetal *c-myb*$^{-/-}$ colon (Zorbas et al, 1999). However, we have subsequently explored this relationship further by studying the *bcl-2*$^{-/-}$ mouse colon during development and at homeostasis. Earlier literature suggested that the absence of Bcl-2 led to a similar colon and small intestinal phenotype to our c-*myb*$^{-/-}$ foetal colon transplants (Kamada et al., 1995; Zorbas et al., 1999). In our experience however, the loss of Bcl-2 does not affect intestinal integrity on a C57Bl/6 background, although most of the other reported defects were very demonstrable. Nor did we find any defect in the developing *bcl-2*$^{-/-}$ colons from embryos that were subjected to the same transplantation procedure used to reveal the *c-myb*$^{-/-}$ phenotype (see Figure 2).

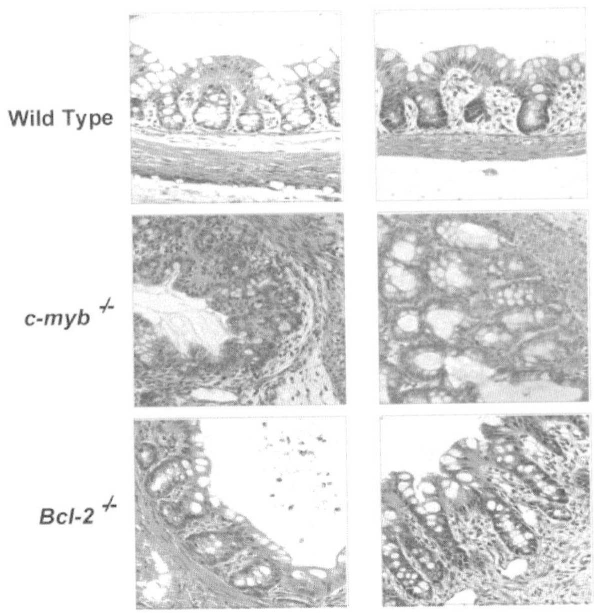

Figure 2

Normal colon development requires *c-myb* expression and in its absence crypt architecture is profoundly disrupted. Foetal colon transplants are shown from 2 wild type, 2 *c-myb*$^{-/-}$ and 2 *bcl-2*$^{-/-}$ mice. Although the *c-myb*$^{-/-}$ embryonic colon shows a deficit in Bcl-2 expression (Zorbas *et al.*, 1999) the absence of *bcl-2* does not affect colon development.

In retrospect, the lack of a gut phenotype in *bcl-2*$^{-/-}$ mice is not surprising as a study by Merritt and colleagues showed that Bcl-2 expression was tightly restricted to a few cells, or perhaps only one cell, within the crypt and that the colon was essentially normal in *bcl-2*$^{-/-}$ mice (Merritt et al., 1995). It is tempting to conclude that the small population of cells with Bcl-2 may represent steady-state stem cells. In view of these data we are currently examining the expression of other Bcl-2 family members in foetal colon in an attempt to explain the elevated apoptosis in *c-myb*$^{-/-}$ colons (Zorbas et al, 1999). Certainly the most promising candidate, Bcl-X_L is abundantly expressed in the *c-myb*$^{-/-}$ colon (data not shown) so loss of expression of this gene is not likely to be responsible for the higher rate of apoptosis in *c-myb*$^{-/-}$ colon transplants (Zorbas et al, 1999).

To directly address the role of *c-myb* in adult colon and other GI tissues where homeostasis has been established it will be necessary to ablate *c-myb* using conditional knock-out strategies, and these are currently underway in our laboratory. Meanwhile we examined the *c-myb*$^{+/-}$ adult mouse for any defects in colon biology, but found none. It is clear that under normal physiological conditions a single functional *c-myb* allele is sufficient for colonic crypt maintenance. But what is the situation when the colonic crypt is stressed by damage following treatments such as γ-irradiation?

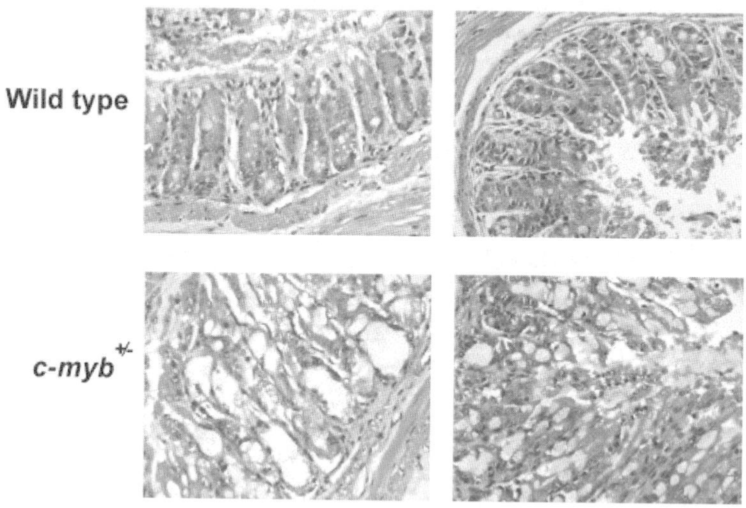

Figure 3

Colon homeostasis is profoundly disrupted when mammals are exposed to high doses of ionizing radiation. Wild type and c-*myb*$^{+/-}$ mice were irradiated with 13 gray of γ-rays and assessed 4 days later. The wild type colon had recovered normal crypt morphology while the *c-myb*$^{+/-}$ mice showed loss of cellularity and crypt morphology suggesting that both *c-myb* alleles are required for a correct response to stress.

The treatment of mice with semi-lethal and lethal doses of cytotoxic drugs or radiation has been a powerful means of revealing differences in the biological role of genes like bcl-2 that are required for cell survival and response to stress (Ogilvy et al., 1999). When wild type and c-myb+/- mice are subjected to a single lethal dose of 13 gray there are no apparent differences in the induction of apoptosis or initial crypt length recovery (RGR, unpublished). Indeed wild type crypts recover quickly from their reduced size with relatively normal crypt morphology although they do exceed the length of steady state crypts (a phenomenon typical of irradiated colon). While the crypts recover in length to some extent in the c-myb$^{+/-}$, colon crypt integrity is profoundly disturbed. By day 4 post-irradiation there is a marked loss of cellularity and crypt structure in the c-myb$^{+/-}$ colons (Figure 3). Usually it is at this time that the c-myb$^{+/-}$ mice become sick. It would therefore appear that both copies of c-myb are required to regulate the exquisite balance between cell proliferation and differentiation, but how this is directly coupled to apoptosis remains unclear.

3. CYTO DIFFERENTIATION

Many studies on haemopoietic cell differentiation have documented a commensurate decline in *c-myb* expression as cells mature and specialise (de Both et al., 1989; Gonda & Metcalf, 1984; Ramsay et al., 1986; Richon et al., 1989). More importantly, it was shown that over-expression of c-Myb blocks differentiation (Clarke et al., 1988). In view of these observations we explored the possibility that c-Myb expression correlates inversely with colonic cell differentiation. This fits well with the spatial distribution of *c-myb* within the colonic crypt where expression is strongest at the base where cells are immature and relatively weak expression at the top of the crypt where cells are about to apoptose before being extruded into the lumen (Rosenthal et al., 1996; Thompson et al., 1998) (see Figure 1). Some colon carcinoma cell lines recapitulate *in vitro* the proliferation, differentiation and apoptosis cycle that takes place in the crypt because they can be induced to differentiate. To directly examine the role of c-Myb in this process stable transfected colon lines were generated that harbour the inducible expression construct, MybER (Schmidt et al., 2000). We also used this construct to replace *c-myb* but in this instance where it declines as the colon cell lines differentiate (Thompson et al., 1998).

In the case of colonic epithelium the natural exogenous differentiation agent of choice is sodium butyrate (NaButyrate) as this short chain fatty acid is found at high concentrations (up to 131 mM) within the colon (Cummings et al., 1987). Using the cell line LIM1215 that responds uniformly to NaButyrate treatment (Whitehead et al., 1986), we examined the effects of Myb-ER on differentiation. One reliable hallmark of colonic epithelia cell differentiation is the expression of alkaline phosphatase (AP) within the membranes of columnar-like cells. In NaButyrate-treated parental LIM1215 cells AP expression can be readily detected by histochemical staining or by enzyme activity assays such that after 3 days exposure to 2 mM NaButyrate at least 30% of cells are stained strongly for AP and enzyme activity has increased 1000% above baseline. However, the activation of the Myb-ER fusion protein by the oestrogen analogue, 4-hydroxy Tamoxifen (4-OHT) largely blocks this differentiation response (Figure 4).

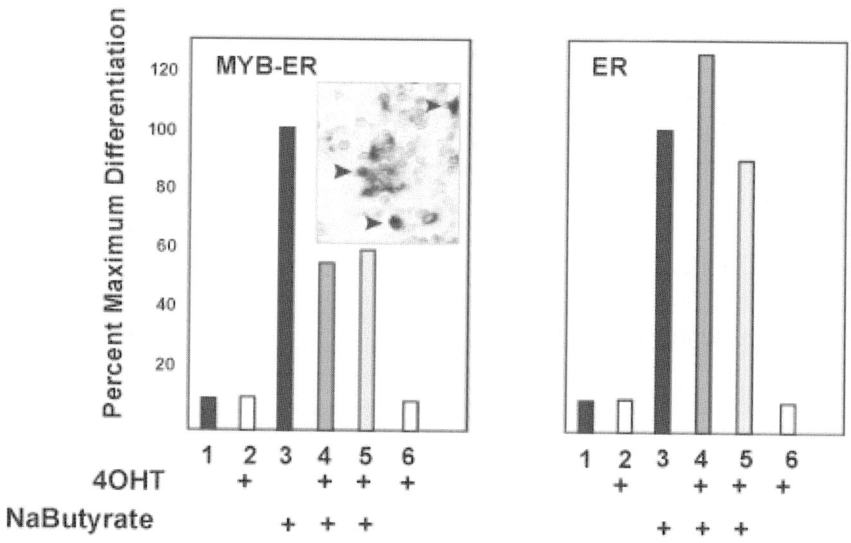

Figure 4

The Myb-ER fusion protein is activated by the addition of 4-OHT allowing it to bind to Myb binding sites in regulatory regions of target genes. Approximately 30% of LIM1215 cells stain positive for AP by 72 hr when treated with NaButyrate (see inset with arrowheads directed to AP staining). However, this differentiation is partially blocked by exogenous Myb-ER expression and activation. Cells stably transfected with Myb-ER (left panel) or ER (right panel) were treated with 4-OHT at -3 hr (lanes 2 and 4) or +24 hr (lanes 5 and 6) or with 2 mM Nabutyrate (lanes 3, 4 and 5). Percentage maximum differentiation was determined from counts of AP positive cells per 1000 cells.

More recently we have investigated the expression of *c-myb* in primary colonocytes grown as colonies in soft agar. As predicted, the cells within the colonies express *c-myb* and other epithelial associated transcripts (Micallef et al, submitted). We have successfully grown colonies from foetal, perinatal and, with less efficiency, adult colon (Micallef et al, submitted). The colonies that have some properties characteristic of bone marrow-derived hematopoietic progenitors but in this instance express the intestinal-specific antigen A33 (Johnstone et al., 2000). Importantly, they also express *c-myb* and *bcl*-2, a profile expected of cells that arise at the base of the colonic crypt. At present the diversity of experiments that can be performed with these cells is limited and has led us to the use of a relatively normal immortalised colon cell line, YAMC (Whitehead et al., 1993).

YAMC cell lines are based on the immorto-transgenic mouse that harbours a temperature sensitive SV40 T-antigen. They are classically derived from young adult mouse colons but can also be generated with reasonable efficiency from foetal colons. In terms of differentiation, these cells respond uniformly to NaButyrate whereby most cells have detectable AP expression following 3 days of treatment (Sicurella and RGR, unpublished) (Figure 5). Using a novel strategy to examine transcriptional pausing of the mouse *c-myb* gene (T. Mantamadiotis and R.G.R., unpublished) we found that *c-myb* expression declines in differentiating YAMC cells as a result an arrest in transcriptional elongation. This mechanism of gene regulation will be discussed in greater detail later. Like immature haemopoietic cells, undifferentiated colonic cells express relatively high levels of *c-myb* that decline as cells mature and specialise. As with leukaemic cells, c-Myb appears to also block colonic cancer cell differentiation.

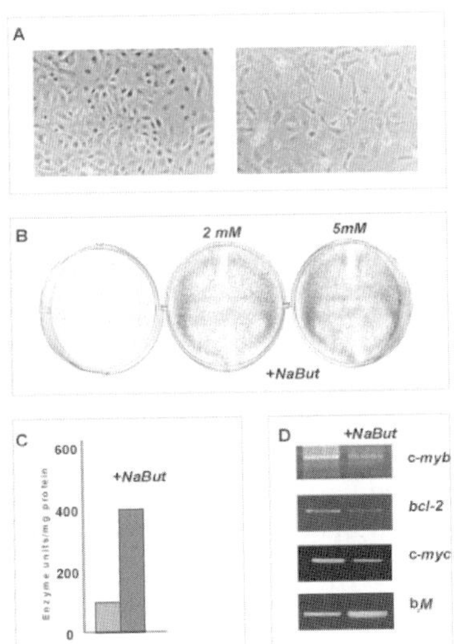

Figure 5

Young adult mouse colon cells also undergo morphological differentiation following addition of NaButyrate (A). They also show a dose response to induction of alkaline phosphatase as measured by dark staining cells (B) or enzyme activity (C). Under these conditions *c-myb*, *bcl-2* and *c-myc* expression is reduced (D).

4. CARCINOGENESIS

The most common GI tumours in humans living in the West arise in the colon, in which the *c-myb* proto-oncogene is frequently over-expressed (Alitalo et al., 1984; Kanei-Ishii et al., 1994; Torelli et al., 1987). The expression of c-myb progressively increases as human colon cancers become more aggressive (Dukes stage B to D) and ultimately metastatic (Figure 6).

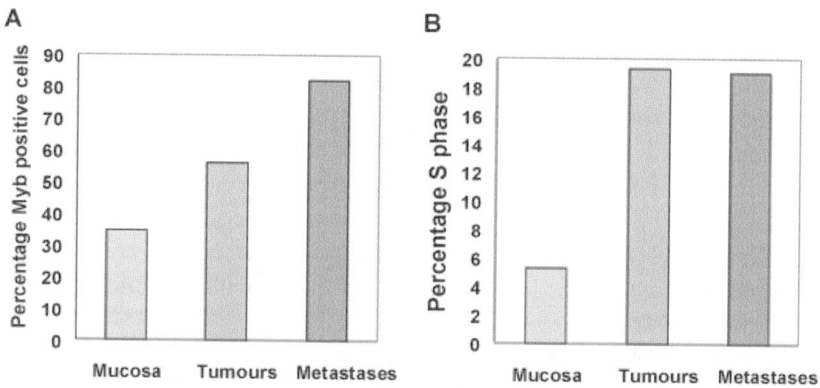

Figure 6

Colon tumours, metastases and matching mucosa were taken from 54 patients and examined for c-Myb protein expression and DNA content by FACS as described (Biroccio *et al.*, 2001; Ramsay *et al.*, 1992). As colon carcinogenesis progresses the number of cells expressing c-Myb increases whereby a large proportion of metastatic tumour cells express the proto-oncogene. Although there is a large increase in c-Myb expression from mucosa to primary tumour to metastasis the proportion of cells in S-phase reaches a maximum in primary tumours. This suggests that the apparent selection for higher c-Myb in metastatic colon cancers is not directly related to increasing cell cycling.

High c-*myb* expression alone is indicative of a poor prognosis for survival and even worse when coupled with over-expression of Bcl-X_L (Biroccio et al., 2001). A recent survey of public databases examined gene expression profiles in epithelial tumours (Su et al., 2001). When we mined this database, a remarkable correlation between high c-*myb* expression and malignancies of the colon, gastro-oesophageus and breast was revealed. Figure 7 shows very consistent over-expression of c-*myb* in colon tumours as we had also reported (Biroccio et al., 2001; Ramsay et al., 1992). We also found some time ago that premalignant colon adenomas or polyps over-express c-*myb* (Ramsay et al, 1992) and recently others have observed the same in the premalignant condition, Barrett's oesophagus as well as in upper GI tract cancers (Brabender et al., 2001). Barrett's oesophagus presents as an inflammatory-like disorder and its presence is predisposing for upper gastrointestinal adenocarcinoma. It is also notable that in patients with long-term ulcerative colitis there is an associated 5-7 fold increased risk with a of colon cancer (Shacter & Weitzman, 2002) with c-*myb* being readily detected in this tissue (Ramsay, unpublished). The high expression of c-*myb* in inflammatory diseases may best reflect the heightened proliferative state of the cells in the affected tissues. Our current thinking is that high c-*myb* levels drive proliferation and block differentiation and afford protection from apoptosis through the regulation of target genes discussed below.

Figure 7

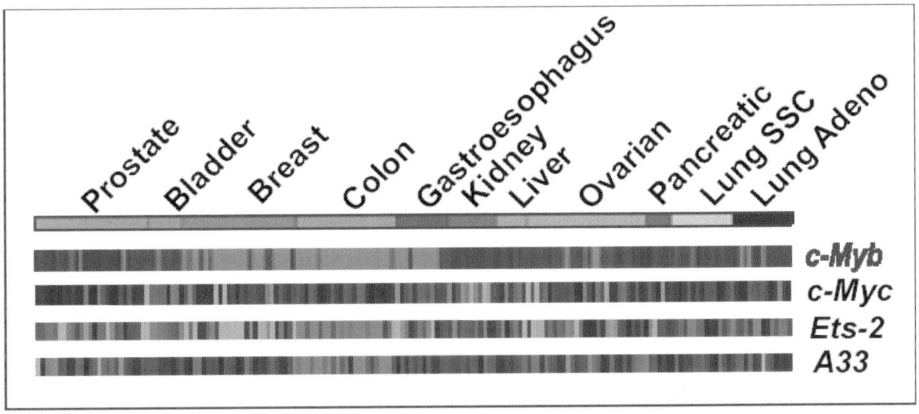

c-Myb over-expression is a feature of haemopoietic malignancies, however a recent survey of 174 epithelial tumours categorised by tissue of origin by Su et al (2001) demonstrated the relatively high expression (Red - high expression; Green - low expression) of *c-myb* in colon, gastroesophageal and breast cancers. For comparison two other transcription factors *c-myc* and *ets-2* and the intestine-specific gene *A33* are shown. (see colour section p. xxxi)

5. TRANSCRIPTIONAL REGULATION OF THE C-MYB GENE

c-Myb over-expression in malignancy is principally due to an increased abundance of *c-myb* mRNA which, in some cases, can be correlated with gene amplification (Alitalo et al., 1984; Kauraniemi et al., 2000). In most instances however, it appears that increased transcription rates are more likely to be responsible for the high levels of *c-myb* mRNA that are observed and understanding how this is achieved may lead to therapeutic approaches that reduce c-Myb. The relative activity of the *c-myb* promoter is indistinguishable from one cell type to another and is operational in many different cell lineages (Campanero et al., 1999). Nevertheless, mature *c-myb* transcripts are restricted in the normal adult tissues to cycling haemopoietic cells and the gastrointestinal epithelium (Thompson & Ramsay, 1995). When cells from these tissues embark upon differentiation *c-myb* transcript levels decline due to a blockade of transcriptional elongation. Our run-on transcription assays in colon cells indicate that *c-myb* down regulation involves transcriptional pausing in intron 1 (Thompson et al., 1997; 1998) in the same way haemopoietic cells down regulate *c-myb* during differentiation (Bender et al., 1987; Watson, 1988).

Like many transcription factors, the half-life of c-Myb protein and *c-myb* mRNA is relatively short compared to most "house-keeping" genes. Protein and mRNA half-lives in "normal" cells are in the order of 120 and 30 minutes respectively with longer half-lives reported in some cancer cell lines (Baer et al., 1992). This short half-life makes *c-myb* mRNA an ideal target for antisense oligonucleotide (ODN) intervention (see below). It appears that *c-myb* transcripts are initiated in many different cell types, even in those that are generally agreed to have minimal to undetectable levels of *c-myb* mRNA (Thompson et al., 1998).

The phenomenon of transcriptional pausing or arrest is not unique to *c-myb*. Many other genes, most notably those encoding transcription factors, employ elongation control. Indeed, an over-representation of transcripts containing up-stream exon(s) is seen for *c-fos* (Mechti et al., 1991) and *c-myc* and *N-myc* (Keene et al., 1999), even in cells that produce abundant mature mRNA. Mapping the sequences of *c-myb* intron 1 that are responsible for pausing has previously been imprecise. We think that the vagaries of precisely defining the mechanism underlying *c-myb* attenuation are due in part to the complex interplay between promoter departure and transcript elongation. Several models of *c-myb* transcriptional regulation have been proposed that are consistent with published data. These include conventional DNA enhancer-binding sites, but we favour the RNA-based model that is described in Figure 8. In this model parallels can be drawn between some bacterial and eukaryotic genes that use transcriptional attenuation as an important mechanism of control. Most notable is the role played by motifs including poly-thymidine (T) tracts that follow sequences which are transcribed to form energetically favoured stem-loops (Xu et al., 1995). We have drawn upon these precedents to identify a sequence motif encoding a putative RNA stem-loop just prior to a long polyT tract that corresponds to the attenuation site within the first intron of *c-myb* (Thompson et al., 1997). The essential features are the proposed RNA polymerase II stalling that occurs at a poly-T tract of 19-20 residues in intron 1. This allows the trailing nascent RNA transcript to form a stem-loop structure that is subsequently bound by cellular proteins. Our current view is that one class of proteins stabilise the attenuated stem-loop and another reactivates elongation. In leukaemic and colon cancer cell lines this process of transcriptional arrest seems to be subverted yet in some circumstances attenuation can be reinstated following the addition of differentiating agents leading to reduced *c-myb* levels, differentiation and apoptosis (Thompson et al., 1998). Agents such as Nabutyrate (Weaver et al., 2000), HMBA and SAHA (Cohen et al., 1999) that inhibit histone deacetylation are capable of inducing or reinstating *c-myb* transcriptional pausing in tumour cells and are therefore of considerable pre-clinical interest.

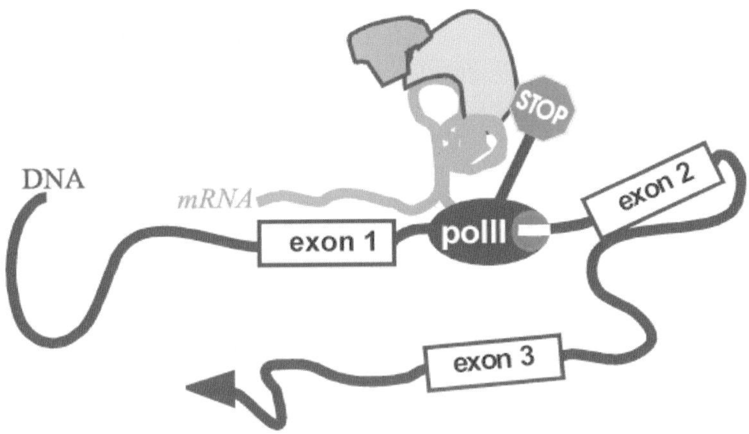

Figure 8

A proposed *c-Myb* RNA stem-loop structure. This is based upon the other annenuator sequences used to arrest transcriptional elongation. This RNA is thought to act as a protein-docking scaffold for cellular proteins. These proteins may either stabilise the attenuated RNA polymerase II (polII) transcript or activate further elongation towards exon 2. The arrest at this sequence is increased in the presence of NaButyrate.

6. C-MYB TARGETS IN CANCER CELLS

Under normal physiological conditions c-Myb regulates genes essential for colonic crypt formation. Current evidence suggests that *c-myb* exerts its oncogenic activity by maintaining expression of both physiological and non-physiological target genes by virtue of its over-expression. Four c-Myb target genes that are likely to be relevant to malignancy are *bcl-2* (Frampton et al., 1996; Taylor et al., 1996; Thompson et al., 1998), *bcl-X_L* (Biroccio et al., 2001), *c-myc* (Cogswell et al., 1993; Nakagoshi et al., 1992; Schmidt et al., 2000) and *cox-2* (Ramsay et al., 2000). c-*myb* is often over-expressed in malignancy and it is also likely that novel target genes might be recruited. Two classes of *c-myb* target genes should therefore be considered: normal physiological targets positively or negatively regulated by *c-myb* and aberrantly recruited targets.

Another important consideration when examining target genes is the role of other transcription factors that operate in partnership with c-Myb. These also fall into two categories: co-operating and antagonistic. For instance, c-Myb and Ets family proteins co-operate, while p53 has been found to antagonize c-Myb activity (Tanikawa et al., 2001). This issue is likely to be important because epithelial cancers are commonly null for wild type p53 and as shown in Figure 7, *Ets-2* and *c-myb* are both abundantly expressed in colon cancers (Su et al., 2001). We have focused on c-Myb regulation of target genes but have never assumed that c-Myb acts alone. Indeed the prospects of any oncoprotein like c-Myb being solely responsible for the maximal activation of a target promoter are remote in colon cancer. Another relevant example of this concept concerns the Wnt/APC pathway. Although more than 80% of human colon cancers show defects in the Wnt/APC pathway, the activation of target genes by the pathway downstream transcription factors, LEF-1 and TCF-4 in isolation is unlikely. Consistent with this principle we have found that c-Myb and the Wnt pathway appear to share target genes.

Two very important examples of colon oncogenes are c-*myc* and *cox-2*, and both appear to be activated by c-Myb and the Wnt pathway. Cox-2 is a key regulator of prostaglandin synthesis and is inducible under a number of physiological stresses. Several lines of evidence strongly suggest that Cox-2 expression is important in colon cancer progression and metastases (Tsujii et al., 1998). In human and animal studies, non-steroidal anti-inflammatory drugs (NSAIDS) reduce the incidence of colorectal cancer by 40-60% (Taketo, 1998). Studies using *cox-2*$^{-/-}$ mice crossed with the GI tract cancer-prone Min (APC mutant) mouse shows a marked reduction in polyp formation and tumour burden, while exogenous over-expression of Cox-2 accelerates colon cancer in mice. We reported that *c-myb* could transactivate the human *cox-2* promoter (Ramsay et al., 2000). The *cox-2* gene has 13 *c-myb* binding sites, eight of which can be bound *in vitro* by recombinant c-Myb and nuclear extracts containing c-Myb. Similarly, *c-myc* over-expression has been reported in colon tumours and on occasion is amplified as in the case of the Colo320 cell line (Schwab et al., 1983). c-Myc is also a transcription factor implicated in cell cycle control and other functions (Dang, 1999; Elend & Eilers, 1999). It had been noted for some time that *c-myc* and *c-myb* are commonly over-expressed in the same tumour biopsies (Rothberg, 1987). Several studies have demonstrated transactivation of *c-myc* reporter constructs by c-Myb, although the level of activation has been modest (eg. 2-4 fold) and appears to be cell type specific (Cogswell et al., 1993; Nakagoshi et al., 1992). In the context of this review the important observation is that TCF-4 also weakly transactivates the *c-myc* promoter (He et al., 1998).

Depending on the tumour status, E-cadherin expression and the combination of the adenomatous polyposis coli protein (APC)-axin and glycogen kinase-3 complex, β-catenin is either directed to proteosome degradation (Hart et al., 1999) or transported to the nucleus where it interacts with TCF/LEFs (Hecht et al., 2000). Activation of TCF-4/LEF-1 is one important downstream consequence of modulation or disruption of the Wnt pathway (Polakis et al., 1999).

There is a strong indication that Wnt pathway stimulation can lead to activation of the *cox-2* gene but attempts to show transcriptional activation in reporter assays in epithelial cells have been mixed (Howe et al., 1999). It has also been reported that the human *c-myc* promoter has two TCF-4 binding motifs that are responsible for *c-myc* transactivation in epithelial cells (Howe et al., 1999).

We have replicated these experiments and found modest transactivation of an extended *c-myc* promoter (Nakagoshi et al., 1992a) by mutant β-catenin and that the *c-myc* promoter has in fact three TCF-4 sites. Indeed there is some dispute as to whether TCF-4 acts best as a transcription factor or more as a histone acetylation-recruiting agent through association with β-catenin (Bieniasz et al., 1998; Hecht et al., 2000). For instance, it is thought that the histone acetyltransferase p300 is recruited by β-catenin and thus indirectly by TCFs (Hecht et al., 2000; Korinek et al., 1997; Morin et al., 1997; Porfiri et al., 1997; Rubinfeld et al., 1997). In the human *cox-2* and *c-myc* promoters here are respectively 13 and 11 c-Myb binding sites and 3 TCF-4 recognition motifs. We have found that these transcription factors work on both reporters co-operatively (D.C. and R.G.R., unpublished). This observation has broad implications for colon carcinogenesis and therapy.

7. PROSPECTS FOR THERAPEUTIC INTERVENTION IN GI CANCERS

From a therapeutic point of view, it is helpful to recognise that mice expressing only one functional *c-myb* allele appear to be perfectly normal and fertile (Mucenski et al., 1991). These mice also appear to have half the expression level of the target gene *bcl-2* in the colon (Zorbas et al., 1999). The aim of reducing high *c-myb* levels in tumours could be achieved without substantial systemic consequences.

One promising approach is to directly target the *c-myb* mRNA using antisense ODNs as pioneered by Gewirtz and colleagues (Gewirtz, 1993; 1996; 1999). In view of the relatively high level of transcription at exon 1 compared to exon 2, it would appear that the use of ODNs designed to hybridise to the translation start codon in exon 1 might not be strategically ideal. The second exon should serve as a better target as mature *c-myb* transcripts containing this sequence are less abundant than the total pool of exon 1 transcripts. An additional advantage of positioning of the ODN target to exon 2 is that drugs that have differentiation-inducing activity and that activate transcriptional pausing in intron 1 would act independently of the ODN and any transcripts which escape transcriptional attenuation would still remain amenable to ODN action at exon 2. We are currently exploiting this strategy of combining antisense ODNs and differentiation agents to test whether these will act together to maximally reduce c-Myb mRNA levels (Ramsay et al, 2003 *in press*).

The more realistic prospect for the future use of antisense *c-myb* ODNs is in combination with cytotoxic chemotherapeutic agents. Results so far indicate that the addition of antisense ODNs does not cause an increase in the toxicity of standard chemotherapy in animals. It seems that treatment with antisense ODNs can render some malignant cells more sensitive to standard chemotherapy agents such as Cytarabine and Methotrexate (Flaherty et al., 2001). While antisense c-*myb* ODNs can retard cancer cell growth both *in vitro and in vivo*, combining ODNs with other approaches to killing or differentiating cells offers many advantages. Examples where antisense *c-myb* ODNs have been used with chemotherapeutic drugs are few but are very supportive of this proposition. For example, we have found that ODNs in combination with Cisplatin were at least additive in their effect on tumour cell line growth in xenografts (Del Bufalo et al., 1996). A separate study showed that Cisplatin-resistant clones were inhibited by the antisense *c-myb* ODN to the same extent as the parental line (Funato et al., 2001). Thus the future prospects of circumventing drug resistance by ODNs *in vivo or ex vivo* has substantial potential.

8. CONCLUSION

There are a number of parallels to be drawn between c-Myb expression in the haemopoietic system and in gastrointestinal tissues. These point to a role for c-Myb in balancing differentiation and cell survival. If c-Myb is over-expressed the first process appears to be blocked and the second is potentiated. The combined consequence of retarded differentiation and protection from apoptosis is to allow for further events in the progression to transformation. Inappropriate regulation of c-Myb target genes and perhaps non-physiological targets feature in GI tumours. These targets are likely to be regulated in co-operation with other transcription factors that are also deregulated in GI cancers. Intervention in GI cancers using antisense ODN based therapies directed to block c-myb expression is a realistic prospect particularly where their strategic design is mindful of the transcriptional regulation of the gene. Certainly, the preclinical data generated in animal studies offer great hope particularly when used in combination with conventional therapies.

REFERENCES

Alexander, R.J., Buxbaum, J.N. and Raicht, R.F. (1992) Oncogene alterations in rat colon tumors induced by N-methyl-N- nitrosourea. Am J Med Sci *303*, 16-24.

Alitalo, K., Winqvist, R., Lin, C.C., de la Chapelle, A., Schwab, M. and Bishop, J.M. (1984) Aberrant expression of an amplified c-myb oncogene in two cell lines from a colon carcinoma. Proc Natl Acad Sci USA *81*, 4534-4538.

Baer, M.R., Augustinos, P. and Kinniburgh, A.J. (1992) Defective c-myc and c-myb RNA turnover in acute myeloid leukemia cells. Blood *79*, 1319-1326.

Baribault, H., Penner, J., Iozzo, R.V. and Wilson-Heiner, M. (1994) Colorectal hyperplasia and inflammation in keratin 8-deficient FVB/N mice. Genes Dev *8*, 2964-2973.

Bender, T.P., Thompson, C.B. and Kuehl, W.M. (1987) Differential expression of c-myb mRNA in murine B lymphomas by a block to transcription elongation. Science *237*, 1473-1476.

Bieniasz, P.D., Grdina, T.A., Bogerd, H.P. and Cullen, B.R. (1998) Recruitment of a protein complex containing Tat and cyclin T1 to TAR governs the species specificity of HIV-1 Tat. EMBO J *17*, 7056-7065.

Biroccio, A., Benassi, B., D'Agnano, I., D'Angelo, C., Buglioni, S., Mottolese, M., Ricciotti, A., Citro, G., Cosimelli, M., Ramsay, R.G., Calabretta, B. and Zupi, G. (2001) c-Myb and Bcl-x overexpression predicts poor prognosis in colorectal cancer: clinical and experimental findings. Am J Pathol *158*, 1289-1299.

Brabender, J., Lord R.V., Danenberg, K.D., Metzger, R., Schneider, P.M., Park, J.M., Salonga, D., Groshen, S., Tsao-Wei, D.D., DeMeester, T.R., Holscher, A.H. and Danenberg, P.V. (2001) Increased c-myb mRNA expression in Barrett's esophagus and Barrett's- associated adenocarcinoma. J Surg Res *99*, 301-306.

Campanero, M.R., Armstrong, M. and Flemington, E. (1999) Distinct cellular factors regulate the c-myb promoter through its E2F element. Mol Cell B*iol 19*, 8442-8450.

Chawengsaksophak, K., James, R., Hammond, V.E., Kontgen, F. and Beck, F. (1997) Homeosis and intestinal tumours in Cdx2 mutant mice. Nature *386*, 84-87.

Clarke, M.F., Kukowska-Latallo, J.F., Westin, E., Smith, M. and Prochownik, E. V. (1988) Constitutive expression of a c-myb cDNA blocks Friend murine erythroleukemia cell differentiation. Mol Cell Biol *8*, 884-892.

Cogswell, J.P., Cogswell, P.C., Kuehl, W.M., Cuddihy, A.M., Bender, T.P., Engelke, U., Marcu, K.B. and Ting, J.P. (1993) Mechanism of c-myc regulation by c-Myb in different cell lineages. Mol Cell Biol *13*, 2858-2869.

Cohen, L.A., Amin, S., Marks, P.A., Rifkind, R.A., Desai, D. and Richon, V.M. (1999) Chemoprevention of carcinogen-induced mammary tumorigenesis by the hybrid polar cytodifferentiation agent, suberanilohydroxamic acid (SAHA). Anticancer Res *19*, 4999-5005.

Cummings, J.H., Pomare, E.W., Branch, W.J., Naylor, C.P. and Macfarlane, G.T. (1987) Short chain fatty acids in human large intestine, portal, hepatic and venous blood. Gut *28*, 1221-1227.

Dang, C.V. (1999) c-Myc target genes involved in cell growth, apoptosis, and metabolism. Mol Cell Biol *19*, 1-11.

de Both, N.J., van der Feltz, M.J., Mooren, A., Vermaas, D., Klaassen, P., Rhijnsburger, E. H. and Kranendonk-Odijk, M.E. (1989) Oncogene expression in Rauscher murine leukemia virus induced erythroid, myeloid and lymphoid cell lines. Leuk Res *13*, 53-64.

de Both, N.J., Wijnhoven, B.P., Sleddens, H.F., Tilanus, H.W. and Dinjens, W.N. (2001) Establishment of cell lines from adenocarcinomas of the esophagus and gastric cardia growing in vivo and in vitro. Virchows Arch *438*, 451-456.

Del Bufalo, D., Cucco, C., Leonetti, C., Citro, G., D'Agnano, I., Benassi, M., Geiser, T., Zon, G., Calabretta, B. and Zupi, G. (1996) Effect of cisplatin and c-myb antisense phosphorothioate oligodeoxynucleotides combination on a human colon carcinoma cell line in vitro and in vivo. Br J Cancer *74*, 387-393.

Elend, M. and Eilers, M. (1999) Cell growth: downstream of Myc - to grow or to cycle? Curr Biol *9*, R936-938.

Flaherty, K.T., Stevenson, J.P. and O'Dwyer, P.J. (2001) Antisense therapeutics: lessons from early clinical trials. Curr Opin Oncol *13*, 499-505.

Frampton, J., Ramqvist, T. and Graf, T. (1996) v-Myb of E26 leukemia virus up-regulates bcl-2 and suppresses apoptosis in myeloid cells. Genes Dev *10*, 2720-2731.

Fu, S.L. and Lipsick, J.S. (1997) Constitutive expression of full-length c-Myb transforms avian cells characteristic of both the monocytic and granulocytic lineages. Cell Growth Differ *8*, 35-45.

Funato, T., Satou, J., Kozawa, K., Fujimaki, S., Miura, T. and Kaku, M. (2001) Use of c-myb antisense oligonucleotides to increase the sensitivity of human colon cancer cells to cisplatin. Oncol Rep *8*, 807-810.

Gavrieli, Y., Sherman, Y. and Ben-Sasson, S.A. (1992) Identification of programmed cell death in situ via specific labeling of nuclear DNA fragmentation. J Cell Biol *119*, 493-501.

Gewirtz, A.M. (1993) Oligodeoxynucleotide-based therapeutics for human leukemias. Stem Cells *11 Suppl 3*, 96-103.

Gewirtz, A.M. (1996) The c-myb proto-oncogene: a novel target for human gene therapy. Cancer Treat Res *84*, 93-112.

Gewirtz, A.M. (1999) Myb targeted therapeutics for the treatment of human malignancies. Oncogene *18*, 3056-3062.

Gonda, T.J., Buckmaster, C. and Ramsay, R.G. (1989a) Activation of c-myb by carboxy-terminal truncation: relationship to transformation of murine haemopoietic cells in vitro. EMBO J *8*, 1777-1783.

Gonda, T.J. and Metcalf, D. (1984) Expression of myb, myc and fos proto-oncogenes during the differentiation of a murine myeloid leukaemia. Nature *310*, 249-251.

Gonda, T.J., Ramsay, R.G. and Johnson, G.R. (1989b) Murine myeloid cell lines derived by in vitro infection with recombinant c-myb retroviruses express myb from rearranged vector proviruses. EMBO J *8*, 1767-1775.

Gordon, J.I. and Hermiston, M.L. (1994) Differentiation and self-renewal in the mouse gastrointestinal epithelium. Curr Opin Cell Biol *6*, 795-803.

Hart, M., Concordet, J.P., Lassot, I., Albert, I., del los Santos, R., Durand, H., Perret, C., Rubinfeld, B., Margottin, F., Benarous, R. and Polakis, P. (1999) The F-box protein beta-TrCP associates with phosphorylated beta-catenin and regulates its activity in the cell. Curr Biol *9*, 207-210.

He, T.C., Sparks, A.B., Rago, C., Hermeking, H., Zawel, L., da Costa, L.T., Morin, P.J., Vogelstein, B. and Kinzler, K.W. (1998) Identification of c-MYC as a target of the APC pathway. Science *281*, 1509-1512.

Hecht, A., Vleminckx, K., Stemmler, M.P., van Roy, F. and Kemler, R. (2000) The p300/CBP acetyltransferases function as transcriptional coactivators of beta-catenin in vertebrates [In Process Citation]. EMBO J *19*, 1839-1850.

Hermiston, M.L., Wong, M.H. and Gordon, J.I. (1996) Forced expression of E-cadherin in the mouse intestinal epithelium slows cell migration and provides evidence for nonautonomous regulation of cell fate in a self-renewing system. Genes Dev *10*, 985-996.

Howe, L.R., Subbaramaiah, K., Chung, W.J., Dannenberg, A.J. and Brown, A.M. (1999) Transcriptional activation of cyclooxygenase-2 in Wnt-1-transformed mouse mammary epithelial cells. Cancer Res *59*, 1572-1577.

Johnstone, C.N., Tebbutt, N.C., Abud, H.E., White, S.J., Stenvers, K.L., Hall, N.E., Cody, S. H., Whitehead, R.H., Catimel, B., Nice, E.C., Burgess, A.W. and Heath, J.K. (2000) Characterization of mouse A33 antigen, a definitive marker for basolateral surfaces of intestinal epithelial cells. Am J Physiol Gastrointest Liver Physiol *279*, G500-510.

Kaestner, K.H., Silberg, D.G., Traber, P.G. and Schutz, G. (1997) The mesenchymal winged helix transcription factor Fkh6 is required for the control of gastrointestinal proliferation and differentiation. Genes Dev *11*, 1583-1595.

Kamada, S., Shimono, A., Shinto, Y., Tsujimura, T., Takahashi, T., Noda, T., Kitamura, Y., Kondoh, H. and Tsujimoto, Y. (1995) bcl-2 deficiency in mice leads to pleiotropic abnormalities: accelerated lymphoid cell death in thymus and spleen, polycystic kidney, hair hypopigmentation, and distorted small intestine. Cancer Res *55*, 354-359.

Kanei-Ishii, C., Yasukawa, T., Morimoto, R.I. and Ishii, S. (1994) c-Myb-induced trans-activation mediated by heat shock elements without sequence-specific DNA binding of c-Myb. J Biol Chem *269*, 15768-15775.

Kauraniemi, P., Hedenfalk, I., Persson, K., Duggan D.J., Tanner, M., Johannsson, O., Olsson, H., Trent, J.M., Isola, J. and Borg, A. (2000) MYB oncogene amplification in hereditary BRCA1 breast cancer. Cancer Res *60*, 5323-5328.

Keene, R.G., Mueller, A., Landick, R. and London, L. (1999) Transcriptional pause, arrest and termination sites for RNA polymerase II in mammalian N- and c-myc genes. Nucleic Acids Res *27*, 3173-3182.

Korinek, V., Barker, N., Morin, P.J., van Wichen, D., de Weger, R., Kinzler, K.W., Vogelstein, B. and Clevers, H. (1997) Constitutive transcriptional activation by a beta-catenin-Tcf complex in APC-/- colon carcinoma. Science *275*, 1784-1787.

Lefebvre, O., Chenard, M.P., Masson, R., Linares, J., Dierich, A., LeMeur, M., Wendling, C., Tomasetto, C., Chambon, P. and Rio, M.C. (1996) Gastric mucosa abnormalities and tumorigenesis in mice lacking the pS2 trefoil protein. Science *274*, 259-262.

Mashimo, H., Wu, D.C., Podolsky, D.K. and Fishman, M.C. (1996) Impaired defense of intestinal mucosa in mice lacking intestinal trefoil factor. Science *274*, 262-265.

Mechti, N., Piechaczyk, M., Blanchard, J. M., Jeanteur, P. and Lebleu, B. (1991) Sequence requirements for premature transcription arrest within the first intron of the mouse c-fos gene. Mol Cell Biol *11*, 2832-2841.

Merritt, A.J., Potten, C.S., Watson, A.J., Loh, D.Y., Nakayama, K. and Hickman, J.A. (1995) Differential expression of bcl-2 in intestinal epithelia. Correlation with attenuation of apoptosis in colonic crypts and the incidence of colonic neoplasia. J Cell Sci *108*, 2261-2271.

Morin, P.J., Sparks, A.B., Korinek, V., Barker, N., Clevers, H., Vogelstein, B. and Kinzler, K.W. (1997) Activation of beta-catenin-Tcf signaling in colon cancer by mutations in beta-catenin or APC. Science *275*, 1787-1790.

Mucenski, M.L., McLain, K., Kier, A. B., Swerdlow, S.H., Schreiner, C.M., Miller, T.A., Pietryga, D.W., Scott, W.J., Jr. and Potter, S.S. (1991) A functional c-myb gene is required for normal murine fetal hepatic hematopoiesis. Cell *65*, 677-689.

Nakagoshi, H., Kanei-Ishii, C., Sawazaki, T., Mizuguchi, G. and Ishii, S. (1992a) Transcriptional activation of the c-myc gene by the c-myb and B-myb gene products. Oncogene *7*, 1233-1240.

Nakagoshi, H., Kanei-Ishii, C., Sawazaki, T., Mizuguchi, G. and Ishii, S. (1992b) Transcriptional activation of the c-myc gene by the c-myb and B-myb gene products. Oncogene *7*, 1233-1240.

Ogilvy, S., Metcalf, D., Print, C.G., Bath, M.L., Harris, A.W. and Adams, J.M. (1999) Constitutive Bcl-2 expression throughout the hematopoietic compartment affects multiple lineages and enhances progenitor cell survival. Proc Natl Acad Sci USA *96*, 14943-14948.

Park,J.G. and Gazdar, A.F. (1996) Biology of colorectal and gastric cancer cell lines. J Cell Biochem Suppl *24*, 131-141.

Polakis, P., Hart, M. and Rubinfeld, B. (1999) Defects in the regulation of beta-catenin in colorectal cancer. Adv Exp Med Biol *470*, 23-32.

Porfiri, E., Rubinfeld, B., Albert, I., Hovanes, K., Waterman, M. and Polakis, P. (1997) Induction of a beta-catenin-LEF-1 complex by wnt-1 and transforming mutants of beta-catenin. Oncogene *15*, 2833-2839.

Potten, C.S. (1998) Stem cells in gastrointestinal epithelium: numbers, characteristics and death. Philos Trans R Soc Lond B Biol Sci *353*, 821-830.

Ramsay, R.G., Friend, A., Vizantios, Y., Freeman, R., Sicurella, C., Hammett, F., Armes, J. and Venter, D. (2000) Cyclooxygenase-2, a colorectal cancer nonsteroidal anti-inflammatory drug target, is regulated by c-MYB. Cancer Res *60*, 1805-1809.

Ramsay, R.G., Ikeda, K., Rifkind, R.A. and Marks, P.A. (1986) Changes in gene expression associated with induced differentiation of erythroleukemia: protooncogenes, globin genes, and cell division. Proc Natl Acad Sci USA *83*, 6849-6853.

Ramsay, R.G., Thompson, M.A., Hayman, J.A., Reid, G., Gonda, T.J. and Whitehead, R.H. (1992) Myb expression is higher in malignant human colonic carcinoma and premalignant adenomatous polyps than in normal mucosa. Cell Growth Differ *3*, 723-730.

Richon, V.M., Ramsay, R.G., Rifkind, R.A. and Marks, P.A. (1989) Modulation of the c-myb, c-myc and p53 mRNA and protein levels during induced murine erythroleukemia cell differentiation. Oncogene *4*, 165-173.

Rosenthal, M.A., Thompson, M.A., Ellis, S., Whitehead, R.H. and Ramsay, R.G. (1996) Colonic expression of c-myb is initiated in utero and continues throughout adult life. Cell Growth Differ *7*, 961-967.

Rothberg, P.G. (1987) The role of the oncogene c-myc in sporadic large bowel cancer and familial polyposis coli. Semin Surg Oncol *3*, 152-158.

Rubinfeld, B., Robbins, P., El-Gamil, M., Albert, I., Porfiri, E. and Polakis, P. (1997) Stabilization of beta-catenin by genetic defects in melanoma cell lines. Science *275*, 1790-1792.

Schmidt, M., Nazarov, V., Stevens, L., Watson, R.J. and Wolff, L. (2000) Regulation of the resident chromosomal copy of c-myc by c-Myb is involved in myeloid leukemogenesis. Mol Cell Biol *20*, 1970-1981.

Schwab, M., Alitalo, K., Klempnauer, K.H., Varmus, H.E., Bishop, J.M., Gilbert, F., Brodeur, G., Goldstein, M. and Trent, J. (1983) Amplified DNA with limited homology to myc cellular oncogene is shared by human neuroblastoma cell lines and a neuroblastoma tumour. Nature *305*, 245-248.

Shacter, E. and Weitzman, S.A. (2002) Chronic inflammation and cancer. Oncology (Huntingt) *16*, 217-226, 229; discussion 230-212.

Su, A.I., Welsh, J.B., Sapinoso, L.M., Kern, S.G., Dimitrov, P., Lapp, H., Schultz, P.G., Powell, S.M., Moskaluk, C.A., Frierson, H.F., Jr. and Hampton, G.M. (2001) Molecular classification of human carcinomas by use of gene expression signatures. Cancer Res *61*, 7388-7393.

Taketo, M.M. (1998) Cyclooxygenase-2 inhibitors in tumorigenesis (part I). J Natl Cancer Inst *90*, 1529-1536.

Tanikawa, J., Ichikawa-Iwata, E., Kanei-Ishii, C. and Ishii, S. (2001) Regulation of c-Myb activity by tumor suppressor p53. Blood Cells Mol Dis *27*, 479-482.

Taylor, D., Badiani, P. and Weston, K. (1996) A dominant interfering Myb mutant causes apoptosis in T cells. Genes Dev *10*, 2732-2744.

Tebbutt, N.C., Giraud, A.S., Inglese, M., Jenkins B., Waring, P., Clay, F.J., Malki, S., Alderman, B.M., Grail, D., Hollande, F., Heath, J.K. and Ernst, M. (2002) Reciprocal regulation of gastrointestinal homeostasis by SHP2 and STAT-mediated trefoil gene activation in gp130 mutant mice. Nat Med *8*, 1089-1097.

Thompson, M.A., Flegg, R., Westin, E.H. and Ramsay, R.G. (1997) Microsatellite deletions in the c-myb transcriptional attenuator region associated with over-expression in colon tumour cell lines. Oncogene *14*, 1715-1723.

Thompson, M.A. and Ramsay, R.G. (1995) Myb: an old oncoprotein with new roles. Bioessays *17*, 341-350.

Thompson, M.A., Rosenthal, M.A., Ellis, S.L., Friend, A.J., Zorbas, M.I., Whitehead, R.H. and Ramsay, R.G. (1998b) c-Myb down-regulation is associated with human colon cell differentiation, apoptosis, and decreased Bcl-2 expression. Cancer Res *58*, 5168-5175.

Torelli, G., Venturelli, D., Colo, A., Zanni, C., Selleri, L., Moretti, L., Calabretta, B. and Torelli, U. (1987) Expression of c-myb protooncogene and other cell cycle-related genes in normal and neoplastic human colonic mucosa. Cancer Res *47*, 5266-5269.

Trainer, D.L., Kline, T., McCabe, F.L., Faucette, L.F., Field, J., Chaikin, M., Anzano,, M., Rieman, D., Hoffstein, S., Li, D.J., et al. (1988) Biological characterization and oncogene expression in human colorectal carcinoma cell lines. Int J Cancer *41*, 287-296.

Tsujii, M., Kawano, S., Tsuji, S., Sawaoka, H., Hori, M. and DuBois, R.N. (1998) Cyclooxygenase regulates angiogenesis induced by colon cancer cells. *Cell* 93, 705-716.

Watson, R.J. (1988) A transcriptional arrest mechanism involved in controlling constitutive levels of mouse c-myb mRNA. Oncogene *2*, 267-272.

Weaver, G.A., Tangel, C.T., Krause, J.A., Parfitt, M.M., Stragand, J.J., Jenkins, P.L., Erb, T.A., Davidson, R.H., Alpern, H.D., Guiney, W.B., Jr. and Higgins, P.J. (2000) Biomarkers of human colonic cell growth are influenced differently by a history of colonic neoplasia and the consumption of acarbose. J Nutr *130*, 2718-2725.

Weston, K.M. (1990) The myb genes. Semin Cancer Biol *1*, 371-382.

Whitehead, R.H., Novak, U., Thomas, R.J., Lukeis, R.E., Walker, F.E. and Jones, J. (1989) A new gastric carcinoma cell line (LIM1839) derived from a young Caucasian male. Int J Cancer *44*, 1100-1103.

Whitehead, R.H., VanEeden, P.E., Noble, M.D., Ataliotis, P. and Jat, P.S. (1993) Establishment of conditionally immortalized epithelial cell lines from both colon and small intestine of adult H-2Kb-tsA58 transgenic mice. Proc Natl Acad Sci USA *90*, 587-591.

Whitehead, R.H., Young, G.P. and Bhathal, P.S. (1986) Effects of short chain fatty acids on a new human colon carcinoma cell line (LIM1215). Gut *27*, 1457-1463.

Wong, M.H., Hermiston, M.L., Syder, A.J. and Gordon, J.I. (1996) Forced expression of the tumor suppressor adenomatosis polyposis coli protein induces disordered cell migration in the intestinal epithelium. Proc Natl Acad Sci USA *93*, 9588-9593.

Wong, M.H., Rubinfeld, B. and Gordon, J.I. (1998) Effects of forced expression of an NH2-terminal truncated beta-Catenin on mouse intestinal epithelial homeostasis. J Cell Biol *141*, 765-777.

Xu, L., Meng, Y., Wallen, R. and DePinho, R.A. (1995) Loss of transcriptional attenuation in N-myc is associated with progression towards a more malignant phenotype. Oncogene *11*, 1865-1872.

Zorbas, M., Sicurella, C., Bertoncello, I., Venter, D., Ellis, S., Mucenski, M.L. and Ramsay R.G. (1999) c-Myb is critical for murine colon development. Oncogene *18*, 5821-5830.

Chapter 20

C-MYB AND CREB FUNCTION IN ADULT NEUROGENESIS

Theo Mantamadiotis, Sally Lightowler, Marijana Vanevski, Mark A. Rosenthal, Nikla R. Emambokus[1], Jon Frampton[1], Robert G. Ramsay
Differentiation and Transcription Laboratory, Peter MacCallum Cancer Centre, St. Andrews Place, East Melbourne, 3002, Australia; [1]Institute of Biomedical Research, The Medical School, Birmingham University, Edgbaston, Birmingham, B15 2TT, United Kingdom.

Abstract: The molecular pathways regulating post-natal neurogenesis are poorly understood. c-Myb is recognised as maintaining haemopoietic and colonic cells in an undifferentiated state, while CREB is known to be important for neuronal cell survival and function. These properties are important for maintaining a neural progenitor population in adult brain. We review the literature that suggests c-Myb and CREB co-regulate a number of factors involved in neurogenesis. Emerging data is discussed, which shows that c-Myb is required for normal postnatal brain development in the mouse, and we consider how these mice will be used to identify factors co-regulated by c-Myb and CREB, specifically involved in neurogenesis.

1. INTRODUCTION

It is no longer thought that the adult brain contains only terminally differentiated, post-mitotic nerve cells that exist throughout the life of an organism, never to be replaced when lost. During embryonic development there is a massive expansion of neurons and glia, balanced with programmed cell death as the brain matures and remodels. As developing brain cells differentiate they migrate toward the region where they will ultimately seek out interactions with other cells and perform their specialised tasks. The only recognised plasticity within the adult brain was thought to be alterations in synapse formation. Over time adult brains of several species were shown to harbour regions of newly generating nerve cells, a process referred to as neurogenesis (Altman and Das, 1965; Altman and Das, 1967; Cayre, et al., 2002; Das and Altman, 1971; Eccles, 1970; Eriksson, et al., 1998; Kranz and

J. Frampton (ed.), Myb Transcription Factors: Their Role in Growth, Differentiation and Disease, 389-397.

Richter, 1975; Morest, 1970; Suginoshita, 1971). Neurogenesis in the adult human brain was formally proven only recently (Eriksson, et al., 1998). The implications of neurogenic potential in mammalian brains is enormous, offering hope to sufferers of neurodegenerative diseases (e.g. Alzheimer's), and brain injury (e.g. stroke, cancer).

The major sites of neurogenesis in the adult mammalian brain are the subventricular zone (SVZ) of the lateral ventricles (Lois and Alvarez-Buylla, 1993) and the sub-granular zone (SGZ) of the dentate gyrus (DG) (Gage, et al., 1998) (Figure 1). The transcriptional machinery necessary for modulating the maintenance of both the neuroblast and the differentiated neuronal population and for regulating the transition between the two is unknown. *In vivo* knockout and transgenic rodent models suggest that transcription factors such as E2F1, mineralocorticoid receptor and CREB (Cooper-Kuhn, et al., 2002; Gass, et al., 2000; Nakagawa, et al., 2002), and signalling molecules such as Shh (Lai, et al., 2003), are involved in the neurogenic process. We propose that c-Myb and CREB (c-AMP Responsive Element Binding protein) are two potent transcription factors that play an important role in regulating neurogenic homeostasis. This view is based upon a number of observations. 1) Stem cells exist in the mammalian brain. 2) A number of genes encoding factors involved in neurogenesis, including *cox2* and *bcl-2* are regulated by CREB and c-Myb. 3) c-Myb expression in adult brain is highest in the dentate gyrus (DG) where neurogenesis occurs. 4) CREB is constitutively active in at least a subset of, and perhaps the entire neuroblast population.

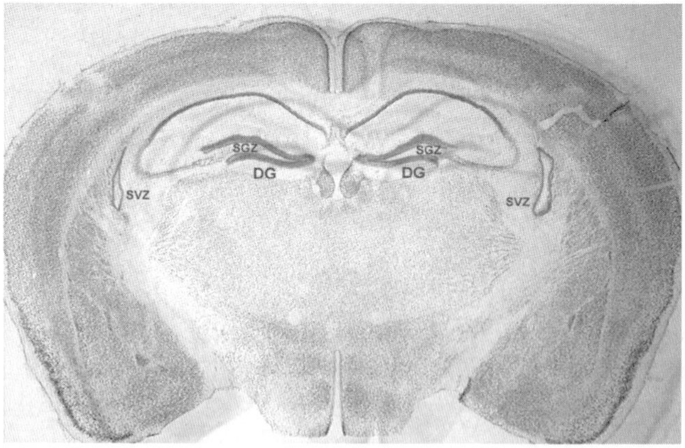

Figure 1

Sites of neurogenesis in the adult brain. A coronal section of a mouse brain at the level of the hippocampus showing the dentate gyrus (DG), the DG sub-granular zone (SGZ) and the lateral ventricle's sub-ventricular zone (SVZ).

2. C-MYB AND CREB IN BRAIN

2.1 c-Myb Function in Brain

To our knowledge there are no published studies addressing c-Myb function in brain development and homeostasis, although its expression in brain has been documented by us (Rosenthal, et al., 1996) and others (Ess, et al., 1999; Shin, et al., 2001) (Figure 2). c-Myb is most highly expressed in the murine embryonic brain when there is extensive neuronal expansion. In the adult rodent brain, c-Myb mRNA is most highly expressed in the hippocampal and neo-cortical regions. Moreover, the site within the hippocampus where expression is highest is in the DG (Shin, et al., 2001), which, as mentioned, is a major site of neural proliferation in adult rodent and human brains. It is therefore tempting to speculate that c-Myb has a role in the maintenance of neural progenitor cells, consistent with its role in the haemopoietic system (Mucenski, et al., 1991) and colon epithelium (Zorbas, et al., 1999).

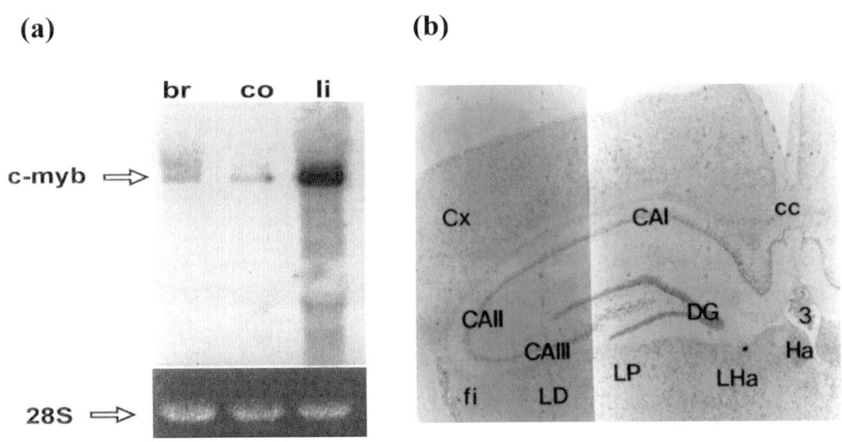

Figure 2

c-Myb expression in brain. a) Northern blot analysis showing the presence of c-Myb mRNA in embryonic mouse brain (br), colon (co) and liver (li). b) *In situ* analysis showing that the DG exhibits the highest levels of c-Myb expression in adult rat hippocampus [(b) taken from Shin et al.,2001]

The aberrant expression of c-Myb in brain pathologies suggests a role for c-Myb in brain tumours. In our recent review, neurological cancer was noted as a malignancy associated with elevated levels of c-Myb (Ramsay, et al., 2003). Thiele et al. (1988) had previously documented the presence of the c-Myb proto-oncogene in several paediatric malignancies of neuroectodermal origin including neuroblastoma, peripheral neuroepithelioma, Ewing's Sarcoma and glial tumour cell lines. Xu et al. (2001) showed that upon administration of anti-sense c-*myb* oligonucleotides in nude mice, C6 glioma cell tumours significantly reduced in size. Rearrangement of the c-Myb gene within the coding region resulting in enhanced transcriptional activity has also been reported in human glioblastoma cells (Welter, et al., 1990).

2.2 CREB Function in Brain

CREB function in neurons has been extensively studied (West, et al., 2002), especially its role in the maintenance of long-term memory (Kandel, 2001). CREB is constitutively and almost ubiquitously expressed in brain and other tissues. However, CREB is not constitutively active but is activated through a number of different pathways, including the cAMP, MAPK, Akt/PKB, and Ca^{2+}/CaM pathways. The penultimate step in these cascades is the activation of a CREB kinase (eg PKA) that phosphorylates residue Ser133. This event allows the recruitment of the CREB Binding Protein (CBP) and other transcription co-factors, as well as RNA polymerase II. Aside from CREB's role in the maintenance of long-term memory, we recently reported that CREB and the related molecule CREM play more fundamental roles in neurons, showing that these factors are required for neuronal survival *in vivo* (Mantamadiotis, et al., 2002).

We recently discovered that although CREB is widely expressed in the adult mouse brain, the active phosphorylated form, pCREB, is constitutively expressed in the DG-SGZ and SVZ, both sites of adult neurogenesis, an observation supported by another recently published study (Nakagawa, et al., 2002). This highly restricted pattern of constitutive CREB activation supports the view that CREB function has a role in neurogenesis. This in turn suggests that constitutive CREB-dependent target gene expression will be confined to this region.

2.3 CREB and c-Myb Target Genes in Brain

We and others have identified c-Myb target genes in colon and haemopoietic cells, showing that *cox2* is regulated by c-Myb (Ramsay, et al., 2000), as is *bcl-2* (Frampton, et al., 1996; Taylor, et al., 1996; Zorbas, et al., 1999). CREB similarly regulates *bcl-2* (Wilson, et al., 1996) and *cox2* (Wardlaw, et al., 2002). *Bcl-2* and *cox2* are thought to play a role in neurogenesis, as Bcl-2 is more highly expressed in the proliferating DG cells, compared with the more differentiated nerve cells (Abe-Dohmae, et al., 1993; Bedard, et al., 2002; Merry, et al., 1994; Vinet, et al., 2002). *Cox2* has also been implicated as a neuro-protective factor, especially in neurodegenerative disorders (Ho, et al., 1999). Moreover, *cox2* appears to be important in regulating neurogenesis, as attenuation of its activity in the brain results in reduced DG neuron proliferation (Kumihashi, et al., 2001; Uchida, et al., 2002).

There are numerous putative CREB target genes in nerve cells, including *cox2, bcl-2, BDNF, nNOS, presenilin, tyrosine hydroxylase, cyclinD1* and *somatostatin* (Mayr and Montminy, 2001; West, et al., 2002). Although CREB phosphorylation correlates in many instances with upregulation of putative target genes, our previous studies on whole-brain mRNA indicate that *in vivo* CREB ablation does not significantly perturb the expression of a number of factors considered as CREB target genes (Mantamadiotis, et al., 2002). This implies that CREB regulation of any one target gene may occur in only a small subset of nerve cells, such as neural progenitor cells.

For the majority of target genes, it is likely that loss of only CREB activity in a nerve cell can be compensated for by other transcription factors that bind to target gene promoters. As the *cox2* and *bcl-2* genes can be transactivated by both c-Myb and CREB, we propose that c-Myb and CREB can co-regulate a number of factors important in regulating neurogenic homeostasis. The neuro-cellular role of c-Myb and CREB may be the regulation of neural progenitor cell survival, proliferation and differentiation, consistent with these roles in other tissues.

2.4 Generation of Brain-Specific Mutant Mice

Previous investigations using knockout mice were hampered by the early lethality of c-Myb and CREB null mutants, precluding studies in adult animals. To study the function of c-Myb and CREB in adult neurogenesis we are using a novel *c-Myb*^{lox} mouse and the previously reported *Creb1*^{lox} mouse to generate mice devoid of c-Myb and CREB in brain. This is achieved by crossing the "floxed" mice to *nestinCre* transgenic mice, which express the Cre recombinase enzyme under the control of the neural progenitor specific *nestin* promoter and enhancer (Tronche, et al., 1999). When expressed within cells, the Cre recombinase enzyme recombines and excises DNA flanked by loxP sequences (Figure 3). The consequences of the loss of each transcription factor with respect to neurogenesis will be studied. Preliminary histological analysis of mice in which c-Myb is lost specifically in the nervous system, shows that indeed c-Myb has a role in mouse brain development. Mice devoid of c-Myb during brain development display reduced brain cellularity and larger than normal lateral ventricles. Most notably, there is obvious abnormal cellularity in the DG-SGZ, suggestive of a defect in the neural progenitor cells residing there.

Generating mice in which both c-Myb and CREB are lost in brain will further allow us to determine if there is genuine co-operativity between these factors in the maintenance of neurogenic target gene expression. Microarray analysis of neurons microdissected from the DG-SGZ will allow us to determine a set of genes whose expression is altered as a consequence of c-Myb and CREB loss. A parallel *in silico* screen of c-Myb and CREB promoter binding sites in the mouse and human genomes, together with the microarray results will allow us to identify a subset of direct c-Myb and CREB target genes which impact upon adult neurogenesis. The ultimate aim of these investigations is to identify some of the factors and pathways involved in neurogenesis *in vivo*, aiding in the identification of drugs which may enhance neural progenitor cell activity, so that neural degeneration and injury can be more effectively treated.

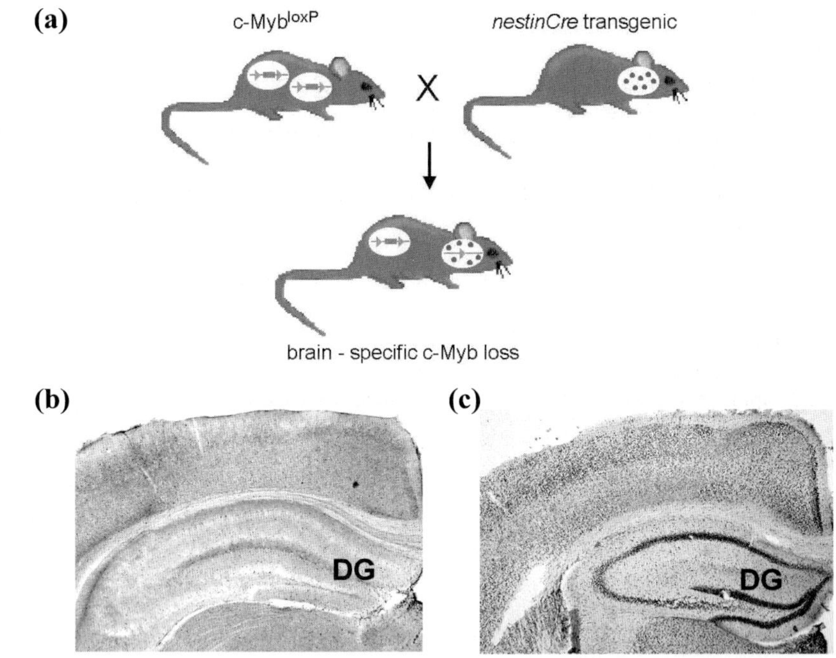

Figure 3

Generation of brain-specific mutant mice using the Cre-loxP recombination system. a) By crossing c-Myb^{lox} mice with *nestinCre* transgenic mice, the floxed *c-Myb* allele (rectangles) flanked by loxP sites (triangles) is recombined by the brain-specific Cre recombinase expression (dark circles) of the *nestinCre* transgene, resulting in the loss of c-Myb. b) The specificity of recombination is illustrated by the loss of CREB protein in $Creb1^{lox;nestinCre}$ hippocampus. c) $Creb1^{lox}$ mice without the *nestinCre* transgene maintain normal CREB levels. DG – dentate gyrus.

REFERENCES

Abe-Dohmae, S., Harada, N., Yamada, K. and Tanaka, R. (1993) Bcl-2 gene is highly expressed during neurogenesis in the central nervous system. Biochem Biophys Res Commun *191*, 915-921.

Altman, J. and Das, G.D. (1965) Autoradiographic and histological evidence of postnatal hippocampal neurogenesis in rats. J Comp Neurol *124*, 319-335.

Altman, J. and Das, G.D. (1967) Postnatal neurogenesis in the guinea-pig. Nature *214*, 1098-101.

Bedard, A., Levesque, M., Bernier, P.J. and Parent, A. (2002) The rostral migratory stream in adult squirrel monkeys: contribution of new neurons to the olfactory tubercle and involvement of the antiapoptotic protein Bcl-2. Eur J Neurosci *16*, 1917-1924.

Cayre, M., Malaterre, J., Scotto-Lomassese, S., Strambi, C. and Strambi, A. (2002) The common properties of neurogenesis in the adult brain: from invertebrates to vertebrates. Comp Biochem Physiol B Biochem Mol Biol *132*, 1-15.

Cooper-Kuhn, C.M., Vroemen, M., Brown, J., Ye, H., Thompson, M.A., Winkler, J. and Kuhn, H.G. (2002) Impaired adult neurogenesis in mice lacking the transcription factor E2f1. Mol Cell Neurosci *21*, 312-323.

Das, G.D. and Altman, J. (1971) Postnatal neurogenesis in the cerebellum of the cat and tritiated thymidine autoradiography. Brain Res *30*, 323-330.

Eccles, J.C. (1970) Neurogenesis and morphogenesis in the cerebellar cortex. Proc Natl Acad Sci USA *66*, 294-301.

Eriksson, P.S., Perfilieva, E., Bjork-Eriksson, T., Alborn, A.M., Nordborg, C., Peterson, D.A. and Gage, F.H. (1998) Neurogenesis in the adult human hippocampus. Nat Med *4*, 1313-1317.

Ess, K.C., Witte, D.P., Bascomb, C.P. and Aronow, B.J. (1999) Diverse developing mouse lineages exhibit high-level c-myb expression in immature cells and loss of expression upon differentiation. Oncogene *18*, 1103-1111.

Frampton, J., Ramqvist, T. and Graf, T. (1996) v-Myb of E26 leukemia virus up-regulates bcl-2 and suppresses apoptosis in myeloid cells. Genes Dev *10*, 2720-2731.

Gage, F.H., Kempermann, G., Palmer, T.D., Peterson, D.A. and Ray, J. (1998) Multipotent progenitor cells in the adult dentate gyrus. J Neurobiol *36*, 249-266.

Gass, P., Kretz, O., Wolfer, D.P., Berger, S., Tronche, F., Reichardt, H.M., Kellendonk, C., Lipp, H.P., Schmid, W. and Schutz, G. (2000) Genetic disruption of mineralocorticoid receptor leads to impaired neurogenesis and granule cell degeneration in the hippocampus of adult mice. EMBO Rep *1*, 447-451.

Ho, L., Pieroni, C.. Winger, D., Purohit, D.P., Aisen, P.S. and Pasinetti, G.M. (1999) Regional distribution of cyclooxygenase-2 in the hippocampal formation in alzheimer's disease. J Neurosci Res *57*, 295-303.

Kandel, E.R. (2001) The molecular biology of memory storage: a dialog between genes and synapses. Biosci Rep *21*, 565-611.

Kranz, V.D. and Richter, W. (1975) Neurogenesis and regeneration in the brain of teleosts in relation to age. (Autoradiographic Studies) Z Alternsforsch *30*, 371-382.

Kumihashi, K., Uchida, K., Miyazaki, H., Kobayashi, J., Tsushima, T. and Machida, T. (2001) Acetylsalicylic acid reduces ischemia-induced proliferation of dentate cells in gerbils. Neuroreport *12*, 915-917.

Lai, K., Kaspar, B.K., Gage, F.H. and Schaffer, D.V. (2003) Sonic hedgehog regulates adult neural progenitor proliferation in vitro and in vivo. Nat Neurosci *6*, 21-27.

Lois, C. and Alvarez-Buylla, A. (1993) Proliferating subventricular zone cells in the adult mammalian forebrain can differentiate into neurons and glia. Proc Natl Acad Sci USA *90*, 2074-2077.

Mantamadiotis, T., Lemberger, T., Bleckmann, S.C., Kern, H., Kretz, O., Martin Villalba, A., Tronche, F., Kellendonk, C., Gau, D., Kapfhammer, J., Otto, C., Schmid, W. and Schutz, G. (2002) Disruption of CREB function in brain leads to neurodegeneration. Nat Genet *31*, 47-54.

Mayr, B. and Montminy, M. (2001) Transcriptional regulation by the phosphorylation-dependent factor CREB. Nat Rev Mol Cell Biol *2*, 599-609.

Merry, D.E., Veis, D.J., Hickey, W.F. and Korsmeyer, S.J. (1994) Bcl-2 protein expression is widespread in the developing nervous system and retained in the adult PNS. Development *120*, 301-311.

Morest, D.K. (1970) A study of neurogenesis in the forebrain of opossum pouch young. Z Anat Entwicklungsgesch *130*, 265-305.

Mucenski, M.L., McLain, K., Kier, A.B., Swerdlow, S.H., Schreiner, C.M., Miller, T.A., Pietryga, D.W., Scott, Jr., W.J. and Potter, S.S. (1991) A functional c-myb gene is required for normal murine fetal hepatic hematopoiesis. Cell *65*, 677-689.

Nakagawa, S., Kim, J.E., Lee, R., Malberg, J.E., Chen, J., Steffen, C., Zhang, Y.J., Nestler, E.J. and Duman, R.S. (2002) Regulation of neurogenesis in adult mouse hippocampus by cAMP and the cAMP response element-binding protein. J Neurosci *22*, 3673-3682.

Ramsay, R.G., Barton, A.L. and Gonda, T.J. (2003) Targeting c-Myb expression in human disease. Expert Opin Ther Targets *7*, 235-248.

Ramsay, R.G., Friend, A., Vizantios, Y., Freeman, R., Sicurella, C., Hammett, F., Armes, J. and Venter, D. (2000) Cyclooxygenase-2, a colorectal cancer nonsteroidal anti-inflammatory drug target, is regulated by c-Myb. Cancer Res *60*, 1805-1809.

Rosenthal, M.A., Thompson, M.A., Ellis, S., Whitehead, R.H. and Ramsay, R.G. (1996) Colonic expression of c-Myb is initiated in utero and continues throughout adult life. Cell Growth Differ *7*, 961-967.

Shin, D.H., Lee, H.W., Jeon, G.S., Lee, H.Y., Lee, K.H. and Cho, S.S. (2001) Constitutive expression of c-Myb mRNA in the adult rat brain. Brain Res *892*, 203-207.

Suginoshita, K. (1971) ^{3}H-thymidine autoradiographic studies on the neurogenesis of the mouse cerebellum. Kaibogaku Zasshi *46*, 289-311.

Taylor, D., Badiani, P. and Weston, K. (1996) A dominant interfering Myb mutant causes apoptosis in T cells. Genes Dev *10*, 2732-2744.

Tronche, F., Kellendonk, C., Kretz, O., Gass, P., Anlag, K., Orban, P.C., Bock, R., Klein, R. and Schutz, G. (1999) Disruption of the glucocorticoid receptor gene in the nervous system results in reduced anxiety. Nat Genet *23*, 99-103.

Uchida, K., Kumihashi, K., Kurosawa, S., Kobayashi, T., Itoi, K. and Machida, T. (2002) Stimulatory effects of prostaglandin E2 on neurogenesis in the dentate gyrus of the adult rat. Zoolog Sci *19*, 1211-1216.

Vinet, J., Bernier, P.J. and Parent, A. (2002) Bcl-2 expression in thalamus, brainstem, cerebellum and visual cortex of adult primate. Neurosci Res *42*, 269-277.

Wardlaw, S.A., Zhang, N. and Belinsky, S.A. (2002) Transcriptional regulation of basal cyclooxygenase-2 expression in murine lung tumor-derived cell lines by CCAAT/enhancer-binding protein and activating transcription factor/Camp response element-binding protein. Mol Pharmacol *62*, 326-333.

Welter, C., Henn, W., Theisinger, B., Fischer, H., Zang, K.D. and Blin, N. (1990) The cellular myb oncogene is amplified, rearranged and activated in human glioblastoma cell lines. Cancer Lett *52*, 57-62.

West, A.E., Griffith, E.C. and Greenberg, M.E. (2002) Regulation of transcription factors by neuronal activity. Nat Rev Neurosci *3*, 921-931.

Wilson, B.E., Mochon, E. and Boxer, L.M. (1996) Induction of bcl-2 expression by phosphorylated CREB proteins during B-cell activation and rescue from apoptosis. Mol Cell Biol *16*, 5546-5556.

Zorbas, M., Sicurella, C., Bertoncello, I., Venter, D., Ellis, S., Mucenski, M.L. and Ramsay, R.G. (1999) c-Myb is critical for murine colon development. Oncogene *18*, 5821-5830.

Chapter 21

THE C-MYB GENE: A RATIONAL TARGET FOR TREATMENT OF HUMAN DISEASES

Susan E. Shetzline and Alan M. Gewirtz
Departments of Internal Medicine, Pathology and Laboratory Medicine, University of Pennsylvania School of Medicine, 421 Curie Boulevard, Philadelphia, PA 19104, United States of America.

Abstract: c-Myb is a nuclear transcription factor that plays a key role in regulating cell survival, differentiation, and proliferation. Aberrant expression or mutated forms of *c-myb* have been associated with human leukaemia as well as with a number of solid tumours, and non-malignant human diseases. Recognition of this association has led to the development of therapeutic strategies focused on inhibiting *c-myb* gene expression at the transcriptional, translational, and protein levels. With regard to the latter, endogenous Myb activity has been abrogated in malignant haemopoietic cells with anti-Myb intracellular single-chain antibodies, as well as by the use of dominant negative *c-myb* expression constructs. To disrupt *c-myb* gene expression, we have developed an approach that utilises reverse complementary, or 'antisense' oligodeoxynucleotides (ODN). Using this therapeutic strategy, we conducted clinical trials to evaluate the effectiveness of *c-myb* targetted ODNs as marrow purging agents. We have also examined the toxicity of systemically administered *c-myb* antisense ODNs. The results from these trials are encouraging, as are models of Myb-targetted therapy in colon cancer, melanoma, and cardiovascular disease.

1. INTRODUCTION

The Myb family of transcription factors consists of A-, B-, and c-Myb. While B-Myb is expressed ubiquitously, A-Myb and c-Myb are expressed predominantly in reproductive tissue and haemopoietic cells, respectively (Weston, 1998). Common to the expression patterns of Myb family members is their association with proliferating tissue, where the expression of Myb proteins declines as cells withdraw from the cell cycle and progress toward terminal differentiation. In addition to the role Myb proteins play in

J. Frampton (ed.), Myb Transcription Factors: Their Role in Growth, Differentiation and Disease, 399-415.

modulating proliferation, they also play a critical role in regulating cell survival and differentiation. Aberrant expression and mutated forms of c-Myb have been associated with leukaemia and a number of different solid tumour cancers. However, actual patient data demonstrating the importance of Myb in leukaemogenesis are not abundant and data implicating the protein in determining clinical disease phenotype are somewhat inferential. This chapter will focus on the important role c-Myb plays in modulating cellular functions and discuss advances in Myb-targetted therapeutics for the treatment of human malignancies.

2. THE THERAPEUTIC RELEVANCE OF MYB BIOLOGY

Located in the human genome on chromosome 6q (Majello et al., 1986), the *c-myb* proto-oncogene is the cellular homologue of *v-myb*, the avian retroviral oncogene that causes acute myeloblastosis leukaemia and erythroblastosis (Lipsick and Wang, 1999; Roussel et al., 1979; Souza et al., 1980). The predominant *c-myb* transcript encodes a ~75-kDa nuclear transcription factor Myb that is predominantly expressed in haemopoietic cells (Kastan et al., 1989). c-Myb is composed of three distinct functional domains: an N-terminal DNA binding domain, a central transactivation domain, and a C-terminal negative regulatory domain (Sakura et al., 1989). Within the DNA binding domain, there are three imperfect tandem repeats (R1, R2, and R3) each consisting of 51-52 amino acids and a conserved tryptophan residue. Together, these tandem repeats form a cluster in the hydrophobic core of the protein which maintains the DNA binding helix-turn-helix structure and recognise a consensus sequence 5'-PyAAC(G/Py)G-3' referred to as the Myb responsive element in its gene targets (Biedenkapp et al., 1988; Howe et al., 1990). Phosphorylation of the DNA binding domain destabilises the c-Myb-DNA complex, which prevents c-Myb from transactivating its gene targets (Andersson et al., 2003; Lüscher et al., 1990; Oelgeschlager et al., 1995). Adjacent to the DNA binding domain is an acidic transactivation domain (Weston and Bishop, 1989) that interacts with the co-activator CREB-binding protein (CBP). When this histone acetyltransferase binds to c-Myb, a bridge forms between c-Myb and the basal transcriptional machinery (Dai et al., 1996; Oelgeschlager et al., 1996). At the C-terminus is the negative regulatory domain that contains a putative leucine zipper and a PEST/EVES motif, which play a critical role in regulating Myb protein activity. When the leucine zipper in the negative regulatory domain of c-Myb interacts with p67, p160, c-Myb, or BS69, c-Myb transactivating activity is inhibited (Favier and Gonda, 1994; Keough et al., 1999; Ladendorff et al., 2001; Nomura et al., 1993). Post-translational modifications such as phosphorylation, acetylation, and ubiquitination occur at the PEST/EVES motif and adjacent residues. Phosphorylation decreases the transactivating activity of c-Myb while acetylation increases its DNA binding affinity and transactivation capacity (Miglarese et al., 1996; Sano and Ishii, 2001; Tomita et al., 2000). Ubiquitination of the c-Myb negative regulatory domain decreases the half-life of the protein by rendering c-Myb a good substrate for degradation by the 26S proteasome proteolytic system (Bies and Wolff, 1997; Feikova et al., 2000). When c-Myb is covalently modified by a small ubiquitin-related protein SUMO-1, its protein stability increases possibly because the ubiquitination sites are masked (Bies et al., 2002). In addition, there is some evidence to suggest the SUMO-1 modifications negatively regulate c-Myb transactivation (Bies et al., 2002).

The C-terminus of c-Myb is deleted in v-Myb and it has been thought to contribute to the proteins transforming ability. The different mutations in v-Myb in the acute myeloblastosis virus and the avian leukaemia virus E26 result in leukaemias of different phenotypes depending on the cell type in which they are expressed (Lipsick and Wang, 1999; Roussel et al., 1979; Souza et al., 1980). That seemingly minor changes in the c-Myb protein may play such an important role in the phenotype of the disease they cause has been commented on (Graf, 1998). Deletion of the N- or C-terminus of c-Myb has also been associated with leukaemias. Retroviral insertion in avian and murine models produced a truncated form of the *c-myb* locus and caused B- and T-cell lymphomas (Kanter and Hayward, 1988; Pizer, 1989; Pizer and Humphries, 1992; Rouzic, 1996) and myeloid leukaemias (Gonda et al., 1987; Nason-Burchenal, 1993; Schmidt et al., 2000; Shen-Ong, 1987), respectively. In TK-6 cells, which were established from a chronic myelogenous leukaemia (CML) patient in T-cell blast crisis, the C-terminus of c-Myb is deleted (Tomita et al., 1998). Truncating the N- and C-termini of c-Myb produces a protein that induces the formation of haemopoietic cells that are more primitive than those produced by N-terminal deletions alone (Press and Ewert, 1994). Taken together, these data indicate that the loss of the N- or C-terminal sequences of c-Myb unmasks its oncogenic potential.

The transforming ability of c-Myb has also been attributed to aberrant gene expression and activating mutations particularly in its DNA-binding domain. Constitutive c-Myb expression in M1 cells, a murine myeloblastic cell line, prevented the induction of the tumour suppressor gene Cdkn2b ($p15^{INK4b}$) and concomitant monocyte maturation (Schmidt et al., 2001). Clinically, amplification of *c-myb* in acute myelogenous leukaemia (AML) patients and over-expression in 6q- syndrome has been reported (Barletta et al., 1987). We have observed elevated *c-myb* gene expression in patients with AML and with acute lymphocytic leukaemia (ALL) compared to normal donors (Shetzline et al., 2002). The mechanism whereby over-expression of c-Myb might be leukaemogenic is uncertain. *In vivo* assays using the truncated form of c-Myb revealed a dramatic increase in its transactivation activity, possibly because it could not interact with its regulatory proteins (Vorbrueggen et al., 1994). Mutations within the DNA-binding domain of c-Myb have been identified (Introna et al., 1990). Functionally, these activating mutations led to transformation of cells with a promyelocyte-like phenotype. Mechanistically, mutations within the DNA-binding domain of c-Myb decreased its DNA-binding affinity for target genes such as *mim-1* (Dini and Lipsick, 1995; Dini, 1993), suggesting that c-Myb in transformed cells regulates a subset of genes. Together, these data suggest that simultaneous loss of the ability of c-Myb to bind DNA and interact with various regulatory proteins is a significant transforming stimulus.

The transformation of haemopoietic cells might also be associated with the ability of c-Myb to regulate cell survival, differentiation, and proliferation. c-Myb binds to and transactivates its target genes through a Myb responsive element (Biedenkapp et al., 1988; Howe et al., 1990). The role of c-Myb in the survival of myeloid and T-cells has been associated with the expression of the anti-apoptotic gene *bcl-2*, where inhibition of endogenous c-Myb activity resulted in a decrease in *bcl-2* expression and concomitant cell death (Frampton et al., 1996; Taylor et al., 1996). During differentiation, c-Myb regulates a number of lineage-specific genes such as *mim-1*, GATA-1, Rag-2, TCRδ, CD4, and neutrophil elastase through cooperation with several transcription factors (Ness and Engel, 1994; Ness, 1996). Like c-Myb, v-Myb regulates the expression of a number of genes. Most notable is the homeobox protein encoding gene *gbx2*. AMV v-Myb transactivates *gbx2*, the protein product of which in turn mediates autocrine growth and monocytic differentiation by transactivation of a myelo-monocytic growth factor gene (cMGF), thus illustrating a mechanism whereby Myb may transform haemopoietic cells (Kowenz-Leutz et al., 1997). Since c-Myb activity reaches a maximum during the G1/S transition of the cell cycle, it has been suggested that it plays an important role in cell cycle progression and proliferation (Thompson et al., 1986). Inhibition of *c-myb* gene expression arrested cell growth in the G1 phase of the cell cycle (Gewirtz et al., 1989; Lyon and Watson, 1996). c-Myb-regulated genes whose products have been implicated in proliferation are DNA topoisomerase IIa (Brandt et al., 1997), *cdc2* (Ku et al., 1993), *c-kit* (Hogg et al., 1997; Ratajczak et al., 1998; Vandenbark et al., 1996), *c-myc* (Cogswell et al., 1993; Evans et al., 1990; Schmidt et al., 2000; Zobel et al., 1992) and myeloblastin (Lutz et al., 2001). *In vivo* studies have demonstrated that c-Myb regulates *c-myc* expression in a murine myeloblastic cell line and protects cells from cytokine-induced cell cycle arrest, indicating that c-Myb is involved in myeloid leukaemogenesis (Schmidt et al., 2000). Our preliminary clinical data indicates that malignant haemopoietic cells are more susceptible to apoptosis than normal haemopoietic cells when exposed to *c-myb* antisense oligodeoxynucleotides (ODN) (Calabretta et al., 1991; Ratajczak et al., 1992a). c-Myb plays a critical role in haemopoietic cells, directly or indirectly, which may contribute to the pathogenesis or maintenance of human leukaemia. Therefore, it is a rational target for therapeutic strategies designed to disrupt gene expression or protein activity.

3. TARGETTED INHIBITION OF C-MYB AND ITS GENE TARGETS

A number of innovative technologies for inhibiting c-Myb protein and *c-myb* gene expression have been developed in recent years. Inhibition of endogenous c-Myb activity at the protein level has been achieved using antibody and dominant negative approaches. An intracellular single-chain antibody (sFv) has been described to achieve a functional knock-out of c-Myb protein (Kasono et al., 1998). Immunofluorescent staining with anti-c-Myb sFv revealed the presence of sFv in the cytoplasmic and nuclear compartments. More importantly, the anti-c-Myb sFv inhibited the transctivation activity of c-Myb, which appears to correlate with the cytotoxic effect of this sFv in the c-Myb positive leukaemia cell line, K562. Interfering with endogenous c-Myb activity has also been achieved by expressing a dominant negative Myb protein. Conditional inhibition of endogenous c-Myb activity by a tamoxifen-inducible dominant negative Myb arrested the cell cycle at the G1/S transition in the haemopoietic cell line K562 and in T cells (Lyon and Watson, 1996; Taylor et al., 1996; Yi et al., 2002). This Myb-dependent attenuation of the cell cycle is often associated with a decrease in *bcl-2* gene expression and apoptosis. Using a microarray approach to identify candidate Myb gene targets, we have also observed a decrease in *bcl-2* gene expression in tamoxifen-treated K562 cells that were engineered to express the inducible dominant negative Myb (Shetzline et al., 2002). However, the greatest change in gene expression in our model system was seen for neuormedin U, a gene whose product is elevated in bone marrow and is involved in G-protein coupled receptor signalling (Fujii et al., 2000; Raddatz et al., 2000; Szekeres et al., 2000). Colony forming assays revealed that exogenously added neuromedin U could rescue K562 cells expressing the dominant negative Myb from cell cycle arrest. Even though our initial observations suggest that neuromedin U stimulates the proliferation of human myeloid leukaemia cells, the role of neuromedin U in haemopoiesis or leukaemogenesis remains to be defined. While the notion of inhibiting c-Myb protein activity is tantalizing, the therapeutic utility of interfering with endogenous c-Myb activity remains to be tested.

Targetted silencing of gene expression has included the use of ribozymes (Eckstein, 1996; James and Turner, 1997; Sullenger, 1995), triple helix forming ODNs (Gunther et al., 1996), antisense ODNs (Gewirtz et al., 1998), and short interfering RNAs (siRNA) (McManus and Sharp, 2002; Shuey et al., 2002). This discussion will focus on the ribozyme and ODN therapeutics because they are in the advanced stages of clinical development. From a therapeutic intervention point of view, ribozymes and ODNs each have advantages and deficiencies. The mechanisms by which ribozymes and particularly ODNs function *in vivo* remain controversial (Wagner and Flanagan, 1997). Both methods involve using the antisense sequence of the mRNA intended for destruction. For the ribozyme, the antisense sequence to the target mRNA flanks the catalytic domain of the ribozyme and functions to guide the ribozyme to the mRNA intended for destruction. Hybridisation of the ODN containing the antisense sequence to its target mRNA forms a DNA:RNA complex that creates a substrate for RNase H, which degrades the mRNA portion of the duplex. Consequently, the expression of the targetted gene is either inhibited or totally abrogated.

The success of inhibiting gene expression relies on efficient delivery and sequence accessibility. Generally, ODNs enter the cell primarily through a combination of adsorptive and fluid-phase endocytosis (Beltinger et al., 1995). Confocal and electron microscopy studies suggest that the bulk of the ODNs enter the endosome/lysosome compartment after internalisation, where most of the material is either trapped or degraded rendering the ODN biologically inactive. However, ODNs can escape intact from these vesicles, enter the cytoplasm, and diffuse into the nucleus, where they gain access to their target mRNA. The mechanism that regulates ODN intracellular trafficking remains poorly understood and requires further investigation. Once the ODN comes in close proximity to its target mRNA, sequence accessibility becomes important. Sequence accessibility is at least in part a function of the physical structure of the target mRNA, which is dictated by internal base composition and associated proteins in the living cell. Attempts to describe the *in vivo* structure of mRNA, in contrast to DNA, have been fraught with difficulties. Accordingly, targetting mRNA is largely a random process, accounting for many experiments in which the addition of an ODN yields no effect on expression. Strategies to address this fundamental problem are presently under development (Ho et al., 1998; Sokol et al., 1998).

4. INHIBITION OF MYB AS A TREATMENT FOR LEUKAEMIA

Attempting to exploit the *c-myb* gene as a target for an antisense ODN therapeutic for leukaemia stems from studies in which we sought to define the role that c-Myb plays in regulating normal haemopoiesis (Gewirtz and Calabretta, 1988). Exposing mononuclear cells from normal bone marrow to *c-myb* antisense ODN resulted in a dose-dependent decrease in colony formation and a decrease in progenitor proliferation. The observed inhibition in cell growth following exposure to the *c-myb* antisense ODN appears to be attributed to a block in the cell cycle at the G1/S transition. In subsequent studies, *c-myb* antisense ODN inhibited the expression of *c-myb* mRNA and protein, indicating that the effects we observed with the *c-myb* antisense ODN were due to an 'antisense' mechanism (Calabretta and Gewirtz, 1991; Calabretta et al., 1991; Gewirtz et al., 1989; Ratajczak et al., 1992a; Ratajczak et al., 1992b; Ratajczak et al., 1998). Our studies also revealed that specific stages in the maturation of haemopoietic cells required c-Myb protein, especially when they are actively cycling (Caracciolo et al., 1990). Taken together, c-Myb appears to play a critical role during normal haemopoiesis, a result which was verified using the technique of homologous recombination (Mucenski, et al., 1991).

As noted earlier, c-Myb is expressed in malignant haemopoietic cells and its expression is elevated in primary haemopoietic cells from ALL and AML patients compared to normal donors (Shetzline et al., 2002). Exposing malignant cell lines to *c-myb* antisense ODN inhibits cell growth (Anfossi et al., 1989), strongly suggesting that c-Myb is also required for leukaemic haemopoiesis. Even though c-Myb is expressed in normal and leukaemic cells, *in vitro* results and *in vivo* data from a SCID mouse model revealed that leukaemic cells are more dependent on c-Myb protein than their normal counterparts (Calabretta et al., 1991; Ratajczak et al., 1992a) and as such these leukaemic cells are preferentially killed following exposure to *c-myb* antisense ODN. The mechanism by which *c-myb* antisense ODN induces cell death in leukaemic cells may be attributed to c-Myb gene targets. For example, in studies designed to define the role of the c-Kit receptor in haemopoietic cells, *c-kit* was determined to be a c-Myb regulated gene (Melotti and Calabretta, 1994; Ratajczak et al., 1992c; Ratajczak et al., 1998; Vandenbark et al., 1996). Because c-Kit is a tyrosine kinase receptor that is critical for haemopoietic cells, we reasoned that dysregulation of this c-Myb regulated gene might be an important mechanism of action of the *c-myb* antisense ODN. In support of this hypothesis, haemopoietic cells that are either deprived of the c-Kit receptor ligand, stem cell factor (SCF) (Yu et al.,

1993) or exposed to *c-myb* antisense ODN (Calabretta et al., 1991; Ratajczak et al., 1992a) undergo apoptosis.

Using a tissue culture based system, we demonstrated that the *c-myb* antisense ODN induced cell death in malignant cell lines and in primary leukaemic cells, suggesting that this ODN might serve as an effective therapeutic against leukaemia. To this end, we conducted clinical trials to evaluate the effectiveness of: (1) phosphorothioate modified *c-myb* antisense ODN as marrow purging agents for chronic phase (CP) or accelerated phase (AP) CML patients (Luger et al., 2002), and (2) intravenous infusion as a means to treat blast crisis (BC) patients or patients with other refractory leukaemias (Gewirtz, 1999). Our pilot marrow purging study revealed that seven out of eight study subjects engrafted. Four out of six CP patients that could be evaluated had 85-100% metaphases that were normal three months following engraftment, suggesting that a significant purge in the marrow had taken place. Five CP patients demonstrated marked, sustained, haematological improvement with essential normalisation of their blood counts. Increasing the marrow purging time from 24 to 72 hours with *c-myb* antisense ODN did not significantly improve efficiency. In fact, it was found that this longer purging period resulted in a poor engraftment in five study subjects due to enhanced toxicity to normal haemopoietic stem cells. Our Phase I systemic infusion study consisted of 18 refractory leukaemia patients. Recurrent dose related toxicity was not observed. However, two idiosyncratic toxicities were noted (one transient renal insufficiency and one pericarditis). One patient survived for more than 14 months with a transient restoration of CP disease. The data from these studies demonstrated that *c-myb* antisense ODN may be administered safely to patients. The clinical benefit received by the patients in each study is uncertain. However, our initial observations suggest that ODN may serve as an effective therapeutic agent for human leukaemia.

5. MYB TARGETTED THERAPEUTICS IN COLON CANCER AND MELANOMA MODELS

In more recent years, *c-myb* expression has been observed in normal human and murine colonic mucosa and its expression is elevated in premalignant adenomatous polyps and carcinomas (Thompson et al., 1998). Treating colon cancer cells with butyrate induces differentiation and apoptosis. To understand the role *c-myb* plays in this differentiation process, electron microscopy, molecular and biochemical analyses were conducted (Thompson et al., 1998). As observed in haemopoietic cells, *c-myb* expression decreased once the colonic cells committed to differentiation and apoptosis, suggesting that c-Myb plays an important role in regulating the delicate balance between differentiation, apoptosis, and proliferation. Associated with the observed decrease in *c-myb* expression was a decrease in *bcl-2* expression, which was identified in haemopoietic cells as a Myb regulated gene. In an independent study, *c-myb* and *bcl-X_L* expression was also elevated in colorectal carcinoma patients (Biroccio et al., 2001). Based on these data, elevated levels of c-Myb in colon cancer cells could lead to persistent expression of anti-apoptotic genes and concomitant protection from programmed cell death. Therefore, inhibition of *c-myb* expression in colon cancer cells might be of therapeutic utility.

Rearrangement of chromosome 6 and alterations in *c-myb* expression have been implicated in the pathogenesis of melanoma (Dasgupta et al., 1989; Linnenbach et al., 1988; Meese et al., 1989; Trent et al., 1989). Using *c-myb* positive human melanoma cell lines, we targetted *c-myb* gene expression with *c-myb* antisense ODN to investigate the biological significance of its expression and to determine the therapeutic potential of disrupting *c-myb* expression in this melanoma model (Hijiya et al., 1994). Unmodified or phosphorothioate-modified *c-myb* antisense ODN inhibited the growth of representative melanoma cell lines in a dose- and sequence-dependent manner. These *in vitro* assays also revealed that the inhibition of cell growth correlated with a specific decrease in the level of *c-myb* mRNA. Infusion of *c-myb* antisense ODN into SCID mice bearing human melanoma tumours transiently suppressed *c-myb* gene expression, but effected long-term growth suppression of transplanted tumour cells. Toxicity of the *c-myb* antisense ODN was minimal.

6. TARGETING MYB AS A THERAPEUTIC INTERVENTION FOR CARDIOVASCULAR DISEASE

Cytokines released from cells in response to tissue injury that promote cell proliferation may also play a pivotal role in the pathogenesis of non-malignant diseases. A prime example is the re-occlusion of coronary arteries that arises following angioplasty procedures performed on patients with artherosclerotic disease. Infusing antibodies against PDGF or bFGF into smooth muscle cells have identified growth factors responsible for their proliferation and migration (Rosenberg, 1993). To identify and define the role of intracellular intermediates in these cellular processes, an antisense ODN approach appeared to be a logical strategy. Given the critical role c-Myb plays in regulating proliferation, *c-myb* antisense ODN was used to target *c-myb* gene expression for disruption in an *in vivo* model (Simons et al., 1992). When *c-myb* antisense ODN was applied locally to rat corotid arterial smooth muscle cells, specific inhibition of cell proliferation was observed. Independent studies have confirmed these observations (Azrin et al., 1997; Pitsch et al., 1996), even though doubt has been raised as to whether the inhibition in the proliferation of smooth muscle cells by their *c-myb* antisense ODN was truly due to an 'antisense' mechanism. Some groups argue that the anti-proliferative effects observed in these smooth muscle cells are due to the presence of four continguous guanosine residues in the ODN sequence employed, and that such 'G-quartets' allow for base stacking and tetraplex formation (Burgess et al., 1995; Castier et al., 1998), which have been shown to inhibit cell proliferation. Regardless of the mechanism, the physiologic role of c-Myb in regulating smooth muscle cell proliferation seems certain (Brown et al., 1992; Kypreos et al., 1998; Simons et al., 1993; Simons et al., 1995). *c-myb* targetted ribozymes have also inhibited the local intimal proliferation of smooth muscle cells (Jarvis et al., 1996). For these reasons, targetted disruption of *c-myb* for the treatment of vascular disease might yet prove to be a useful manoeuver.

7. CONCLUSIONS

Many malignant diseases remain difficult to treat, and are often incurable. Furthermore, they often affect individuals during their most productive years. Treating patients with highly toxic chemotherapeutic agents and radiation is very debilitating, and can at times result in the iatrogenic death of patients. Oncogene-targetted therapeutics have become an attractive pharmaceutical prospect because they may allow for the development of effective, relatively non-toxic therapeutic agents that can be used to treat various diseases. A number of different strategies for treating various human malignancies have evolved. We have focused on developing RNA targetted ODNs for the purposes of 'silencing' genes of etiologic importance to the disease of interest. c-*myb* is an appropriate target for this approach in leukaemia and possibly other tumour types. Because *c-myb* is expressed in both normal and malignant cells, more rational targets might be the transcriptional targets of c-Myb or its cooperating proteins, especially if cancer specific partners are identified. This approach might also benefit patients with non-malignant diseases characterised by hyperproliferation. Regardless of the disease, improving the efficacy of RNA targetted drugs will depend on solving fundamental problems in 'drug' delivery and physical accessibility of sequence within the targetted RNA species.

REFERENCES

Andersson, K.B., Kowenz-Leutz, E., Brendeford, E.M., Tygsett, A.H., Leutz, A. and Gabrielsen, O.S. (2003) Phosphorylation-dependent down-regulation of c-Myb DNA binding is abrogated by a point mutation in the v-myb oncogene. J Biol Chem *278*, 3816-3824.

Anfossi, G., Gewirtz, A.M. and Calabretta, B. (1989) An oligomer complementary to c-myb-encoded mRNA inhibits proliferation of human myeloid leukaemia cell lines. Proc Natl Acad Sci U S A. *86*, 3379-3383.

Azrin, M.A., Mitchel, J.F., Bow, L.M., Pedersen, C.A., Cartun, R.W., Aretz, T.H., Waters, D.D. and McKay, R.G. (1997) Local delivery of c-myb antisense oligonucleotides during balloon angioplasty. Cathet Cardiovasc Diagn *41*, 232-240.

Barletta C, P.P., Kenyon LC, Smith SD, Dalla-Favera R. (1987) Relationship between the c-myb locus and the 6q-chromosomal aberration in leukaemias and lymphomas. Science *235*, 1064-1067.

Beltinger, C., Saragovi, H.U., Smith, R.M., LeSauteur, L., Shah, N., DeDionisio, L., Christensen, L., Raible, A., Jarett, L. and Gewirtz, A.M. (1995) Binding, uptake, and intracellular trafficking of phosphorothioate-modified oligodeoxynucleotides. J Clin Invest *95*, 1814-1823.

Biedenkapp, H., Borgmeyer, U., Sippel, A.E. and Klempnauer, K.H. (1988) Viral myb oncogene encodes a sequence-specific DNA-binding activity. Nature *335*, 835-837.

Bies, J. and Wolff, L. (1997) Oncogenic activation of c-Myb by carboxyl-terminal truncation leads to decreased proteolysis by the ubiquitin-26S proteasome pathway. Oncogene *14*, 203-212.

Bies, J., Markus, J. and Wolff, L. (2002) Covalent attachment of the SUMO-1 protein to the negative regulatory domain of the c-Myb transcription factor modifies its stability and transactivation capacity. J Biol Chem *277*, 8999-9009.

Biroccio, A., Benassi, B., D'Agnano, I., D'Angelo, C., Buglioni, S., Mottolese, M., Ricciotti, A., Citro, G., Cosimelli, M., Ramsay, R.G., Calabretta, B. and Zupi, G. (2001) c-Myb and Bcl-x overexpression predicts poor prognosis in colorectal cancer: clinical and experimental findings. Am J Physiol. *158*, 1289-1299.

Brandt, T.L., Fraser, D.J., Leal, S., Halandras, P.M., Kroll, A.R. and Kroll, D.J. (1997) c-Myb trans-activates the human DNA topoisomerase IIalpha gene promoter. J Biol Chem *272*, 6278-6284.

Brown, K.E., Kindy, M.S. and Sonenshein, G.E. (1992) Expression of the c-myb proto-oncogene in bovine vascular smooth muscle cells. J Biol Chem *267*, 4625-4630.

Burgess, T.L., Fisher, E.F., Ross, S.L., Bready, J.V., Qian, Y.X., Bayewitch, L.A., Cohen, A.M., Herrera, C.J., Hu, S.S. and Kramer, T.B., et al. (1995) The antiproliferative activity of c-myb and c-myc antisense oligonucleotides in smooth muscle cells is caused by a nonantisense mechanism. Proc Natl Acad Sci U S A *92*, 4051-4055.

Calabretta, B. and Gewirtz, A.M. (1991) Functional requirements of c-myb during normal and leukaemic haemopoiesis. Crit Rev Oncog *2*, 187-194.

Calabretta, B., Sims, R.B., Valtieri, M., Caracciolo, D., Szczylik, C., Venturelli, D., Ratajczak, M., Beran, M. and Gewirtz, A.M. (1991) Normal and leukaemic haemopoietic cells manifest differential sensitivity to inhibitory effects of *c-myb* antisense oligonucleotides: An *in vitro* study relvant to bone marrow purging. Proc Natl Acad Sci U S A *88*, 2351-2355.

Caracciolo, D., Venturelli, D., Valtieri, M., Peschle, C., Gewirtz, A.M. and Calabretta, B. (1990) Stage-related proliferative activity determines c-myb functional requirements during normal human haemopoiesis. J Clin Invest *85*, 55-61.

Castier, Y., Chemla, E., Nierat, J., Heudes, D., Vasseur, M.A., Rajnoch, C., Bruneval, P., Carpentier, A. and Fabiani, J.N. (1998) The activity of c-myb antisense oligonucleotide to prevent intimal hyperplasia is nonspecific. J Cardiovasc Surg (Torino). *39*, 1-7.

Cogswell, J.P., Cogswell, P.C., Kuehl, W.M., Cuddihy, A.M., Bender, T.M., Engelke, U., Marcu, K.B. and Ting, J.P. (1993) Mechanism of *c-myc* regulation by c-Myb in different cell lineages. Mol Cell Biol *13*, 2858-2869.

Dai, P., Akimaru, H., Tanaka, Y., Hou, D.X., Yasukawa, T., C., K.-I., Takahashi, T. and Ishii, S. (1996) CBP as a transcriptional coactivator of c-Myb. Genes and Development *10*, 528-540.

Dasgupta, P., Linnenbach, A.J., Giaccia, A.J., Stamato, T.D. and Reddy, E.P. (1989) Molecular cloning of the breakpoint region on chromosome 6 in cutaneous malignant melanoma: evidence for deletion in the c-myb locus and translocation of a segment of chromosome 12. Oncogene *4*, 2101-2105.

Dini PW, E.J., Lipsick JS. (1995) Mutations in the DNA-binding and transcriptional activation domains of v-Myb cooperate in transformation. J Virol *69*, 2515-2524.

Dini PW, L.J. (1993) Oncogenic truncation of the first repeat of c-Myb decreases DNA binding in vitro and in vivo. Mol Cell Biol *13*, 7334-7348.

Eckstein, F. (1996) The hammerhead ribozyme. Biochem Soc Trans. *24*, 601-604.

Evans, J.L., Moore, T.L., Kuehl, W.M., Bender, T. and Ting, J.P. (1990) Functional analysis of c-Myb protein in T-lymphocytic cell lines shows that it *trans*-activates the c-myc promoter. Mol Cell Biol *10*, 5747-5752.

Favier, D. and Gonda, T.J. (1994) Detection of proteins that bind to the leucine zipper motif of c-Myb. Oncogene *9*, 305-311.

Feikova, S., Wolff, L. and Bies, J. (2000) Constitutive ubiquitination and degradation of c-myb by the 26S proteasome during proliferation and differentiation of myeloid cells. Neoplasma *47*, 212-218.

Frampton, J., Ramqvist, T. and Graf, T. (1996) v-Myb of E26 leukaemia virus up-regulates *bcl-2* and suppresses apoptosis in myeloid cells. Genes Dev *10*, 2720-2730.

Fujii, R., Hosoya, M., Fukusumi, S., Kawamata, Y., Habata, Y., Hinuma, S., Onda, H., Nishimura, O. and Fujino, M. (2000) Identification of neuromedin U as the cognate ligand of the orphan G protein-coupled receptor FM-3. J Biol Chem *275*, 21068-21074.

Gewirtz, A.M. and Calabretta, B. (1988) A c-myb antisense oligodeoxynucleotide inhibits normal human haemopoiesis in vitro. Science *242*, 1303-1306.

Gewirtz, A.M., Anfossi, G., Venturelli, D., Valpreda, S., Sims, R. and Calabretta, B. (1989) G1/S transition in normal human T-lymphocytes requires the nuclear protein encoded by c-myb. Science *245*, 180-183.

Gewirtz, A.M., Sokol, D.L. and Ratajczak, M.Z. (1998) Nucleic acid therapeutics: state of the art and future prospects. Blood *92*, 712-736.

Gewirtz, A.M. (1999) Myb targetted therapeutics for the treatment of human malignancies. Oncogene *18*, 3056-3062.

Gonda TJ, C.S., Sobieszczuk P, Holtzman D, Adams JM. (1987) Generation of altered transcripts by retroviral insertion eithin the *c-myb* gene in two murine monocytic leukaemias. J. Virol. *61*, 2754-2763.

Graf, T. (1998) Leukemogenesis: small differences in Myb have large effects. Curr Biol *8*, R353-R355.

Gunther, E.J., Havre, P.A., Gasparro, F.P. and Glazer, P.M. (1996) Triplex-mediated, in vitro targeting of psoralen photoadducts within the genome of a transgenic mouse. Photochem Photobiol. *63*, 207-212.

Hijiya, N., Zhang, J., Ratajczak, M.Z., Kant, J.A., DeRiel, K., Herlyn, M., Zon, G. and Gewirtz, A.M. (1994) Biologic and therapeutic significance of MYB expression in human melanoma. Proc Natl Acad Sci U S A *91*, 4499-4503.

Ho, S.P., Bao, Y., Lesher, T., Malhotra, R., Ma, L.Y., Fluharty, S.J. and Sakai, R.R. (1998) Mapping of RNA accessible sites for antisense experiments with oligonucleotide libraries. Nat Biotechnol. *16*, 59-63.

Hogg, A., Schirm, S., Nakagoshi, H., Bartley, P., Ishii, S., Bishop, J.M. and Gonda, T.J. (1997) Inactivation of a c-Myb/estrogen receptor fusion protein in transformed primary cells leads to granulocyte/macrophage differentiation and down-regulation of c-kit but not c-myc or cdc2. Oncogene *15*, 2885-2898.

Howe, K.M., Reakes, C.F. and Watson, R.J. (1990) Characterization of the sequence-specific interaction of mouse c-myb protein with DNA. EMBO J *9*, 161-169.

Introna M, G.J., Frampton J, Nakano T, Ness SA, Graf T. (1990) Mutations in v-myb alter the differentiation of myelomonocytic cells transformed by the oncogen. Cell *63*, 1289-1297.

James, H.A. and Turner, P.C. (1997) Ribozymes. Methods Mol Biol. *74*, 1-9.

Jarvis, T.C., Alby, L.J., Beaudry, A.A., Wincott, F.E., Beigelman, L., McSwiggen, J.A., Usman, N. and Stinchcomb, D.T. (1996) Inhibition of vascular smooth muscle cell proliferation by ribozymes that cleave c-myb mRNA. RNA *2*, 419-428.

Kanter MR, S.R., Hayward WS. (1988) Rapid induction of B-cell lymphomas: insertional activation of *c-myb* by avian leukosis virus. J. Virol. *62*, 1423-1432.

Kasono, K., Piche, A., Xiang, J., Kim, H.G., Bilbao, G., Johanning, F., Nawrath, M., Moelling, K. and Curiel, D.T. (1998) Functional knock-out of c-myb by an intracellular anti-c-Myb single-chain antibody. Biochem Biophys Res Commun. *251*, 124-130.

Kastan, M.B., Stone, K.D. and Civin, C.I. (1989) Nuclear oncoprotein expression as a function of lineage, differentiation stage, and proliferative status of normal human haemopoietic cells. Blood *74*, 1517-1524.

Keough, R., Woollatt, E., Crawford, J., Sutherland, G.R., Plummer, S., Casey, G. and Gonda, T.J. (1999) Molecular cloning and chromosomal mapping of the human homologue of MYB binding protein (P160) 1A (MYBBP1A) to 17p13.3. Genomics *62*, 483-489.

Kowenz-Leutz, E., Herr, P., Niss, K. and Leutz, A. (1997) The homeobox gene GBX2, a target of the myb oncogene, mediates autocrine growth and monocyte differentiation. Cell *91*, 185-195.

Ku, D., Wen, S.C., Engelhard, A., Nicolaides, N.C., Lipson, K.E., Marino, T.A. and Calabretta, B. (1993) c-Myb transactivates cdc2 expression via Myb binding sites in the 5'-flanking region of the human cdc2 gene. J Biol Chem *268*, 2255-2259.

Kypreos, K.E., Nugent, M.A. and Sonenshein, G.E. (1998) Basic fibroblast growth factor-induced decrease in type I collagen gene transcription is mediated by B-myb. Cell Growth & Differentiation *9*, 723-730.

Ladendorff, N.E., Wu, S. and Lipsick, J.S. (2001) BS69, an adenovirus E1A-associated protein, inhibits the transcriptional activity of c-Myb. Oncogene *20*, 125-132.

Linnenbach, A.J., Huebner, K., Reddy, E.P., Herlyn, M., Parmiter, A.H., Nowell, P.C. and Koprowski, H. (1988) Structural alteration in the MYB protooncogene and deletion within the gene encoding alpha-type protein kinase C in human melanoma cell lines. Proc Natl Acad Sci U S A *85*, 74-78.

Lipsick, J.S. and Wang, D.-M. (1999) Transformation by v-Myb. Oncogene *18*, 3047-3055.

Luger, S.M., O'Brien, S.G., Ratajczak, J., Ratajczak, M.Z., Mick, R., Stadtmauer, E.A., Nowell, P.C., Goldman, J.M. and Gewirtz, A.M. (2002) Oligodeoxynucleotide-mediated inhibition of c-myb gene expression in autografted bone marrow: a pilot study. Blood *99*, 1150-1158.

Luscher, B., Christenson, E., Litchfield, D.W., Krebs, E.G. and Eisenman, R.N. (1990) Myb DNA binding inhibited by phosphorylation at a site deleted during oncogenic activation. Nature *344*, 517-522.

Lutz, P., Houzel-Charavel A, Moog-Lutz C, Cayre YE. (2001) Myeloblastin is a Myb target gene: mechanisms of regulation in myeloid leukaemia cells growth-arrested by retinoic acid. Blood *97*, 2449-2456.

Lyon, J.J. and Watson, R.J. (1996) Interference of Myb transactivation activity by conditional dominant negative protein: functional interference in a cytotoxic T-cell line results in G1 arrest. Gene *182*, 123-128.

Majello, B., Kenyon, L.C. and Dalla-Favera, R. (1986) Human c-myb protooncogene: nucleotide sequence of cDNA and organization of the genomic locus. Proc Natl Acad Sci U S A *83*, 9636-9640.

McManus, M.T. and Sharp, P.A. (2002) Gene silencing in mammals by small interfering RNAs. Nat Rev Genet. *3*, 737-747.

Meese, E., Meltzer, P.S., Witkowski, C.M. and Trent, J.M. (1989) Molecular mapping of the oncogene MYB and rearrangements in malignant melanoma. Genes Chromosomes Cancer *1*, 88-94.

Melotti, P. and Calabretta, B. (1994) Ets-2 and c-Myb act independently in regulating expression of the haemopoietic stem cell antigen CD34. J Biol Chem *269*, 25303-25309.

Miglarese, M.R., Richardson, A.F., Aziz, N. and Bender, T.P. (1996) Differential regulation of c-Myb-induced transcription activation by a phosphorylation site in the negative regulatory domain. J Biol Chem *271*, 22697-22705.

Mucenski, M.L., McLain, K., Kier, A.B., Swerdlow, S.H., Schreiner, C.M., Miller, T.A., Pietryga, D.W., Scott, W.J.J. and Potter, S.S. (1991) A functional c-myb gene is required for normal murine fetal hepatic haemopoiesis. Cell *65*, 677-689.

Nason-Burchenal K, W.L. (1993) Activation of c-myb is an early bone-marrow event in murine model for acute promonocytic leukaemia. Proc Natl Acad Sci U S A *90*, 1619-1623.

Ness, S.A. and Engel, J.D. (1994) Vintage reds and whites: combinatorial transcription factor utilization in haemopoietic differentiation. Curr Opin Genet Dev. *4*, 718-724.

Ness, S.A. (1996) The Myb oncoprotein: regulating a regulator. Biochim. Biophys. Acta *1288*, F123-F139.

Nomura, T., Sakai, N., Sarai, A., Sudo, T., Kanei-Ishii, C., Ramsay, R.G., Favier, D., Gonda, T.J. and Ishii, S. (1993) Negative autoregulation of c-Myb activity by homodimer formation through the leucine zipper. J Biol Chem *268*, 21914-21923.

Oelgeschlager, M., Krieg, J., Luscher-Firzlaff, J.M. and Luscher, B. (1995) Casein kinase II phosphorylation site mutations in c-Myb affect DNA binding and transcriptional cooperativity with NF-M. Mol Cell Biol *15*, 5966-5974.

Oelgeschlager, M., Janknecht, R., Krieg, J., Schreek, S. and Luscher, B. (1996) Interaction of the co-activator CBP with Myb proteins: effects on Myb-specific transactivation and on the cooperativity with NF-M. EMBO J *15*, 2771-2780.

Pitsch, R.J., Goodman, G.R., Minion, D.J., Madura, J.A.n., Fox, P.L. and Graham, L.M. (1996) Inhibition of smooth muscle cell proliferation and migration in vitro by antisense oligonucleotide to c-myb. J Vasc Surg *23*, 783-791.

Pizer E, H.E. (1989) RAV-1 insertional mutagenesis: disruption of the c-myb locus and development of avian B-cell lymphomas. J. Virol. *63*, 1630-1640.

Pizer ES, B.T., Humphries EH. (1992) Activation of the c-myb locus is insufficient for the rapid induction of disseminated avian B-cell lymphoma. J. Virol. *66*, 512-523.

Press RD, R.E., Ewert DL. (1994) Overexpression of C-terminally but not N-terminally truncated Myb induces fibrosarcomas: a novel nonhaemopoietic target cell for the myb oncogene. Mol Cell Biol *14*, 2278-2290.

Raddatz, R., Wilson, A.M., Artymyshyn, R., Bonini, J.A., Borowshy, B., Boteju, L.W., Zhou, S., Kouranova, E.V., Nagorny, R., Guevarra, M.S., Dai, M., Lerman, G.S., Vaysse, P.J., Branchek, T.A., Gerald, C., Forray, C. and Adham, N. (2000) Identification and Characterization of two neuromedin U receptors differentially expressed in peripheral tissues and the central nervous systems. J Biol Chem *275*, 32452-32459.

Ratajczak, M.Z., Hijiya, N., Catani, L., DeRiel, K., Luger, S.M., McGlave, P. and Gewirtz, A.M. (1992a) Acute- and chronic-phase chronic myelogenous leukaemia colony-forming units are highly sensitive to the growth inhibitory effects of c-myb antisense oligodeoxynucleotides. Blood *79*, 1956-1961.

Ratajczak, M.Z., Kant, J.A., Luger, S.M., Hijiya, N., Zhang, J., Zon, G. and Gewirtz, A.M. (1992b) In vivo treatment of human leukaemia in a scid mouse model with c-myb antisense oligodeoxynucleotides. Proc Natl Acad Sci U S A. *89*, 11823-11827.

Ratajczak, M.Z., Luger, S.M., DeRiel, K., Abrahm, J., Calabretta, B. and Gewirtz, A.M. (1992c) Role of the KIT protooncogene in normal and malignant human haemopoiesis. Proc Natl Acad Sci U S A. *89*, 1710-1714.

Ratajczak, M.Z., Perrotti, D., Melotti, P., Powzaniuk, M., Calabretta, B., Onodera, K., Kregenow, D.A., Machalinski, B. and Gewirtz, A.M. (1998) Myb and ets proteins are candidate regulators of c-kit expression in human haemopoietic cells. Blood *91*, 1934-1946.

Rosenberg, R.D. (1993) Vascular smooth muscle cell proliferation: basic investigations and new therapeutic approaches. Thromb Haemost. *70*, 10-16.

Roussel, M., Saule, S., Langrow, S., Langrow, C., Rommens, C., Berg, H., Graf, T. and Stehelin, D. (1979) Three types of viral oncogenes of cellular origin for haematopoietic cell transformation. Nature (London) *281*, 452-455.

Rouzic E, P.B. (1996) Retroviral insertional activation of the c-myb proto-oncogene in a Marek's disease T-lymphoma cell line. J. Virol. *70*, 7414-7423.

Sakura, H., Kanei-Ishii, C., Nagase, T., Nakagoshi, H., Gonda, T. and Ishii, S. (1989) Delineation of three functional domains of the transcriptional activator encoded by the c-myb protooncogene. Proc Natl Acad Sci U S A *86*, 5758-5762.

Sano, Y. and Ishii, S. (2001) Increased affinity of c-Myb for CREB-binding protein (CBP) after CBP-induced acetylation. J Biol Chem *276*, 3674-3682.

Schmidt M, K.R., Haviernik P, Bies J, Maciag K, Wolff L. (2001) Deregulated c-Myb expression in murine myeloid leukaemias prevents the up-regulation of p15(INK4b) normally associated with differentiation. Oncogene *20*, 6205-6214.

Schmidt M, N.V., Stevens L, Watson R, Wolff L. (2000) Regulation of the resident chromosomal copy of c-myc by c-Myb involved in myeloid leukemogenesis. Mol Cell Biol *20*, 1970-1981.

Shen-Ong GL, W.L. (1987) Moloney murine leukaemia virus-induced myeloid tumours in adult BALB/c mice: requirement of *c-myb* activation but lack of *v-abl* involvement. J. Virol. *61*, 3721-3725.

Shetzline, S.E., Dowd, K.J., Neely, R.J., Choi, J.K., Zou, S., Nakata, Y., Swider, C.R. and Gewirtz, A.M. (2002) Neuromedin U: A novel gene target of the c-Myb proto-oncogene in normal and mailgnant haemopoietic cells. Blood *100*, 742a.

Shuey, D.J., McCallus, D.E. and Giordano, T. (2002) RNAi: gene-silencing in therapeutic intervention. Drug Discov Today *7*, 1040-1046.

Simons, M., Edelman, E.R., DeKeyser, J.L., Langer, R. and Rosenberg, R.D. (1992) Antisense c-myb oligonucleotides inhibit intimal arterial smooth muscle cell accumulation in vivo. Nature *359*, 67-70.

Simons, M., Morgan, K.G., Parker, C., Collins, E. and Rosenberg, R.D. (1993) The proto-oncogene c-myb mediates an intracellular calcium rise during the late G1 phase of the cell cycle. J Biol Chem *268*, 627-632.

Simons, M., Ariyoshi, H., Salzman, E.W. and Rosenberg, R.D. (1995) c-myb affects intracellular calcium handling in vascular smooth muscle cells. Am J Physiol. *268*, C856-868.

Sokol, D.L., Zhang, X., Lu, P. and Gewirtz, A.M. (1998) Real time detection of DNA.RNA hybridization in living cells. Proc Natl Acad Sci U S A *95*, 11538-11543.

Souza, L.M., Briskin, M.J., Hillyard, R.L. and Baluda, M.A. (1980) Identification of the avian myeloblastosis genome. J. Virol. *36*, 325-336.

Sullenger, B.A. (1995) Colocalizing ribozymes with substrate RNAs to increase their efficacy as gene inhibitors. Appl Biochem Biotechnol. *54*, 57-61.

Szekeres, P.G., Muir, A.I., Spinage, L.D., Miller, J.E., Butler, S.I., Smith, A., Rennie, G.I., Murdock, P.R., Fitzgerald, L.R., Wu, H., McMillan, L.J., Guerrera, S., Vawter, L., Elshourbagy, N.A., Mooney, J.L., Bergsma, D.J., Wilson, S. and Chambers, J.K. (2000) Neuromedin U is a potent agonist at the orphan G protein-coupled receptor FM3. J Biol Chem *275*, 20247-20250.

Taylor, D., Badiani, P. and Weston, K. (1996) A dominant interfering Myb mutant causes apoptosis in T-cells. Genes and Development *10*, 2732-2744.

Thompson, C.B., Challoner, P.B., Neiman, P.E. and Groudine, M. (1986) Expression of the c-myb proto-oncogene during cellular proliferation. Nature *319*, 374-380.

Thompson, M.A., Rosenthal, M.A., Ellis, S.L., Friend, A.J., Zorbas, M.I., Whitehead, R.H. and Ramsay, R.G. (1998) c-Myb down-regulation is associated with human colon cell differentiation, apoptosis, and decreased Bcl-2 expression. Cancer Res *58*, 5168-5175.

Tomita, A., Towatari, M., Tsuzuki, S., Hayakawa, F., Kosugi, H., Tamai, K., Miyazaki, T., Kinoshita, T. and Saito, H. (2000) c-Myb acetylation at the carboxyl-terminal conserved domain by transcriptional co-activator p300. Oncogene *19*, 444-451.

Tomita A, W.T., Kosugi H, Ohashi H, Uchida T, Kinoshita T, Mizutani S, Hotta T, Murate T, Seto M, Saito H. (1998) Truncated c-Myb expression in the human leukaemia cell line TK-6. Leukaemia *12*, 1422-1429.

Trent, J.M., Thompson, F.H. and Meyskens, F.L.J. (1989) Identification of a recurring translocation site involving chromosome 6 in human malignant melanoma. Cancer Res *49*, 420-423.

Vandenbark, G.R., Chen, Y., Friday, E., Pavlik, K., Anthony, B., deCastro, C. and Kaufman, R.E. (1996) Complex regulation of human c-kit transcription by promoter repressors, activators, and specific *myb* elements. Cell Growth & Differentiation *7*, 1383-1392.

Vorbrueggen G, K.F., Guehmann S, Moelling K. (1994) The carboxyterminus of human c-myb protein stimulates activated transcription in trans. Nucleic Acids Res *22*, 2466-2475.

Wagner, R.W. and Flanagan, W.M. (1997) Antisense technology and prospects for therapy of viral infections and cancer. Mol Med Today *3*, 31-38.

Weston, K. and Bishop, J. (1989) Transcriptional activation by the v-myb oncogene and its cellular progenitor, c-myb. Cell *58*, 85-93.

Weston, K. (1998) Myb proteins in life, death, and differentiation. Current Opinion in Genetics & Development *8*, 76-81.

Yi, H.K., Nam, S.Y., Kim, J.C., Kim, J.S., Lee, D.Y. and Hwang, P.H. (2002) Induction of apoptosis in K562 cells by dominant negative c-myb. Exp Hematol. *30*, 1139-1146.

Yu, H., Bauer, B., Lipke, G.K., Phillips, R.L. and Van Zant, G. (1993) Apoptosis and haemopoiesis in murine fetal liver. Blood *81*, 373-384.

Zobel A, K.F., Vorbrueggen G, Moelling K. (1992) Trans-activation of the human c-myb gene by c-Myb. Biochim. Biophys. Res. Commun *186*, 715-722.